DATE DUE

JAN 24 2007		
BROWSED		

GAYLORD   PRINTED IN U.S.A.

Carl Friedrich Gauss

Titan of Science

Carl Friedrich Gauss

The full-length portrait by R. Wimmer
in the Deutsches Museum in Munich (1925)

Carl Friedrich Gauss
Titan of Science

by

G. Waldo Dunnington, PhD

with additional material
by
Jeremy Gray and Fritz-Egbert Dohse

Published and Distributed by
The Mathematical Association of America

By G. Waldo Dunnington

Carl Friedrich Gauss: Inaugural Lecture on Astronomy and Papers on the Foundations of Mathematics

Carl Friedrich Gauss: Titan of Science

© 1955 by G. Waldo Dunnington
Previously published by Hafner Publishing, New York

Reprinted 2004 by
The Mathematical Association of America (Incorporated)

Additional material by Jeremy Gray (Introduction to Dunnington's *Gauss: Titan of Science*, Introduction to Gauss's Diary, Gauss's Diary, and Commentary on Gauss's Diary) and by Fritz-Egbert Dohse (biography of Guy Waldo Dunnington)
©2004 by
The Mathematical Association of America (Incorporated)

ISBN: 0-88385-547-X
Library of Congress Catalog Card Number: 2003113540

Current Printing (last digit):
10 9 8 7 6 5 4 3 2

MEMORIAE
VIRI INCORRUPTISSIMI
CAROLI FRIDERICI GAUSS
ANIMAE AMOENISSIMAE
ET
INGENII FOECUNDISSIMI
BIOGRAPHUS EJUS
TOTO PECTORE
HUNC LIBRUM
DEDICAT

Yet though thy purer spirit did not need
The vulgar guerdon of a brief renown,
Some little meed at least—some little meed
Our age may yield to thy more lasting crown.
For praise is his who builds for his own age;
But he who builds for time must look to time for wage.
— Grant Allen

SPECTRUM SERIES

Published by

THE MATHEMATICAL ASSOCIATION OF AMERICA

Committee on Publications
Gerald L. Alexanderson, *Chair*

Spectrum Editorial Board
Gerald L. Alexanderson, *Editor*

Robert Beezer
William Dunham
Michael Filaseta
Erica Flapan
Eleanor Lang Kendrick
Ellen Maycock
Russell L. Merris

Jeffrey L. Nunemacher
Jean Pedersen
J. D. Phillips, Jr.
Marvin Schaefer
Harvey J. Schmidt, Jr.
Sanford Segal
Franklin Sheehan

John E. Wetzel

The Spectrum Series of the Mathematical Association of America was so named to reflect its purpose: to publish a broad range of books including biographies, accessible expositions of old or new mathematical ideas, reprints and revisions of excellent out-of-print books, popular works, and other monographs of high interest that will appeal to a broad range of readers, including students and teachers of mathematics, mathematical amateurs, and researchers.

777 Mathematical Conversation Starters, by John dePillis
All the Math That's Fit to Print, by Keith Devlin
Carl Friedrich Gauss: Titan of Science, by G. Waldo Dunnington, with additional material by Jeremy Gray and Fritz-Egbert Dohse
The Changing Space of Geometry, edited by Chris Pritchard
Circles: A Mathematical View, by Dan Pedoe
Complex Numbers and Geometry, by Liang-shin Hahn
Cryptology, by Albrecht Beutelspacher
Five Hundred Mathematical Challenges, Edward J. Barbeau, Murray S. Klamkin, and William O. J. Moser
From Zero to Infinity, by Constance Reid
The Golden Section, by Hans Walser. Translated from the original German by Peter Hilton, with the assistance of Jean Pedersen.
I Want to Be a Mathematician, by Paul R. Halmos
Journey into Geometries, by Marta Sved
JULIA: a life in mathematics, by Constance Reid
The Lighter Side of Mathematics: Proceedings of the Eugène Strens Memorial Conference on Recreational Mathematics & Its History, edited by Richard K. Guy and Robert E. Woodrow
Lure of the Integers, by Joe Roberts
Magic Tricks, Card Shuffling, and Dynamic Computer Memories: The Mathematics of the Perfect Shuffle, by S. Brent Morris
The Math Chat Book, by Frank Morgan
Mathematical Apocrypha, by Steven G. Krantz
Mathematical Carnival, by Martin Gardner
Mathematical Circles Vol I: In Mathematical Circles Quadrants I, II, III, IV, by Howard W. Eves
Mathematical Circles Vol II: Mathematical Circles Revisited and *Mathematical Circles Squared,* by Howard W. Eves
Mathematical Circles Vol III: Mathematical Circles Adieu and *Return to Mathematical Circles,* by Howard W. Eves
Mathematical Circus, by Martin Gardner
Mathematical Cranks, by Underwood Dudley
Mathematical Evolutions, edited by Abe Shenitzer and John Stillwell

Mathematical Fallacies, Flaws, and Flimflam, by Edward J. Barbeau
Mathematical Magic Show, by Martin Gardner
Mathematical Reminiscences, by Howard Eves
Mathematical Treks: From Surreal Numbers to Magic Circles, by Ivars Peterson
Mathematics: Queen and Servant of Science, by E.T. Bell
Memorabilia Mathematica, by Robert Edouard Moritz
New Mathematical Diversions, by Martin Gardner
Non-Euclidean Geometry, by H. S. M. Coxeter
Numerical Methods That Work, by Forman Acton
Numerology or What Pythagoras Wrought, by Underwood Dudley
Out of the Mouths of Mathematicians, by Rosemary Schmalz
Penrose Tiles to Trapdoor Ciphers ... and the Return of Dr. Matrix, by Martin Gardner
Polyominoes, by George Martin
Power Play, by Edward J. Barbeau
The Random Walks of George Pólya, by Gerald L. Alexanderson
Remarkable Mathematicians, from Euler to von Neumann, Ioan James
The Search for E.T. Bell, also known as John Taine, by Constance Reid
Shaping Space, edited by Marjorie Senechal and George Fleck
Sherlock Holmes in Babylon and Other Tales of Mathematical History, edited by Marlow Anderson, Victor Katz, and Robin Wilson
Student Research Projects in Calculus, by Marcus Cohen, Arthur Knoebel, Edward D. Gaughan, Douglas S. Kurtz, and David Pengelley
Symmetry, by Hans Walser. Translated from the original German by Peter Hilton, with the assistance of Jean Pedersen.
The Trisectors, by Underwood Dudley
Twenty Years Before the Blackboard, by Michael Stueben with Diane Sandford
The Words of Mathematics, by Steven Schwartzman

MAA Service Center
P.O. Box 91112
Washington, DC 20090-1112
800-331-1622   FAX 301-206-9789

Foreword

It has been observed by many writers that there is no such thing as an impartial biography. If this be true, then the author must plead guilty of a bias in favor of Gauss. Yet Gauss is not a controversial figure, and hence decisions for or against him have been rare. In 1925 I was struck by the fact that there was no full-scale biography of Gauss. It seemed incredible that a man of his stature had never received detailed biographical treatment. Various scholars have commented on the fact. Several planned such a biography but never carried out their purpose. The reason is not hard to find, for it is difficult to make the life of a mathematician "readable." I hope that I have at least partially succeeded, but it is with some trepidation that I lay this work before the public. Had I known in 1925 the magnitude of the task, I would probably never have started.

Yet the search has been sweet and rewarding. I have "lived intimately" with Gauss for almost thirty years, and there have been many pleasant by-products of my research. I have found all that I sought except the death mask and three of the four Petri deathbed pictures. If it is acceptable to spend part of a lifetime in the study of Shakespeare or Bach, why should it be thought strange to do the same for the greatest mathematician of modern times, a man whose life exhibits so many charming facets?

The present volume is derived almost entirely from contemporary sources, both manuscript and printed works. The letters and collected works of Gauss have been a rich mine of information. Reminiscences of his friends and students have been useful. His descendants have been very kind in allowing me to use letters, pictures, and other material. I have drawn freely from what I found. The nearest approximation to a biography of Gauss is the memorial monograph by his friend Sartorius (1856). Any biographer must use it. For some aspects of his life it is the only source. I have profited by the learning of those who have covered the ground before me.

My purpose has been to set Gauss off against the times in which he lived, to show him as a man and scientist. It has been frequently observed that biography is more difficult to write than history. My aim has not been mere entertainment, but a full record of the life and achievements of Gauss.

The most pleasant part of my searches occurred in the year I spent in his home at the Göttingen observatory, several weeks of which I spent at Brunswick amid the scenes of his boyhood. One cannot help enjoying the local flavor. Work in manuscripts and accumulated memorabilia conveys a vital spark which is lost in the printed work. It enables one to reconstruct the past.

Probably all the evidence on Gauss is in; it is not likely that any new vital information will turn up, particularly since the great destruction of World War II. Year after year, the notes pile up, but I hope I have not yielded to vanity and exhibited too many of my diggings to the public. The chapters vary considerably in length, but that is inevitable in a biography; a novelist is not bothered by such a problem. There are certain natural dividing points in a human life, and the biographer cannot change them.

It is hoped that the Appendixes will serve as a place of reference to primary sources. The reader will observe that Chapter XXIV and Appendix I are largely quotation, since it seems advisable to allow Gauss and certain authorities to speak for themselves. An effort has been made to keep footnotes to a minimum. In a work of this character it has not been deemed advisable to cite in footnotes bibliographical references to published letters. These are arranged chronologically in the published correspondence and can be easily found by anyone who is interested. The published correspondence is listed completely in the bibliography.

FOREWORD

The list of acknowledgments in a work of this kind, whose preparation has extended over a number of years, is necessarily a long one. I trust that I shall not omit any person whose name belongs here and I regret that a number of those named below have not lived to see the appearance of this biography in print.

The Gauss family has been extremely helpful in the undertaking. First and foremost should be mentioned the late William T. Gauss of Colorado Springs, Colorado. For many years he eagerly collected all that he could find touching on the life and achievements of his grandfather; he very kindly allowed me to use as much of his collection as I desired. He displayed a more ardent interest in his ancestor than any other descendant. His daughter, Miss Helen W. Gauss, did not lag behind him in interest; she went over the early chapters with me and made many valuable suggestions. The late Miss Anne D. Gauss of St. Charles, Missouri, furnished much information through correspondence, and sent a number of pictures, as did the late Mrs. Ida H. Gauss of St. Louis, Missouri. Carl August Adolph Gauss of Hamlin, Germany, wrote me his childhood reminiscences of his grandfather and added many firsthand incidents which would have been otherwise very difficult (or impossible) to obtain; his son, Dr. Carl Joseph Gauss of Bad Kissingen, has kindly supplemented his father's aid. The late Miss Virginia Gauss and her brother Eugene of Columbia, Missouri, were generous enough to give me their father's set of Gauss' collected works. Other descendants who have given aid are Mrs. J. Paul Annan of Shreveport, Louisiana, Professor Henry F. Gauss of the University of Idaho, Philip W. Gauss of Port Arthur, Texas, and the late Matthew J. Gauss of St. Charles, Missouri.

Special thanks are due Mrs. Carl Mirbt of Göttingen, who allowed me to use the manuscript record of her grandfather Wagner's conversations with Gauss in his last days. The late Mrs. Elisabeth Stäckel of Heidelberg graciously placed at my disposal all of her husband's Gauss notes. I am under great obligation to the late Mrs. Charlotte Hieb of Rübeland in the Harz, who gave me the original of the Petri daguerreotype showing Gauss on his deathbed, as well as other items; she was the widow of Georg Hieb, who was the founder of the Gauss Museum in Brunswick.

Geheimrat Dr. Bruno Meyermann of the Göttingen observatory, and his wife, gave me innumerable leads and hints, thus expediting the work and making my year's stay in their home (the Gauss apartment) a most pleasant and memorable one. I am also peculiarly indebted to a leading Gauss authority, the late Dr. Heinrich Mack, librarian of Brunswick, who displayed great enthusiasm for the work, gave me copies of his own publications, and guided me into the Gaussiana under his charge. The editors of Gauss' *Collected Works,* Dr. Martin Brendel of Freiburg im Breisgau, and the late Dr. Ludwig Schlesinger of Giessen, as well as the latter's widow, turned over to me a large quantity of valuable material.

Others who gave special assistance of various sorts are Dr. A. Wietzke of Bremen, Rudolf Borch of Brunswick, Dr. Heinrich Schneider of Harvard University, the late Dr. G. A. Miller of the University of Illinois, the late Dr. Friedrich Hesemann of Göttingen, the late Dr. Karl R. Berger of Hamlin, and the late Dr. Harald Geppert of Giessen.

The following list includes those to whom I am indebted for answering specific questions, giving me copies of their publications, or otherwise aiding me, though to a lesser extent than those mentioned above: Dr. Götz von Selle of Göttingen, Dr. Andreas Galle of Potsdam, Professor R. C. Archibald of Brown University, Miss Marthe Ahrens of Stettin, Dr. F. E. Brasch of the Library of Congress (Smithsonian Division), Dr. Wilhelm Lorey of Frankfurt am Main, Dr. Clemens Schaefer of the University of Cologne, Dr. Otto Spiess of Basel, Dr. Karl Metzner of Berlin, Rolf Erb and Mrs. Lucie Noack of Dresden, the late Dr. Erich Bessel-Hagen of Bonn, Dr. J. E. Hofmann of Tübingen, Mrs. Hildegard Leidig of Berlin, the late Baron August Sartorius von Waltershausen of Gauting (Bavaria), Dr. Harald Elsner von Gronow of Berlin, Dr. Günther Reichardt of Berlin, the late Dr. Alfred Stern of Zurich, Friedrich Sack of Brunswick, Dr. W. Jaeger of Berlin-Friedenau, Dr. H. Ludendorff of Potsdam, Professor Richard Courant of New York University, Dr. Eduard Berend of Geneva, the late Dr.

Johannes Joachim of Göttingen, Major Fritz von Lindenau of Berlin-Schlachtensee, and, finally, a grandson of the mathematician A. F. Möbius (a pupil of Gauss), Dr. M. Möbius of Frankfurt am Main, as well as the three sons of Ernst Schering: Dr. Harald Schering of Hanover, Dr. Carl Schering of Darmstadt, and Dr. Walther M. Schering of Berlin.

Grants from the Alexander von Humboldt Foundation and the Carl Schurz Memorial Foundation enabled me to spend a year in Göttingen and Brunswick on the research. A special grant from the Deutsche Forschungsgemeinschaft (Notgemeinschaft der Deutschen Wissenschaft) covered the cost of reproducing a number of pictures. More recently a special allowance from the Northwestern State College of Louisiana made it possible for me to spend a summer completing the writing of the work.

Thanks of a unique nature must go to Miss Rosemary Johnston of Mansfield, Louisiana, who took on the onerous task of typing the entire manuscript, and to Irene Crawford Wagner, M.A., of Natchitoches, Louisiana, who compiled the index.

G. WALDO DUNNINGTON

Northwestern State College
Natchitoches, Louisiana
December 7, 1954

Contents

Foreword		ix
Introduction to Dunnington's *Gauss: Titan of Science* by Jeremy Gray		xix
Guy Waldo Dunnington by Fritz-Egbert Dohse		xxvii
I	Introduction: Family Background	3
II	The Enchanted Boyhood	11
III	Student Days	23
IV	The Young Man	32
V	Astronomy and Matrimony	49
VI	Further Activity	67
VII	Back to Göttingen	85
VIII	Labor and Sorrow	90
IX	The Young Professor: A Decade of Discovery, 1812–1822	100
X	Geodesy and Bereavement: The Transitional Decade, 1822–1832	113
XI	Alliance With Weber: Strenuous Years	139
XII	The Electromagnetic Telegraph	147
XIII	Magnetism: Physics Dominant	153
XIV	Surface Theory, Crystallography, and Optics	163
XV	Germination: Non-Euclidean Geometry	174
XVI	Trials and Triumphs: Experiencing Conflict	191
XVII	Milestones on the Highways and Byways	208
XVIII	Senex Mirabilis	225
XIX	Monarch of Mathematics in Europe	250
XX	The DOYEN of German Science, 1832–1855	265
XXI	Gathering Up the Threads: A Broad Horizon	280
XXII	RELIGIO SCIENTIAE: A Profession of Belief From the Philosopher and Lover of Truth	298

XXIII	Sunset and Eventide: Renunciation	318
XXIV	Epilogue	
	1. Apotheosis: Orations of Ewald and Sartorius	331
	2. Valhalla: Posthumous Recognition and Honors	335

Appendixes

A	Estimates of His Services	343
B	Honors, Diplomas, and Appointments of Gauss	351
C	The Will of Gauss	356
D	Children of Gauss	361
E	Genealogy	376
F	Chronology of the Life of Carl F. Gauss	391
G	Books Borrowed by Gauss From the University of Göttingen Library During His Student Years	398
H	Courses Taught by Gauss	405
I	Doctrines, Opinions, Theories, and Views	411
	Bibliography	
	1. Publications of Gauss	420
	Collected Works	430
	2. About Gauss	
	Manuscripts and Related Material	431
	Books	431
	Pamphlets	434
	Articles	435
	An Introduction to Gauss's Mathematical Diary by Jeremy Gray	449
	Gauss's Mathematical Diary	469
	Commentary on Gauss's Mathematical Diary by Jeremy Gray	485
	Annotated Bibliography	497
	Index	507

List of Illustrations

Frontispiece:
Carl Friedrich Gauss: full-length portrait by R. Wimmer in the Deutsches Museum in Munich (1925)

Following page 130:
The birthplace of C. F. Gauss
The Gauss coat-of-arms
Silhouette of Gauss in his youth
Bust of Gauss by Friedrich Künkler (1810)
The Collegium Carolinum in Brunswick
The Schwarz portrait of Gauss (1803)
Minna Waldeck, second wife of Gauss
Portrait of Gauss by S. Bendixen (1828)
A sketch of Gauss by his pupil J. B. Listing
The observatory of the University of Göttingen
The courtyard of the Göttingen observatory as it appeared in Gauss' time
Gauss' personal laboratory in the observatory as he left it
The Gauss-Weber telegraph (Easter, 1833)
Gauss' principal instrument, the Repsold meridian circle
Gauss and Weber
Biermiller's copy (1887) of the Jensen portrait of Gauss (1840)
Ritmüller's portrait of Gauss on the terrace of the observatory
Gauss in 1854
Gauss about 1850
Portrait of Wolfgang Bolyai by János Szabó
Johann Friedrich Pfaff

Following page 322:
Heinrich Christian Schumacher
Heinrich Wilhelm Matthias Olbers
Friedrich Wilhelm Bessel
Alexander von Humboldt
Johann Franz Encke
Johann Benedikt Listing

The Philipp Petri daguerreotype of Gauss on his deathbed
The copper memorial tablet in Gauss' death chamber, given by King George V of Hanover (1865)
The grave of Gauss in St. Albans Cemetery in Göttingen
The bust of Gauss by C. H. Hesemann (1855) in the University of Göttingen library
Joseph Gauss
Heinrich Ewald
Minna Gauss Ewald, a watercolor by L. Becker (1834)
Wilhelm Gauss
Eugene Gauss
Therese Gauss and her husband, Constantin Staufenau
Therese Gauss, a sketch by J. B. Listing
Schaper's monument to Gauss in Brunswick (1880)
Bust of Gauss by W. Kindler (in the Dunnington Collection)
Bust of Gauss by Friedrich Küsthardt in Hildesheim
The Gauss–Weber monument in Göttingen, by Hartzer (1899)
The Gauss monument by Janensch, in Berlin
The Gauss Tower on Mount Hohenhagen, near Dransfeld, dedicated 1911
The Brehmer medal of Gauss (1877)
The Eberlein bust of Gauss in the Hohenhagen Tower
The bust of Gauss by F. Ratzenberger (1910), in the Dunnington Collection
Commemorative postage stamp issued for the Gauss Centennial, 1955

Introduction to Dunnington's Gauss: Titan of Science

Jeremy Gray

Dunnington's biography of Gauss was first published almost 50 years ago, and remains unrivalled for its combined breadth, depth, and accuracy. It was the product of three decades of labour on substantially all the known sources, and it is regrettable that it has been so long out of print. Gauss continues to reward the attention of historians—there have been at least five other biographies since 1953 (those by Wussing [1976], Reich [1977], Bühler [1981], Hall [1970], and Worbs [1955]) as well as an extensive, thoroughly-researched essay by K.O. May [1972]) —and some of their discoveries are noted in the text as we proceed, but Dunnington's book remains the core of any attempt to understand Gauss and his work. Walter Kaufmann Bühler, one of these later biographers, rightly called Dunnington's book 'by far the most important' of the major biographies of Gauss (Bühler [1981], p. 166).

Apart from its merits as a biography, to which I shall return, Dunnington's study is notable for the wealth of information it contains in its numerous appendices. There is a useful chronology of his life and a good bibliography of his work. There is a list of the books Gauss read at university, which is very helpful in sorting out what he learned and when. There is a remarkable genealogy of his descendants reaching to 1953 (the family tree is currently maintained by Susan Chambless, herself a Gauss descendant, at `http:homepages.rootsweb.com/~schmblss`).

xix

There is a lengthy bibliography of writings about Gauss. The Gauss scholar may now also consult the extensive work by Uta Merzbach: *Carl Friedrich Gauss: a Bibliography*, published in 1994. This lists the full publication details of all Gauss's works, locates all his surviving letters, and records such information as the names of everyone known to Gauss, whether a predecessor or a contemporary, and where their names were mentioned by him. It also brings up to date the list of works about Gauss. The simplest way to keep up with studies on Gauss may well be to navigate the Web, starting perhaps with the Web site of the British Society for the History of Mathematics (http://www.dcs.warwick.ac.uk/bshm.html) but one should also consult the *Mitteilungen der Gauss-Gesellschaft*, which carries many articles devoted to Gauss (their home page can be found at http://www.math.uni-hamburg.de/math/ign/gauss/gaussges.html). A select bibliography has been added to this book, but unlike those by Dunnington and Merzbach it does not aim to be complete but merely to point out some recent additions to the literature which the reader of this book may welcome.

Any account of someone's life and work founders, in the last analysis, on our inability to grasp the creative process. Just as novelists leave behind works that are no simple record of the life and cannot be used to supplement the documentary record in any easy way (if at all) mathematicians and scientists do not simply respond to circumstances and turn them into theorems and discoveries. Much of the life Gauss lived could have been lived by anyone without there being so much to show for it. In many ways, Gauss led the life of a competent astronomer, no different from a number of his contemporaries. Nothing in his situation can account for his profound originality. Dunnington rightly establishes that the circumstances of Gauss's life did dispose him to some topics rather than others. His astronomical work was part of an extensive German enterprise, often well-funded and with good new instruments. His survey of the State of Hanover in the 1820s likewise saw him securely placed in the world of useful work. Both activities were congenial to someone with Gauss's remarkable ability as a calculator, not to mention his capacity for sheer hard work. But the circumstances cannot explain how Gauss came to invent the method of least squares and rediscover the lost asteroid Ceres, or discover the concept of intrinsic curvature and open a new field in the study of geometry. To his credit, Dunnington does not speculate where the evidence is inevitably lacking.

Dunnington, instead, put a great deal of careful effort into his account of the German political and intellectual scene in Gauss's day.[1] Those who knew Gauss, who helped him on his way in his teens and as a young man, and who were his friends in later life, are described and located in their various milieus. The prosaic but essential side of Gauss's activity is well described, from the details of equipment, its purchase and maintenance, to the conduct of the famous survey and the work on the electric telegraph. At the same time, every aspect of Gauss's scientific and mathematical work is discussed in detail.

Nonetheless, Dunnington's book has become dated in two significant ways. A modern reader is likely to feel that he became too close to his subject, and that he did not always explain the mathematics with sufficient clarity. It is hard not to be impressed, even over-impressed, with Gauss's achievements, but while some have found his personality less attractive, Dunnington comes over as Gauss's best friend, excusing and explaining away all the imperfections of his hero. As he endearingly admits in the opening words of the Foreword, he 'must plead guilty to a bias in favor of Gauss.' The entries in the index for aspects of Gauss's personality are, in their entirety: sensitive, noble in bearing, thoroughly conservative, not a utilitarian, slow in passing judgement, dislike of travel, wise investor, aristocratic, thorough, despised pretence, religious in nature, practical in concept of religion, kind but austere, accepted misfortune. Some are undoubtedly correct—Gauss not only left his family a lot of money but rescued the university widows' fund—but unless one would spurn the label 'thoroughly conservative' or holds different religious views, there is little here that one would not wish for oneself. Bühler, by contrast, noted that 'Early in his adult life, Gauss severed most

---

[1] I am grateful to Menso Folkerts for his expert opinion on the generally high quality of Dunnington's account of German history, which is largely endorsed by Bühler in his more recent book.

of the 'meaningful' social and emotional ties a man could have.' (Bühler, p. 13), which is not such an attractive aspect, and May commented that 'Those who admired Gauss most and knew him best found him cold and uncommunicative' (*Dictionary of Scientific Biography*, p. 308). The result of Dunnington's bias is a book which, although it was written in 1953, is oddly Victorian in tone. While the reader may easily make allowance for this, and no attempt has been made to mitigate the effect, he or she may also wonder what is missing. How reliable are the facts presented here, how fair, has a full picture been presented or a flattering half-view?

It has become inevitable that we doubt the anecdotes about the young Gauss. They were written down only late in life, they derive from the fond but perhaps inaccurate memories of Gauss and his mother. They exaggerate, but such were Gauss's prodigious abilities that they came to be believed. Today, we find it easy to strike them down, instinctively reacting against stories that we see as pandering to the romantic ideal of the genius. Did Gauss really learn to read and to do elementary arithmetic with so little help? Bühler, in his biography of Gauss, wrote simply that 'many of these anecdotes cannot be substantiated, nor are they of serious interest' [1981, p. 5]. And perhaps they are fairy tales, even if they can be traced back to Gauss himself, as Bühler admitted in a footnote. May, on the other hand, found the famous stories convincing. But even if the evidence is not unimpeachable, even if it is at times unreliable, it is held in question today as much for the message it carries as for its lack of documentation.

The feeling that Dunnington is partisan lingers throughout the book. What sort of a colleague was he among the astronomers? When he surveyed Hanover? During the affair of the Göttingen Seven, when two of his close friends were dismissed from their jobs at the University of Göttingen for refusing to take an oath of allegiance to the reactionary Duke of Cumberland? There is no one, true view of a person. There are only multiple, often contradictory views, and Dunnington's is partial and perhaps inevitably so; there is a general tendency for biographers to sympathise with their subjects. But in fact, most biographers concur that it was among astronomers that Gauss was most at home. He felt useful, they in turn were impressed by his observational, and still more by his computational, skills. But they were not over-awed. With Bessel, Olbers, Schumacher and others Gauss enjoyed a long correspondence and friendships that he was denied elsewhere. In the long survey of Hanover, Gauss climbed the mountains, endured the bad weather and the sleepless nights that every one else had to endure, and generally impressed everyone as a team player, albeit an important one. And in the miserable affair of the Göttingen Seven, Gauss was prudent, less reactionary than some of his circle, and in any case incapable of rescuing Weber, Ewald, and the others from the consequences of their actions. No one at the time seems to have expected him, or wanted him, to do much more than he did, or much differently. Dunnington's hero is no one's villain.

The problem is that the same haze diffuses over the scientific achievements. Was there no one doing comparable work at the time? Had Abel really only come one-third of the way that Gauss had managed in creating a theory of elliptic functions? What, precisely, did Gauss know about non-Euclidean geometry? Is Gauss the chief architect of the theory and method of least squares in statistics? Dunnington is not given to outright exaggeration, but contemporary mathematicians and scientists are kept in the shade unless they are also close friends or colleagues of Gauss, and mentioned as often for their personal views as the qualities of their work. This not only misses an opportunity to make Gauss's work stand out more clearly by comparison with the best of what else was being done at the time, it hinders the readers' chances of assessing Gauss's influence. This may be why much of the subsequent literature on Gauss omits reference to Dunnington's book, not only in cases where they have little to add, but on the more interesting occasions when they disagree.

By keeping to himself rather than putting himself out socially, by his carefully honed writing style, which Crelle compared to 'gruel', Gauss minimised his influence on his contemporaries. He could be generous, as with his praise of Eisenstein and Sophie Germain (no prejudice against women in this instance at least) but he could be withholding too, as he surely was with the discoverers of non-Euclidean geometry. Part of

Dirichlet's lasting legacy is undoubtedly that he made Gauss's number theory comprehensible to the small but important audience of German research mathematicians. As a result, number theory became a major concern of the major mathematical nation of the nineteenth century, but to whom is the credit due, Gauss or Dirichlet? The question is barely tackled here. One has the impression that nonetheless German mathematicians gravitated around Gauss, or perhaps his published work, but that French mathematicians did not, but the biographer would have had to have shifted focus considerably to bring such matters into discussion.

It was pointed out by May that Gauss's motto 'Pauca, sed matura' (Few, but ripe) tends to obscure the fact that Gauss wrote a very great deal. There are two substantial books and 12 volumes in his *Werke*, and even if one allows for the generous commentaries, and the fact that quite a bit of it was written for the desk drawer, that is a lot of pages. The bulk of it is made up of astronomical work, much of it about asteroids, surveying, and error analysis. This has given rise to a controversy about whether Gauss should be regarded as primarily a mathematician, and if so a pure or an applied mathematician, or as a scientist. The controversy is at once enjoyable and fruitless. The sciences and mathematics advance in different ways and even at different speeds. Gauss was acclaimed in his day for his work on the orbits of asteroids, an exciting topic then but one of less interest today, so much so that some recent commentators have seen fit to explain why Gauss bothered. His work on the electric telegraph was pioneering in its day, but—as Dunnington described—it did not lead to a great technological advance. He won a prize for his work on theoretical cartography, which laid down the theory that has guided work in the subject ever since, but which was by no means the deepest part of his work on differential geometry. Rightly or wrongly, it seems that one cannot win fame as a 'geodesist'. The mathematics, on the other hand, that Gauss did, not all of which is described here, is a secure part of modern mathematics in a number of different areas, and it was appreciated by the best of his younger contemporaries.

It is therefore to be regretted that Dunnington's account of Gauss's mathematics has become, and perhaps always was, a little indistinct.

There is, of course, an admirable account by Felix Klein in his *Entwickelung der Mathematik*, and the lengthy essays on different aspect of Gauss's mathematics that adorn volume X.2 of the *Werke* are classics of German historical scholarship. Guided by these, and in the belief that accounts of anything can be given which are clear at least to those with the right sort of background knowledge, I have written an Appendix on Gauss's mathematical work, concentrating on those topics that Dunnington himself discussed. I have also appended a translation, with commentary, of Gauss's mathematical diary, an interesting document in itself and which I hope will interest a wider readership than it had on its original publication.

Throughout the book there are occasions when the reader will wish Dunnington had done things differently. I have resisted the temptation to rewrite the book. It is Dunnington's *Gauss* that is worth reprinting. Were it not to pass that test one would be driven to contemplate co-authorship across the tomb, and while there are successful examples of such collaboration I did not wish to attempt one. It is better to have in print the considered view of someone who spent nearly thirty years on the task, I believe, than to tinker nervously with it, even when accuracy and the highest of current standards may be on one's side. There are remarkably few footnotes, for example. This makes it difficult, if not impossible, to determine the sources of Dunnington's information. Short of spending a year doing someone else's research, or hiring a fact checker from the *New York Times*, there is nothing to be done. But there are occasions when later writers have brought up information that changes the picture Dunnington presented in significant ways. I have taken up these matters in the annotated bibliographical essay which will be found at the end of the book, and added footnotes, where I note a number of recent and valuable accounts of the mathematical and scientific work have appeared.[2]

---

[2] I am very happy to acknowledge here the contributions that were sent in by many members of the list on history of mathematics maintained by Julio Gonzalez Cabillon, and to thank him in particular for the way he has so expertly and convivially managed his valuable task.

Curiously enough, Gauss seems much better remembered among mathematicians and historians of mathematics than among other groups of his intellectual descendants, and to be better remembered for his mathematics and statistics than for any other aspects of his work. At least, searches of the literature, Web searches, and the like revealed much more scholarship on these matters than any other, except perhaps on the minutiae of his life. Some of this perception is doubtless due to my own limitations, some perhaps to the longer memory mathematicians have for their subject than scientists do. After all, much of what Gauss did in number theory is suitable today for advanced undergraduate courses if not still a topic for graduate school, but Gauss's work on statistics has become just one tool among many, astronomers search distant galaxies for new science (and the asteroids more for threats to our survival), and geodesy has been transformed by satellites and computers. His lasting contribution to electro-magnetic theory is the mathematics of potential theory, not the physics or the technology inherent in his achievements.

Accordingly, additional material on number theory, algebra, function theory, differential geometry, topology, and non-Euclidean geometry, has been collected together in the above-mentioned Appendix, whereas the commentaries I have added to Dunnington's book updating it on astronomy, surveying and geodesy, magnetism and the telegraph, and statistics will be found after the Appendix in the annotated bibliographical essay, organised by topic. These commentaries should be thought of as a lightly annotated bibliography, in which some entries appear without further comment by me. As noted above, I do not claim completeness here.

Guy Waldo Dunnington

Fritz-Egbert Dohse

It may seem somewhat strange that it was an American professor of German, G. Waldo Dunnington, who undertook the enormous task of compiling resource material for the first comprehensive biography of one of the greatest mathematicians of all time: Carl Friedrich Gauss. The author was born in Bowling Green, Missouri, in 1906, though his ancestry goes back to Colonial Virginia, Maryland and Massachusetts. As a twelve-year-old, he learned the basic principles of mathematics from a charming young teacher, Miss Minna Waldeck Gauss, a granddaughter of Gauss' third son Eugen. Besides teaching mathematics to her class, she also told stories from time to time about her famous great-grandfather. Intrigued, young Waldo asked his teacher for a book about the great scientist, in order to learn more about him. When Miss Gauss responded that no full-length biography existed either in English or German, the boy declared that someday he would write one. Thus began a lifelong fascination with Gauss.

Waldo Dunnington completed his primary and secondary education in Missouri. After high school, he enrolled at Washington and Lee University in Lexington, Virginia, where he earned A.B. and M.A. degrees in German and also took courses in mathematics. Once more struck by the fact that no full-length biography of Gauss was available, he decided then and there to fulfill the promise he had made to his schoolteacher Minna Gauss. In the foreword, he describes how he pursued that goal.

At the University of Illinois, Urbana, he earned a doctorate in German, with a minor in English philology. His dissertation about Jean Paul, the favorite poet of Gauss, was entitled 'The Relationship of Jean Paul to Karl Philipp Moritz'. He taught German Literature and occasionally the History of Mathematics at various colleges and universities: first in St. Louis, then in Kansas City; later in La Crosse, Wisconsin and, from 1946 to his retirement in 1969, at Northwestern State University in Natchitoches, Louisiana. During World War II, Dunnington served in the US Army. In the fall of 1945 he was briefly assigned to the Nuremberg Trials as an interrogator and interpreter.

Dr. Dunnington published numerous articles, in both English and German, in foreign language and mathematical journals. From 1936 to 1945 he served as associate editor of the *National Mathematics Magazine*, which became *Mathematics Magazine* in 1947, heading a department for the humanism and history of mathematics. In 1937, the Louisiana State University Press published his monograph on Carl Friedrich Gauss. His other contributions have appeared in journals such as the *Monatshefte für deutschen Unterricht*, *Jean Paul Blätter*, *The Scientific Monthly*, *The Open Book*, *Scripta Mathematica*, and *The American Mathematical Monthly*, as well as in the *Encyclopædia Britannica*. He was well liked as a guest speaker, delivering one of the memorial speeches at Gauss' gravesite in 1955, on the occasion of the 100th anniversary of the death of the scientific genius. He was a good teacher, a stickler for detail, a living lexicon of Gauss genealogy, well read and amusing in conversation, loved and respected by his many students.

Upon his death in 1974, Dr. Dunnington left a collection of Gauss memorabilia (including several Gauss letters), assembled in a Gauss Archive and Museum, to Northwestern State University where he had served as an archivist after his retirement in 1969.

Dr. Dunnington was also active in the Lutheran Church as a lay preacher and chairman of the board of elders. During the memorial service for Dunnington in the small church he had loved, the minister recalled his life and work as man and scholar, quoting extensively from the eulogy delivered by Heinrich Ewald on the occasion of Gauss' death. Dr. Dunnington was buried in his home town of Bowling Green, Missouri, under a red granite tombstone bearing the following inscription on a tetrahedron crowned by a sphere: "G. Waldo Dunnington, Ph.D., born January 16, 1906, died April 10, 1974. Professor of German 1928–1969. Biographer of C. F. Gauss. 'Jetzt kann mein Geist recht freudig rasten, komm, sanfter Tod, und führ mich fort' (Now my soul can rest in peace, come, gentle death, and lead me away)".

Since 1995 an annual presentation, the Dunnington-Gauss Award, has been made to the outstanding student of mathematics at Northwestern State University. The purpose of the award is to honor the author of this book and to remind all those present of the 'Titan of Science,' Carl Friedrich Gauss. If today the English-speaking world knows more about Gauss, it is in large measure the result of Dr. Dunnington's work and devotion to a noble cause.

CARL FRIEDRICH GAUSS:

Titan of Science

CHAPTER ONE

Introduction: Family Background

During the years 1813–1832 Germany was in possession of her three greatest geniuses: Goethe, Gauss, and Wagner. To be sure, the three were in various stages of development. Goethe was in his declining years, Gauss at the height of his fame, and Wagner was not yet in the ascendant. The life and works of Goethe have been fully treated, as has the life of Wagner. The sublime attainments of these two men are fairly well known to the general reading public, but what can be said about the lofty achievements of Gauss in the more abstract fields of thought? The average reader outside the scientific field can scarcely tell who he was! Let it be said at this point that no apology need be made for the above grouping.

A student of Heinrich Ewald[1] once signified his intention of translating into English the brief monograph on Gauss by Wolfgang Sartorius von Waltershausen, but even this small task was never pushed through to completion. In recent years, especially in 1927, many scholars in Germany have again become interested in Gauss. Nothing more than brief magazine articles and encyclopedia notices have appeared about him in English. The author has had the conviction for some years that a comprehensive life story of Gauss should be presented to the reading public, in English, one that would not be too technical, and yet one complete enough to give a true picture of Carl Friedrich Gauss as man and scientist, and to give his philosophy of life.

---

[1] Renowned theologian and Orientalist who was a son-in-law of Gauss.

The life of this great thinker offers no tragic moments and scenes of excitement and disaster such as we find in Kepler's or Galileo's life. His life is marked by the same proportion of happiness and sorrow that one finds in the life of an ordinary human being. Yet we venture to say that his life offers more interesting features than does Newton's.

What distinguishes him are his fruitful contributions to science, made over a period of more than fifty years. He worked in a sphere which was not accessible to the layman and which even the most gifted scholars could reach only with the greatest effort. Thus he was little known to the general public. We see him striding on alone from one original discovery to another.

Accurate information about the ancestors of Gauss has recently been given by Rudolf Borch after a search of church records and account books. The family name Gauss, Gaus, and in variant forms Goss, Goess, Gooss, Goes, and Goos, is found very frequently in the region north of Brunswick as far as Meine and Kalberlah. Between the years 1500 and 1600 it occurs especially often from Gross-Schwülper to Essenrode.[2] Hänselmann has succeeded in establishing the fact that Hinrich Gooss in Völkenrode was the great-grandfather of Carl Friedrich. Rudolf Borch has located the father of this Hinrich Gooss in Wendeburg (supposedly he was born about 1600 and had migrated from Hanover), where a Hans "Gauss" was living between 1630 and 1660. The form of the name, therefore, had changed. This Hans Gauss had two sons and several daughters; the sons were named Heineke and Henrick (or Henrich). The latter got by marriage a small farm in Völkenrode.

Hinrich Gauss was married three times and had twelve children, four by each marriage. The sons of the third marriage were called Engel or Engelke, Jürgen, and Andreas. Engel Goos had a son named Heinrich Engel, who died in 1843 at the age of ninety-four. He knew of his relationship with Carl Friedrich and visited him at the observatory in Göttingen. Jürgen (Goos) Gauss

---

[2] The name is also common in Württemberg, where it may have no connection with the Hanoverian branch.

had a son, Gebhard Dietrich, the father of Carl Friedrich. The fate of a son of Heinrich Engel Goos and the fate of Andreas Goos are unknown. The small farm of Hinrich Gooss went to his son of the second marriage, who also was called Hinrich, and through a daughter of his to the family named Gremmel.

At Völkenrode, which is about an hour and a half from Brunswick in a northwesterly direction, the above-mentioned Hinrich Gooss was united in marriage to Anna Grove, the widow of Hinrich Wehrtmann. The church records reach back to 1649, but the name Goss does not occur there before this entry. The bride was a resident of Völkenrode and the owner of a small farm. She had no children by her first husband; of the second marriage, which lasted until her death twelve years later, there were two daughters. Having become a widower in 1695, Hinrich Gooss married on July 16, 1696, Ilse Geermanns, who bore him a son and three daughters, but died after nine years. On November 24, 1705, he married Katharine Lütke, who bore him during the next twelve years three sons and a daughter.

After Hinrich's death on October 25, 1726, his younger son of the same name, born in 1690 of the first wife, took over the farm. Hans, who was six years older, was unfit or had no desire to take on himself the burdens of the place. According to the peasant custom, he served his brother as a farm hand, dying single in 1739. Hinrich, who survived him until 1772, left four daughters, one of whom was married to Konrad Gremmel. She inherited the farm.

Henry, the half brother of Hans and Hinrich, by the second marriage, was born in 1700 and died prematurely in 1724. Of the brothers of the third marriage, Engel's younger brothers, Jürgen and Andreas, had to seek their living in strange towns. Andreas disappears in the darkness; his fate is unknown. The tracks of Jürgen, however, lead us into the city of Brunswick.

A protocol in the "new citizens' book" reveals that Jürgen presented papers on January 23, 1739; with that step he became established in Brunswick. Just before leaving he had married in Völkenrode Katharine Magdalene Eggelings from Rethen, a

Hanoverian village a few hours distant from there. These people might have belonged to their kin, those from Rethen who later at many christenings named Engel Gooss as godfather. But there was also an Anton Gooss from Rethen, who in 1709 got his second wife from Völkenrode. For this reason, Hänselmann supposes that Rethen was the original home of the elder Hinrich Gooss, although its church records do not begin until 1692.

Jürgen Gooss had registered as a day laborer at the city hall in Brunswick, where a note is made of him; he is called a claymason and a street butcher. Both these trades at that time belonged to domestic functions in the flat country. For a country worker who was seeking his living in town, by his own hand, they were closely related; even their union in one person was natural. They could be exchanged according to the time of year; the work of the street butcher began when that of the clay mason left off, and at that time, when the town houses in Brunswick did their own butchering even more generally than later, members of their guild earned substantial salaries.

In the year of Jürgen Gooss' entry on October 29, 1739, a certain Peter Hoyer, who acted as witness for him at the granting of citizenship, sold to Jürgen Gaus, as the name now runs, his house on the Ritterbrunnen "therefore and of such manner that purchasers shall pay to him yearly *in termino* Michaelmas 5 thalers and 10 groschen unreminded, and with it, as long as he, Peter Hoyer, shall continue to live, properly and stably, but after his death shall have and hold such house exclusively and hereditarily."

It was the house No. 10 of the Ritterbrunnen, a narrow building of only two windows' breadth, as it stood there two centuries later. People called it the "gingerbread slice." The married couple prospered there fourteen years; their three sons and one daughter were born in the house. In no wise had Jürgen paid too much for it; when it was sold in 1735 it brought 217 thalers.

Jürgen Gauss now bought another house, 1550 Wendengraben, later 30 Wilhelmstrasse. Of the price of 900 thalers he lacked 500, and the mayor, Wilmerding, held a mortgage on it.

In order to be able to pay the rest in cash, Jürgen had to borrow back 100 thalers of the proceeds on the old house. Thus the fruit of his own fourteen years' labor was a mere 85 thalers. For the next twenty-one years of his life his best efforts achieved a decrease of only 200 thalers in this mortgage.

Consumption ended his toilsome life on July 5, 1774, after his wife, just fifty-nine years old, had succumbed to a "bilious fever" on April 3. The daughter mentioned had died as a child. The eldest son, Gebhard Dietrich, born on February 13, 1744, had helped his father in business and on April 28, 1768, had married Dorothea Emerenzia Warnecke. Her name is also given as Sollerich or Sollicher. She brought him a dowry of 150 thalers, and on January 14, 1769, bore him a son, Johann Georg Heinrich. On April 27, 1775, Gebhard Dietrich made an agreement with his brothers Peter Heinrich and Johann Franz Heinrich concerning the father's estate in this wise: the house was to be his for 800 thalers, including the mortgage of 400 thalers to Mayor Wilmerding and 150 thalers to his wife (nee Sollicher).

Gebhard Dietrich was able to pay his brothers their share out of his own means, supplemented by a loan of 125 thalers, which he again received from Mayor Wilmerding. The wiping out of the debt on his portion was the goal he pursued unceasingly and after twenty-five years attained. When he sold his house on June 5, 1800, for 1,700 gold thalers there were no claims against it except the maternal inheritance of his son by the first marriage and that brought by his second wife.

Carl Friedrich Gauss knew of the existence of his two uncles but was of the opinion that they died before his father. It has been impossible to establish any facts about Peter Heinrich. It is quite possible that the families of the late Mr. Emil and Mr. Theodore Gaus in Brooklyn, New York, belong to his line. Their father came to America from Brunswick in the year 1854. Johann Franz Heinrich had descendants, among them a son, who died in 1798 at the age of twenty. Otherwise, his family continued only in the female line and through children of his wife's second marriage.

Gebhard Dietrich's first wife died on September 5, 1775, of consumption, aged thirty. On April 25, 1776, he married Dorothea Benze, a daughter of Christoph Benze, deceased, who had been a stonemason in Velpke, a nearby village. The marriage record states on April 16 that she brought to her husband, besides a bed and "eventual linen," one hundred thalers as a "true" dowry; Dorothea was born on June 18, 1743; she had no special schooling, could not write, and could scarcely read. Gebhard Dietrich was the second in his family; he bore the title of master of waterworks, but carried on various occupations and had assisted his father. During the last fifteen years of his life his only occupation was gardening. He also assisted a merchant in the Brunswick and Leipzig fairs. Because he wrote and calculated very well, he was placed in charge of the accounts and receipt of money in a large burial insurance company. Gebhard was an absolutely upright, respectable, and really honest man; but in his home he was quite domineering, often rough and uncouth, hence Gauss' childlike heart could not join itself to him in full confidence and trust, although no misunderstanding ever arose from this, because the son very early became entirely independent of him.

The birth of Carl Friedrich occurred in the house (later a museum and marked with a tablet over the front door) on Wilhelmstrasse[3] on April 30, 1777. His mother could not remember the exact day of birth, according to his own story. She only knew that it was on a Wednesday eight days before Ascension. This was the occasion of his later giving the formula by which one can calculate the day on which Easter falls in any year. His half brother, Johann Georg Heinrich, early left the home in order to learn a trade, then "wandered" as was customary and came back to Brunswick in 1794. A dangerous eye trouble rendered it necessary for him to give up his trade, but the father would not tolerate an idler, and since it was too late to begin any other business, he had to become a soldier. In that position he also continued to help his father, but withdrew from military service in 1806, and

---

[3] Destroyed in an air raid October 15, 1944.

when Gebhard Dietrich died in 1808, Johann Georg Heinrich took over his duties, which he performed until his death on August 7, 1854. Georg married twice; by the first wife he had a daughter, Caroline Magdalene Dorothee, and a son, Georg Gebhard Albert. The daughter married Eduard Wilhelm Bauermeister. Georg Gebhard Albert had a tin shop; he had been in Munich a long time when he visited his uncle in Göttingen. His son Georg Christian Albert also became a tinman and had his shop on Weberstrasse in Brunswick; he died in 1907. He lost two sons in infancy; his daughter Albertine was Mrs. Böttger and lived at Altewiekring 41, in Brunswick. She, in turn, had two sons, born in 1899 and 1901.

Gauss' maternal grandfather, Christoph Benze, was a stonemason in the small village Velpke, near Brunswick. As a consequence of working on sandstone, he suffered from the customary pulmonary consumption, from which he died at the age of thirty. He was survived by a daughter, Dorothea, and a younger son, Johann Friedrich. The son took up weaving, in which he soon attained artistic damask-weaving, without further guidance, and as a whole revealed an extraordinarily intelligent, shrewd mind. Gauss as a small boy thought a great deal of him, and later this feeling increased as he guided him in conversation on stimulating matters and thereby recognized his unusual talents and capacities. He always bewailed the uncle's untimely death December 2, 1809, with the declaration: "A born genius was lost in him."

The daughter, Dorothea, moved about 1769 from Velpke to Brunswick and married Gebhard Dietrich Gauss in 1776. She was a woman of natural, clever understanding, of unpretentious, happy spirit and a strong character. Her great son was her only child, her pride! She clung to him with the deepest love just as he did to her up to her last hour. Possessed of sound health, although in the last four years totally blind, she reached the unusually advanced age of ninety-seven and died at her son's home in the observatory, where she had lived for twenty-two years, on April 19, 1839.

The founder of the Benze line, Andreas Benze, was a contemporary of the elder Hinrich Gooss. Of his twin sons born on February 4, 1687, one, also christened Andreas, was survived by two daughters and two sons, the eldest of whom, Christoph Benze, married Katharina Marie Krone; he died on September 1, 1748, in the seventh year of his marriage.

Dorothea served as a maid for seven years before she married Gebhard Dietrich. From the son we learn that this marriage was not a very happy one, "chiefly because of external circumstances and because the two characters were not compatible." He boasts of his mother as "a very good, excellent woman." Of his father: "in many respects worthy of esteem and really esteemed, but in his home was quite domineering, uncouth, and unrefined"; also "he never possessed his complete confidence." This is indication enough of the fact that this child belonged more to the mother than the father in mental affinity. Dorothea had a younger brother, Christoph Andreas, born on June 11, 1748, several months before the father's death; of him we know very little.

CHAPTER TWO

The Enchanted Boyhood

Up to the end of his life Gauss loved to recall numerous episodes of his early childhood. These anecdotes reveal unmistakably occasional sparks of genius. He remembered them correctly and knew how to lend them rare charm by his lively, happy way of narrating; never did the slightest deviation occur in the retelling of them.

His memory went back to very earliest childhood, when he had once been near death. The previously mentioned Wendengraben (now Wilhelmstrasse) on which his parents were living, although later walled over, was once an open canal connected with the Ocker, abundantly filled with water in the spring. The little boy, unguarded, was playing on it and fell in, but was rescued just before drowning, as though destined by Providence for high scientific accomplishment.

Even in his earliest years Gauss gave extraordinary proofs of his mental ability. After he had asked various members of the household about the pronunciation of letters of the alphabet, he learned to read by himself, we are told, even before he went to school, and showed such remarkable comprehension of number relationships and such an incredible facility and correctness in mental arithmetic that he soon attracted the attention of his parents and the interest of intimate friends.

Gauss' father carried on in the summer what we would call today a bricklayer's trade. On Saturdays he was accustomed to give out the payroll for the men working under him. Whenever a man worked overtime he was, of course, paid proportionately more. Once, after the "boss" had finished his calculations for each man and was about to give out the money, the three-year-old boy got up and cried in childish voice: "Father, the calculation is wrong," and he named a certain number as the true result. He had been following his father's actions unnoticed, but the figuring was carefully repeated and to the astonishment of all present was found to be exactly as the little boy had said. Later Gauss used to joke and say that he could figure before he could talk.

Gauss entered the St. Katharine's Volksschule in 1784, after he had reached his seventh year. Here elementary instruction was offered, and the school was under the direction of a man named J. G. Büttner. The schoolroom was musty and low, with an uneven floor. From the room one could look on one side toward the two tall, narrow Gothic spires of St. Katharine's Church, on the other toward stables and the rear of slums. Here Büttner, the whip in his hand, would go back and forth among about two hundred pupils. The whip was recognized by great and small of the day as the *ultima ratio* of educational method, and Büttner felt himself justified in making unsparing use of it according to caprice and need. In this school, which seems to have had the cut and style of the Middle Ages, young Gauss remained for two years without any incident worth recording.

Eventually he entered the arithmetic class, in which most pupils remained until their confirmation, that is, until about their fifteenth year. Here an event occurred which is worthy of notice because it was of some influence on Gauss' later life, and he often told it in old age with great joy and animation.

Büttner once gave the class the exercise of writing down all the numbers from 1 to 100 and adding them. The pupil who finished an exercise first always laid his tablet in the middle of a big table; the second laid his on top of this, and so forth. The problem had scarcely been given when Gauss threw his tablet on the table and said in Brunswick low dialect: *"Ligget se"* (There 'tis). While the other pupils were figuring, multiplying, and adding, Büttner went back and forth, conscious of his dignity; he cast a sarcastic glance at his quick pupil and showed a little scorn.

In the end, however, he found on Gauss' tablet only one number, the answer, and it was correct. But the young boy was in a position to explain to the teacher how he arrived at this result. He said: "100+1=101; 99+2=101, 98+3=101, etc., and so we have as many 'pairs' as there are in 100. Thus the answer is 50×101, or 5,050." Gauss sat quietly, firmly convinced that his problem had been correctly solved, just as he later did in the case of any completed piece of work. Many of the other answers were wrong and were at once "rectified" by the whip.

Büttner now considered that the proper thing to do was to order a better arithmetic book[1] from Hamburg, in order to give it to the boy whose accomplishments astonished him and soon brought him face to face with the fact that there was nothing more for him to teach the boy.

The Velpke relatives shook their heads, and of course an early death was prophesied for him, according to the old popular belief that heaven's favorites must perish young. The Brunswick neighbors, the father's customers, and others were impressed. His talent had already been shown in his fourth year. In the living room, in which he stayed a great deal with his mother, was hanging an old-fashioned calendar, and soon the little boy could read all the numbers on it. But when the relatives were called together to witness this feat, he got along very poorly, not because he could not read the numbers, but because of the nearsightedness with which he was troubled.

Either Büttner or Bartels[2] even called the boy's father in to talk about the boy's education. His questions as to how to get

---

[1] Remer's *Arithmetica* or Hemeling's *Arithmetisches kleines Rechenbuch*.

[2] Bartels was born in Brunswick on August 12, 1769, and entered the Collegium Carolinum in 1788; he became professor of mathematics at Reichenau in Switzerland, later at the University of Kasan in Russia, and finally at Dorpat, where he died on December 19, 1836, while retired on a pension. His daughter married the astronomer Otto Struve. He published essays on the theory of functions (1822), an essay on the analytic geometry of space in the St. Petersburg Academy *Reports* (1831), lectures on mathematical analysis (Vol. I, 1833). Besides this, he translated Bailly's *History of Astronomy* into German.

means for continued study were met with the rejoinder that assistance of patrons in high position would be won. Thus the rather stubborn father agreed that the boy would no longer have to spin a certain amount of flax every evening. It is said that the elder Gauss, when he arrived home after the conversation, carried the spinning wheel into the back yard and later chopped it up for kindling wood in the kitchen.

Mathematical books now took the place of the spinning wheel in the evening hours; this was attended to by Johann Christian Martin Bartels (1769–1836), the son of Heinrich Elias Friedrich Bartels, a pewterer, living on the Wendengraben. He assisted Büttner, the duties of this office being to cut pens for the smaller boys and to help them in their writing by erasing all the flourishes. A close relationship soon grew up between Bartels, who was himself interested in mathematics, and this remarkable neighbor's child. Teacher and pupil became intimate friends; they studied mathematics together, with such zeal that Bartels himself decided to devote his life to this branch. Thus Gauss in his eleventh year came into independent possession of the binomial theorem in its complete generality and became acquainted with the theory of infinite series, which opened for him the way to higher analysis.

We owe much to Bartels for the service which he performed in informing several persons of high rank in Brunswick about the ability of young Gauss, particularly E. A. W. Zimmermann.[3]

One day Zimmermann ordered Bartels to bring the young boy to him. News of unusual talent had already reached his ears; schoolmates had set the report in circulation. Professor Hellwig, the new mathematics teacher at the Katharineum, had handed

---

[3] Eberhard August Wilhelm Zimmermann (1743–1815) had been full professor of mathematics, physics, and natural history at the Collegium Carolinum since 1766. He had resumed his lectures in 1789 after two years' traveling in England, France, and Italy. This was a short time after Bartels had entered the Carolineum. In 1786 Zimmermann received the title of councilor and in 1796 was raised to the nobility by the emperor; in 1802 he was appointed privy councilor by the Duke, Carl Wilhelm Ferdinand. Zimmermann was highly respected as a scholar and writer, and stood well in the Duke's home. He was a man of attractive nature, lovable, possessed of insight and great humanity.

back Gauss' first written work with the remark that it was superfluous that such a mathematician should continue to appear in his classes.

According to Gauss' own statement it was almost against his father's will that he left Büttner's school. With the aid of "older" friends, among whom were Bartels and doubtless also the philologist Johann Heinrich Jakob Meyerhoff (1770–1812), whom we shall meet later, he had mastered by private study the elements of the ancient languages. In every other respect he was far in advance of persons of his age. Two years later the Katharineum began to take on new life under the direction of Conrad Heusinger.

The Duchess once found young Gauss in the yard of the palace, absorbed in a book. At first skeptical, she soon found in the course of her conversation with him that the little boy understood what he was reading. Very much astonished, she told the Duke to have the boy summoned. When the lackey reached the Gauss home, he was first sent to the older brother, Georg, with his message; but Georg, weeping, strove against the lackey's entrance. After being corrected, he rejoined that it concerned his brother, the "good-for-nothing" who always "stuck his nose in a book." After Carl became a world-famous man and the industrious brother an ordinary laborer, Georg is alleged to have said: "Yes, if I had known that, then I would be a professor now; it was offered to me first, but I didn't want to go to the castle." Georg Gauss was by no means as silly as he is here pictured. Shortly after the incident with the lackey, he went on his period of "wandering." On winter nights Gebhard Dietrich would make the boys go to bed early to save light and heat. In his attic room Carl Friedrich would take a turnip, hollow it out, and roll a wick of rough cotton for it; some fat furnished the fuel, and by the dim light thus obtained he studied half the night until cold and exhaustion forced him to seek his bed.

The surroundings of the Duke delighted the modest, somewhat bashful fourteen-year-old boy, and the tactful Duke, conscious of the fact that he had a very unusual individual before him, knew how to win his love and how to use the means which were necessary for his further education. Young Gauss appeared a bit awkward, but the chief thing of importance was that the Duke quickly and clearly recognized his ability. Gauss departed, enriched in several ways. He received his first logarithmic tables from Geheimrat Feronçe von Rotenkreuz, the minister of state.[4] Assisted by the Duke, he entered the Collegium Carolinum in 1792.

"Five thalers to Councilor Zimmermann for a mathematical [instrument] case bought from the mechanic Harborth for a young man named Gauss." This expenditure, by an order of June 28, 1791, from the special account of the ducal chamber, reveals the first actual trace of the Duke's interest in Gauss. From the chamberlain's accounts it seems that on July 20 ten thalers were to be furnished him yearly, and Zimmermann was to be paid for further expense. On June 12, 1792, came the clause: "These payments are to continue as long as he shall attend the Carolineum." The designation "tuition" cannot be taken in its genuine sense (these are entered under that heading) because Gauss had a "free place" and according to the monthly attendance lists was an "extra free-pupil." There is no doubt that the Duke gave him many other funds and considerable support from his own private means.

The friendship between Gauss and Zimmermann lasted until the death of the latter on July 1, 1815. The son wrote thus about his father's death, under date of March 16, 1816: "My father died last year at the moment when the body of his majesty the Duke was being interred,[5] so much was he overpowered by feelings of melancholy (as is disclosed in a letter half completed by him) that he suffered a stroke. Unfortunately I was just then absent and neither physician nor surgeon to be had. When help came to him after two hours, alas it was too late."[6]

---

[4] Schulze, *Sammlung von Tafeln*.
[5] Duke Friedrich Wilhelm fell at Quatre Bras on June 16, 1815.
[6] Heinrich Mack published two of Gauss' letters to Zimmermann; these are dated November 22, 1797, and December 24, 1797. P. Zimmermann pub-

When Gauss entered the Collegium Carolinum in 1792, it was at the zenith of its fame. Hänselmann states that there were two viewpoints which were important in the plan on which Duke Karl had founded the institution in 1745. At that time there were no institutions of higher learning for those who wished to study subjects not taught in the four faculties of a university. This new creation was to fill a gap between Gymnasium and university. Future officers, architects, engineers, mechanics, merchants, and farmers were to find the opportunity to equip themselves with an education grown universal to meet the higher demands of life, and at the same time with the elements of their own fields of specialization. Ancient and modern languages; Christian dogma and morals; philosophy; universal, ecclesiastical, and literary history; statistics; civil and canon law; mathematics, physics, and natural history; anatomy; German poetry and oratory; the theory of the beautiful in painting and sculpture; exercises in drawing and painting, in music, dancing, fencing, and riding, in turning and glass polishing—all this and more were embraced in the curriculum. But what gave the Carolineum its really attractive standing was the spirit in which they were presented. Not at the purely practical demands of life alone was the scheme of this course directed; the pupils were taught to regard themselves as carriers of the new culture, of the freer and nobler education of the taste and the heart. Zachariä, Gärtner, Ebert, K. A. Schmidt, and a group of colleagues were dedicating their best powers to the Carolineum. The day of this newer culture was just then dawning in Germany. In spite of occasional disappointments, such as could not have been avoided in so high an undertaking, these men might have said that that goal was not an idle one; the course started was the true one.

Distinguished and useful men of various positions in life, both native and foreign, were proud of what they had taken away from Brunswick. Academic teachers like Gellert, Ernesti, Kästner,

---

lished three other of Gauss' letters to Zimmermann, dated October 19, 1796, May 20, 1796, and November 16, 1803, in the Grimme Natalis Co. *Braunschweiger Monatsschrift* (1921), pp. 753 *et seq.*

and Heyne avowed that the youths prepared there were distinguished by thorough knowledge, diligence, and worthy morals. Whenever and wherever the new culture of *bon-sens* and good taste was pursued, the Carolineum was recognized as one of its worthiest nurseries. In the last decade of the eighteenth century there went forth from this institution a number of graduates who were to gain a certain amount of fame; besides Gauss and Bartels there were Ide, Illiger, and Dräseke, the well-known pulpit orator. (We shall return to Ide and Illiger later.) All were born in Germany, and all the children of poor parents, with the exception of Illiger, whose father was a merchant.

When Gauss entered, most of the famous teachers had passed away. Only the last one, Ebert, died on March 15, 1795, not long before Gauss left there. But excellent successors worked in their place. Johann Joachim Eschenburg (1743–1820) took Zachariä's place in 1777 as professor of philosophy and belles-lettres; Johann Ferdinand Friedrich Emperius was professor of Greek, Latin, and English and after Schmidt's death also professor of religious instruction; after 1787 August Ferdinand Lueder was professor of history and statistics in the position vacated by Remer, who was called to Helmstedt. In 1810 he became A. L. von Schlözer's successor in Göttingen. Except for Zimmermann these were the more important of those who saw Gauss sitting in their classes at that time.

On February 18, 1792, Gauss signed himself in the register of the Carolineum *462 Johann Friedrich Carl Gauss, of Brunswick.* Later he never used this name Johann; on all his writings one finds *Carl Friedrich Gauss.* While in this institution he completed his knowledge of the ancient languages and learned the modern languages. During the four years he remained there he was occupied with abstruse mathematical investigations and studies. The essential foundations of his fund of knowledge were deepened and broadened before he left Brunswick.

He seems to have studied carefully at that time the works of Newton, Euler, and Lagrange. Especially did he feel himself attracted by the great spirit of Newton, whom he revered and whose method he fully mastered. During the last year in his native city Gauss discovered the "method of least squares," that method of bringing observations into a calculation so that the unavoidable observational errors affect the result as little as possible, and so that the deviations of the value finally obtained from the observational results are a minimum in any single case and as a whole. Daniel Huber in Basel seems to have arrived at the same method. Adrien-Marie Legendre (1752–1833) also discovered it and published it in 1805 in his *Nouvelles méthodes pour la détermination des orbites,* while the Gaussian deduction did not appear in print until 1809. Thus according to the custom then in vogue Legendre gained the right of priority; the date of publication of a discovery passed for the date of discovery itself. Adrain, in the United States, is also said to have deduced the method in 1808, an article on the law of probability of error appearing in *The Analyst,* a periodical published by him at Philadelphia.

About the year 1750 certain indirect observations in astronomy led to observation equations, and the question as to the proper manner of their solution arose. Boscovich in Italy, Mayer and Lambert in Germany, Laplace in France, Euler in Russia, and Simpson in England proposed different methods for the solution of such cases, discussed the reasons for the arithmetical mean, and endeavored to determine the law of facility of error. Simpson, in 1757, was the first to state that positive and negative errors are equally probable; and Laplace, in 1774, was the first to apply the principles of probability to the discussion of errors of observation. Laplace's method for finding the values of q unknown quantities from n observation equations consisted in imposing the conditions that the algebraic sum of the residuals should be zero, and that their sum, all taken with the positive signs, should be a minimum. By introducing these conditions, he was able to reduce the n equations to q, from which the q unknowns were determined. This method he applied to the deduction of the shape of the earth from measurements of arcs of meridians, and also from pendulum observations.

In his publication of 1805 Legendre proposed the principle of least squares as an advantageous and convenient method of adjusting observations. He called it *méthode des moindres quarrés,* showed that the rule of the arithmetical mean is a particular case of the general principle, deduced the method of normal equations, and gave examples of its application to the determination of the orbit of a comet and to the form of a meridian section of the earth. Although Legendre gave no demonstration that the results thus determined were the most probable or best, his remarks indicated that he recognized the advantages of the method in equilibrating the errors.

Adrain showed from the law of probability of error that the arithmetical mean followed, and that the most probable position of an observation point in space is the center of gravity of all the given points. He also applied it to the discussion of two practical problems in surveying and navigation.

In 1795 Gauss deduced the law of probability of error and from it gave a full development of the method. To Gauss is due the algorism of the method, the determination of weights from normal equations, the investigation of the precision of results, the method of correlatives for conditional observations, and numerous practical applications. Few branches of science owe so large a proportion of subject matter to the labors of one man.

The method thus thoroughly established spread rapidly among astronomers. The theory was subjected during the following fifty years to rigid analysis by Encke, Gauss, Hagen, Ivory, and Laplace, while the labor of Bessel, Gerling, Hansen, and Puissant developed its practical applications to astronomical and geodetic observations. During the period since 1850, the literature on the subject has been greatly extended. The writings of Airy and De Morgan in England, of Liagre and Quetelet in Belgium, of Bienaymé in France, of Schiaparelli in Italy, of Andrä in Denmark, of Helmert and Jordan in Germany, and of Chauvenet and Schott in the United States have brought the method to a high degree of perfection in all its branches and have caused it to be universally adopted by scientific men as the only proper method for the discussion of observations.

There falls to the part of Gauss the service of having rendered the method reliable against all exceptions and of having equipped it so comfortably for use that its utility could not be unfolded until the twenties when he published his "polished" supplementary essays on the subject. From that time on no further application of observations was to be thought of.

Gauss' fundamental paper on this subject is "Theoria combinationis observationum erroribus minimis obnoxiae," 1821, and is given in Volume IV of his *Works*. Woodward wrote:

No single adjunct has done so much as this to perfect plans of reduction, and to give definiteness to computed results. The effect of the general adoption of this method has been somewhat like the effect of the general adoption by scientific men of the metric system; it has furnished common modes of procedure, common measures of precision, and common terminology, thus increasing to an untold extent the availability of the priceless treasures which have been recorded in the century's annals of astronomy and geodesy.[7]

At school in 1792 or 1793 Gauss also investigated the law of primes, i.e., of their rarer and rarer occurrence in the series of natural numbers. Euler he studied for the rich content; Newton and Lagrange for the form presented. Newton's rigor of proof influenced him, and Lagrange's work on the theory of numbers interested him.

Gauss stayed at the Collegium Carolinum well into the fourth year; on August 21, 1795, the order was issued from the ducal office to arrange "that 158 thalers yearly shall be paid to the student named Gauss, going to Göttingen, for assistance and that he be informed of this, as well as that the 'free table' is open to him in Göttingen." His stipend was raised to 400 thalers in 1801 and in 1803 to 600 thalers, besides free apartments.

On October 11 the young scientist left Brunswick. He indicated later on the occasion of his receiving the doctorate that Göttingen University (the Georgia Augusta) was chosen because

---

[7] R. S. Woodward, "The Century's Progress in Applied Mathematics," *Bulletin of the American Mathematical Society*, 2d series, Vol. VI, No. 4 (N. Y., 1900), pp. 149–50.

of the great wealth of mathematical literature in its library. The Duke made no objection, thus furnishing proof of his lively interest in the development of the young man's talent. On this occasion high social circles in Brunswick spoke a great deal of Gauss. It was believed that the Duke added a considerable, specified amount to the stipend from his own pocket. Even then the augurs of the court were setting up a horoscope for the young man, which was well suited to recommend him to the attention of those who listened to such things.

CHAPTER THREE

Student Days

The same age saw the founding of three German universities: Halle, Göttingen, and Erlangen. Göttingen, founded in 1737 by King George II of England, soon took the lead among all German universities and held it until the end of the nineteenth century. George II, from whom comes the name Georgia Augusta, endowed the institution in a princely manner.

Göttingen was equipped, with Halle as a model, by its first curator, Gerlach Adolph von Münchhausen, though it differed from its model in several respects. The legal and political faculty stood forth prominently rather than the theological, as in Halle. *Lehrfreiheit* ruled from the very first. Its theologians followed historical and critical studies rather than controversy. In the first decades there were men like Albrecht von Haller, to whom Göttingen owes the establishment of its medical and scientific school, and the great philologist Johann Matthias Gesner, the former colleague of Johann Sebastian Bach in Leipzig at the St. Thomas School, who founded in Göttingen the first philological seminar in Germany. Great jurists like Georg Ludwig Böhmer labored there. Johann Stephan Pütter made Göttingen an outstanding school for the study of civil law.

German youth from every corner of the land, especially the West and South, people of rank, princes and counts, in fact almost everybody who had an interest in general culture soon streamed into Göttingen. Foreign countries, England above all, took notice of Göttingen. Students came from northern Europe. The faculty included such men as Gatterer, Achenwall, Schlözer,

Spittler, and later Heeren, so that at the end of the eighteenth century one could almost speak of a Göttingen historical school. These men exerted a very strong influence.

That was the foundation for the activity of the man to whom Göttingen owes the most, except perhaps Münchhausen: Christian Gottlob Heyne, the successor of Gesner.[1] Heyne was a pure classicist; the ancient humanities and classical literature, he thought, were the means of every nobler "training of the mind" for the true, the good, and the beautiful. His lectures were attended not only by philologists but by students of all faculties, especially lawyers. Besides, he was a splendid administrative officer. The fate of the university was tied up with him until the close of the eighteenth century. Just as Hollman kept the university from the deeper shocks of the impact of the French occupation at the time of the Seven Years' War, so Heyne, during the Westphalian period, knew how to sweep away the danger of Göttingen's being transformed according to a French model.

Into this atmosphere came young Gauss. When he reached Göttingen, he secured a room at Gothmarstrasse 11. Worries about the future had not left him. He was still undecided whether to follow philology or mathematics. The former offered a surer and speedier prospect of livelihood. For a time, therefore, he was among the auditors of Heyne, who attracted him personally more than did Kästner,[2] whom he dubbed "the leading mathematician among poets and leading poet among mathematicians." N. Fuss once remarked that Euler's *Tentamen novae theoriae musicae* contained "too much geometry for musicians and too much music for geometers."

---

[1] Heyne was born on September 25, 1729, and died on July 14, 1812. The biography of him by Arnold Heeren, the historian, his son-in-law, appeared in 1813. For a good impersonal account see that by Friedrich Leo (1901).

[2] Abraham Gotthelf Kästner was born at Leipzig on September 27, 1719, became docent (1739), and professor (1746) of mathematics at the university there. He went to Göttingen in 1756 as professor and died there June 30, 1800. His *Geschichte der Mathematik* (4 vols.; Göttingen, 1796–1800) was the first work on this subject by a prominent author. He wrote also on equations, geometry, hydrodynamics, and various other branches. His works and memoirs are by no means brilliant and his career, as a whole, was rather mediocre.

At the same time there was another young man from Brunswick striving toward the same goal as Gauss. Johann Joseph Anton Ide, born on January 26, 1775, was the son of a mill inspector who died young and left his family in reduced circumstances. Even in the lifetime of his father the boy had attended the orphange school; on recommendation of the inspector Jenner he was received in 1788 into the instruction department of the Freemasons' lodge, which at the time was under the direction of Hellwig, who discovered in him, as he himself reported, "an eminent genius for the mathematical sciences" and urged him to devote himself to the study of them. But, remarked Hellwig further, "as it is becoming easy for many a young man in this splendid world, even lacking natural abilities, industry, and good rearing, to acquire assistance through various channels, so it was hard for Ide." Indeed, he found a ready reception in the Martineum, and, after Hellwig had become teacher of mathematics and natural history at the reorganized Katharineum, at this school also. In 1794 he got a "free place" at the Carolineum, where he was attracted to Zimmermann and Lueder. But mathematics and its related areas belonged to those sciences "whose culture unfortunately deserved no special assistance, in the eyes of high patrons." For a while the efforts of those two men were in vain. The twenty-five thalers yearly from Count Veltheim were worthy of thanks, but of course did not go very far. There was nothing left except to approach the Duke again. Almost contrary to expectation, this step was of consequence. The Duke promised the necessary means; a half year after Gauss, Ide was able to enter the University of Göttingen at Easter, 1796, where he continued his studies for five years.

Ide belonged to the small circle to which Gauss limited his social intercourse during his student days. He was the only one of the older acquaintances at Göttingen until Arnold Wilhelm Eschenburg (born on September 15, 1778) arrived in 1797. The latter was a son of the Carolineum professor already mentioned; he and Gauss had been close friends there. Eschenburg matriculated to study law and finance. In 1800 after the completion of his studies Eschenburg became a lawyer in Brunswick, and a year later secretary of the lower court. He was appointed cabinet secretary to the Duke in 1805. It was said to be almost incredible how much he influenced the Duke in favor of Gauss while in this position. Eschenburg died in 1861 as governmental councilor and treasurer at Detmold.

The lectures of the brilliant physicist Georg Christoph Lichtenberg (1744–1799) seem to have offered Gauss considerable stimulation, for he called him "the leading adornment of Göttingen." Also Carl Felix Seyffer (1762–1822), after 1789 the assistant professor of astronomy, should be mentioned; Gauss was in friendly association with him, and they carried on a correspondence after Seyffer had moved to Munich.

Ide was called to Moscow in 1803 as professor of mathematics and died there in 1806 as a result of the Russian climate. To the circle of Gauss' student friends belong also Heinrich Wilhelm Brandes (1777–1834), later professor of mathematics at Breslau and Leipzig, and Johann Albrecht Friedrich Eichhorn of Wertheim (1779–1856), a lawyer who was Prussian minister of religion from 1840 to 1848. Gauss frequently came in touch with Johann Friedrich Benzenberg (1777–1846) and later assisted him with his work *Experiments on the Laws of Gravity, the Resistance of the Air, and the Rotation of the Earth* (Hamburg, 1804).

By far the most intimate friend of Gauss at this time was Wolfgang Bolyai von Bolya, descendant of an old Hungarian noble family whose records reach back into the thirteenth or fourteenth century of Magyar history. He was born on the Bolya estate, which lies about fifteen miles north of Hermannstadt; his mother was Christine Vajna of Pava. His father, Caspar, took him in 1781 to the Evangelical Reformed College at Nagy-Enyed. The quiet, introspective boy seldom took part in the games of his schoolmates; indeed, he had to be forced to do so. On the other hand, he had special aptitude for languages, poetry, and mental arithmetic. At the various festivities of the college he was presented as the child prodigy.

At the college Bolyai became a close friend of the cultured

Baron Simon Kemény. About 1790 the two went to Klausenburg and lived in the house of the famous professor of theology Michael Szathmáry, who managed for a while to interest Bolyai exclusively in theology. George Méhes was the instructor in mathematics there. In Klausenburg, Bolyai's eyes were injured by the explosion of some gunpowder, which he had himself prepared, so that for a long while he could read only with the greatest difficulty.

It was at that time the fashion for sons of the Hungarian nobility to attend German universities for their higher education. An older, talented student acted as "mentor." Thus Bolyai accompanied Baron Simon Kemény, who went to Jena in the summer of 1796. In Vienna, Bolyai got sick and had to remain behind. Here he saw the Artillery School and was so carried away that he wanted to enter a military career. Such enthusiasm was a notable characteristic of his. A letter from Kemény caused him to drop this plan and go to Jena and from there to Göttingen, where he arrived in September, 1796.

Bolyai wrote thus:

We [he and his fellow countryman Baron Simon Kemény] went to Göttingen, where Kästner and Lichtenberg were able to stand us, and I became acquainted with Gauss, at that time a student there. Even today I am a friend of his, although I am far from being able to compare myself with him. He was very modest and didn't make much showing; not three days as in the case of Plato, but for *years* one could be with him without recognizing his greatness. What a shame, that I didn't understand how to open up this silent "book without a title" and read it! I didn't know how much he knew, and after he saw my temperament, he regarded me highly without knowing how insignificant I am. The passion for mathematics (not externally manifested) and our moral agreement bound us together so that while often out walking we were silent for hours at a time, each occupied with his own thoughts.

Gauss once said that Bolyai was the only one who understood how to penetrate into his views on the foundations of mathematics.

In discussing his studies of Euler and Lagrange at this time, Gauss later wrote: "I became animated with fresh ardor, and by treading in their footsteps, I felt fortified in my resolution to push forward the boundaries of this wide department of science." Then on March 30, 1796, the nineteen-year-old student made a discovery which determined above all else his future career. That evening remained vividly impressed on his memory. In the *Allgemeine Literaturzeitung* of April, 1796, appeared this notice:

It is known to every beginner in geometry that various regular polygons, viz., the triangle, tetragon, pentagon, 15-gon, and those which arise by the continued doubling of the number of sides of one of them, are geometrically constructible.

One was already that far in the time of Euclid, and, it seems, it has generally been said since then that the field of elementary geometry extends no farther: at least I know of no successful attempt to extend its limits on this side.

So much the more, methinks, does the discovery deserve attention . . . that besides those regular polygons a number of others, e.g., the 17-gon, allow of a geometrical construction. This discovery is really only a special supplement to a theory of greater inclusiveness, not yet completed, and is to be presented to the public as soon as it has received its completion.

CARL FRIEDRICH GAUSS
Student of Mathematics at Göttingen

It deserves mentioning, that Mr. Gauss is now in his 18th year, and devoted himself here in Brunswick with equal success to philosophy and classical literature as well as higher mathematics.

18 April, 1796 E. A. W. ZIMMERMANN, *Prof.*

This was his first publication, and Gauss always considered the discovery one of his greatest. No mathematician in two thousand years had thought of it. It proved to be a rich field for the discoverer. He told Bolyai that this polygon of seventeen sides should adorn his tombstone. That could not be carried out, but there was a design on the side of the base of the Brunswick monument to him.[3]

---

[3] On the back of the monument's base was a seventeen-pointed star, because the stonemason Howaldt said that a polygon of seventeen sides would be mistaken by everyone for a circle.

This reminds one of the tomb of Archimedes, which bore the figure of a

In his youth the lad was less interested in geometry than in algebra, according to Sartorius. What equation could be closer to him than $x^p = 1$, whose roots were so closely connected with the problem of circle division? He opened his scientific diary with the discovery of March 30. The realization that the division of a circle into p equal parts depends on the solution of the equation $x^p = 1$ is due to Cotes and Demoivre; it was first explained and firmly established by Euler in 1748. Gauss was acquainted with the later work of Vandermonde, as shown by his letter of October 12, 1802, to Olbers. The roots of the equation

$$x = \frac{x^p - 1}{x - 1} = 0,$$

if we set

$$r = \cos\frac{2k\pi}{p} + \sqrt{-1}\,\sin\frac{2k\pi}{p}$$

are represented by the powers $r, r^2, r^3 \ldots r^{p-1}$, where k has the values $0, 1, 2 \ldots n-1$; in this manner the points of the regular n-gon are determined by the complex quantities just given. This figure is inscribed in a circle of radius unity. The quantities $\cos 2k\pi/p$ and $\sin 2k\pi/p$ are the rectangular Cartesian coordinates of the points concerned, if one chooses the center of the circle as the starting point and if the axis of abscissae is laid through the point for which $x = 0$.

The above values show that, if p is presupposed to be an uneven prime, the powers $1, g, g^2 \ldots g^{p-2}$ of a primitive root g (mod. p), without respect to order, are congruent to the residues $1, 2, 3 \ldots p-1$, which is the peculiar relation that in the cyclic

---

sphere inscribed in a cylinder. Cicero found the tomb buried under rubbish when he was in Syracuse. Jacques Bernoulli, from his study of the logarithmic spiral $r = a^\theta$, directed that this curve should be engraved on his tombstone with the words EADEM MUTATA RESURGO, and we are told that the visitor to the cloisters at Basel may still see the rude attempt of the stonecutter to carry out his wish.

Gauss' companion, Bolyai, said that no monument should stand over his grave, only an apple tree, in memory of the three apples: the two of Eve and Paris, which made hell out of earth, and that of Sir Isaac Newton, which elevated the earth again into the circle of heavenly bodies.

order, $r, r\varepsilon, r\varepsilon^2 \ldots r\varepsilon^{p-2}$ every one is the equal rational function (viz., the gth power) of the above. With this insight the original viewpoint was won, and from it arose the entire Gaussian theory of circle-division equations.

When Gauss went to Göttingen, J. Wildt (1770–1844) was studying mathematics; he published in 1795 a paper with the title "Theses quae de lineis parallelis respondent." Although Gauss did not become very intimate with Wildt and Kästner, he corresponded with Seyffer until the latter's death. It was in Seyffer's home that Gauss and Bolyai met accidentally, as the latter's son Johann related. Wolfgang spoke freely of the ease with which he could handle mathematics. Shortly afterward, while out walking, he met Gauss on the old town fortification. They approached each other. Wolfgang mentioned, among other things, his definition of the straight line, and several ways of proving Euclid's eleventh axiom. Gauss, surprised and delighted, broke in with the laconic words: "You are a genius; you are my friend!"

Gauss gave to Bolyai the tablet on which he had made the discovery of the 17-gon, as a souvenir; also a pipe. In 1797 the two made a trip to Brunswick by foot to visit Gauss' parents. When Gauss was not in the room, Frau Dorothea asked Bolyai whether Gauss would amount to anything. Bolyai replied: "The first mathematician in Europe," and she burst into tears.

On September 28, 1798, Gauss returned to Brunswick, while Bolyai remained until June 5, 1799. Their letters of this period are extremely interesting from a personal side. Bolyai tells of their last meeting in very touching terms:

When he left a year earlier, and desired that we see each other once more, he wrote that I was to set the time and place [outside of Göttingen]. I decided on Clausthal; and we appeared punctually at the same time, on foot. The professor of astronomy [Seyffer], who was with Napoleon at Austerlitz and then became his engineer colonel, and others accompanied me by foot to the next village. On leave-taking I wept like a child; I went back against my will, but finally gained control of myself. From the last hill, from which Göttingen was still visible, I looked back once again. That picture of the farewell has remained impressed on me forever.

STUDENT DAYS

I accompanied him [Gauss] in the morning, Saturday, May 25, 1799, to the peak of a small mountain toward Brunswick. That feeling of seeing each other for the last time is indescribable. Even a word about tears is ineffective. The Book of the Future is closed. Then we parted with a dying, farewell handshake, almost without words, with the distinction that he, led by angels of the temple of fame and glory, returns to Brunswick, and I, much less worthy, yet calm with a good conscience, go to Göttingen, although pursued afterward as a martyr of truth by an army of the many less worthy ones.

The two friends never saw each other again, but later we shall come to their correspondence on scientific and personal matters. On Thursday, May 23, 1799, Ide, to whom Bolyai had just related the episode of his meeting with Gauss in Klausthal, wrote to Gauss:

If I reckon properly, then our Bolyai arrives [in Göttingen again] at the right time to be able to take some part in the pleasures of the rifle meeting. (Saturday, I believe, the king is led around the house three times with noisy playing and to the sound of bells; afterward follow firecrackers and a riotous parade.) He will in all probability attend this, but only as a philosopher, who on such occasions finds material with which to institute observations on human follies. This is so much his rule of habit, as I have discovered from several cases, that it is difficult for him to miss any of these worldly affairs, not that he wants to enjoy them along with the others, but in order to strengthen his tranquillity of mind. Recently there was a turbulent student uprising again, at which we were accidentally spectators. I went home about ten o'clock because the affair lasted so late into the night, but couldn't persuade him to go along, not because he would have still been willing to be in the action (for we were both empty-handed), but in order to be able to philosophize even further on the idleness of the deed, which he had done the whole time over and over while I was by him.

Unfortunately there is extant no picture of Gauss during his student days. The first we have of him (which is reproduced in this book) was made by Schwarz in 1803.

CHAPTER FOUR

The Young Man

When Gauss returned to his parents in Brunswick on September 28, 1798, at the close of the summer semester, the future was still uncertain. He hoped that the Duke would not withdraw his assistance. Zimmermann did his best to get a promise for him. Shortly after Gauss' arrival Zimmermann had inquired in writing if the Duke would be willing to talk to his protégé. Months passed without an answer, and no opportunity presented itself. It was not likely that Gauss himself would be able to hasten the decision with a request for an audience, because Carl Wilhelm Ferdinand had suffered from considerable abuse of this indulgent custom and now seldom admitted anyone unless personally summoned. Gauss felt keenly the fact that he could not discuss his case personally with the Duke. He wrote Bolyai on September 30, 1798:

Brunswick.

DEAR BOLYAI:
   I reached here last Tuesday. On the second day of my trip I had to travel a good while in the rain. This, together with the fact that I left Monday with a half-empty stomach and spent the entire night *traveling* in the open, though not indeed on foot, brought on a little indisposition from which I have slightly recovered. Now the dear native air has already completely cured me again. Except for Zimmermann, I haven't yet seen any of my older friends. I am planning to visit the duke in a few days. Thus I know very little that is certain about my future fate: if I meanwhile infer your sentiments from my own, then even this item will not be uninteresting to you. Of my duke I have reason to hope that he will still continue his assistance until I receive a regular job. I lost a

very lucrative occupation. A Russian envoy[1] is staying here; I was to have instructed his two young, very spirited daughters in mathematics and astronomy. But because I arrived too late, a French immigrant has already taken over the assignment.

But another affair, very attractive to me, is awaiting me.

Major General von Stamford,[2] whom I have often mentioned to you as an excellent man, intelligent connoisseur, and warm friend of mathematics, desires to go through certain parts of it with me. What parts and on what basis I do not know, because I haven't visited him yet. I think that this will be adequate for my subsistence, and that almost my entire time will be my own. That is the most important news that I can write you now. My future landlord seems to be a good, honest man: I have seen my room only at night: it appears quite comfortable and suitable to me. In about eight days . . . when my riding trousers are done . . . I plan to take a trip to Helmstedt. Then I hope to be able to write you of my other circumstances. And so, until then, farewell, good Bolyai,

GAUSS

P.S. I have used the Latin letters on account of your eyes and because you once told me that you would rather have them than the *Gothik* [i.e., cursive].

My address: Charles Frederic Gauss, *Candid. en philosophie* (or even nothing at all) at Mr. Schröder's on Wendenstrasse. I need not say that all tidings from you will be welcome to me.

Remember me to my acquaintances there: Ide, Simonis, Eichhorn, Seyffer, Lichtenberg, Kaestner, Persoon, or whomsoever you see.

Because you are in danger of traveling with Murhard,[3] I consider it

---

[1] The Russian mentioned in this letter was Count Murawjeff, envoy in Nether-Saxony.

[2] Franz Karl von Stamford (1742–1807) was successively an officer in the Brunswick, Prussian, and Dutch Service. He wrote *An Attempt to Present the Fundamentals of the Differential and Integral Calculus, Without Introducing the Idea of Infinitesimals* (Berlin, 1784). He received an ambassador's post in Berlin and Gauss' hopes again were not fulfilled. Several other disappointments of this kind occurred as to similar offers.

[3] Friedrich Wilhelm August Murhard (1799–1853) received his Ph.D. at Göttingen in 1796 and had been Privatdozent in mathematics there. His *Bibliotheca mathematica* (5 vols.; Leipzig, 1797–1805), valuable even today, is a famous proof of his far-reaching knowledge of literature. In 1798 he left Göttingen on a long journey which led him to Asia Minor. He plagiarized and dedicated the book to the emperor, was arrested in Hungary for idle babblings about government and religion, and made a translation of Lagrange's *Mecanique analytique*. Later Murhard led a varied and adventurous life as a political and journalistic writer.

my duty to tell you of his behavior here, which may show you that it is not without danger to join with such a man on a trip. Save for this reason I would not even mention such a name in a letter to you, even as I do not make mention here of a single one of any of his other windy chatterings.

Here in Brunswick he hires a man who drives his (Murhard's) carriage to Helmstedt with his own horses. When they get there he asks him (without paying him) if he doesn't want to come again two days later in order to drive him back to Brunswick again, when he is then to receive double pay. The coachman agrees. When he comes again in two days Murhard has been gone a long time. (*En passant*: When I went through Nordheim Monday night, Murhard came directly into the tavern; he appeared very disturbed and said that he had just traveled from Helmstedt to Goslar, and planned to go to Göttingen the next day. What became of him there you probably know already.)

He goes with Secr. Brückmann,[4] whose father's famous mineralogical cabinet he had inspected, into a loan library and takes out several books. Thereupon, when he is to leave something as security, as is the custom, he calls on his companion, Mr. Brückmann, to identify him. But M. didn't return the books, and Mr. Brückmann (who doesn't know M.) has to pay for them.

On November 29, the following observations occur in another letter to Bolyai:

I am living for the most part now on credit, because my financial prospects are all shattered. . . . I have been in Helmstedt and found a very good reception with Pfaff as well as with the custodian of the library. Pfaff came up to my expectations. He shows the unmistakable sign of the genius, of not leaving a matter until he has dug it out as far as possible. With great kindness he offered me the use of his library and I am going to write him in a few days to request various books.

Greet all my acquaintances. But mention to no one what I have told you of my circumstances, and should anyone inquire about the matter, then say that you only know in general that I have good, although not entirely certain, prospects, which in fact is quite true. And visit me as soon as you can.

GAUSS.

An improvement in his circumstances soon came, as the following extracts from a letter of December 30, 1798, to Bolyai show:

---

[4] Urban Friedrich Benedict Brückmann (1728–1812), professor of anatomy and ducal physician in Brunswick, had a special collection of precious stones.

Several favorable changes have occurred in my situation since my last letter. Indeed I have not yet talked with the duke himself, but he has explained that I am to continue receiving the sum which I enjoyed in Göttingen (which amounts to 158 thalers annually and is now fairly adequate to my needs). He desires further that I become a doctor of philosophy, but I am going to postpone it until my *Work* is done, when I hope I can become one without costs and without the usual harlequinery.

About the middle of December, 1799, Gauss returned to Helmstedt again in order to use its library. He was cordially received by the librarian Bruns, as well as by the professor of mathematics Johann Friedrich Pfaff (1765–1825), whose acquaintance Gauss had already made during his stay in Helmstedt in October, 1798. He rented a room in Pfaff's home and furnished it himself, but studied so strenuously and incessantly that the others in the house saw him for only a few hours in the evening. He and Pfaff would then take walks "to the spring and to Harpke." The topics of conversation were usually mathematical in nature. On such occasions it is believed that Gauss gave out much more than he received.

After a brief introductory paragraph Gauss wrote to Bolyai on December 16, 1799:

Do not allow me to spoil very many lines in explaining the cause of my somewhat belated answer [to your last letter]. The chief one is that I did not know for certain until October that I would not yet take my trip to Gotha as intended then; and I desired to be able to write you only when certain what sort of a place I have now temporarily substituted for it. That I can do now; it is Helmstedt, where I arrived a few days ago and from where I write you this letter. I now narrate from there in chronological order.

You remember that I had already sent in a thesis to the philosophical faculty at Helmstedt, when we saw each other in Clausthal for the last time, as candidate for the title of doctor. This affair has progressed since then, and the faculty conferred this title on me on July 16 without burdening me with most of the formalities heretofore customary. Our good prince has taken over the costs of it. The paper is printed and was already finished in August. I know no way, at once safe and convenient, of sending it to you, but you will probably be able to get it easily through the bookstore; hence I am writing the complete title. It is: Demonstratio nova theorematis, omnem functionem algebraicam rationalem integram unius variabilis in factores reales primi vel secundi gradus resolvi posse, auctore Carolo Frederico Gauss, Helmstadii apud C. G. Fleckeisen, 1799. 5 folios in quarto, with a copperplate print. The title indicates quite definitely the chief purpose of the essay; only about a third of the whole, nevertheless, is used for this purpose, the remainder contains chiefly the history and a critique of works on the same subject by other mathematicians (viz. d'Alembert, Bougainville, Euler, de Foncenex, Lagrange, and the encyclopedists . . . which latter, however, will probably not be much pleased) besides many and varied comments on the shallowness which is so dominant in our present-day mathematics.

Certainly this pamphlet will interest you at least as the first attempt of your friend. To my knowledge, published reviews of it haven't yet appeared anywhere. Up until now I have distributed some thirty copies, partly to mathematicians, partly to those whom I owed a debt of courtesy. Until now I have lacked an opportunity to send one to France. Of private comments which have come to my knowledge only that of General von Tempelhoff[5] in Berlin is especially important to me and gladdened me the more, because he is one of the best German mathematicians, and especially because my criticism touched him as the author of a compendium. From third hand I have found out that he thus passes opinion: (his own words) "that Gauss is an absolutely hopeless mathematician; he doesn't yield a hand's breadth of ground; he has fought bravely and well and holds the battlefield completely." . . .

Since I shall probably not enter the fetters of a position very soon and had too slight assistance in Brunswick, I made the decision to betake myself here to Helmstedt for a while, and I shall probably remain here until Easter; you can send your letters at your pleasure either here or to Brunswick, for I have arranged that all letters directed to me there are forwarded at once.

I am staying here at Professor Pfaff's whom I esteem as an excellent geometer as well as a good man and my warm friend; a man of an innocent, childlike character, without any of the violent emotions which so dishonor a man and are widespread among scholars. Since I haven't been here eight days, I can't decide yet how well I shall be pleased otherwise; the place itself is terrible, the environs are praised; one must do without many comforts of life; the tone among the students as a whole is said to be rather rough; among the professors with whom I have become acquainted there are well-bred men.

You write me of my portrait; I shall certainly send it to you, but for

---

[5] Georg Friedrich von Tempelhoff (1737–1807), who is mentioned in the letter, was a Prussian artillery officer, member of the Berlin Academy of Sciences, and published a textbook on algebra in 1773.

several reasons not now. Also you will prefer a later picture of me, when my looks will have changed, for my present features are, I hope, as fresh in your memory as yours are in mine.

This letter will hardly reach you this year; tell me in your next one when you received it; the last day of December will at least be the last day which we call seventeen hundred (if micrological exegetes now postpone the end of the century one more year) and will be especially sacred to me. Note that when it is midnight here for us, midnight is already an hour past with you. On such festive occasions my mind passes into a loftier mood, into another spiritual world; the partitions of the room disappear, our filthy, paltry world with everything that appears so big to us, makes us so happy and so unhappy, disappears, and I am an immortal pure spirit united with all the good and noble who adorned our planet and whose bodies space or time separated from mine, and I enjoy the higher life of those greater joys which an impenetrable veil conceals from our eyes until death . . . do not cease to love your constant friend

C. F. G.

In 1797 Gauss discovered a new proof of Lagrange's theorem, and at the close of this year two of his letters addressed to Zimmermann show that he was working out the *Disquisitiones arithmeticae*. These are probably the first reports of a work which was to put Gauss in the ranks of the greatest mathematicians of all time. The printing of it was begun in Brunswick by the printer Kircher, who had recently acquired ownership of a print shop as an adjunct of the public-school system under the superintendency of Joachim Heinrich Campe. In 1799 he sold it to Friedrich Vieweg, Campe's son-in-law, in order that he might move back to Goslar, where he also had a printing shop, previously owned by Duncker, which he had gotten in 1783 by marriage. In Goslar he completed the printing of the *Disquisitiones,* although the work took much longer than Gauss had anticipated. Of unusual interest is the part which Meyerhoff[6] took in this book—the correction of the Latin.

---

[6] Johann Heinrich Jakob Meyerhoff (1770–1812) became in 1794 collaborator, and in 1802 director, of the gymnasium in Holzminden. He was thoroughly grounded and trained in the ancient and modern languages. As a Göttingen student he had won a golden prize medal for a Latin dissertation on the Phoenicians. Yet mathematics was rather foreign to him.

The above is striking enough, if one considers how little Gauss needed to mistrust his own proficiency in this respect. According to Moritz Cantor, Gauss wrote a classical Latin, giving rise to the expression that Cicero, if he could understand the mathematics of it, would have censured nothing in the Gaussian Latinity, except perhaps several customary incorrect modes of expression which Gauss used purposely. But it was Latin just the same and therefore attractive and stimulating to only a narrow circle of readers. Referring to Meyerhoff's work, Gauss wrote:

Of course I understand that it cannot be an especially attractive work for Mr. Meyerhoff, since he does not seem to be sufficiently acquainted with mathematics, in order to look on it just as reading. Thus the word *algorithmus* was unknown to him. Only on a single point must I take the liberty of disagreeing with him. I well know that *si* with the subjunctive is not good Latin; but modern mathematicians seem to have made for themselves the rule of constantly using the subjunctive in hypotheses and definitions; I do not remember an example of the opposite, and in Huyghens, who according to my notion writes the most elegant Latin and whom I purposely, therefore, have imitated, I find the subjunctive continually in these cases. I open at random and find *Opera,* p. 156, *Quodsi fuerit;* p. 157, *Si sit, si fiat, si agitetur;* p. 158, *si suspendatur;* pp. 188 *seqq.* are examples by the dozen. Therefore, since in this instance the desire to be a genuine Roman would be only purism (which as far as I am concerned would be less allowable, because I am well minded not to be so in any case) and the thing is not at all absurd in itself, I went with the current. I hope Mr. M. will not take offense at me. What was incomprehensible to him in the *accedere possunt,* p. 5, I have not been able to guess; I have therefore let it stand. The passage p. 7, which previously ran thus: *Si numeri decadice expressi figurae singulae sine respectu loci quem occupant addantur,* Mr. M. misunderstood, because he probably didn't know that *figurae* means numbers; he took *numeri* for the nominative plural and *figurae* for the dative singular and on that account suggested to me that *singulus* is not wrong; but just for this reason a mathematician will probably not construe it incorrectly, chiefly because it doesn't make sense; nevertheless I have now arranged the words somewhat differently by this time.

It was probably for Zimmermann's sake that Gauss was so careful of his Latin, for as the latter's protégé he must have been anxious that his first work should in no wise lack the stamp of completion. His preliminary reports had so aroused the Duke's high expectations that he assumed the cost of publication and the book was gratefully dedicated to him when it finally appeared in 1801.

For Zimmermann as well as for Gauss it would have been painful to a high degree if the critics had found any well-grounded defects in it. Meyerhoff was the proper person to relieve them of these cares. Zimmermann had known him from the Carolineum. Meyerhoff had entered at the same time with Bartels, and through him may perhaps have come into touch also with Gauss. How much the classic form of the *Disquisitiones* owes to his polishing is hard to establish.

The *Disquisitiones arithmeticae,* which is perhaps Gauss' principal work, contains many important researches, one of which, known as the celebrated Fundamental Theorem of Gauss, or the Law of Quadratic Reciprocity of Legendre, of itself alone, R. Tucker wrote, "would have placed Gauss in the first rank of mathematicians." In March, 1795, Gauss discovered by induction, before he was eighteen years old, the proposition that -1 quadratic residue of prime numbers of the form $4n+1$ is a quadratic nonresidue of primes of the form $4n+3$ and worked out the first proof of it, which he published the following year. He was not satisfied with this, but published other demonstrations resting on different principles, until the number reached six. The proposition which he had discovered appeared to its author to be one of no ordinary beauty; and as he conjectured that it was connected with others of still greater value and generality, he applied (as he himself assures us) all the powers of his mind to find the principles upon which it rested and to establish its truth by a rigid demonstration. Having fully succeeded in this aim, he felt himself so completely fascinated by this class of researches that he found it impossible to abandon them. He was thus led from one truth to another, until he had finished the larger part of the first and most original, if not the greatest, of his works, before he had read the writings of any of his precursors in this department of science, notably those of Euler and Lagrange. He had, however, been anticipated in enunciating the theorem, although in a more complex form, by Euler. Legendre had unsuccessfully tried to prove it. R. Tucker remarked that the question of priority of enunciation or of demonstrating by induction is in this case a trifling one; any rigorous demonstration of it involved apparently insuperable difficulties.

The subsequent study of the arithmetical researches of these great masters of analysis could hardly fail to expose Gauss to the chagrin which young men of premature and creative genius have so often experienced, of finding that they have been anticipated in some of their finest speculations. The crowning result of his labors was, as is well known, the complete solution of binomial equations, and a most unexpected achievement in placing the imaginary on a firm basis. He was the first to use the symbol i for $\sqrt{-1}$, giving it the interpretation of a geometric mean between $+1$ and -1. This convention rapidly spread into general use. Of his theorem on binomial equations he made two other distinct demonstrations in December, 1815, and January, 1816. But these works, though he was the first in the field on the subject, gave him little fame. Lagrange seems not to have heard of the first one; and Cauchy, whose subsequent demonstrations have been preferred in textbooks, received in France all the praise due to a first discoverer.

The fruits of the Helmstedt sojourn were manifold. His doctoral dissertation embodied the complete solution of binomial equations, and it was upon this that the philosophical faculty awarded him the Ph.D. degree, in 1799. Doctoral dissertations, even of the greatest scholars, rarely exhibit anything of more than passing value. But not so with Gauss. The foundation of the whole theory of equations is formed by the proposition that every expression placed as a sum of powers of one and the same unknown with positive, integral exponents, in order, is factorable into factors of the first or second degree of that unknown. The existence of this fundamental theorem of algebra had long been known. Many writers had published supposedly rigorous proofs of it. Gauss proceeded to show, in his first section, a model of historical and critical presentation, that all those early "proofs" were only pseudo proofs, were only miscarried attempts, and in its second section his thesis presented an indisputable proof of the theorem, of faultless, dogmatic sharpness. The proofs of 1815 and 1816 were just as rigorous. In 1849, at the golden-anniversary celebration of his doctorate, almost the last paper which he ever published gives still another proof. This is essentially the elaboration of a thought indicated in the thesis and presents the same proof in altered form. Which of these proofs is to be regarded as the most characteristic, and the most beautiful, is a matter of taste. It speaks highly for the four proofs when first this one and then that one receives the highest praise from various writers.

In dedicating his *Disquisitiones arithmeticae* to the Duke of Brunswick, Gauss acknowledged in very touching terms the wise and liberal patronage which had not only provided for the expenses of his work but also enabled him to exchange the humble pursuits of trade for those of science. The work itself, its author assures us, assumed many changes of form in its progress to maturity, as new views presented themselves from time to time to his mind; but, as is well known, the course which is followed in the discovery of new truths is rarely that which is most favorable to their clear exposition, more especially when it has been pursued in solitude, with little communication with other minds. The peculiar terminology which Gauss employed in the classification of numbers and their relations, and which is so completely embodied in the enunciation and demonstration of nearly every proposition that it can never be absent from the mind of the reader, rendered the study of this work so laborious and embarrassing that few persons have ever mastered its contents. The book is one of the standard works of the nineteenth century and has never been really surpassed or supplanted. Not an error has ever been detected in it. But though the character of Gauss' subjects tempts few readers, though his own severe brevity renders these subjects even more difficult than they need be, yet the young reader of Euclid may be brought into contact with Gauss so as to understand the tone of his genius in a manner which would be utterly impossible in the case of Newton, or Lagrange, or Euler.

Even Legendre, who had written so much and so successfully on the same subject, and who, in the second edition of his *Théorie des nombres* (1808) makes the great discovery of what this book contains the occasion not merely of special investigation but of the most emphatic praise, complains of the extreme difficulty of adapting its forms of exposition to his own; while the writers of the *Biographie des contemporains* in a notice of the author at a much later period, when he had established many other and almost equally unquestionable claims to immortality, quote an extract from a report of a commission of the Institute of France, to whom it was referred in 1810, in which it is said, "that it was impossible for them to give an idea of this work, inasmuch as everything in it is new, and surpasses our comprehension even in its language." The biographers then proceed to stigmatize the book as full of puerilities, and refer to its success, including its translation into two languages, as ground for the assumption that "charlatanism sometimes extended even to the domain of mathematics." It is rather amusing that twelve years later we read in this work: "Suffice it to say that his works are esteemed in general by the most distinguished mathematicians and that they recommend themselves, rather by their exactitude than by clarity, by their precision and elegance of style."

Another section in the *Disquisitiones* involves the theory of the congruence of numbers, or the relation that exists between all numbers that give the same remainder when they are divided by the same number. As Gauss was working on the book a new problem arose; it was much longer than originally planned and hence the funds promised by the Duke would not cover the cost of publication. He decided to close with the seventh section, to lay aside the eighth, and also to shorten it in many places. Even so, the publication suffered a long interruption until one day the Duke inquired about it and learned the difficulty. That was enough. The Duke supplied the needed funds, and in the summer of 1801 the final sheets of the folio left the press of Gerhard Fleischer in Leipzig. The sale of the book was greater than had originally been hoped for, but the author received almost no profit, because the bookseller in Paris, to whom most of the copies had been consigned, went bankrupt. Gauss was not the man to dwell on the misfortune.

The dedication, dated July, runs thus:

> I account it the highest joy, most gracious Prince, that you allow me to grace with your exalted name this work which I offer to you in fulfillment of the holy obligations of loyal love. For if your Grace had not opened up for me the access to the sciences, if your unremitting benefactions had not encouraged my studies up to this day, I would never have been able to dedicate myself completely to the mathematical sciences to which I am inclined by nature. Indeed, the observations, some of which are presented in this volume . . . the fact that I could undertake them, continue them for several years, and publish them, I owe only to your kindness which freed me from other cares, and permitted me to devote myself to this work. Your magnanimity has pushed aside all the obstacles which delayed publication.

He also refers to the Duke's understanding of the coherence of all the sciences, and to his wise insight, which did not deny assistance to those sciences which appear of little use in practical life and are criticized as abstract. This was not a phrase of flattery, and no one was more deeply convinced of the truth of the expression than Gauss himself.

The Duke's long delay in granting Gauss an annuity on his return to Brunswick is explained by the fact that his state treasury was close to bankruptcy, depleted as it was by his father's too generous spending. Tongues wagged in criticism of such gratuities as impractical and unwarranted. But in the end the Duke's own better judgment settled the matter, and young Gauss' financial worries were at least temporarily removed.

One can easily see now how the publication of the work was delayed for four years by various circumstances. Some of the material in the first six sections relates back many centuries to Diophantus. The presence of some old theorems is accounted for by the fact that the publications of the Berlin and St. Petersburg academies were not available to Gauss. Two things stand out: the doctrine of quadratic forms, and the name "determinants," which

Gauss first introduced. "Mathematics," he said, "is the queen of the sciences and the theory of numbers the queen of mathematics." If this is true, one may continue by calling the *Disquisitiones* the Magna Charta of the theory of numbers. Gauss published memoirs on biquadratic residues in 1817 and 1831. After his death, an eighth section of the *Disquisitiones* treating of congruences of higher degree was found and published.

Lagrange, who was called by Napoleon the high pyramid of the mathematical sciences, wrote to the young scholar shortly after the *Disquisitiones* were published: "Your *Disquisitiones* have with one stroke elevated you to the rank of the foremost mathematicians, and the contents of the last section [theory of the equations of circle division] I look on as the most beautiful analytical discovery which has been made for a long time."

Laplace, who was the author of the famous *Mécanique céleste* and before whose impressive greatness everyone in the Paris academy bowed down, is said to have exclaimed: "The Duke of Brunswick has discovered more in his country than a planet: a superterrestrial spirit in a human body!" This was just after the young Brunswick mathematician had made his first important astronomical contribution. Several years later Laplace recommended to Napoleon the Conqueror especial consideration for the university city of Göttingen, because "the foremost mathematician of his time dwells there."

Gauss' own opinion of the *Disquisitiones,* uttered in his closing years, is not without interest: "The *Disquisitiones* belong to history, and in a new edition, to which I am not disinclined, but for which I now possess no leisure, I would change nothing, with the exception of the misprints; I would merely like to attach the eighth section, which indeed was essentially worked out, but did not appear at the time in order not to increase the cost of printing the book." Gauss wrote to Bolyai that he hoped to have published as a second volume of the work many supplementary pieces. These investigations as a matter of fact were later published in bulletins of the Royal Society of Göttingen and the university's periodicals. The complete list of Gauss' writings, exclusive of larger works, contains 124 papers. It is singular that Gauss devoted so few memoirs to subjects of a purely algebraic character; except for a rather unimportant paper on "Descartes' Rule of Signs" in *Crelle's Journal* (1828), his only algebraic papers relate to the doctoral thesis.

Very soon the *Disquisitiones* went out of print; Eisenstein, the brilliant mathematician who died very young, could not get possession of a copy of the original. On account of the scarcity of copies, other pupils of Gauss had the courage to copy the work by hand from beginning to end. Just as certain clergymen go about with their prayer book, so many a great mathematician of the nineteenth century was always accompanied by a much-read copy of the *Disquisitiones,* taken from the original volume. Nevertheless, it remained a book with seven seals.

By a mischance the thoroughly rigorous and simple proof of Lagrange's theorem, which Gauss found in 1797, was never published. Gauss communicated it to Pfaff, who forwarded it to Hindenburg; Hindenburg soon died, and the manuscript never appeared again. Gauss never published this proof, of which a transcript is extant, because he later found that Laplace had applied a similar method. The theorem ran thus: "If p and q are positive uneven prime numbers, p has the same quadratic character with regard to q that q has with regard to p; except when p and q are both of the form $4n+3$, in which case the two characters are always opposite, instead of identical." Legendre's unsuccessful attempt to prove this appeared in the *Memoirs* of the academy of Paris for 1784. Gauss was not content with once vanquishing the difficulty; he returned to it again and again in the fifth section of the *Disquisitiones* and there obtained another demonstration based on entirely different but perhaps still less elementary principles. In 1875 Kronecker showed that Euler was the real discoverer of this law, in that he enunciated it before Legendre.

Gauss rightfully called this theorem the fundamental one in quadratic residues, because it not only forms the real nucleus in the theory but also has fundamental meaning in the later portions of the theory.[7] He could not foresee that scientists would furnish forty proofs of it! He did say that proofs would have to be created more *ex notionibus* rather than *ex notationibus*. Dirichlet succeeded in reducing Gauss' eight cases to two. Gauss stated that he struggled for a year with one little point.

It is no wonder that he should have felt a sort of personal attachment to a theorem which he had made so completely his own, and which he used to call the "gem" of the higher arithmetic. It would be impossible to exaggerate the importance of the influence which this theorem has had on the later development of arithmetic, and the discovery of its demonstration by Gauss must certainly be regarded (as it was by Gauss himself) as one of his greatest scientific achievements. The fifth section ("these marvelous pages") abounds with subjects each of which has been the starting point of long series of important researches by later mathematicians. It is curious that wonderful researches only alluded to in this section first saw the light sixty-three years later in the second volume of the collected edition of his works. Until the time of Jacobi the profound researches of the fourth and fifth sections were almost completely ignored. The seventh section on the theory of circle division and the theory of equations was received with great enthusiasm.

Very noteworthy is a passage (Article 335) where he observes that the principles of his method are applicable to many other

---

[7] The chronology of the eight Gaussian proofs of the fundamental theorem, dating them according to the time of their publication and counting them as Gauss does, is—

Proof I (*Disquisitiones*, Article 135 *et seq.*; *Works*, Vol. I, p. 104):	1801
Proof II (*Disquisitiones*, Art. 262, *Works*, Vol. I, p. 292):	1801
Proof III (Royal Soc. of Göttingen, *Works*, Vol. II, p. 1):	1808
Proof IV (Royal Soc. of Göttingen, *Works*, Vol. II, p. 9):	1811
Proofs V & VI (Royal Soc. of Göttingen, *Works*, Vol. II, p. 47):	1818
Proofs VII & VIII (posthumous, *Works*, Vol. II, p. 234):	1863

Proof I dates from April 8, 1796, and Proof II from June 7, 1796. Proofs VII and VIII (really one full proof) are at the latest September 2, 1796. Proof IV dates from May 15, 1801; the remainder of the proofs originated after 1805, but the exact date of the discovery of each is uncertain.

functions besides circular functions, and in particular to the transcendents dependent on the integral

$$\frac{dx}{\sqrt{1-x^4}}$$

and have equal importance with respect to the lemniscate, just as the trigonometrical functions are related to the circle. We know from his diary that on March 19, 1797, he investigated the equation which is used in the division of the lemniscate and on March 21 established the fact that this curve may be geometrically divided into five equal parts by means of a straightedge and compasses. This almost casual remark (as Jacobi observed) shows that Gauss before 1801 had already examined the nature and properties of the elliptic functions and had discovered their fundamental property, that of double periodicity. From May 6 to June 3, 1800, he formulated the general concepts in this field. Nothing of this was published; Gauss was ahead of his time. Abel and Jacobi found the doubly periodic or elliptic functions in the twenties. The elliptic modular functions which Gauss mastered in 1800 were fully developed only in recent times.

It was in the course of an astronomical investigation that Gauss arrived at some elliptic integrals, the evaluation of which he was able to effect by means of a transformation included in Jacobi's series of transformations. It is said that this distinguished analyst was induced by his knowledge of the fact to seek (after his own discoveries were completed) an interview with the great mathematician who had thus intruded, prematurely as it were, into one of the deepest recesses of his own province. Jacobi submitted his various theorems to Gauss' inspection, and they were met, as they successively appeared, by others of corresponding character and import produced from Gauss' manuscript stores. The interchange concluded with an intimation that Gauss had many more in reserve. Such an anticipation of discoveries which totally changed the aspect of this difficult department of analysis constituted no infringement of the rights which Jacobi undeniably secured by priority of publication. Wide circulation was given to this story. The story is confirmed by the papers on elliptic transcendents, published in the third volume of Gauss' *Collected Works*.

The *Disquisitiones* were to have included an eighth section; at first a complete theory of congruences was planned, but later Gauss appears to have proposed to continue the work by a more complete discussion of the theory of circle division. Manuscript drafts on each of these subjects were found among his papers; the first of them is especially interesting, as it treats of the general theory of congruences from a point of view closely allied to that later taken by Evariste Galois, Serret, and Richard Dedekind. This draft appears to belong to the years 1797 and 1798.

Euler had enriched the theory of numbers with a multitude of results, relating to Diophantine problems, to the theory of the residues of powers, and to binary quadratic forms; Lagrange had given the character of a general theory to some, at least, of these results by his discovery of the reduction of quadratic forms and of the true principles of the solution of indeterminate equations of the second degree. Legendre, with many additions of his own, had endeavored to arrange as many as possible of these scattered fragments of the science into a systematic whole in his *Essai sur la théorie des nombres* (1799). But the *Disquisitiones* were already in press, and in the *Essai* what was new to others was already known to Gauss.

The remarkable interpretation of the arithmetical theory of positive binary and ternary quadratic forms is found in Gauss' review of the *Works* of L. Seeber (1831).[8] The two important memoirs on the theory of biquadratic residues appeared in 1825 and 1831. In the second of these two papers Gauss gives a theorem of biquadratic reciprocity between any two prime numbers, no less important than the quadratic law: "If p_1 and p_2 are two primary prime numbers, the biquadratic character of p_1 with regard to p_2 is the same as that of p_2 with regard to p_1." This theorem itself and the introduction of imaginary integers upon which it depends are memorable in the history of higher arithmetic for the variety of the researches to which they have given rise.

---

[8] Gauss, *Collected Works*, II, 188.

CHAPTER FIVE

Astronomy and Matrimony

The discovery of Ceres (Ferdinandea) made by Joseph Piazzi (1746–1826) in Palermo on New Year's Day, 1801, did not become known through the German newspapers until May, and the first reasonably accurate reports about it were given by the *Monatliche Correspondenz zur Beförderung der Erd- und Himmelskunde* in its June number. The editor was Baron Franz Xavier G. von Zach (1754–1832), lieutenant colonel and director of the Seeberg observatory near Gotha, and this journal served as the collecting point for important new geographical and astronomical reports. Piazzi had dispatched on January 24 letters to Bode, director of the Berlin observatory, to Oriani in Milan, and to Lalande in Paris reporting that he had discovered a very small comet without tail and envelope. In February Lalande informed von Zach about it, without accurately indicating the place in the heavens, so that von Zach awaited further information. The letters of Piazzi to Oriani and Bode did not reach their destinations until April; the one to Bode was seventy-one days on the way. Meanwhile Piazzi had been able to pursue the object only up to February 11. In his letters Piazzi gave only two observed locations, those of January 1 and 23, rounded off in whole minutes, and only noted that from January 10 to 11 the receding motion went over into right motion; also he added in his letter to Oriani that he supposed that it was a planet, while he spoke in the letters to Bode and Lalande only of a comet.

Both Bode and Oriani immediately gave the new report to von Zach, who promptly brought out in the June number of the *Monatliche Correspondenz* an exhaustive article "on a long supposed, now probably discovered, new major planet of our solar system between Mars and Jupiter." Bode communicated the discovery to the Royal Prussian Academy of Sciences, and saw to its publication in several newspapers.

On the occasion of a "little astronomical trip to Celle, Bremen, and Lilienthal" about which he published an exhaustive diary in the issues of the *Monatliche Correspondenz* in 1800–1801, von Zach had decided, along with five other astronomers (Schröder, Harding, Olbers, and probably von Ende and Gildemeister) who met in Lilienthal, to found an "exclusive society of twenty-four practical astronomers scattered over Europe" who were to plan the study of the supposed planet between Mars and Jupiter by simultaneous correction of the star catalogues. Piazzi was also among the twenty-four, but had not yet received the invitation to take part in the society. Bode, as well as Oriani, held fast to the belief that the new object was a planet, moving between Mars and Jupiter, and even von Zach agreed with this view, which would lead them beyond the superficial calculation of a circular orbit. He therefore tried a somewhat sharper computation of a circular orbit which exhibited a noteworthy similarity to the comet of 1770; the semi-major axis is the only element which shows similarity in both orbits. He also questioned whether both objects were not perhaps identical; certain doubts as to the nature of the new planet were, therefore, continually arising. Piazzi spoke of it in a later letter as only a comet, and even the Paris astronomers appear to have valued the discovery lightly.

Meanwhile Bode had written Piazzi requesting an accurate record of his observations. He received no satisfactory answer, and an extensive correspondence on the subject of the discovery sprang up between Bode, von Zach, and Olbers (who had learned of it from the newspapers), and also Burckhardt in Paris. Just at this time Lalande arrived in Paris, and received a letter from Piazzi which gave the observations more accurately, but with the request not to publish them prematurely. Lalande however shared them with Burckhardt on the same conditions, and Burckhardt with the German astronomers.

On the basis of these more accurate observations Burckhardt computed an ellipse; his attempts to represent the observations by a parabola failed. The corresponding elements are in the July issue of the *Monatliche Correspondenz,* in which von Zach gave in all monthly periodicals "Continued Reports about a New Major Planet."

The complete observations of Piazzi from January 1 to February 11 were finally published in the September issue of the *Monatliche Correspondenz,* after Piazzi had sent them with several corrections to Bode, Lalande, and Oriani, and thus they reached Gauss.

In the October issue of the *Monatliche Correspondenz* von Zach recorded that from about the middle of August until the close of September attempts were made by almost all astronomers to find the planet as it again emerged from the rays of the sun, but in vain; bad weather was also prevalent at this time. The elliptic orbit calculated by Burckhardt was unreliable, not so much because the observed portion of the orbit was quite small (which was then considered by astronomers as the chief difficulty) as because it deviated from an arbitrary attraction in the region of perihelion. The problem of determining a completely unknown planetary orbit from observations had arisen previously only in the case of Uranus, where a circular orbit could be reckoned more accurately through reference to Bode's discovery of the much earlier observations of Flamsteed (1690) and Tobias Mayer (1756). Olbers had also begun the calculation of an elliptical orbit, but with little hope of success, since he considered the preliminary calculation of a circular orbit to be fundamental. He gave the elements of such an orbit as follows:

If the new planet had gone through its aphelion before January 1, then its heliocentric rapidity is always increasing, and even its geocentric lengths must be greater in August and September than according to the circle hypothesis. But if it went through its perihelion in February, then it later decreased the heliocentric velocity and its geocentric lengths must be smaller in August and September than according to the circle hypothesis. Since it is not known now which of the two cases obtains, it is safer, in future searching for the star, to hold as basic those cases from the circle hypothesis which cannot deviate very much from the true ones and to keep the mean of both possible cases.

Olbers, just as Burckhardt, falsely assumed that the planet stood at the time of its discovery not far from perihelion or aphelion, while Gauss later showed that it was about halfway between.

Piazzi wrote a brief essay in which he made careful communications about the first discovery and the later observations. He also included in this his own calculations of one of the circular orbits as well as those sent to him by Oriani and the orbits computed by the other astronomers. To this paper is attached a corrected list of his observations. In the November issue of the *Monatliche Correspondenz,* von Zach gave an exhaustive review of this work "which probably could not come into the German bookstores either soon or easily." He printed opposite the observations that had appeared in the September issue the corrected observations, which contain a correction of the right ascension for February 11 of about 15 minutes.

With respect to Olbers' proposal to base the preliminary calculation for rediscovery of the planet on a circular orbit, von Zach computed an ephemeris for November and December "in order to perform a little service for all astronomers and lovers of the subject who want to busy themselves with the search for the star."

During the interim Gauss, who received the *Monatliche Correspondenz* in Brunswick, had silently given himself over to the problem; the interest in the new planet caused him to lay aside temporarily his purely mathematical researches and his theory of the moon. In his diary notes, entries 119 and 120 show that he was working out the method in 1801 (September and October). His earliest notes on Ceres date from the first part of November, 1801, but lack clearness.

One can see that Gauss, as soon as he took up this work, created new practical methods for orbit determination. He would not limit himself to a certain hypothesis by experiment, but systematically sought the orbit which would fit the observations as well as possible: if the Piazzi observations embrace only forty-one days, then there must be an ellipse which fits them, and which approximates the predicted places for the rediscovery. It was a matter, therefore, of finding an ellipse that was free from all arbitrary assumptions. On the first pages of his manual for November, 1801, we find this problem completely solved, although in a less complete form than in the *Theoria motus*. In a brief manuscript, "Summary Survey of the Methods Applied in the Determination of the Orbits of Both New Planets," Gauss collected his earliest methods, and sent this to Olbers on August 6, 1802; it was returned to him in November, 1805. Shortly after the appearance of the *Theoria motus*, von Lindenau got it, supposedly on a visit with Gauss, and published it with Gauss' consent in the *Monatliche Correspondenz* for September, 1809.

The determination of the orbit of a celestial body from three given observations is an almost impossible task; an explicit solution is not achievable, because the observed places and the relations to the determining elements of the orbit are very involved. The solution of the problem is indicated by approximations, and hence the setting up of an almost unlimited number of methods is possible; these are distinguished by more or less important factors.

It is to be expected that Gauss thoroughly investigated the field to which these methods apply; but it will also be understood that he was not able at the time to carry through this work on his first orbital calculations because of the need to speed calculation of the single case now pressing. It was important that the orbit of the new planet should be known as soon and as accurately as possible so that the planet might be rediscovered in the heavens. Thus Gauss explained that the first orbital calculations rest on an important new fundamental thought to which the arbitrariness of older methods no longer applies. But in this particular achievement one does not find the consummate perfection and delicate elaboration of the methods of the *Theoria motus*.

This first fundamental thought consists in the setting up of an equation between the distances of the planet from the sun and from the earth in the mean observation; it is too involved to be given here. Gauss wrote of it: "This formula is the most important part of the entire method and its main foundation." He tells us that he came on his fundamental formula in a very bizarre way.

As Gauss explained in Article 2 of the "Summary Survey," the first computation of the completely unknown orbit of a celestial body from three observations rests on the solution of two different problems: first, to find an approximate orbit in any sort of way; second, so to correct this orbit that it "satisfies" the observations as well as possible.

If the orbit is even approximately known, then the first problem vanishes.

His older calculations for the determination of Ceres' orbit are (so far as preserved) dated November, 1801. Gauss numbered the systems of elements which he found in further correction of the orbit.

Von Zach and many others had sought Ceres according to the Gaussian ephemeris in December, but under the most unfavorable weather and without success. In February, however, von Zach could finally announce in his *Monatliche Correspondenz* the lucky rediscovery of the planet. He had informed Gauss by letter on January 17, 1802. During the night of December 31 to January 1, von Zach made sure that a suspicious star observed by him was definitely Ceres. On January 1, Olbers also discovered the planetoid, its location agreeing exactly with the Gaussian ephemeris. Gauss seems to have first heard of Olbers' rediscovery of Ceres through the newspapers. He wrote to him on January 18 in order to get Olbers' observations; with this letter began the correspondence which shows how busy Gauss was with new corrections of the Ceres orbit. The correspondence ripened into a lifelong, intimate friendship between Gauss and Olbers.[1]

---

[1] Heinrich Wilhelm Matthias Olbers was born October 11, 1758, at Arbergen on the Weser, a village near Bremen, where his father was a pastor. He studied medicine at the University of Göttingen during the years 1777–1780, at the same time attending Kästner's lectures on mathematics. In 1770, while watching by the sickbed of a fellow student, he devised a method of calculat-

The discovery of the planet Ceres introduced Gauss to the world as a theoretical astronomer of the highest order. He was able to calculate in one hour the orbit of a comet, for which task Euler, using the old methods, had used three full days. Through such close application Euler later lost the sight of one eye. "To be sure," said Gauss, "I would probably have become blind also, if I had been willing to keep on calculating in this manner for three days." The planet Uranus had been discovered twenty years before Ceres, when near opposition; this was a critical position, which at once gave a near approximation to the elements of its orbit. A stationary elongation of Ceres, though less fertile in its results, was sufficient to assign her such a place between Mars and Jupiter as was required to satisfy Bode's singular law. Kepler had found a planet to be lacking, as he sought to round out one of that series of cosmic speculations which had guided him to the discovery of his laws. The complete determination, however, of the elements of a planet's orbit from three geocentric longitudes and latitudes—or from four of the first and two of the

---

ing cometary orbits which made an epoch in the treatment of the subject, and is still extensively used. This important discovery was published by Baron von Zach with the title *Ueber die leichteste und bequemste Methode die Bahn eines Cometen zu berechnen* (Weimar, 1797). A table of 87 orbits was attached, increased to 178 by Encke in the second edition (1847), and to 242 by Galle in the third (1864 and 1894).

About the end of 1781 Olbers settled as a physician in Bremen; in June, 1785, he married Dorothee Elisabeth Köhne, who died in less than one year, having fourteen days previously given birth to a daughter, Doris, who married the lawyer Dr. Christian Focke, and died in 1818. She had six children, and through her Olbers had seven great-grandchildren during his lifetime. In 1789 Olbers married Anna Adelheid Lürssen (born in 1765); one son, Senator Georg Heinrich Olbers, was born on August 11, 1790, and survived his father, who died March 2, 1840. Anna Lürssen Olbers died on January 23, 1820.

Olbers practiced medicine actively for forty years in Bremen and retired on January 1, 1823. He is said never to have slept more than four hours, the major portion of each night being given over to astronomy. The upper part of his house was fitted up as an observatory. He gave special attention to comets, and that of March 6, 1815, (period, seventy-four years) was named for him, in memory of its discovery by him.

His bold hypothesis of the origin of the minor planets by the disruption of a primitive large planet *(Monatl. Corr.* VI, 88), although later discarded,

second in those cases where the latitudes are evanescent or small —was still a new problem, already solved only in the case of comets moving in parabolic orbits. Newton, whom we can thank for its first solution, had pronounced it to be *problema omnium longe difficillimum*.

Gauss devoted one memoir to the beautiful demonstration of a very remarkable proposition that the secular variations which the elements of the orbit of a planet would undergo from another planet which disturbs it are the same as if the mass of the disturbing planet were distributed into an elliptical ring coincident with its orbit, and in such a manner that equal masses of the ring would correspond to portions of the orbit described in equal times.

In marked contrast to the publication of the *Disquisitiones*, this rediscovery of Ceres was a spectacular accomplishment. It was Piazzi's discovery which gave Gauss the opportunity of revealing, in most impressive form, his remarkable mathematical superiority over all his contemporaries. This particular work also

---

was strengthened by the finding of Juno by Harding and of Vesta by himself, in the precise regions of Cetus and Virgo where the nodes of such supposed planetary fragments should be situated. Olbers was deputed by his fellow citizens to assist at the baptism of the King of Rome on June 9, 1811, and was a member of the *corps legislatif* in Paris 1812-1813, besides being a member of many learned societies. He received decorations from the governments of various countries, and a monument was erected in Bremen to his memory. In 1828, the largest ship which had ever left Bremen (up to that time) was named *Olbers*, for him, and carried a thousand German emigrants to Brazil. It was wrecked in 1837.

He was the eighth of sixteen children. His father became preacher of the Domkirche in Bremen in 1760 and died in 1772. Olbers attended the Gymnasium in Bremen; his preference for astronomy showed itself when he was fourteen years of age. At the age of nineteen he observed the solar eclipse of 1777. At Göttingen he computed the orbit of the comet of 1779. His doctor's thesis of 1780 was entitled *De oculi mutationibus internis*. In 1781 he made a trip to Vienna, where he gained admission to the observatory under Maximilian Hell, and for the first time, on August 17, 1781, saw the planet Uranus, discovered that year by Herschel, even before the Vienna astronomers saw it.

His home in Bremen was at Sandstrasse 16. He had a large medical practice and at ten o'clock in the evening would withdraw to his private observatory. A monument was erected to his memory on the wall-promenade in 1850.

had the effect of greatly improving his personal affairs in a financial way, and made possible the establishment of his own home. Brunswick had no observatory, very likely not even a telescope (worthy of the name) within its walls. This young, little-known scholar, without the external aids and instruments of astronomers, was possessed of an inner vision so far-reaching and a mathematical genius so marvelously piercing that he was able from calculations at his desk to locate the missing orbit of the lost asteroid so accurately that now the work of retracing and rediscovery by men equipped with the telescope could no longer fail.

Further observations of Ceres by Olbers led him in 1802 to the discovery of a second planet in the immediate neighborhood of Ceres, and, with the privilege accorded him as discoverer, he. named it Pallas. In 1804 a third planetoid was discovered by Ludwig Harding, and named by him Juno. A natural consequence of Gauss' calculation of the orbit of Ceres was that all calculations of these new celestial bodies now devolved upon him. In this he had no rival. Gauss devoted especial care and work to the Olbers planetoid, Pallas, making close investigation and calculations of its movements and perturbations, as a result of which he came to call it his favorite.

The year 1801 was truly epoch-making for Gauss. In later years he recalled that so many new discoveries and important ideas came to him, as sources of scientific truths, that he could not control them. Many of these occupied his time during the remainder of his life, while others were lost in the press of other affairs. The works of Lagrange and Laplace, which appeared at the time, gave him material for investigation in the field of celestial mechanics. He clearly realized how these men had attained their discoveries by means of Newton's creative ideas, and his admiration for Newton reached new heights. "Newton remains forever the master of all masters!" he exclaimed. The deeper he penetrated into mathematics, the more fully he was persuaded that its true meaning lies in its application to practical life and natural science. Thus we explain his turning to theoretical and practical astronomy. He had made use of the "astronomical instruments" at the Collegium Carolinum, but we do not know what they were. It was there that he may have gotten his first practice in observing.

An anecdote relates that Gauss' attention was first directed to Ceres in the course of a conversation with Zimmermann. Piazzi rejoiced more than anyone else at the rediscovery of Ceres: *"Faites, je vous en prie, mes compliments et mes remerciments à M. Gauss, qui nous a épargné beaucoup de peine et de travail, et sans lequel peut-etre il ne m'aurait pas réussi de vérifier ma decouverte!"* Bode, Schröder, Mechain, Maskelyne, and others expressed themselves in a similar manner. Gauss' joy was inexpressibly great. He knew the full value of his accomplishment and found it natural that specialists laid emphasis on it. Yet he remained modestly in the background and said that had it not been for Newton's *Principia* he would not have been able to establish the new method. The best thing in the whole affair, he said, was the confirmation of the Newtonian hypothesis of universal gravitation.

The first open recognition Gauss received came when the Academy of Sciences in St. Petersburg elected him a corresponding member on January 31, 1801. Many honors quickly came to him from other learned societies. The Russian minister of state, Nikolaus von Fuss (1755-1826), wrote him officially to announce his election, and their continued correspondence presently matured into a warm friendship. This led to the Russian government's subsequent attempt to call Gauss to the St. Petersburg observatory. This was in part due to high admiration for the *Disquisitiones*. The last letter of N. von Fuss to Gauss, dated March 24, 1824, announced his nomination as foreign member of the academy. N. von Fuss died at St. Petersburg on January 24, 1826, in his seventy-first year, and his death genuinely grieved Gauss.

In the summer of 1801, Zimmermann received a call from St. Petersburg. This was very enticing, for it meant greater leisure for literary activity than his triple professorship at the Carolineum allowed him. The Duke outbid this offer, raising Zimmermann to the rank of privy councillor and releasing him from all the duties of his former position. Thus Zimmermann was retained for Brunswick. Filling the two chairs he vacated at the Carolineum was not to be easy, however, for placing mathematics and natural history in one person's hands demanded care. Even Hellwig, who had the first "say" as to this succession, urgently requested that one person be appointed for each of these branches. He proposed two excellently suited persons. Count Hollmannsegg, just returned from a four years' journey in Portugal, had several weeks previously decided to reside in Brunswick as long as Hellwig lived, in order to collaborate with him in arranging the rich natural-history collections which he had brought home and which he hoped to complete from the sources open to him. An earlier pupil of Hellwig, Johann Carl Wilhelm Illiger, took part in this work. He had written a recognized work, his *Naturhistorische Terminologie,* for which the philosophical faculty at Kiel had conferred on him a Ph.D. *honoris causa.* Assisted financially by the Duke, he was teaching privately in Brunswick. On November 14, 1801, Hellwig wrote in a "Promemoria": "Without much commotion there has been established by this league in Brunswick a council to which will be called, in different branches of natural history, German, French, Italian, and northern research men to the complete satisfaction of those concerned." In the interest of science, therefore, as well as of the country, it seemed highly desirable to keep Illiger in Brunswick, and for that purpose the best opportunity occurred at Zimmermann's withdrawal. Hellwig wished to see Illiger made professor of natural history. For the professorship of mathematics he recommended no one but Ide.

This other protégé of the Duke had also justified the hopes placed in him. He had made a name for himself in 1800 with his *Theorie der Weltkörper unseres Sonnensystems und ihrer elliptischen Figur, nach Herrn Laplace frei bearbeitet;* in 1801 there came his *System der reinen und angewandten Mechanik fester Körper.* In September, 1801, he received a Ph.D. at Helmstedt, and at the beginning of the winter semester attended academic lectures in Göttingen. Although he could not match Gauss in creative depth of mind, it was agreed that he was better qualified as a teacher. "It would be wrong," wrote Hellwig, "to pass over our talented and brilliant Gauss in silence. But that I am more for Ide in the holding of this position is settled for me by his well-known excellent teaching capacities."

Hellwig's plans were shattered. He himself became Zimmermann's successor in both departments. Illiger became his son-in-law nine years later and died in 1813 as professor and director of the Zoological Museum in Berlin.

Toward the end of the year 1801 the Duke granted Gauss a yearly income of four hundred thalers in order to retain him in Brunswick and in recognition of the *Disquisitiones*. When Zimmermann announced this to him, he exclaimed: "But I certainly haven't earned it, I haven't done anything for the country yet." He now decided to buy a sextant at his own cost and put it to practical use for his country.

The Duke sent out the order on January 25, 1803, as follows:

Then to Dr. Gauss of this city, who has refused a call to St. Petersburg, an increase of 200 thalers besides an extra wood allowance of 4 cords of beech and 8 cords of fir wood and instead of the free apartment, until he can receive such *in natura*, a reimbursement of 50 thalers annually has been granted. It is so ordered that said increase is to be paid, without deduction of the first quarterly payment from Christmas last year, in quarterly installments, besides the lodging compensation of 50 thalers, out of that fund from which he collects his present salary.

Thus Gauss formally entered the service of the Duke without any definite official duties, a position which suited his inclinations and needs. There was always the possibility that an observatory would be established, such as Duke Ernst II of Gotha had erected on Seeberg and had placed under the direction of von Zach.

Gauss was subjected to some criticism because of his loyalty to the Duke, which prevented him from accepting any offers of a position away from Brunswick. Even his father expressed doubt, but his mother would always defend her Carl Friedrich. Self-confident and enjoying the companionship of a small group of friends, he paid little attention to his critics.

Presently he was to be made happy by a circumstance in which he found all that had been lacking. In the course of a letter to Bolyai on December 3, 1802, Gauss wrote:

How much I rejoice over your domestic bliss. I embrace your wife, who gives my friend the sweetest jewel of life. You write me I must not allow myself to be misled by your example, and unfortunately it is a sure thing that whosoever marries enters a lottery where there are many blanks and few prizes. May heaven grant that if I should one day take the leap, I do not draw a blank. . . . On the whole I have slight connection with Göttingen. Just two weeks ago the Society of Sciences named me a corresponding member. . . . May the dream which we call life be a sweet one for you, a foretaste of the true life in our real home, where the fetters of the slothful body, the barriers of space, the scourges of earthly passions and the mockery of our petty needs and desires no longer oppress the awakened spirit. Let us courageously and without grumbling bear the burden to the end, but never lose sight of that higher goal. Joyfully shall we then lay down the burden when our hour strikes and see the thick curtain fall.

Again on June 20, 1803, to Bolyai:

I wish you good luck, a thousand times, in your son. Truly only he who is a father has full citizen's rights on the earth. You now have in your hand the first links of fate's chain of eternal beings continuing on into the infinite. An important and serious, but sweet, calling. May your son some day bless you as the establisher of his well being! . . . Astronomy and pure mathematics are the magnetic poles toward which the compass of my mind ever turns. I am going now to Bremen for a visit with Dr. Olbers, whose friendship I have cultivated up to this time by letters without knowing him personally.

Gauss' mother had worked for a tanner named Ritter, before she married Gebhard Dietrich in 1776. Two of that name, Friedrich Behrend and Georg Karl Ritter were the godfathers of Carl Friedrich. As a child Gauss had been frequently in the Ritter home. Every Christmas he had found a present there and had otherwise enjoyed himself in the family. After his return from Göttingen he again took part in the gatherings at Ritter's house. The happy tone of the group, unassuming but not without some of the better culture then found in the Brunswick middle class, had a strong charm for Gauss in moments of recreation.

In this circle he became acquainted in 1803 with Johanna

Elisabeth Rosina Osthoff, the daughter of Christian Ernst Osthoff (1742–1804) and Johanna Maria Christine Ahrenholz (1747–1821). Her father was a master tanner and owned the house at what is now Leopoldstrasse 3, a man of moderate means according to the standards of his position. Johanna, born May 8, 1780 (St. Martin's Day), was the only child and her parents' darling. An education in complete contrast to that which Gauss had known had developed in her all the gifts of a happily endowed nature. She was not a dazzling beauty and her letters now and then show a lack of proper schooling. But she was cheerful, unusually kind, happy as a child, charmingly roguish, and gifted with innate understanding. To all who met her she appeared as a creature of light, we are told. Unfortunately no picture of her is extant, but her only daughter, Minna, is said to have been the very image of her.

In a letter dated June 28, 1804, after a paragraph on Borstorf apples and on the whereabouts of old Göttingen friends and professors, and after recommending certain mathematical works, condemning certain astrological superstitions, and commenting on how he enjoyed Hungarian wines, Gauss reported to Bolyai about Johanna:

Since my last letter to you I have made a lot of new acquaintances, some of them extremely interesting. I was in Bremen and Lilienthal for one month with Olbers, one of the most lovable of all men whom I know, and with Schröter, and was four months in Gotha. Here also the circle of my acquaintances and friendships was considerably widened. The most beautiful however, is the friendship of a splendid girl, exactly such a girl as I have always desired for a life companion. A wondrously fair madonna countenance, a mirror of spiritual peace and health, kind, somewhat romantic eyes, a perfect figure and size (that is something), a clear understanding and an intelligent conversation (that is also something), but a quiet, happy, modest, and chaste angelic soul which can harm no one, that is the best. Ostentatious display of flirting and inordinate passion are foreign to her. But I shall not give free rein to my feelings toward this comely creature until I see hope that I can make her just as happy as she deserves. A one-sided happiness is none at all.

Heaven and Hell! I freeze before the picture which you paint of women. What sombre demon guided your pen when it wrote:

"Trust no girl even if she appears as clear as a ray of light, her heart like a crystal spring with a lovely pure crystal bottom, gentle as the soft, cooling evening air in sultry summer. Trust not. The white snow passes away and leaves after it a black muddy filth."

Those features are not my Johanna's (her name the same as your son John's). But your terrible picture can never be applied to her. Be advised, therefore, of even more. I have already known her a year. Indeed I was struck by her quiet virtues as soon as I first saw her, but always observed her very coolly from afar and only recently have I approached her. My conviction of the excellence of her heart is not the result of blinding by passion, but of the most unrestrained observation.

But I stop. If we become more closely related, then you shall find out more of her at the proper time; besides, if heaven wills otherwise, a picture of her has no interest for you.

After a year of active courtship Gauss realized that the possession of her was his highest earthly desire, and on July 12, 1804, opened his heart to her in the following letter, which must always be classed with the gems of German amatory literature:

My very dear friend, accept favorably the fact that in this letter I pour out my heart to you, about an important matter, regarding which I have found no proper opportunity to speak till now.

Finally, let me say from the fullness of my heart, that I have a heart for your silent angelic virtues, an eye for the noble features which make your face a true mirror of these virtues. You, dear modest soul, are so far removed from all vanity that you yourself do not realize your own value; you don't know how richly and kindly heaven has endowed you. But *my* heart knows your worth—O! more than it can bear with ease. For a long time it has belonged to you. You won't repel it? Can you give me yours? Dear, can you grasp the proffered hand, and do it gladly? My happiness hangs on the answer to this question. Indeed, at present, I can't offer you riches or splendor. Still, dear, I cannot have mistaken your beautiful soul—you are certainly as indifferent to riches and splendor as I am. But I have more than I need for myself alone, enough for two young people to start a carefree, agreeable life, regardless of my prospects for the future. The best that I can offer you is a true heart full of the warmest love for you.

Ask yourself, beloved friend, whether this heart completely satisfies you, whether you can respond with equally sincere feelings, whether you can contentedly make the journey of life hand in hand with me, and decide soon.

I have placed before you, darling, the desires of my heart in artless, but candid words. I could have done it in entirely different words. I could paint for you a portrait of your charms, which, although it would be nothing more than the truth, you would have received as flattery; with burning colors I could make for you a picture of my love—to be sure, I would only need to give rein to my feelings—a portrait of the bliss or disconsolation which await me ever after you have accepted or rejected my desires. But I didn't want to do that. At least, don't mistake the purity of my unselfish love. I don't want to bribe your decision. In the gravest concern of your life you must not allow any extraneous considerations to influence you. You are not to bring a sacrifice to my happiness. Your own happiness alone must guide your decision. Yes, dearest, so warmly do I even love you, that only possession of you can make me happy, provided you are in agreement with me.

Dearest, I have exposed to you my inner heart; passionately and in suspense am I waiting for your decision. With all my heart,

Yours,
C. F. GAUSS.

Three months passed before an answer was received. Not that Johanna was still uncertain of her feeling for him. She had loved him for a long time; but the loftiness of his personality and the fame which surrounded him had led her to repress her own humble desires. And later something came to her ears which must have aroused doubt. Busy gossip had spread it abroad that another young lady, well educated and of great wealth, was engaged to Gauss. It is not known just how the report started.

For many days Gauss wondered why Johanna was hesitating. A very different sort of scrupulous compunction also engaged his attention. The report was noised abroad that Napoleon would sell the southern portions of Hanover, including Göttingen, to the Elector of Hesse. If this occurred, there was a strong possibility of his being called to Göttingen University. Moreover, things in Brunswick had taken a turn which made his future there look very uncertain, dependent as he was on the bounty of the Duke. In this state of affairs, uncertain of the future, he doubted whether he had the right to tie up the fate of Johanna irrevocably with his own.

Nevertheless the two became engaged on November 22, 1804; and three days later he wrote to Bolyai:

Do apply yourself somewhat to practical astronomy. It is, according to my notion, next to the joys of the heart and the contemplation of truth in pure mathematics, the sweetest enjoyment which we can have on earth.

A second, even more important reason, however, why I delayed my answer somewhat, was that I didn't want to send it until I could tell you something of myself. I can do that now, dearest Bolyai. For three days that angel, almost too celestial for this earth, has been my fiancée. I am superabundantly happy. You wish that the picture which I sketched for you would be a good one. It is not accurate; it says far too little. Her cardinal trait is a quiet devout soul without one drop of bitterness or sourness. Oh, she is much better than I. I had only one scruple. Not the fear of a negative answer. No, she was always kind to me, conscious as I am of my shortcomings; indeed if one becomes better acquainted with most young men, their frivolity and heartlessness, one cannot help taking fresh courage and gaining more confidence in oneself. But it seemed to me that her devout heart was too inaccessible to earthly love, that she could not accept my most patient love, that perhaps she could not respond and that she could not make me really happy. But God, how glad it makes me, because the crust of ice now melts away from her heart in the heat of celestial love, when she chastely snuggles up to my bosom, and her soulful eye, her warm, silken hand, and her delicately formed mouth with maidenly, bashful purity speak nothing but love. Life stands before me like an eternal spring with new glittering colors; not until now do I understand with full clarity the beautiful words of our incomparable poet Jean Paul:

"Wie zwei Selige vor Gott schauen einander in die Augen und in die Herzen: sie reden nicht, um sich anzublicken: sie erheben die Augen, um durch den Freudentropfen durchzuschauen, und senken sie nieder, um ihn mit den Augenlidern abzutrocknen."

Oh, I had never hoped for this bliss; I am not handsome, not gallant, I have nothing to offer except a candid heart full of devoted love; I despaired of ever finding love.

"*Puissances du Ciel: j'avais une âme pour la douleur, donnez m'en une pour la félicité*," I would like to exclaim with Rousseau's St. Preux.

How much it affects me, good fellow, that you are not entirely happy. Your noble soul deserves it so very much. May the dissonances which seemed to be bringing discord into your life concert only have seemed to be such, and may they soon resolve themselves into pure consonants of an eternal, celestial harmony.

I purposely interrupted this letter for several days in order to write you in a really cool mood concerning your scientific communications: but how can I come to such! Every new day gives me new assurances of my happiness, new proofs of how much the good pure soul loves me.

If anything could increase my happiness even more, it would be the discovery that she loved me first, even earlier than I loved her. Our first acquaintanceship in the summer of 1803 covered only a period of a few weeks (because I soon afterward left for Gotha) and was not renewed until April of this year. Although she had made such a favorable impression on me even at the beginning, I put off at that time any serious thought of marriage, partly because I had the opportunity of meeting her only a few times, and partly because of other conditions. In vain I have so often endeavored, therefore, to recall the day when I saw her for the first time. What a pleasant surprise for me, that she herself could tell me this and every following day when I saw her. Do think always of your friend, dear Bolyai, when he celebrates his happiest days on July 27 and November 22. On the latter day she became mine before God, when she will become mine before the world, certain other important conditions must determine. These will soon be decided, and I shall inform you immediately after our plans have matured.

On October 9, 1805, Gauss and Johanna were married at St. Katharine's Church. They had their apartment at the time in Ritter's house, where Gauss had been living as a bachelor, now Steinweg 22.

CHAPTER SIX

Further Activity

In all his major writings Gauss is a notable representative of that type of scholar and investigator that Wilhelm Ostwald designated as *Klassiker*. In view of the general incomprehensibility of the *Disquisitiones* at the time of their publication, one may well imagine his pleased surprise three years later when he received a certain letter from Paris. He wrote to Olbers: "Recently I have had the joy of receiving a letter from a young Parisian geometer, Le Blanc, who is familiarizing himself enthusiastically with the higher arithmetic and gives proofs that he has penetrated deeply into my *Disquis. Arith.*" This word of recognition and praise is significant, since Gauss was conservative in such matters. His pleasure over this and following letters from the same correspondent would have been still greater if he had had an inkling that this "young geometer Le Blanc" who was offering additions to his own book was a woman.

In the late fall of 1806, when Brunswick was under French rule, Gauss apparently had planned to leave the city, but finally decided to remain. On November 27, a French officer named Chantel, chief of a batallion, entered the room where Gauss and his wife were. His general, Pernety, who was besieging Breslau and was busy with the encampment, sent him at the instance of "Demoiselle" Sophie Germain in Paris to inquire after Gauss' health and if necessary to offer his protection. *"Il me parut un peu confus,"* Chantel reported to his general shortly after this scene. Of course Gauss was perplexed and astonished by this visitor, for he knew neither a General Pernety nor a "Demoiselle"

Sophie Germain[1] in Paris. In all Paris, he explained to the officer, he knew only one lady, Madame Lalande, the wife of the famous astronomer. When the officer asked further whether he wanted to write a letter to Sophie Germain in Paris and give it to him for forwarding, Gauss did not know whether to answer with yes or no. Under these circumstances he merely thanked the officer and his general for the kind attention shown him.

Not until three months later did Gauss discover through Denon who Sophie Germain really was. She saw fit to tell him herself. He wrote to Olbers, "That Le Blanc is a mere assumed name of a young lady, Sophie Germain, certainly amazes you as much as it does me." Bolyai jokingly put these words in a letter

---

[1] Sophie Germain was born on April 1, 1776, the daughter of a well-to-do Parisian family. She was just thirteen years old when the Revolution broke out and would frequently take refuge in her father's library during those days. One day Montucla's *History of Mathematics* fell into her hands. Here she read of Archimedes' death. The story made a deep impression on her. In order to study Newton and Euler, she learned Latin without a teacher's instruction. When the Ecole Polytechnique was opened, she smuggled her own mathematical papers in among those of the male students under the pseudonym of Le Blanc, and the professor, no less a person than Lagrange, found the work of this Le Blanc so praiseworthy that he became her adviser in studies, when he found out who the author was. Thus she became a city celebrity; great scholars visited her at her father's home. In 1799 appeared Legendre's *Theory of Numbers,* and when Gauss' work appeared nearly three years later she was prepared to present to him the results of her own research in the subject, as a result of study of the two volumes. She again used the pseudonym, because at that time a learned woman was hardly taken seriously. But when Brunswick was taken by the French and she became worried about the safety of the young mathematician there, "Le Blanc" was laid aside, and she appealed to General Pernety, who was a friend of her family.

Sophie Germain's most important accomplishments are in the field of the theory of numbers. To be sure, she reaped the greatest honors with a problem in the theory of surfaces and mathematical physics, for which the Paris Academy of Sciences had set up a prize by order of Napoleon. A mathematical theory of elastic surfaces was sought, a result of Chladni's physical experiments. Sophie Germain received the prize. Her paper has today only historical value. She died of cancer, on June 27, 1831, aged fifty-five. When the matter of honorary degrees came up in 1837 at the centenary celebration of the University of Göttingen, Gauss regretted exceedingly that Sophie Germain was no longer alive. "She proved to the world that even a woman can accomplish something worth while in the most rigorous and abstract of the sciences and for that reason would have well deserved an honorary degree," he said.

of his to Gauss: "You once wrote me of a Sophie in Paris; if I were your wife, I would not be too pleased. Write me more of her."

One of the fruits of Gauss' sojourn in Helmstedt was the famous Easter formula. According to his own story his mother could not tell him the exact day on which he was born; she only knew that it was a Wednesday, eight days before Ascension Day. That started him on his search for the formula. His first article on this subject was published by von Zach in the *Monatliche Correspondenz*, II (August, 1800), 121. In this, the cyclic calculation of the Easter date is reduced to purely analytic processes. The simplicity of the method is extremely remarkable.

The method originally applied only to the Julian and Gregorian calendars but was presently extended to the Passover in the Jewish calendar. Gauss made note of this extension in his diary on April 1, 1801. It was published by von Zach in the *Monatliche Correspondenz* (May 5, 1802), p. 435. Chevalier Cisa Gresy gave the first proof of the Gaussian rule for the Jewish calendar in the *Correspondance astronomique*, I (1818), 556.

The involved process was designed to prevent Easter and the Passover from occurring at the same time.[2] But since no cyclic calculation can accurately agree with the course of the moon, this is impossible. The event will recur for centuries; in the nineteenth century it took place in 1805 and 1825; in the present century this occurs in 1903, 1923, 1927, 1954, and 1981. The last occurrence will be in the year 7485.

Following is the original Gaussian method of calculating the date of Easter:

In the Julian calendar:	$m = 15$	$n = 6$
In the Gregorian calendar,		
from 1700 to 1799:	$m = 23$	$n = 3$
from 1800 to 1899:	$m = 23$	$n = 4$

---

[2] Ch. Z. Slonimsky so expanded the latter formula that it gives any required information about the Jewish calendar of any year, in *Crelle's Journal*, Vol. 28 (1844), p. 179. M. Hamburger gave the first thorough proof in *Crelle's Journal*, Vol. 116 (1896), p. 90.

Divide the year by 19, and call the remainder	a
Divide the year by 4, and call the remainder	b
Divide the year by 7, and call the remainder	c
Divide $(19a + m)$ by 30, and call the remainder	d
Divide $(2b + 4c + 6d + n)$ by 7, and call the remainder	e

Hence, Easter Sunday is $(22 + d + e)$ March. The fact that April 25 was Easter in 1734 and 1886 eluded the notice of Gauss, von Zach, and Delambre. Gauss published a second paper on the method in 1807 (September 12) in the *Braunschweigisches Magazin*. In 1816 he published a correction which applies to his method from 4200 on; Dr. P. Tittel, then in Göttingen, had called his attention to the need for this.

Gauss' first major undertaking in the astronomical field is the working out of a theory of the moon, found among his papers in notebook form, and is printed in Volume VII of his *Works* (1906), pages 633 *et seq*. It dates from 1801, according to diary note 120 ("Theoriam motus Lunae aggressi sumus").

Up to 1788 Mayer's Lunar Tables were in use for the computation of the Nautical Almanac Ephemerides. Meanwhile, Mason of the Board of Longitude had been commissioned to correct them further. After 1789, therefore, the Mason Tables took the place of the original Mayer Tables. As the need of further correction made itself felt, the Paris academy in 1798 set up the prize question: "To determine from a large number of the best, reliable, old and new lunar observations, at least 500 in number, the epochs of the mean length of the apogee and of the ascending nodes of the moon's orbit."

Bürg worked over this task, using more than 3,000 observations, which he compared with the Mayer Lunar Tables, and in 1800 he received the prize (as did Bouvard, who had also worked out a paper). Bürg continued his researches on lunar motion, and Laplace also began to formulate his theory of the moon about this time. In 1800 the Paris academy set up a new prize for the fulfillment of the following conditions:

(1) To determine from the comparison of a large number of good observations the value of the coefficients of lunar

disparity most accurately, to give more accurate and more complete formulas for the length, breadth, and parallax of this body, than those on which the previously used lunar tables are based.

(2) To devise lunar tables from these formulas with adequate ease and reliability for the calculation.

The division of the prize was limited to no certain time.

Herein, perhaps, is to be sought the motive that led Gauss to take in hand the working out of lunar tables. He deduced as fundamental equations the differential equations of the reciprocal shortened radius vector, of the mean length (or time), and of the tangents of the breadth, and used the true length as the independent variable. His fundamental equations are, therefore, similar to those of Clairaut, d'Alembert, and those set up later by Laplace, Plana, and others, but not to those of Euler.

The integration is carried out by approximations, in which the developments progress according to powers of the eccentricity and the slope of the tangents. Gauss develops the divisors of integration according to powers of the ratio of the mean motions of the moon and earth. On that account the form given to the results agrees essentially with those of Plana's later theory (1832). Gauss compared the result of the first approximation with the values of Tobias Mayer. Meanwhile he soon gave up the entire work again, finishing only the perturbations of the breadth. The sudden breaking off is explained in a letter to Schumacher dated January 23, 1842: "In the summer of 1801, I had just set myself the task of executing a similar work on the moon. But scarcely had I begun the preparatory theoretical work (for this it is, which is alluded to in the preface of my *Theoria motus*) when news of Piazzi's Ceres observations drew me in an entirely different direction."

Exhaustive reports on the progress of the researches of Laplace and Bürg are in the *Monatliche Correspondenz* (1800–1802) and this may have been the reason that Gauss did not later resume his own work. Laplace's results appeared in 1802 in Book III of the *Mécanique céleste;* Bürg received in 1803 the new prize, while the printing of his lunar tables was postponed until 1806. In 1803 Gauss copied the results of Laplace's investigations from Book III of *Mécanique céleste* and added a few notes, which seem to relate to the tables of Mason and the results of Bürg's researches then known.

The Duke had planned to have an observatory built in Brunswick for Gauss, and this plan would have been executed had it not been for the disturbing consequences of the French Revolution and the Napoleonic troubles, in which Carl Wilhelm Ferdinand was no indifferent onlooker.

Shortly after the rediscovery of Ceres under Gauss' direction, there came another welcome opportunity to apply his method of planetary calculation and to add to his stature among specialists.

Olbers, while observing the constellation Virgo on March 28, 1802, had seen a star of about the seventh magnitude, which was not to be found on the star chart or in the star catalogue. He hoped that he had discovered a new planet. In order to be certain, he took a star chart and marked the exact location of this body in the heavens. The next day he awaited nightfall with great excitement, eager to ascertain whether or not this was a fixed star. The evening sky on March 29 was perfect for observing. He repeated the measurements exactly and convinced himself that the new celestial body was a planet. It had moved from its position and showed an advance of 10 minutes in right ascension and a difference of 19 minutes in declination. Before publishing this discovery he observed the star on April 3, the weather being very favorable, and had the joy of being more strongly convinced that he was dealing with a new planet.

Then Olbers wrote to von Zach in Seeberg. The letter arrived on the morning of April 4 and von Zach, by his own observation, convinced himself on the same day of the correctness of his friend's discovery. Olbers continued his observation of the new body and sent the data to his young friend Gauss with the request that he calculate the orbit. Gauss had the problem finished on April 18; it is said that the actual calculation required three hours. Previously people had marveled when Euler performed such a feat in three weeks, when others required months. Gauss' work was distinguished by accuracy as well as speed. The notable fact was revealed that the orbit of the body (like that of Ceres) lay between Mars and Jupiter, and that both planets possessed the same period of revolution. This new planet presented one peculiarity; while Mars and Jupiter remained near the ecliptic and did not go beyond the zodiac, this new body passed the old limit considerably.

Olbers, with the privilege accorded him as discoverer, named the planetoid Pallas and expressed the conviction that still others would be discovered. John Herschel was moved to stigmatize the hope thus: "This may serve as a specimen of the dreams in which astronomers, like other speculators, occasionally and harmlessly indulge." Today the number of known planetoids is in excess of a thousand, and at least the first fifty were discovered under the supposition of Olbers that "the connected orbits would have their nodal points in Virgo and Pisces."

Certain people of that day, not so well trained as John Herschel, felt themselves called upon to treat scientific activity with levity and scorn. Thus von Zach told in April, 1801, of receiving a letter from a far corner of the earth in which someone made sport of the versatile endeavors of the astronomers and gave the well-intended advice that it was now time to refrain from building of air castles. Concerning this utterance, as petty as it was senseless, von Zach wrote:

We cannot restrain ourselves from quoting here an excellent passage from a letter of our Dr. Gauss which indicated the noble qualities and attitude of this worthy scholar. "It is scarcely comprehensible," writes Gauss, "how men of honor, priests of science, can reveal themselves in such a light. As for me, I look on such incidents only as tests of whether I work for my own sake, or for the sake of the subject concerned." These are the *onera* of fame, and Gauss will experience more in the course of time since he is just entering upon his literary career. But with such a type of mind as his, with such a consciousness and striving to work only for science, these burdens will never oppress him; they will neither put him out of tune with his age, nor embitter his life. We admonished him, therefore, to persist and abide steadfastly in these noble maxims, which we also would do very well to remember, and to recall our always sprightly, happy, and worthy old patriarch and teacher in the following moral-political-mathematical calculus:

> Résultat d'un Calcul mathématico-politique et moral,
> par le Citoyen La Lande, Doyen des Astronomes
>
> Il y a mille millions d'habitants sur la surface de la terre.
> Sur ces mille millions de têtes
> Que de méchants, de foux, de bêtes,
> Mais nous ne pouvons les guérir,
> Il faut les plaindre, et les servir.

The above verse was copied in his notebook by Gauss from the *Monatliche Correspondenz* under the heading "Cereri Ferdinandeae Sacrum" and shows that Gauss remembered with good humor those who felt themselves clever or ingenious enough to sneer at his scientific accomplishments. The book contains his calculations of Ceres' orbit and comparisons with the observations of others.

In 1804 Olbers wrote to Gauss: "From a few lines in your last letter, I can almost believe that you are under the beneficent attraction of some beautiful star whose compelling force (to put it in a few words) could soon influence you to change your confirmed bachelorhood for the state of matrimony." This obviously refers to Gauss' future wife. Later we shall see the strange connection between these asteroids and Gauss' family life.

On June 22, 1802, Gauss made his first visit to Bremen, where he remained for three weeks with Olbers. Together they visited Johann Hieronymus Schröter (August 31, 1745—August 29, 1796), who was an alumnus of Göttingen, 1764–1768, and an intimate friend of the Herschel family. Although very poor in his student days, Schröter set up a private observatory in Lilienthal in 1782. In April, 1800, Carl Ludwig Harding (September 29, 1765—August 31, 1834) of Lauenburg went to Lilienthal. He had already studied theology, and was now to act as tutor for Schröter's ten-year-old son and also as astronomer at the observatory. On September 1, 1804, he discovered the asteroid Juno. It was on this Bremen visit in 1802 that Gauss became acquainted with him. A portraitist then in Bremen, named Schwarz, made the only picture of Gauss which we have of him in youth. It was a pastel crayon sketch and remained in the Olbers family. Olbers writes of it to Gauss on August 21, 1803: "Above all, my warmest, heartiest thanks for the priceless gift of your portrait, so attractive to me, which our Schwarz brought me. It is astonishingly good and strikes in the same way all who see it and have known you. Schwarz has outdone himself: not a one of your features is neglected." Gauss received as return gift a portrait of Olbers in 1805, equally successful and executed by the same artist. This gift was indescribably dear to him all his life, and at his death it passed into the hands of his physician, Dr. Wilhelm Baum, then to Professor Ernst Schering, and in accordance with the latter's request it became the property of the Göttingen observatory in 1897.

On November 8, 1802, Gauss observed with his meager instrument (a two-foot achromatic lens by Baumann) the solar transit of Mercury. When Juno was discovered by Harding in June, 1804, he observed it with a mirror telescope by Ahort.

On August 26, 1803, Gauss met with von Zach on the Brocken, where powder signals were given for the purpose of longitudinal determinations, and the first week of September the two went to Gotha, where Gauss wanted to perfect himself in practical astronomy. On December 7 he returned to Brunswick, and also spent some time with von Zach at the Seeberg observatory. The following year he had the pleasure of again seeing his friend Olbers at Bad Rehburg (near Hanover); the latter had written him on July 6, 1804: "I am going to Bad Rehburg on August 1 and will remain there fourteen days. In a single day you could come from Brunswick to Rehburg. What a satisfaction, what delight for me, if your good genius would persuade you about this time to refresh itself at this romantically attractive place." For Gauss these three little trips were always among the happiest memories of his youth, especially since those years closed a rich epoch in his life.

Later a large part of Gauss' time was claimed by orbital corrections and calculations of Pallas (April, 1802), Juno (September,

1804), and Vesta (March, 1807). The first orbit for Vesta was computed by Gauss in only ten hours.

Not until 1805 did he begin to develop the perturbations of Ceres. He finally grew tired of this mechanical calculation, as shown in a letter of May 10, 1805, to Olbers: "The methods according to which I had begun to compute the perturbations of Ceres I have given up again. The dead mechanical calculating (entirely too much) which I saw ahead of me turned me away."

Olbers mentioned to Gauss that he was acquainted with a young man named Bessel,[3] who was acting as clerk in Kuhlenkamp's office and who in his leisure hours and at night zealously studied astronomy and showed talent. Gauss did not meet this young man during his first stay in Bremen. When they met later, Gauss made a deep impression on Bessel and gave him new inspiration. An intimate friendship of forty-two years duration thus had its beginning. The copy of the *Disquisitiones* which Gauss presented to Bessel was studied diligently. It never got far from him and was so badly worn that it had to be rebound. Bessel always looked up to Gauss as a teacher, while the latter looked on

---

[3] Bessel was the son of Karl Friedrich Bessel, who studied law in Göttingen and held various governmental positions, dying in 1829 or 1830. The mother was Charlotte Schrader, daughter of a pastor in Hausberge, near Minden. She died in 1814. Franz (usually written Friedrich) Wilhelm Bessel was the second of nine children, born on July 21, 1784, at Minden. He attended the gymnasium in Minden and left it when fourteen years of age "because he could not enter into friendship with Latin." After some private instruction, he was in the employ of A. G. Kuhlenkamp and Sons in Bremen as a clerk, from January 1, 1799, to March 19, 1806. As he intended to go to sea on a merchant vessel, he studied navigation and thus was led to astronomy. His first work was published in von Zach's *Monatl. Corr.* in 1804, "Computation of Harriot's and Torporley's Observations of the Comet of 1607." In 1806 he went to take Harding's place as inspector of the Lilienthal observatory on the Worpe, near Bremen. While there he wrote some valuable memoirs, especially the "Investigation of the True Elliptic Motion of the Comet of 1769." He remained there until March 27, 1810. At Easter, 1810, he moved to Königsberg to become full professor of astronomy and take charge of the erection of a new observatory. His sister, Augusta Dorothea Amalie, accompanied him. In 1812 Bessel married Johanna Hagen, daughter of a well-known Königsberg professor of chemistry. She was born March 20, 1794 and died about 1885. He had one son, Karl Wilhelm (June 16, 1814–October 26, 1840), and three daughters.

Bessel, with his natural genius, as one of the greatest astronomers. "No one," he said, "made the nature of the heavens his mental property as did Bessel, no one had so much aptitude in observing, in the practical use of his instruments; Olbers performed great, very great services for astronomy as a whole, but his greatest service lies in the fact that he properly recognized Bessel's talent for astronomy from its first germination, that he won and educated it for science."

All German astronomers of that day gathered around Olbers. He was the moving spirit of the whole science. Lichtenberg said of him, "If only everyone accomplished in work what Olbers does in his recreation!" Littrow characterized his activity with the words: "Observations and discoveries went forth from Olbers' observatory, i.e., from his living room, every single one of which would make even the greatest observatory immortal forever."

On August 21, 1806, Gauss received full citizen's rights of the world, as he would say, for on that day his wife Johanna presented the happy father with a son. Ceres, the goddess of harvest, had given him harvests also. She first had made possible for him the founding of his own home. The first fruits of this young marriage were to be connected in name with the history of the Ceres discovery. In happy memory of this, Gauss invited Ceres herself, the founder of his domestic fortunes, to stand for this child in the figure of her discoverer, Piazzi, and at St. Katharine's Church on August 24 the newborn child was christened Joseph. At the time Gauss himself had no suspicion that this was the first link in a little chain to which various additional links would be added in years following, a chain through which this phase of astronomical history would be woven directly into his family. Joseph long remained his father's favorite. The young father sent Bolyai this report about the baby: "Talents are visibly developing in the little chap, he is the darling of all who know him. A geometer is scarcely in him, he is too wild, too playful, I should say."

At the beginning of the year 1804, Harding informed Gauss that there was a ten-foot telescope mirror for sale, priced at thirty pistoles. Its excellence was recommended. The Duke heard of this and gave orders to buy the instrument for the observatory he intended to erect. It arrived the latter part of April and Gauss was at once authorized to attend to the mounting. The optical pieces were ordered from Schröder in Gotha and the mechanical workings from Rudloff in Wolfenbüttel.

But this first step toward the fulfillment of Gauss' eager desires brought in its train a series of annoyances and disappointments, destined to embitter the last hours of his stay in Brunswick. Both manufacturers severely tried his patience; Schröder did not furnish his order of goods until August of the following year; Rudloff almost a year later. Rudloff demanded 750 thalers. Scarcely had the Duke's munificence removed this worry when it was discovered that the instrument did not have the desired power and did not accomplish what it should. Gauss was convinced that it must go through a mechanic's hands again, and decided to let the matter rest. Not until September, 1807, when Harding was visiting him, did Gauss discover the difficulty, and realize that the mirror was just as excellent as had been represented.

Harding took the mirror with him to Lilienthal for repolishing. Soon it was reported that the repairs were successful and that since there were few others in Germany like it, it would be an adornment to any observatory. Gauss was glad the purchase of the mirror was justified.

While Gauss was leading a quiet life with his family in Brunswick, the political horizon was ominously disturbed. Napoleon's power became entrenched in western Germany, and although the worst was to be feared, no agreement for defense could be reached between Prussia and Austria. These threatening circumstances turned attention toward Russia as an ally. The Duke of Brunswick was accordingly commissioned by the court at Berlin on January 30, 1806, with a diplomatic mission to St. Petersburg. This seems to have been of no avail. Many questioned him there about the young astronomer Gauss, urging him to yield his claim and allow Gauss to accept the second call to St. Petersburg. Having returned to Brunswick on March 24, he raised Gauss' salary to six hundred thalers. This fact was communicated to Gauss on April 30, his birthday. One day in May Gauss called to see the Duke and thank him for the kindness. This was the last time that they saw each other; the friendship had covered fourteen years and had brought Gauss great opportunity. During the next few months the political storm prevented the Duke from thinking of his protégé.

After the battle of Jena a court of punishment was set up; many of the highest officers were imprisoned, with long or life sentences, many condemned to the arquebuse. With such a corps of officers, a victory against Napoleon was precluded at the outset. Prussia began the war at a time when she was completely isolated. King Friedrich Wilhelm III could have freed his forces more than once, at favorable moments, but under the influence of such diplomats as Haugwitz and Lucchesini, whom the Baron vom Stein called blockheads and knaves, he delayed until too late, when withdrawal was impossible. The King had left Austria in the lurch and she now had her revenge. The prospective Russian auxiliary troops were only on paper, while the French army had 200,000 reliable soldiers. The Prussian generalissimo, Duke Carl Wilhelm Ferdinand of Brunswick, had only 57,000 men under his command, among them many raw recruits, foreigners, vagabonds from various countries, and good-for-nothings who could be held together only by the stick; Poles from newly acquired South Prussia deserted in the garrisons and still more on the field. The equipment was deficient, the extremely heavy guns shot poorly, and the commissary left much to be desired. Many regimental commanders drew on their own funds. There were batteries which had never trained, had never practiced hitching and unhitching a gun carriage. No one knew the roads, the footpaths, and bridges. The result was a mad hurly-burly which caused them to shoot their own men. The general chief of staff, Colonel von Massenbach, an unfit babbler, erred continually like the Austrian general, Mack, in commands, orientation, and so forth. This Massenbach was quartermaster general for the commander of the second army, the Prince of Hohenlohe—Ingelfingen, who had 23,000 men at his disposal. Since the Prince was jealous of the Duke, he was receptive to the intrigues of Massenbach and set up opposition to the commands of superiors. Any carefully planned, systematic collaboration of the two armies was thus rendered impossible. General Rüchel with his 27,000 men did not arrive at the proper time. Hence Blücher later reported: "A reserve was never more inactive than ours at Auerstedt."

These facts explain the disaster which overwhelmed the Prussian army on October 14, 1806; her leader, Duke Carl Wilhelm Ferdinand of Brunswick, should receive the least blame for it. He was seventy-one years old, but in no wise stupid and frail, as has been claimed. An eyewitness gave the following account of his last days:

The rapid movements and progress of a never resting enemy made the advance of the Prussian army more and more necessary. Every moment had now become valuable, and the headquarters moved daily farther on from Erfurt to Blankenhayn, from there on October 12 to Weimar, and then on the twelfth to the fatal Auerstedt. The duke set up quarters at the local nobles' court, the king established his close by in another house and here they awaited the event which *could* honorably rescue a great state, but could also, perhaps, disrupt it. It was a great, frightful game!—a struggle and a massing of men so large that for centuries history had not seen one like it. The duke, always active and busy, almost never left the circle of officers surrounding him. Fieldmarshal von Möllendorf, Colonels von Kleist and von Scharnhorst, and adjutants surrounded him until late at night. They discussed the positions of the opposing armies, the next moves, the break-up of the battle, the next day's hopes. The duke was grave, reflective, and self-contained, but nevertheless talkative. The companionable circle around him was serious and thoughtful. It was a great, awful evening. The future and its possibilities strained courage, hopes, and apprehensions. Opposing them was a general whose genius had been admired for several years and whose high talents in war were recognized by the whole world. "The fourteenth of October," said the duke on the night before the battle, "has been an unfavorable day for me several times," and named the historical reasons for his utterance. Nevertheless it is certain that his spirit was out of tune, disturbed, distrustful, and badly affected several days before the battle. Probably the death of bold Prince Ludwig of Prussia had made a deep impression on him, as it did on the whole army. [Prince Louis Ferdinand died in a skirmish near Saalfeld on

October 10, because he was not willing to flee or surrender.] Even before the battle so much ground had been lost that the conquerors had already gained noticeable advantages for themselves by their advance. That heroic spirit of the Prussian army, which had made itself so immortal in the Seven Years' War, seemed no longer to be the same; even generals with a pusillanimous and vehement sincerity questioned their morning victory, and Prussian defense tactics aroused opposition and irresolution. In short, the duke himself felt grave forebodings.

Around midnight on October 13, before the battle, the duke retired, exhausted by the continuous military demands on him. He reclined on his couch this time, just as he had done for several turbulent days, wearing all his uniform with girdle, rapier, and spurs. Similarly clothed and ever ready for decamping lay also his entire retinue in the adjoining rooms. Scarcely had it struck three o'clock when he was awakened by the voices of his officers in the next room. By 4 A.M. he was active and ready for the great business of the day.

At 4:30 even the king came into the duke's quarters. Again a council of war was held and now for the last time. The king repeatedly advised withdrawal, but soon after six o'clock, in the first shimmering morning light, the duke hastened to the attack on the battlefield. A thick autumn haze slowed the movements of the troops, and a portion of the advance corps fell upon enemy troops, and paid with a heavy loss. Not until nine o'clock did the sun begin to shine through the haze. The struggle was now more general and decisive. The emperor's army, already tasting victory, pressed forward swiftly under a terrible musket fire from several sides. Several regiments of Prussians remained brave, others were only figures in this fantastic struggle. At the critical moment when everything depended on a powerful and decisive move, the duke was blinded by a musket ball, as he was giving orders before the grenadier battalion from Hanstein. The ball had gone in above the right eye, shattered the bone in the nose, and had pushed the left eye from its socket. His uniform was colored with blood, In this indescribably sad condition the wounded general lay for some minutes on the ground, and thought of being compelled to end his life here, until several Prussian officers hastened by and placed him on an officer's horse because his own had escaped. A musketeer got on behind him in order to support his back, while two others went on each side to prevent any swaying. He was brought back to Auerstedt; his people watched his return with deep sorrow. "I am a poor blind man, bring me to rest!" he said dolefully to a rifleman who accompanied him. In Auerstedt his blinded eyes, already very swollen, were bandaged, and an able military physician named Völker accompanied him on the order of the king, on his entire journey.

It was a consolation to him to learn that the enemy was sparing his land. His hope that an avenger would arise in his son, Friedrich Wilhelm, was fulfilled. "The Black Duke" was one of the outstanding figures in the War of Liberation, and fell at Quatre Bras on June 16, 1815.

A deputation was sent to Napoleon, then in Halle, to request clemency for the Duke and permission for him to die in peace among his own people. The deputation was spitefully received and brutally treated. They heard the Duke overwhelmed with the bitterest reproaches, and Napoleon himself jeered at the conduct of the war. The French army was soon to march in on Brunswick, and then they would be subjects of France.

Arrangements were made to protect the Duke from ignominious imprisonment at the hand of the enemy. The plan was to flee with him to England. Gauss had intended to flee also. He always told the following story with touching sadness. Living at the time at Steinweg 22, just opposite the castle gate, Gauss looked out his window on the morning of October 25 and saw a long carriage drawn by two horses leaving the castle yard. The gate opened and it moved on out and toward the Wendenthor, so slowly and gloomily that several thought it contained a corpse. The wounded and dying Duke was within, on his flight to Altona, so that he might at least die in freedom. With stricken heart and in grieving silence Gauss saw the departure of his patron and friend. He was as though annihilated. A serious expression came over his face. He was quiet and speechless and bore this great sorrow without words and sounds of complaint. This same seriousness and quietude was prevalent among all the people. Although Gauss said very little, his thoughts and emotions were the more aroused by this fact. Poignant grief gained control of him, accompanied by rancor against the invaders of Germany, in whom he also hated the enemy of his beloved prince. Personal reasons for hatred of Napoleon soon arose.

After several days' journey the carriage with the dying duke arrived in Altona. The journey could not be continued—the condition of the Duke had become so much worse that care was taken merely to let him die in peace. He died on November 10, 1806, in a tiny hovel at Ottensen, "driven from the land of his fathers and pursued and mocked by his arrogant foes." His resting place is not far from the linden tree which shaded the grave of the German poet Klopstock. Referring to the connection between Gauss and the Duke, von Zach had written to the former on January 27, 1803: "You will see to it that his great name is written even in the heavens."

Gauss remained in Brunswick with his small family during the winter, while Germany was in subjection from the Rhine to the Niemen. He was not popular with many townsmen and was envied on account of his independent position, but with the Duke gone conditions were completely altered. Hence, every mishap with reference to the Lilienthal telescope mirror bothered him.

For some time Dr. August Heinrich Christian Gelpke, assistant principal of the Martineum, had been holding astronomical lectures. He was a comical man; in later years whenever his auditors indefatigably asked him as a joking question who were the three greatest astronomers, he named Kepler and Laplace; his modesty forbade him to name the third. It was this man who wrote to a member of the provisional government, Privy Councilor von Wolffradt, on October 10, 1807, requesting the ten-foot Newtonian telescope for use in his lectures, "which [telescope] Dr. Gauss possesses only as a piece of property borrowed from his majesty the late duke, as I have heard, and who is soon leaving here for Göttingen. . . . In the possession of this instrument, I shall attempt to make myself more and more useful to the institution and thereby more and more worthy of your excellency's favor and affection."

On the same day the council of the Carolineum approved such use of the telescope and directed Gauss to give it up. The council was commissioned to look after the proper and safe mounting of the instrument. Gauss explained that he was not personally interested in the matter and was ready to obey the order, but felt called upon to remonstrate against the intended use of it. His idea was that it should be turned over to Pfaff for use at the University of Helmstedt and on rare occasions for other use, and that it was too fine an instrument, too expensive, for beginners to use. We must remember that the telescope was still in Lilienthal for repairs. Gauss' advice was not heeded by Wolffradt, and thus Gauss departed from Brunswick with some ill feeling.

CHAPTER SEVEN

Back to Göttingen

Gauss received the call to Göttingen on July 9, 1807, while he was visiting Olbers, and accepted it only after careful counseling with his older friend. On his return trip from Bremen he visited Ernst Brandes (1758–1810), an official in Hanover, and became acquainted with A. W. Rehberg, secretary of the cabinet. The country was already occupied by the enemy, although certain officials were still loyal to the Hanoverian government. Gauss arrived in Göttingen just at a time when the old government was in complete dissolution, and the new Westphalian government set up by the French was not yet organized. In the confusion the formal oath of office was never administered to him. He was employed by the Hanoverian government, but this omission never made him less loyal to his country or to the university.

The arrival in Göttingen is best described by a letter of Gauss' wife to her intimate friend, Dorothea Müller Köppe (1780–1857) in Brunswick. This letter is dated December 6, 1807; they had arrived November 21:

> At last I find time to entertain myself with my darlings. Oh, how I long to get word of you! We all arrived here safe and sound. The journey was very hard on me, for I was uncomfortable as long as the trip lasted. As soon as I left the coach I was as happy as a fish. All our personal effects were already unpacked through the kindness of Prof. Harding, so that on our arrival at about three o'clock in the afternoon a warm room and a cup of tea greeted us. Nevertheless such a trip with "sack and pack" is most frightful and tiresome, as one can imagine; I believed the packing job was wonderful, but since I arrived there in fine shape, I wished for the sake of the fun that you had seen this chaos.

The first five days there was no end of hay and straw, now finally I am in order—with our apartment we are nothing less than satisfied. It is all connected, to name one objection. Our living room is the most tolerable; little dirty halls, a smoky draughty kitchen, old phlegmatic landlords, these are incidentally several merits, and not very apt to make the stay here pleasant for me. In the first eight days I didn't see a soul except Harding, because it was impossible for us to make visits before Friday, when we visited 50 or 60 different families in the space of an hour, to be sure, without having heard or seen anyone. O the ridiculous people! A week ago today in spite of the bad weather we began daily to visit one or two families in a more reasonable manner (with body and soul), we are received everywhere very politely, indeed by several very cordially (confidentially, Gauss seems to stand in great respect here), also we now have visits daily from all the different people, among whom many a one is very interesting. I haven't been in high society yet, but the people here seem to me to be trustworthy; I haven't yet been able to make close friendship with anyone. [Omitted here are several lines bewailing the absence of friends in Brunswick, and inquiring about the price of coffee and sugar.] We are all well, my Joseph is becoming a splendid chap. He has been running about alone since two days before our departure. But to be sure this always looks awfully perilous. He gladdens us, and is my idol.

Now farewell, beloved friend, and write soon and much to your ever loving friend

HANCHEN GAUSS.

The old observatory was located on the street Klein-Paris (now Turmstrasse) in an old fortification tower, and this apartment which Hanchen mentions in the letter has not been located. In April, 1808, the family moved to the large house on the corner of Turmstrasse and Kurzestrasse 15. As a student in Göttingen, Gauss had lived at Gothmarstrasse 11 and on Geismarstrasse at Volbaum's. The last address at which he lived before he moved to the new observatory is not known. Many of the family letters of this period are filled with references to Joseph, their only child at the time.

Gauss had just settled in Göttingen and had not yet drawn a cent as director of the observatory when a war contribution was ordered by Napoleon in the form of a compulsory loan for the newly created kingdom of Westphalia. The burden on the university was very heavy and Gauss' portion was set at two thousand francs. One day he received a letter from Olbers with this sum, but felt he should not accept it and sent the money back to Bremen. Shortly afterward came a letter from Laplace stating that he had paid in the two thousand francs direct at Paris. Later Gauss paid back the French mathematician, with interest on the money, from his own funds. From Frankfort on the Main he received an anonymous gift of a thousand florins,[1] which in later years he learned was a gift of the prince primate, Baron Karl Theodor Anton Maria von Dalberg (1744–1817), grand duke of Frankfort. This sum he kept as coming from a public purse, a tribute to his work rather than to personal friendship.

Writing under date of December 24, 1807, Frau Johanna thanked her mother for sausage and Christmas gifts sent to little Joseph, who had been ill. His celebration was to be postponed one week because of this and also because a little table and chair were not yet ready. He had been walking for some time and knew everyone, but did not talk yet. A discussion of servant problems followed.

On January 6, 1808, Johanna wrote to Dorothea Köppe again, expressing her homesickness for Brunswick and begging Dorothea not to reveal to her mother how very ill Joseph had been. She mentioned having made friends with the family of Thomas Christian Tychsen (1758–1834), professor of theology and oriental languages in Göttingen, and being especially attracted to his twin daughters Cäcilie and Adelheid, aged fourteen; also the family of Arnold Hermann Ludwig Heeren (1770–1842), professor of history, and the wife of Friedrich Stromeyer (1776–1835), professor of medicine. She mentioned a *thé dansant* and a concert but had absented herself from these and other social affairs because of her pregnancy. She alluded to Gauss' astronomical observations and said he usually came to bed at one o'clock in the morning.

On February 29, 1808, Gauss wrote his parents that his wife Johanna had given birth that morning at six o'clock to a daughter

---

[1] Sent through the Bethmann Bank.

not as dainty and pretty as Joseph, but well formed, strong, and healthy. The garments which Joseph wore during his first six months fit her perfectly, but the caps which Johanna had knitted were all too small. He regretted that the child had only one birthday in four years. She was named Wilhelmine, in honor of Olbers, who had already promised to act as godfather. In the family she was always called Minna, and grew up to be the very image of her mother. The child was described as loving, clever, goodhearted, pure, open, and happy. Humboldt wrote to her father that she was beautiful.

The young mother was very ill for three weeks after the child's birth. In describing the child, she admitted that the father thought it too fat and large for a girl baby.

On April 11, 1808, Johanna's mother wrote from Brunswick to say that Gebhard, Gauss' father, had been taken seriously ill on Wednesday, April 6, that he had given up all hope of recovery and had made his will. She also mentioned the fact that her intended visit to Göttingen was postponed because of high water.

Then came another letter dated April 14, 1808, saying that Gauss' father had died that day at 11:30 A.M. at his home, Aegidienstrasse 5.

In a later letter to Dorothea Köppe, Johanna described some of the social functions which she and Gauss attended in Göttingen. But there always seemed to be a note of homesickness in her letters. She danced, but Gauss could not induce her to play cards, because, she said, she would rather devote that time to the children.

In the winter semester 1808-1809 Schumacher[2] went to Göt-

---

[2] Heinrich Christian Schumacher, son of Andreas Anthon Friedrich Schumacher, was born at Bramstedt in Holstein on September 3, 1780. In 1813 he married Christine Magdalene von Schoon. From 1817 on he directed the triangulation of Holstein, and later a complete geodetic survey of Denmark (completed after his death). For the survey an observatory was established at Altona, and Schumacher resided there permanently, chiefly occupied with the publication of ephemerides (eleven parts, 1822-1832) and of the Journal *Astronomische Nachrichten*, of which he edited thirty-one volumes. He died at Altona on December 28, 1850.

His nephew, Christian Andreas Schumacher (1810-1854), was associated

tingen to study astronomy under Gauss. He kept notes on their conversations together, which he called "Gaussiana." These have been very useful to later scholars. The intimate friendship between Gauss and Schumacher continued up to the latter's death in 1850. Their correspondence, published 1860–1865, fills six volumes. We shall have occasion to refer to Schumacher continually in connection with Gauss' scientific work.

When Alexander von Humboldt returned to Paris in 1804 from his American voyage, he heard the name of Gauss highly praised in scientific circles. Their first exchange of letters began in 1807. He and his brother Wilhelm had studied in Göttingen earlier than Gauss, but the three later became lifelong friends, the correspondence continuing until the death of Gauss. Another friend was Johann Georg Repsold, born September 19, 1770, in Wremen on the Weser. He was fire chief in Hamburg, and in an old artillery building had equipped an observatory with instruments which he had built himself. Repsold was long a friend of both Gauss and Schumacher. In June, 1809, he visited Gauss in Göttingen. His correspondence with Gauss relates largely to the purchase of his meridian circle for the Göttingen observatory. Repsold died in 1830 while still active, and his sons developed the workshop into a well-known firm for astronomical instruments.

---

with the geodetic survey of Denmark from 1833 to 1838 and afterward (1844–1845) improved the observatory at Pulkowa.

H. C. Schumacher's son Richard (1827–1902) was his assistant from 1844 to 1850 at the Altona observatory. Having become assistant to Carlos Guillelmo Moesta (1825–1884), director of the observatory at Santiago, in 1859, he was associated with the Chilean geodetic survey in 1864. Returning in 1869, he was appointed assistant astronomer at Altona in 1873, and afterward at Kiel.

CHAPTER EIGHT

Labor and Sorrow

Gauss was now ready to publish his second major work, and Olbers referred him to Friedrich Christoph Perthes (1772–1843) of Hamburg and Gotha, the best-known German publisher at that time. Perthes had written Olbers that he had exhausted available funds on Johann Müller's *Universal History;* later, however, he was glad to contract for the Gaussian work. It would be interesting to know what the author received for this work, which was originally written in German but published in Latin at the request of Perthes under the title: *Theoria motus corporum coelestium in sectionibus conicis solem ambientium.*[1]

Florian Cajori thus described the *Theoria motus*: "... a classic of astronomy, which introduced the principle of curvilinear triangulation; containing a discussion of the problems arising in the determination of the movements of planets and comets from observations made on them under any circumstances. In it are found four formulae in spherical trigonometry, now usually called 'Gauss' Analogies.'"[2]

---

[1] Carl Haase (December, 1817–March 23, 1877) translated the Latin original into German in 1864 with the title *Theorie der Bewegung der Himmelskörper, welche in Kegelschnitten die Sonne umlaufen,* and published it at Hanover in 1865. The English translation was published in 1857 at Boston by Rear Admiral Charles Henry Davis, assistant in the U.S. Coast Survey in 1842 and later superintendent of the Naval Observatory.

[2] There has been some confusion as to the discoverer of these formulas. It seems that Delambre discovered them in 1807, but did not publish them until 1809 in the *Connaissance des temps,* p. 443. They were discovered independently by Gauss, and are often called "Gauss' equations." Both systems may be proved geometrically. The geometric proof is the one originally given

The discovery of Ceres had suggested to Gauss' inventive mind a great variety of beautiful contrivances for computing the movement of a body revolving in a conic section in accordance with Kepler's laws. In the *Theoria motus* the author gives a complete system of formulas and processes for computing the movement of a body revolving in a conic section and then explains a general method for determining the orbit of a planet or comet from three observed positions of the body. The work concludes with an exposition of the method of least squares. This work exhibits in a very remarkable degree the effects of Gauss' severe system of revision, in the skillful adaptation and reduction of methods and formulas, and in the careful estimate of the circumstances under which they may be most advantageously employed. We find in it no evasion of difficulties, and no resort to methods of approximation only, when the means of accurate determination are at hand. His aim was in every instance to obtain results of the same order of correctness with the observations upon which they were founded.

The *Theoria motus* will always be classed among those great works the appearance of which forms an epoch in the history of the science to which they refer. The processes detailed in it are no less remarkable for originality and completeness than for the concise and elegant form in which the author has exhibited them. Indeed, it may be considered as the textbook from which have been chiefly derived those powerful and refined methods of investigation which characterize German astronomy and its representatives of the nineteenth century, Bessel, Hansen, Struve, Encke, and Gerling. It is a curious fact that the date of the preface to this immortal work is exactly two centuries later than the date of Kepler's equally renowned work *Praefatio de stella Martis* (March 28, 1609–1809.) A few numerical errors have been found in the calculations of the *Theoria motus,* but these were nearly

---

by Delambre. It was rediscovered by Professor Crofton in 1869, and published in the *Proceedings* of the London Math. Soc., Vol. II. (See Casey's *Trigonometry,* p. 41.) Carl Brandon Mollweide (1774–1825) of Leipzig had also published them before 1809.

all corrected in the authorized edition of Gauss' *Collected Works*, (Vol. VII, 1906). The entire work, of course, is based on Newton's law of universal gravitation. It was forty years before the methods of the *Theoria motus* became the common possession of all astronomers. About that time frequent discoveries of new celestial bodies compelled them to master its methods.

Of the numerous fundamental works on astronomy planned by Gauss, the *Theoria motus* is the only one he presented to his contemporaries. It treats exclusively the elliptical and hyperbolic motion in so far as it concerns the determination of an orbit from observations, which leads to the conjecture that he may have postponed further investigations on this subject with the idea that his friend Olbers had successfully solved the problem of parabolic orbital determination. In the *Theoria motus* he applies methods for determining a hyperbolic orbit similar to those for an elliptic orbit. But he gives in the *Theoria motus* a numerical example only for the calculation of the elements from two radii vectores, the included angle and the time between them. In a letter of January 3, 1806, he pointed out to Olbers a certain case where the latter's method for a parabolic orbit is not applicable. Later, in 1813, he found a correction for the method and in 1815 finally busied himself intensively with the parabolic-orbit problem. He not only gave Lambert's equation a different form, but also developed a series of other important relations. The determination of a circular orbit (which Gauss characterized as easy and simple) was given in 1871 by Klinkerfues, his pupil, in a textbook. Gauss used the method of least squares in making the corrections as given in Articles 175–186 of the *Theoria motus*.

The appearance of the *Theoria motus* soon aroused among savants recognition and admiration for its author. Prince Primate Dalberg, already warmly interested in Gauss, sent him a golden medal; another was sent by the Royal Society in London. Sartorius tells us that Gauss became a member of all learned societies "from the Arctic circle to the Tropics, from the Tajo to the Ural, including the American societies."[3]

---

[3] See Appendix B.

In the year 1810 Gauss received from the Institut de France a new distinction: *la médaille fondée par Monsieur Lalande, pour le meilleur ouvrage ou l'observation astronomique la plus curieuse.* This prize he never allowed to be paid, because he did not want to accept any money from France. Part of the money was used by Delambre, secretary of the institute, and Sophie Germain, in purchasing a pendulum clock which was sent to Gauss and used in his room for the remainder of his life.

On September 2, 1809, Gauss wrote to his friend Karl Köppe (1772–1837) in Brunswick that he had received a call to the University of Dorpat, but rejected it because of the climate and the precarious, disorderly conditions there. Efforts were also being made at this time to call Gauss to Leipzig.

For the third time, on September 10, 1809, Gauss became a father. This child was named Ludwig in honor of Harding, the discoverer of Juno, but was always called Louis in the family; the father's phrase was *der arme kleine Louis.* Olbers sent his congratulations on September 20: "May heaven make little Ludwig, in future years, the equal of his father in mind and heart." Neither of Olbers' expressed wishes was fulfilled. Little Louis died suddenly on March 1, 1810, only five months of age. He was buried on the family lot in St. Albans cemetery in Göttingen. Even before this, however, the great scholar had a much heavier affliction.

The birth of Minna in 1808 had been difficult and caused Frau Johanna much suffering. As a result of the birth of Louis, she died on October 11, 1809. The grief-stricken husband wrote to Olbers: "Last night at eight o'clock I closed the angel eyes in which for five years I have found a heaven." Johanna was buried on October 14 at the family lot in the St. Albans cemetery in Göttingen. Immediately thereafter, Gauss went on a trip to Olbers in Bremen and Schumacher in Altona, returning by way of Brunswick to be with old friends there.

In 1927, Carl August Gauss of Hamlin, Germany, found a lamentation for the deceased wife among his grandfather's papers. There are actual traces of tears on the paper, and the writing covers two and one-half pages of folio paper. It is divided into two parts, the second being dated October 25, (Bremen). The first part is written with a coarser pen and darker ink than the second; the place where it was written is uncertain. This beautiful tribute shows Gauss to be a master of the German language, and it should be added that translation does not do it full justice:

Beloved spirit, do you see my tears? As long as I called you mine, you knew no pain but mine, and you needed for your happiness nothing more than to see me happy! Blessed days! I, poor fool, could look on such happiness as eternal, could blindly presume that you, O once incarnate and now again a newly transfigured angel, were destined to help me bear through my whole life all of its petty burdens. How could I have deserved you? You did not need earthly life in order to become better. You entered life only in order to be an example for us. Oh, I was the happy one whose dark path the Unsearchable allowed your presence, your love, your tenderest and purest love to illumine. Did I dare to consider myself your equal? Dear being, you yourself didn't know how unique you were. With the meekness of an angel you bore my faults. Oh, if it is granted to the departed to wander about invisible but still near us poor creatures in life's darkness, desert me not. Can your love be transitory? Can you take it away from the poor soul, whose highest good it was? O thou best one, remain near unto my spirit. Let your blessed tranquillity of mind, which helped you bear the departure from your dear ones, communicate itself to me; help me to be more and more worthy of you! Oh, for the dear pledges of our love, what can take the place of you, your motherly care, your training, only you can strengthen and ennoble me to live for them, and not to sink away in my sorrow.

25 OCTOBER.—Lonesome, I sneak about among the happy people who surround me here. If for a few moments they make me forget my sorrow, it comes back later with double force. I am of no value among your happy faces. I could become hardened toward you, which you don't deserve. Even the bright sky makes me sadder. Now, dear, you would have left your bed, now you would be walking at my arm, our darling at my hand, and you would be rejoicing in your recovery and our happiness, which we would each be reading in the mirror of the other's eyes. We dreamed of a more beautiful future. An envious demon—no, not an envious demon, the Unsearchable, did not will it so. O Blessed One, already you see clearly now the mysterious purposes which are to be attained through the shattering of my happiness. Aren't you permitted to infuse several drops of consolation and resignation into my desolate heart? Even in life you were so overrich in both. You loved me

so. You longed so much to stay with me! I should not yield too much to grief, were almost your last words. O, how am I to begin shaking it off? O, beseech the Eternal—could he refuse you anything?—only this one thing, that your infinite kindheartedness may always hover and float, living, before me, helping me, poor son of earth that I am, to struggle after you as best I can.

It is illustrative of Gauss' character that in this period, so humiliating to Germany and so full of his own personal grief, he proved himself a genuine son of his native land. He guarded and cherished his native language and German science. With the fascination which mathematics held for Napoleon and the admiration which Laplace voiced concerning Gauss, he could have attained the greatest honors, distinctions, and material advantages possible at that time. Nevertheless he considered the chief purpose of his life to be the fulfillment of his scientific endeavors. The French were occupying Germany and seemed to have no desire to leave. It was therefore to their interest not to neglect the University of Göttingen in their new Kingdom of Westphalia. Doubtless for this reason Gauss was annoyed no further.

In practical astronomy, which along with theoretical astronomy was revolutionized at the beginning of the nineteenth century, little had been accomplished in Göttingen until the arrival of Gauss in 1807. The new observatory at this time consisted only of the foundation, although its completion had been planned. Due to the disturbed and distressing conditions of the times there seemed to be little hope of seeing the plans fully realized. Gauss therefore worked for some time at the old observatory, where Tobias Mayer had been active and had made some noteworthy contributions to astronomy. It was located in an old slate-covered tower which had served in the Middle Ages as a defense for the inner town wall. Until recently parts of it could be seen on Turmstrasse.

After a long delay the Westphalian government appropriated in 1810 the sum of 200,000 francs for the further construction of the observatory, which amount was to be distributed over a period of five years until the completion of the structure. After various architects had worked on the plans, it was finally built in the Doric style under the direction of a contractor named Müller, essentially according to the wishes of the astronomers.

In 1814 the exterior of the observatory was finished—with the interior work yet to be done. The two apartments were ready in 1816; Gauss with his family took the west, and Harding the east, wing. The new instruments did not arrive from Munich until 1819 and 1821. What was then usable from the old observatory included Bird's six-foot wall quadrant, with which old Tobias Mayer had carried on his observations for his catalogue of fixed stars, and William Herschel's ten-foot mirror telescope, which served Harding a long time in the comet observations. The other instruments ordered from Lilienthal were scarcely used, except to show the starry heavens to visitors. The great mirrors of the Schröter telescope demanded a large scaffold, and it was originally planned to set up this structure between the west wing and Geismar-Landstrasse; but when the new instruments arrived from Munich and required Gauss' complete attention, the plan was forgotten, and only the two large metal mirrors of twenty-foot focal length are now extant. Wilhelm Schur found them in 1886 in a strongly oxidized condition and later had them set up in the west meridian hall on blocks of wood.

In 1814 the excellent Fraunhofer heliometer with an objective of 1.5-meter focal length and an aperture of 76 millimeters arrived in Göttingen. Gauss was very much interested in this instrument at first and made several observations with it, but along with other instruments of similar construction from Berlin, Breslau, and Gotha, it was not used until the transit of Venus on December 8, 1874 (on Auckland Island), and again in 1882. The instruments of Repsold were altered for this purpose and set up in a revolving tower on the terrace of the observatory. Here the heliometer was used regularly after 1886.

After the new observatory was completed in the fall of 1816 and Gauss had moved into his rooms, he made a trip to Munich in order to inspect the new instruments ordered there. Before they reached Göttingen, the observatory had already received the

Repsold meridian circle.[4] This instrument, the second oldest of all meridian circles, was previously at the private observatory of Johann Georg Repsold at Elbhöhe (Stintfang) in Hamburg near the place which was later the site of the German Oceanic Observatory. H. C. Schumacher used it in 1804 for the observation of circumpolar stars. When Repsold was compelled by war conditions to give up his observatory, the instrument remained several years in his workshop and was purchased by Gauss for the new Göttingen observatory at Schumacher's suggestion, after Repsold had made several alterations, including a new triple objective constructed according to Gauss' formula. The instrument was eagerly used by Gauss for the first two years in observing Polaris; later he gave a method for determining its elevation. However, when the two new meridian instruments arrived—the meridian circle and the passage instrument of Reichenbach (from Munich) —Gauss no longer used the Repsold. He turned it over to Harding to get positions on it for his star atlas, and occupied himself principally with the new meridian circle.

The observatory in Göttingen lies to the south of the old town wall about five minutes' walk from the Geismar-Tor[5] and consists of a main building, 120 feet in length from east to west and 45 feet wide, on whose west and east ends two wings extend to the north, each with two-story apartments for the families of the professors. These two, together with the main building, form a court opening to the north. To the south, where the observatory presents a façade of beautiful Doric columns, three large double doors open on a terrace 200 by 40 feet, partly laid with concrete and elevated six feet above the yard. The apartments with their gardens join it at both ends. This terrace furnishes room for transportable instruments and also for two iron revolving towers with the Fraunhofer heliometer and little Merz refractor, and can also be used for teaching purposes. The view from the terrace has been cut down by the growth of the city beyond the limits of the

---

[4] Olaus Römer of Copenhagen had in the seventeenth century constructed a meridian circle of wood, but it was lost in a fire.
[5] Removed in 1945.

old town fortification wall out to the horizon, yet it has been possible to keep free and open the view from the Reichenbach meridian circle. To the east a strip of the horizon is lost by the Hainberg, on whose slope the observatory lies. The main building contains in the middle a rotunda with two opposite portals opening on the court and on the terrace. At the north portal the staircase leads to the dome room over this hall. To the east and west one goes from the hall out into two equal-sized meridian halls of 30 feet in a north-south direction, 20 feet east-west, 25 feet high; then come two large halls on the east and west, one serving as a lecture room and the other as a library. These rooms have doors to the terrace and the apartments so that all rooms are connected with each other. The rotunda in the center of the entire structure was covered over at the construction of the observatory by a revolving dome made in Göttingen. Not until 1888 was a new dome, 30 feet in diameter, set up by Grubb of Dublin.

To the joy and surprise of astronomers the great comet of 1811 emerged unexpectedly from the blue sky. Gauss saw it for the first time on August 22, deep in the evening twilight. Actual observations could not be carried out until later, partly because of the cloudy weather and partly because the view from the north side of the old observatory was cut off by buildings. But in early August, when the result of the observations made by von Zach arrived, Gauss calculated the parabolic elements of the comet according to several of these data, and calculated its orbit. The comet's reappearance proved these calculations to be entirely correct, as was also his prediction that it would exhibit greater brilliance after its transit through the solar region. The peoples of Europe saw in the comet only a chastisement of heaven, a prelude of the burning of Moscow and an omen of Napoleonic dominion.

Napoleon's great army, however, soon lay buried in the icy fields of Smolensk and along the Beresina. Germany armed herself. Her ally, the Cossacks, for whom there was great sympathy, soon came to her aid. A Cossack officer was naïve enough, after he had allowed Gauss to show him the observatory, to demand as a perpetual memento the only chronometer there, which had been presented by King Jérôme. After the fall of Napoleon when the "good old days" returned to the country, Gauss enjoyed at all times the good will of the kings and the governing board of the university. The new observatory, construction of which had been ratified by the Westphalian government, as stated, was soon completed after a fashion. It was in this provisional condition when the director moved into his apartment in the fall of 1816.

CHAPTER NINE

The Young Professor: A Decade of Discovery, 1812-1822

Gauss spent exactly four years of happiest married life with Johanna, whom he loved most passionately. That he married so soon after her death appeared strange to relatives and friends. The same sort of thing is to be found today; especially do relatives of the first wife have that feeling. Yet it is easy to understand why a bereaved husband takes such a step. It is a well-known fact that men who have had an unusually happy married life remarry in most cases soon after the death of the wife because the isolation is unbearable. Such was the case with Gauss. It was even worse for him, because he was robbed of the serenity and peace of mind so necessary for successful scientific work. Moreover, the knowledge that he was compelled to give a new mother to his children, the real legacy of his youthful love, was dominant in his thought. They sorely needed a mother's loving care and discipline.

His choice fell on Minna Waldeck, the youngest daughter of the Göttingen professor of law Johann Peter Waldeck (1751–1815) and his wife, Charlotte Augusta Wilhelmine Wyneken (1765–1848). Minna's full name was Friederica Wilhelmine, although she always used the short form. She was born April 15, 1788, and, although eight years younger than Gauss' first wife, had been her best friend in Göttingen. Just before Gauss courted her, she had broken off an engagement and was thus in a somewhat depressed mood. His courtship did not immediately remove this condition. It was not easy for a girl of her type to shift her affections quickly to another. When she finally did accept the offer of marriage, it was due in a considerable degree to parental influence. Gauss was very welcome as a son-in-law to the aristocratic Hofrat Waldeck and his wife.

Reason thus played a part in this marriage, more so in the case of Minna than in the case of Gauss. The engagement (April 1, 1810) was put to a hard test, but eventually the marriage proved to be a very happy one. It was easy for Gauss to win quickly the full love of Minna. He felt profound gratitude for her domestic qualities and her loving care of his children, as though they had been her own. The children in turn loved her—a proof of her desire to make her husband and stepchildren happy. However, it must be admitted that Minna did not possess that hearty, happy nature so characteristic of Johanna. Gauss consulted her parents before he addressed her. Encouraged by them, he wrote on March 27, 1810, to Minna:

With beating heart I write you this letter upon which the happiness of my life depends. When you receive it, you will already know my wishes. How will you, best one, receive this? Shall I not appear to you in a disadvantageous light, because not half a year after the loss of a mate so well loved, I already think of a new union? Will you therefore consider me fickle or even worse?

I hope you will not. How could I have the courage to seek your heart if I did not flatter myself that I stand too well in your opinion for you to think me capable of any motive for which I should have to blush.

I honor you too much to wish to conceal from you that I have only a divided heart to offer you, in which the image of the dear transfigured spirit will never be extinguished. But if you knew, you who are so good, how much the departed one loved and honored you, you would understand that in this important moment, when I ask you whether you can make up your mind to accept the place she has left, I see her vividly before me, smiling joyfully upon my wishes and wishing me and our children happiness and blessing.

But, dearest, I will not bribe you in the most serious affair of your life. That a departed one would look down upon the fulfillment of my wishes with sincere joy; that your mother, whom I have informed of it (she herself will tell you what has induced me to do so)—that your father, who through your mother knows of it, both approve of my intentions and hope for the happiness of us all from it; that I, to whom you were dear from the first moment when I made your acquaintance, should become supremely happy through it, all this I mention only for the purpose of begging you, of imploring you, to give this no consideration, but to consider only your own happiness and your own heart. You deserve a happiness completely pure, and you must not allow yourself to be influenced by any subordinate considerations which lie outside of my personality, of whatever kind they may be. Let me also confess to you quite frankly that however modest and easily satisfied I am in my claims on life otherwise, there can not be for me any halfway condition in the narrowest, closest, domestic relation, and there I must be either extremely happy or very unhappy; and even the union with you would not make me happy if through it you too would not become entirely so.

I probably have had for some time a share in your good wishes: ask yourself, my dearest, whether you are able to give me more. If you should find that you could not, do not hesitate to pronounce judgment over me. But I would not be able to find words to express my happiness if you allowed me to call myself by a still more beautiful name, than that of your warmest friend

CARL FRIEDRICH GAUSS

The wedding is entered on the records of St. John's Church in Göttingen under date of August 4, 1810. At this time Gauss desired his mother to live with him, but she did not do so until 1817. Minna belonged to an aristocratic family; her descendants still treasure the Waldeck crest. This new marital bliss was not of long duration. As early as 1818 Minna's health began to fail. The same year she went to the spa in Pyrmont for treatment. In 1820 and 1824 she tried Ems. Letters of this period leave no doubt that she was suffering from consumption. There were short intervals of improvement, but the disease progressed almost incessantly. Minna bore Gauss three children: (1) Peter Samuel Marius Eugenius, always called Eugene, who was born July 29, 1811; (2) Wilhelm August Carl Matthias, always called Wilhelm, who was born October 23, 1813; and (3) Henriette Wilhelmine Caroline Therese, who was born June 9, 1816. The daughter was always called Therese. Later these children of the mathematician will enter our story more fully.

Minna became practically a semi-invalid and thus could not

take the social position to which she was entitled. The liberal allowance of money given to her was useless, as far as social life was concerned. Some of it was spent in collecting silverware; she had a cabinet in her room where she kept her rare pieces. Gauss took his coffee with Minna and (after 1817) his mother every afternoon, his daughter Therese sitting close to him. In 1848, the year Minna's mother died, a number of her relics, as well as some pieces of the silver, were sent to Eugene and Wilhelm, who had migrated to the United States. Letters, family mementos, and some of Minna's jewelry were also included. Unfortunately the ship sank and all these souvenirs were lost.

A gold medal issued to Gauss by George V of Hanover was in later years converted into spectacle rims by Eugene, who also unfortunately burned most of the letters his father had sent him. Wilhelm used to tell of King George's visit to Gauss' home. Great preparations were made for his entertainment. Two of the mathematician's sons, dressed in black velvet suits, stood at the doorway as the king approached and dropped flowers in his path. The descendants treasure Gauss' small cup and saucer bearing the inscription, *Meinem guten Vater*. They also have a tiny covered china dish decorated with dainty gold-painted china flowers, which served Gauss as a match box. They use under glass in a tray a piece of petit point embroidery which decorated a pillow in Gauss' home. The bright reds, blues, and yellows of the gay design have remained fresh and attractive.

Minna died on September 12, 1831, at the age of forty-three. More must be said later concerning her death, which was undoubtedly hastened by grief over her son Eugene's behavior and his sudden migration to America.

In April, 1810, Gauss had a call to the Royal Academy of Sciences in Berlin; this was engineered by Wilhelm von Humboldt. The marriage to Minna strengthened his ties in Göttingen and contributed much to his decision to reject the Berlin offer. This offer was renewed in 1821 and 1824; it hung fire until 1826. Yet in the beginning he did consider accepting, because in Berlin he would have had full time for research and observation—freedom from teaching duties. However, he enjoyed the companionship of young scholars, and in a sense felt well repaid for his teaching when he later saw substantial scientific accomplishment among his students. Their testimony refutes the charge that Gauss was not a good teacher.

On January 30, 1812, Gauss presented to the Royal Society of Sciences at Göttingen his "Disquisitiones generales circa seriem infinitam":

$$1 + \frac{\alpha \beta}{1 \cdot \gamma} x + \frac{\alpha(\alpha+1)\beta(\beta+1)}{1 \cdot 2 \cdot \gamma(\gamma+1)} x^2 + \\ + \frac{\alpha(\alpha+1)\beta(\beta+1)(\beta+2)}{1 \cdot 2 \cdot 3 \gamma(\gamma+1)(\gamma+2)} x^3 + \cdots$$

This memoir was published in the *Commentationes* of the Royal Society, Vol. II (1813); the German translation by Dr. Heinrich Simon did not appear until 1888. The Latin text was published in Gauss' *Collected Works* (Vol. III, 1866), however, and included a continuation found in his papers after his death. This memoir is important because it founded a section of mathematical literature in which later such names as Kummer, Weierstrass, and Riemann were prominent. From the manuscript it appears that Gauss had originally intended to present this paper in November, 1811. Thus Gauss opened the critical or modern period of research on infinite series. Pfaff called this one the "hypergeometric series"; Euler had studied it, but Gauss was the first to master it. For the first time the convergence (validity) of an infinite series was adequately investigated.

This decade of his life was an especially productive one for Gauss. His work of this period covered a wide range of subjects in approximately two dozen memoirs. A number of them deal with difficult problems in mathematical physics, while others cover theoretical astronomy as well as observations, particularly of the asteroids and comets. Several papers had as a subject astronomical instruments recently acquired, and optical problems connected therewith. One finds also topics in the theory of numbers, and logarithms; he gave, too, a second and third proof of the fundamental theorem of algebra.

The years 1816 and 1817 mark the close of Gauss' work in the field of theoretical astronomy. His later astronomical activity belongs to the area of observational and spherical astronomy, while the emphasis shifted to related domains. A fairly complete record of activity at the Göttingen University Observatory is given by the observation books still to be found in its library catalogue. This manuscript bears the title "Diary of Astronomical Observations at the Observatory in Göttingen." It is in two volumes and a small quarto notebook in the Gauss papers. In it most results of observations are entered in chronological order. While the notebook covers only the time from January 1 to July 31, 1808, the first volume of the diary does not begin until November 3, 1808. With some gaps, this diary runs to April, 1822. After that date Gauss kept special records of his observations, mostly of comets and planets, on separate sheets of paper; these papers are not complete.

The regular observer was Karl Ludwig Harding (1766–1834), who was called to Göttingen in 1805 as assistant professor of astronomy and official observer of the observatory and in 1812 was promoted to the rank of full professor. In addition, he taught elementary astronomy and navigation. In the above-mentioned diary the observations of the sun and fixed stars are his. Harding mentioned in an article published in August, 1810, that the Göttingen observatory was not properly equipped with instruments. To supplement this the King of Westphalia granted Harding four thousand francs for his celestial atlas, and from March to June, 1811, one finds observations by Gauss, since Harding was at that time in Paris. The planets, the moon, Vesta, and Ceres were frequently observed by Gauss. In addition, various observations from 1809 to 1818 were carried out by sixteen of Gauss' pupils.

The principal clock was the Shelton, a gift of George III, king of Great Britain and elector of Hanover. It had been in use since 1770. It is now in the east meridian hall and is still running. Gauss praised it in 1816 and 1818, but by 1819 he was dissatisfied with it.

Gauss always liked to use the sextant and spent many hours in determining the elevation and azimuth of terrestrial objects. In October, 1810, he journeyed to Gotha for a chronometric determination of longitude. During the period 1808–1812 he published five shorter memoirs in the field of spherical astronomy.

In the early days most of the high-quality astronomical instruments were of English make, and practically all of the Göttingen instruments were imported from England. However, there was a mechanic in Stuttgart named Baumann, who had studied for a long time under Ramsden in England, and soon Gauss gave him an order. But Germany had another instrument producer, also a pupil of Ramsden, who was becoming well known. This was an artillery captain, Georg von Reichenbach (1772–1826), in Munich. He was doing the best work of this kind to be found in Germany, and Gauss now began to purchase instruments from him. The results were highly pleasing; the first instrument arrived on November 26, 1812, and it is noticeable in the diary that Gauss' interest in observing was immediately stimulated. His efforts were directed, not to accumulating a mass of observations, but to making a small number of extremely accurate ones. From this time on almost all the diary entries are in Gauss' handwriting. Apparently Harding's observations were no longer entered in the diary. Gauss' astronomical work of this period is indicated in his correspondence with Bessel, one of the leading observers of that time. Their letters go into great detail; Bessel had considerable influence on Gauss in these matters and stimulated his interest in the theory of dioptrics in later years. On May 23, 1814, the observatory received a fine achromatic heliometer produced by Joseph Fraunhofer (1787–1826) in Munich. The observatory already possessed an older heliometer.

In 1810 a special fund was set up for the various departments of Göttingen University; 1,750 francs were earmarked annually for the observatory. Current expenditures, including the allowance for Gauss' rent, absorbed about half of that amount. By application to the governing board of the university Gauss got funds for a stand to support the new heliometer, which arrived in April, 1815. After it was mounted, he tried some measurements of the diameters of Venus, Mars, and the ring of Saturn, as well as of the comet of 1815. In May, 1817, he sent the heliometer back to Fraunhofer for improvements. It was finally returned in November, 1817, but he never made real use of it.

In the winter of 1815–1816 the Göttingen observatory received a collection of astronomical instruments which it had purchased from the estate of Johann Hieronymous Schröter (1745–1816), founder of the Lilienthal observatory. These instruments had been purchased with the stipulation that Schröter could use them as long as he lived. When the Kingdom of Westphalia annexed Lilienthal, Schröter wished to disregard the sale and to ship the instruments to France. The French wanted to have some of the instruments, but realized a second sale would be illegal. Gauss felt so strongly about the matter that he expressed himself only orally!

On November 25, 1811, Laplace wrote to Gauss saying that he would do what he could. Meanwhile, nothing happened until Lilienthal was captured and burned on December 12, 1812, by the French. Schröter had to flee and never recovered from this blow. Fortunately the instruments were rescued. After the restoration of the Kingdom of Hanover Schröter arranged their transfer to Göttingen before his death.

Repsold had constructed a meridian circle which interested Gauss as early as 1810. Finally it was purchased for Göttingen and arrived on Friday, April 10, 1818. Incidentally Gauss visited the Mathematical Society in Hamburg in 1818 and became an honorary member of it. Repsold's sons developed his workshop into a well-known astronomical instrument firm.

Gauss planned to purchase from Reichenbach a second meridian circle. This plan led to his trip to Bavaria in the Easter vacation of 1816. He left Göttingen on April 18 and was absent five weeks. The letters written to Minna at this time are extremely interesting. Gauss was not a frequent traveler, and it is probable that he enjoyed this trip more than any other he ever took. The exquisite Bavarian Alps made a strong appeal to him. Twelve days were spent in Munich and Benediktbeuern, where Fraunhofer's optical work was carried on. His companions on this trip were his ten-year-old son Joseph and Dr. P. Tittel, one of his pupils. Gauss had hesitated to leave Minna at this time, since it was only several weeks before the birth of his daughter Therese. The primary purpose of this trip was to make the personal acquaintance of Fraunhofer and Reichenbach, as well as Joseph von Utzschneider (1763–1840), partner of Fraunhofer in the optical company, and Traugott Lebrecht Ertel (1778–1858), owner of Reichenbach's firm, and to discuss with them the ordering of two new meridian instruments of the best construction.

Gauss journeyed via Gotha, where he spent several days with his friend Bernhard August von Lindenau (1799–1854), premier of Saxony and director of the Seeberg observatory. Lindenau furnished his coach to Gauss, who was not satisfied with the slowness of his Göttingen coachman. At Berchtesgaden he visited the salt mines. Benediktbeuern was some hours distant beyond Munich, and Gauss found the optical shop set up in an old abbey belonging to Utzschneider. In addition, these men had as assistant a Munich mechanic named Liebherr. The workers were at the time busy with meridian instruments for Warsaw, Ofen, and Turin.

On April 26 Gauss wrote Minna from Munich:

Yesterday evening about eight o'clock we arrived here in good shape . . . Now first of all a little account of the trip. On Sunday, early, we left Seeberg, in Lindenau's coach and with horses we hired, to drive through the Thuringian Forest where the roads were still completely covered with ice. In summer this region must be romantically beautiful. These horses brought us to Meiningen, where we immediately took post-horses for the all-night drive over the fine Bavarian highway, reaching Würzburg next morning rather weary. Here we refreshed ourselves with a midday meal. As night approached we continued on our way, again to drive all night and into the next day as far as Augsburg, where we spent the night. Thursday morning we looked around a little and by midday were again on our way. Those last eight and a half miles over an incomparably beautiful road took seven and a half hours. And so we came yesterday to beautiful Munich.

I was rather exhausted by the journey, but rest has set everything right again, and today I feel as well as in Göttingen. We are lodged in a very good inn. Early this morning Reichenbach came to see me, having already learned of my arrival. I have spent the greater part of the day with him and have accepted his very friendly and urgent invitation to stay at his home. Tomorrow we will move there, also Tittel. . . . I have also become acquainted with Utzschneider, and on Tuesday we will go with him to his estate in Benedictbeuern on the Tyrolean border. Reichenbach is a very gracious man who overwhelms me with kindness; his house is in the suburbs, has an extremely pleasing location and bears the stamp of great affluence.

A second letter is headed "Reichenhall, 36 hours beyond Munich, Sunday evening, May 11, 1816," and gives an account of the return journey:

At last I am on the way home. After twelve very pleasant days in Munich, in which I include the trip to Benedictbeuern, I have come this far with Reichenbach, who had to come here on business. After seeing something of this vicinity and of the nearby Berchtesgaden as well as the extremely interesting salt mines and the incomparably beautiful region, tomorrow will find me on my way back to Göttingen. We do not return to Munich, but take the nearer way to Regensburg and Nürnberg, then Gotha for a couple of days. . . . I write this at midnight with my eyes almost closing, since today we made the excursion to Berchtesgaden, and there in the underground salt mines we were constantly on the move.

The tour of inspection also took Gauss to the Munich observatory and that of Father Placidus Heinrich in Regensburg. The entire trip lasted exactly thirty-six days, of which eight were required for the trip to Munich via Mühlhausen, Gotha, Meiningen, and Augsburg. Sixteen days were required for the return trip, because Gauss detoured via Reichenhall, Landshut, Regensburg, and Nürnberg.

An official report on this trip was given by Gauss to the university board of curators on June 5, 1816, just four days before Therese's birth. In it he made his recommendations on the purchase of instruments. Success crowned his efforts; he received the Reichenbach passage instrument in November, 1818, and the meridian circle in August, 1819.

The Repsold meridian circle arrived, as mentioned above, on April 10, 1818. The following day Repsold himself arrived and was a house guest of Gauss until the evening of the nineteenth. During his visit the Shelton clock was cleaned and regulated. On May 1, 1818, Gauss began a new diary entitled "Diary of Observations at the Repsold Meridian Circle." This volume lay unnoticed at the observatory until after 1927. From May 1 to 31 Gauss observed twenty-one days—mostly principal stars, on some days the sun. Occasionally he complained about weather conditions as well as the fact that, in one direction, fruit trees in adjacent yards and, in another, the wooded Hainberg obscured his view. He suggested the possibility of chopping down some of the trees. On the first of June, 1818, Repsold, returning home on his trip, again visited Gauss and made further improvements on the instrument.

Gauss set up a rather extensive plan of observation which in large part he completed, but did not publish his results except for occasional observations of planets. In the catalogue of the Göttingen observatory is Gauss' notebook entitled "Calculations and Notes Concerning the Observations at the Meridian Circles." On the first pages of this book is an unfinished list of 316 stars, predominantly circumpolar. His observations fall almost without exception in the time from noon or early afternoon to the late evening hours and include daily ten to fifteen stars. From the end of September to the end of October, 1818, Gauss was in Lüneburg to make measurements on St. Michael's Church tower as a connecting point with the Danish triangulation. From his return until the summer of 1819 he was observing as many as thirty stars daily, as well as Jupiter, the sun, and the comet of 1819.

The Reichenbach passage instrument was set up in September, 1818, and Gauss began to devote all his attention to it. He stopped using the Repsold circle and began to use more and more the Reichenbach meridian circle, which was set up on October, 1819. It is noteworthy that Gauss seldom permitted Harding or one of the pupils to use these instruments. When he handled the Reichenbach meridian circle, Gauss wore gloves, according to the story told by Encke, his pupil. Encke and Nicolai, another pupil, were allowed to take down notes when Gauss was observing. He made use of a town mechanic who rejoiced in the name of Philipp Rumpf.

In the years 1819–1822 at Gauss' suggestion the right ascensions of the moon and several moon stars were observed by Nicolai in Mannheim, Soldner in Munich, and Encke in Seeberg for the determination of differences in longitude. The observations were continued for many years, although Gauss withdrew in July, 1820.

On the first of May, 1820, he obtained a Liebherr clock, which had been sorely needed, as evidenced by the fact that he had borrowed one from Repsold. Even the Liebherr clock soon caused trouble by running fast, and Rumpf was away for several months! Fortunately a Hardy clock was presented to the observatory in 1826 by the Duke of Sussex.

Gauss was made happy in August, 1819, by a four-and-a-half-day visit from his old friend Olbers. At the moment he was especially vexed at Heidelbach, Harding's father-in-law, who owned the orchard just south of the observatory. The owner would not allow the fruit trees to be cut down and refused compensation for these obstructions to observation. In June, 1819, Gauss was in Lauenburg for the setting up of a Ramsden zenith sector.

One should not imagine that Gauss' work of this period was confined to practical astronomy. He was considering magic squares, quadratic residuals, and a new method of determining integrals by approximation (1814), founded on Newton's method for that purpose. There was one memoir of more than common interest, devoted to the demonstration of a very remarkable proposition in planetary theory that the secular variations which the elements of the orbit of a planet would experience from another planet which disturbs it are the same as if the mass of the disturbing planet were distributed into an elliptic ring coincident with its orbit, in such a manner that equal masses of the ring would correspond to portions of the orbit described in equal times.

In the course of the last investigation Gauss arrived at some elliptic integrals, the evaluation of which he was enabled to effect by means of a transformation which is included in one of the series immortalized by the name of C. G. J. Jacobi (1804–1851), the brilliant German-Jewish mathematician. Somewhat later Gauss wrote to Schumacher (1828) that he had anticipated Abel and Jacobi by a quarter of a century, but that his principle of not publishing anything unless it was a completed work of art (form and content), had robbed him of priority rights in these difficult theories.

In the years with which this chapter is concerned Gauss enunciated a number of theorems on the so-called arithmetico-geometric mean. As a matter of fact, he had been working on elliptic and lemniscate integrals as early as his student days. Posthumous fragments of the year 1808 show that the 1828 letter to Schumacher was correct in its statement. The use of i for $\sqrt{-1}$ goes back to early work of Gauss (1801), but he did not lend his authority to placing the imaginary on a firm basis until the appearance of a memoir in 1831. The names "complex" and "lateral number" are credited to him, but others gave the first geometric interpretation of the imaginary.

Gauss' name has been incorrectly given to a table of addition and subtraction logarithms which he produced and published in 1812. However, he scrupulously referred to Z. Leonelli, whose idea it was and who actually published such a table in 1802. The so-called Gaussian logarithms are arranged to give the logarithms of the sum and difference of numbers whose logarithms are given. Gaussian logarithms are intended to facilitate the finding of the logarithms of the sum and difference of two numbers, the numbers themselves being unknown, but their logarithms being known, wherefore they are frequently called addition and subtraction logarithms.

CHAPTER TEN

Geodesy and Bereavement: The Transitional Decade, 1822-1832

Gauss' interest in geodesy must be traced back to the year 1794 when he discovered the method of least squares. In it he found a method of logically combining magnitudes which involve accidental errors. For Gauss the method of least squares was at once one of the bridges which led from pure to applied mathematics. He visualized his highest ideal in activity in both directions. In a letter to Olbers he expressed it thus: "The most refined geometer and the perfect astronomer—these are two separate titles which I highly esteem with all my heart, and which I worship with passionate warmth whenever they are united."

Originally Gauss did not attach great importance to the method of least squares; he felt it was so natural that it must have been used by many who were engaged in numerical calculations. Frequently he said he would be willing to bet that the elder Tobias Mayer had used it in his calculations. Later he discovered by examining Mayer's papers that he would have lost the bet. Legendre anticipated Gauss in the matter of publication, and Gauss showed Legendre's study to his fellow students, including Bolyai. The interesting thing is that Gauss discovered the method at such an early age. Yet he did not underestimate the practical importance of its use. In June, 1798, he adapted it to the principles of the calculus of probabilities, and as early as 1802 used it in astronomical calculations. Legendre used the name *méthode des moindres carrés,* and by adopting this nomenclature Gauss showed that he did not feel hurt because he was anticipated. His first publication on the subject occurred in Book II, Section III, of the *Theoria motus.*

On February 15, 1821, Gauss presented to the Royal Society of Göttingen his memoir *Theoria combinationis observationum erroribus minimis obnoxiae, pars prior,* in which he gave what he felt to be the only proper connection of the method of least squares with the calculus of probabilities. He stated that he started from the same viewpoint as Laplace, but that he used a different mode of development. On February 2, 1823, he transmitted to the Royal Society *pars posterior* of the above memoir. His *Supplementum theoriae combinationis observationum erroribus minimis obnoxiae,* presented to the Royal Society on September 16, 1826, was a direct result of his practical work in geodesy.[1]

Very little is known about Gauss' first experience in practical geodetic work. During his student years he had no experience in observing, other than a few exercises with the mural quadrant and the reflecting sextant. A letter written in 1797 shows that he did have a certain interest in observing, and realized that he had no training in drawing, architecture, or mechanical arts. In a letter dated February 21, 1802, the astronomer von Zach, speaking from his own experience, wrote that he regarded Gauss' near-sightedness as a hindrance, and alluded to danger to the eyes from solar and sextant observations. However, he let Gauss borrow a sextant, a clock, and a telescope. With the little telescope Gauss began to practice occasionally—primarily as a pastime, but also in preparation for his future vocation. In years 1803–1805 he began practicing with the sextant in an area around Broitzen, five kilometers from Brunswick. At this time he had visions of one day engaging in a large triangulation; an essay of Beigel on the trigonometric survey of Bavaria attracted his attention.

In the summer of 1803 Gauss again came in contact with a councilor of the Superior Court of Appeals named von Ende, who was making determinations of latitude and longitude in Brun-

---

[1] Börsch and Simon edited a German text of Gauss' memoirs on the method of least squares (Berlin, 1887); J. Bertrand published a French text of these memoirs (Paris, 1855).

swick and vicinity. Von Ende, who died in 1816 in Cologne, had a small observatory in Celle. In August, 1803, von Zach was on the Brocken and was giving almost daily signals for the determination of longitude. In Brunswick, Helmstedt, and Wolfenbüttel the work was turned over to Gauss and von Ende. From solar observations Gauss determined the time with a chronometer belonging to the Duke of Gotha.

On August 27, 1803, Gauss journeyed from Brunswick to Brocken, met von Zach there, and then accompanied him to Gotha in order to confer with von Ende and Professor Bürg of Vienna. For three months he was a busy guest of the Seeberg observatory; on December 7 he and von Zach returned to Brunswick, where for ten days von Zach was his guest.

Gauss' interest in this work was evidenced by the fact that he calculated observations made by Baron Friedrich Ferdinand Carl von Müffling (1775–1851), who in 1821 became chief of the general staff of the Prussian army. The latter had already participated in the trigonometric survey of Westphalia, made for military purposes in 1797–1802 by Colonel Lecoq. Lindenau now joined this project. With his sextant Gauss took part in this triangulation, and his calculation gave him a number of places from Brocken to Franconia which he thought were accurate enough for the needs of a map. In August, 1804, Gauss joined Olbers in the spa Rehburg for several days. The two friends measured some angles between Hanover, Brocken, and Minden, from whose approximately known geographic location Gauss calculated the latitude and longitude of Georgplatz, a small knoll in Bad Rehburg.

In the years 1803–1805 the French Colonel Epailly was directing surveys in the occupied zone, and his triangles, especially in the southern part of the Electorate of Hanover, were later of great use to Gauss in the measurement of an arc of meridian for setting up a network of triangles, although their configuration had an unfavorable influence. In 1805 Epailly went to Brunswick to make an observation from the tower of St. Andrew's Church; Gauss used the opportunity to study his work and to familiarize himself with the instruments. Epailly became ill in Brunswick.

Soon after settling in Göttingen in November, 1807, Gauss secured from Harding a ten-inch Troughton sextant. In 1808 he published his *Methodum peculiarem elevationem poli determinandi explicat D. Carolus Fridericus Gauss*. With this method, the sextant and a clock would give good results. It required less preparatory calculation than a method he had previously published. In August, 1810, Gauss journeyed to Münden. One purpose of this trip was to find the effect of traveling on the running of a chronometer. On this occasion he determined on Freytagswerder at the confluence of the Werra and Fulda the latitude and difference in longitude from Göttingen. On February 19 and 20, 1812, Gauss observed star occultations for the determination of the longitude of Göttingen. In the same year on the Hanstein he determined the angles between Göttingen, Brocken, and Boineburg (two Müffling points).

On June 8, 1816, Schumacher wrote Gauss that the King of Denmark had appropriated the necessary funds for the measurement of an arc of meridian from Skagen to Lauenburg and a measurement of longitude from Copenhagen to the west coast of Jutland. He asked Gauss for an improvement in the usual methods and inquired whether he would be interested in extending the survey through Germany to the Bavarian triangles. Gauss replied on July 5, 1816, with enthusiasm and expressed hope of future activity in this field in Hanover, also offering his aid in calculating the principal triangles. He saw an advantage for Schumacher in the fact that Denmark had already been trigonometrically surveyed, although he felt there would be a difficulty inherent in Denmark's flat terrain.

Gauss felt that such operations would help to clear up questions connected with the irregularity of the earth's figure by furnishing additional empirical data. He was especially interested in making fundamental determinations in this field and also in the astronomical constants. His earlier interest in geodesy can be traced in part to a desire for accomplishment, in gratitude to the Duke of Brunswick. He was later greatly interested in von Zach's survey of recently acquired lands authorized by the King of Prussia and realized that after the elaborate results in France and

England a survey would be relatively unimportant unless carried out on a large scale. In 1816 Gauss received a copy of Epailly's triangles.

At this time Gauss was considering as a prize question the problem of so projecting a given surface on another given surface that the projection would be similar to the original in the smallest parts. A special case occurs when the first surface is a sphere and the second a plane. The stereographic and mercator projections are particular solutions. The general solution, to include all particular solutions, would cover every kind of surface. In the Göttingen Royal Society of Sciences, Gauss had the opportunity of proposing a prize question only once every twelve years. He saw a chance of generalizing this theory and giving a geometric description of a large part of Europe by means of his geodetic measurements if others would also communicate their results to him. Thus his first idea of conformal projection can be dated early in the year 1815.

On the occasion of Gauss' visit to Lindenau on the Seeberg in the fall of 1812, geodetic questions were discussed. His paper of 1813 on the theory of the attraction of homogeneous ellipsoids was a fruit of this visit. On September 25, 1814, the astronomer Encke, Gauss' pupil, accompanied him to Seeberg for another visit with Lindenau. The fertility of Gauss' thinking during this decade is comparable to that of his youth, 1790–1800. Encke recorded that he was as though intoxicated by the glorious theories which Gauss enunciated in the evening after dinner.

In the autumn of 1816 Gauss moved into the new observatory. His spare time was claimed by research in theoretical astronomy, non-Euclidean geometry, and the theory of numbers; terrestrial magnetism and the theory of surfaces were already beginning to occupy his attention. It is understandable that he did not immediately agree to join Schumacher's survey project. Gauss lacked practical experience in negotiating with the authorities on such a matter and in gaining the cooperation of experts; in addition he did not have trained assistants. Such a survey was expensive, and he disliked to request an appropriation at the same time as he was seeking funds to equip the new observatory. Even salaries were not always paid promptly. At this juncture Schumacher applied directly to Baron Karl Friedrich Alexander von Arnswaldt (1768–1845) in Hanover, curator of the University of Göttingen from 1816 to 1838. Von Arnswaldt proved to be very cooperative and at once inquired about all details, including the matter of assistants. In September, 1816, when von Arnswaldt was returning from a water cure in Wiesbaden, he stopped in Göttingen and requested Gauss to hand in a memorandum on the proposed survey. In early December Gauss received from Schumacher information on the cost of the Danish survey, including the duration of the measurements, the personnel, and related topics.

By the end of July, 1817, actual operations had begun in Denmark, and in November the triangulation of $1\frac{1}{2}$ degrees of the arc of meridian from Hamburg to the Island of Alsen was complete. In 1818 when Schumacher was ready to measure his southernmost triangles, he again requested Gauss to participate in the connecting measurements, if possible in September, and sent him a sketch of the triangles already measured and of the planned connection. Gauss received this letter too late to discuss the plan with the curator. He had not yet received an answer to his memorandum on participation in the survey. Again Schumacher took the necessary steps and announced to Gauss at once that the continuation of the survey through Hanover was not only approved, but also that von Arnswaldt would do everything to promote it. In September, 1818, Gauss announced that the premier had commissioned him to undertake the survey in Lüneburg. He asked Schumacher about the most suitable tower in Lüneburg. His health had not been as good as usual that summer, but it soon improved and he looked forward with pleasure to the journey. He bought a coach and took along the twelve-inch Borda circle as well as the eight-inch theodolite in order to see how the results would agree. Shumacher sent his assistant, Captain Caroc, who was busy with a survey in Lauenburg at the time. On this account Ursin (Georg Frederik Krüger) was ordered to assist Gauss. Later Schumacher himself came to Lüneburg with an eight-inch Repsold theodolite. The tower of St. Michael's Church in Lüneburg was used to measure the two angles Hamburg-Hohenhorn and Hohenhorn-Lauenburg. This fine, big tower was of solid stone, and its centering was easy and sure.

During these observations Gauss had his first stimulus for the invention of the heliotrope; this was the reflection of the sun from a window in St. Michael's tower in Hamburg, a disturbance in his observing. On advice of H. C. Albers, a geodesist, Gauss stayed in Lüneburg at the Schütting inn on the market place, some distance from St. Michael's Church. On October 9, 1818, he determined on the bastion before a Lüneburg gate the angles between the various towers and the position of Schumacher's theodolite. Gauss sent a report to Hanover on the Lüneburg measurements and requested a larger theodolite, but did not receive an answer.

Schumacher journeyed to London in April, 1819, to receive Ramsden's zenith sector at Woolwich observatory. He used the opportunity to interest Sir Joseph Banks, president of the Royal Society and a man of unlimited influence, in the Hanoverian survey. He also stimulated the Danish envoy in London, Privy Councilor von Bourke, to negotiate on the survey with Count Münster, minister of Hanoverian affairs in London. Count Münster was rather surprised that Gauss had not applied directly to him, and requested an immediate memorandum on the costs. Gauss therefore sent Münster a document on the meaning and usefulness of the Hanoverian survey, and on advice of Schumacher informed von Arnswaldt of this letter, at the same time requesting the latter to approve participation in the observations in Lauenburg. On the first of June Schumacher personally conferred with von Arnswaldt and succeeded in getting a ministerial directive authorizing Gauss to go to Lauenburg and to receive all necessary advances of money. By the end of June the sector was set up in Lauenburg and Gauss participated in the observations, from which he returned home very suddenly on July 18, weakened by the summer heat. So great was the haste that he left behind various articles of clothing.

Early in 1820 Gauss again felt ready to make observations; he knew that eventually these data would be published and therefore wished to have a competent scientific witness present for the most important operations as an aid to accuracy. Soon afterward King Frederick VI of Denmark ordered his envoy in London to request of the Hanoverian government Gauss' presence in the Danish survey, including the observations in Lauenburg. Bessel attached little importance to Gauss' geodetic work and thought it required too much of his time. Such activity, he thought, should be carried on by one of lower mathematical stature. Gauss replied to Bessel:

> All the measurements in the world are not worth *one* theorem by which the science of eternal truths is genuinely advanced. However, you are not to judge on the absolute, but rather on the relative value. *Such a value is without doubt possessed by the measurements by which my triangle system is to be connected with that of Krayenhoff, and thereby with the French and English. And however low you estimate this work, in my eyes it is higher than those occupations which are interrupted by it. I am indeed far removed *here* from being master of my time. I must divide it between teaching (to which I have always had an antipathy, which is increased, though not caused, by the feeling of throwing my time away, an everpresent concomitant of this activity) and practical astronomical work. I have always enjoyed it so much; you will agree with me, that, when one does without all real help in numerous petty affairs, the feeling of losing one's time can only be removed when one is conscious of pursuing a *great important* purpose. But you have made that difficult for the rest of us, since you have anticipated us and supplied most [astronomical] desires in such a model manner. Nothing remains for the rest of us except now and then to glean.
>
> What do I have for such work, on which I myself could place a higher value, except fleeting *hours of leisure?* A character different from mine, less sensitive to unpleasant impressions, or I myself, if many other things were different from what they actually are, would perhaps be able to gain more from such hours of leisure than I, in general, can. As things are, I must not reject an undertaking, which, although connected with a thousand complaints and perhaps acting as an irritant to my strength, is really useful, which of course could be carried out by others, while I myself under more favorable circumstances would do something better, but which, if I do not take it on myself, would definitely not be carried out; finally, I must not conceal from you a matter which *somewhat* balances the inequality which exists between my salary—the same in 1824 as was scheduled in 1810 under Jérôme—

and the needs of a large family. [Gauss wanted Bessel to treat this as confidential.]

In the spring of 1820, as a result of Count Münster's letter from London, the King approved the continuation of the survey through the Kingdom of Hanover. This cabinet order of George IV, king of Great Britain and Hanover, went to the ministry on May 9, 1820, and the ministry notified Gauss on June 30, 1820. On November 1, 1820, he reported to the ministry on his participation in the survey at Braak, twelve kilometers northeast of Hamburg, which lasted from September 12 to October 25. On the return trip Gauss found out through von Arnswaldt that Count Münster had been negotiating with the Duke of Wellington (as Great Master of the Ordnance) on the question of transferring the Ramsden zenith sector, which Gauss valued highly for this work. Actually, it was returned to London in 1827 by Gauss' assistant, Captain Müller. The Danish survey as planned was not completed in Schumacher's lifetime. There was a long delay from 1824 to 1838; his death occurred in 1850.

By the close of 1820 Gauss had begun many preparations for the survey. There were numerous difficulties which often put him in a bad temper. The Göttingen observatory was chosen as a beginning point; he found a good view on a hill five kilometers north, just beyond Weende, the closest village. Difficulty was encountered in getting Epailly's data on ninety-four triangles surveyed in 1804–1805, which by a detour connected southern Hanover with Hamburg. Epailly's points were difficult to locate. The Prussian general staff furnished results of von Müffling's survey of Hessen, Thuringia, and Brandenburg, which was planned to connect with the French, Bavarian, and Austrian network. Another irritation was the delay in getting the proper instruments. Two artillery officers, Captain G. W. Müller and Lieutenant F. Hartmann of the city of Hanover, were appointed his assistants.

First reconnaissance and preliminary measurements in the neighborhood of Göttingen began in the middle of April, 1821. A French signal tower had once stood on the mountain Hohenhagen, 508 meters above sea level, not far from Dransfeld, a small village fifteen kilometers distant. Because of the fine view this spot proved to be suitable as a main point of triangulation. Müller and Hartmann found the Hils near Ammensen to be a good point for an open view northward to the city of Hanover and even farther. Hartmann was assigned the building of a signal tower on the Hohenhagen, and Müller had the same task on the Hils, as well as the reconnaissance of Kruksberg at Lichtenberg in the Brunswick area.

A unified reconnaissance of the survey area was not made and evidently was not planned. Later this caused Gauss many unpleasant hours, although one does not have to look far for the reasons why it was not made: poor travel facilities, slow means of communication, mediocre maps, poor accommodations in villages, expense, and technical difficulties in building reconnaissance structures. Gauss had to familiarize his assistants with the scientific instruments and often performed himself what he did not entrust to them. The principal reason for lack of reconnaissance is found in Gauss' desire to see some results of the survey and to be able to present them to the proper authorities. His method of procedure in theoretical work sometimes worked to a disadvantage in practical geodetic work.

Gauss' invention of the heliotrope was the circumstance which stimulated him to accelerate the survey. At first he convinced himself on a photometric basis that sunlight reflected by a small mirror can be seen at considerable distances. Next he experimented with a mirror attached to a theodolite. In the *Astronomical Yearbook* for 1825 he wrote: "With a somewhat favorable condition of the atmosphere there are no longer any limits for the sides of a triangle, except such as are set by the earth's curvature."

In connection with the heliotrope the Göttingen *Almanach* attributed to Gauss the statement that this discovery would be greater than that of America, if with this instrument we could communicate with our neighbors on the moon. Gauss did write: "With 100 mirrors united, each of 16 square feet area, one would be able to send good heliotrope light to the moon. It is a shame that we can't send out such an apparatus with a detachment of 100 people and several astronomers to give us signals for determination of longitude."

To be sure that his plans for a heliotrope would be carried out, Gauss turned the work over to Philipp Rumpf, inspector and mechanic of the Göttingen observatory. Work began in the spring of 1821.

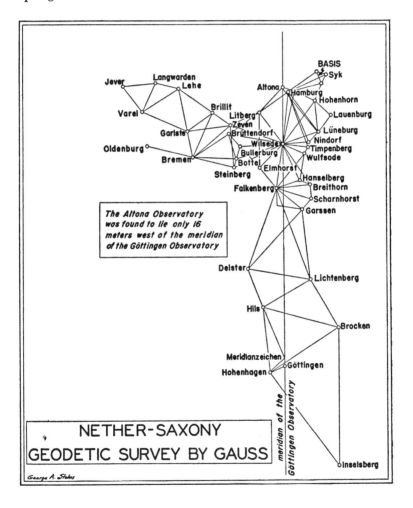

Gauss found that a mirror about the size of a calling card would suffice to reflect sunlight with the brightness of a star of the first magnitude, if one assumes mean atmospheric absorption. His first calculations on this were based on the work of Bouguer, and he was surprised at the smallness of the required mirror.

Meanwhile Gauss had written to Repsold in Hamburg and sent him a sketch of the planned heliotrope, requesting him to undertake the construction of one. In May, 1821, a Hamburg resident traveling through Göttingen told Gauss that one of the instruments was ready and the other one would be in production immediately. In July, 1821, Rumpf completed his first Gaussian heliotrope. An instrument of this form is now in the Geophysical Institute of Göttingen University. In the simple form, this instrument consists of a plane mirror, 4, 6, or 8 inches in diameter, which may be rotated about a horizontal or a vertical axis. This mirror is at the station to be observed, the sun's rays reflected by it impinging on the distant observing telescope. To the observer it appears to be a star of the first or second magnitude. In later years an improved heliotrope was introduced by the geographer and engineer Bertram. Gauss' friends were rather reticent in making suggestions to him out of their experience with the simple heliotrope. Before its completion Gauss made use of a makeshift instrument which he called the viceheliotrope.

Encke came now from the Seeberg observatory, where he had become associate director; from July 19 to 29, 1821, he and Gauss made successful observations and experiments with the heliotrope. Encke was on the Inselsberg and Gauss on the Hohenhagen, a distance of eighty-five kilometers. As a theoretician Gauss was quite naturally enthusiastic about his first invention in the practical field; he even visualized its possibilities as a mode of signaling in time of war.

The actual triangulation covered the years 1821–1823. A large number of points were determined, the principal ones being Hohenhagen, Brocken, Inselsberg, and Hils. For practical reasons the city of Hanover was dropped as a main point of triangulation although its exact location was determined. Gauss' measurements on the Hils occupied the time from August 7 to 27, 1821. Accompanied by Hartmann, Gauss went on September 2 to Brocken, the last triangle point covered in 1821, from which the bearings of Lichtenberg, Hils, Hohenhagen, and Inselsberg were to be taken. The signal tower on the Hohehagen had burnt; Hils had such a tower, but in view of the great distances heliotropes were used. The weather turned out to be so bad that light seldom came from Hohenhagen, and Inselsberg had sunshine only a quarter of an hour in fourteen days. No better weather was to be expected in October, hence Gauss returned on October 3 because of the expected arrival of the King, who planned to visit the library, the observatory, and the riding school. The field work had required five and a half months, and the relatively cool weather that summer had benefited Gauss' health. From November 22 to December 14, 1821, Gauss was in Altona to get the Ramsden zenith sector; Rumpf accompanied him in order to look after the packing.

In June, 1821, Schumacher founded the *Astronomische Nachrichten* and invited Gauss to collaborate. He told Schumacher that he would, as much as circumstances allowed, but warned him never to expect long contributions. The paper of 1822 on the application of the calculus of probability to a problem in practical geometry was intended for Schumacher. In the letter accompanying this paper Gauss showed his condescension by remarking very bluntly that he thought it was trivial.

The survey reached its most difficult stage as it approached the Lüneburg Heath. There were almost no hills, and forests frequently obstructed the view. On April 28, 1822, Gauss again went out on reconnaissance. He found a usable point an hour and a half northeast of Celle on a plateau near Garssen, where a French signal tower had stood. This point could be connected not only with Deister, but also with Lichtenberg. Gauss turned to Falkenberg, from which he could see Deister, Garssen, and Lichtenberg. Movement north from Garssen and Falkenberg offered "unspeakable difficulties." However, he did have the satisfaction of setting up two triangles in the heart of the heath: Wulfsode-Hauselberg-Wilsede and Wulfsode-Hauselberg-Falkenberg.

On June 1, 1822, Gauss returned home, somewhat weakened by the summer heat and the rigors of field work, to postpone further activity until he had more personnel and better instruments. He then spent fourteen days in preparation and decided to use his son Joseph as third assistant. On June 17, 1822, Gauss arrived in Lichtenberg and began measurements which ended on July 8. A moor fire in the direction of Falkenberg caused some disturbance for a week.

This summer (1822) three heliotropes were always used, as well as a heliotrope-style mounted mirror for telegraphing. From July 6 to 16 Deister was the center of activity. Joseph gave signals from Garssen, while Hartmann made the other stations visible. Müller, by cutting through woods, opened the way from Falkenberg to Wilsede. On July 18 Gauss went to Garssen, from which Falkenberg could be seen. On August 4 he went in the direction of Falkenberg and took lodging in Bergen, being forced to remain there five weeks. By the last of August Gauss was heartily tired of the operation.

On September 7 Gauss left Falkenberg and went to Hauselberg. There he could not receive any mail, and in a letter to Schumacher he humorously described his living quarters at Barlhof near Wilsede:

The sojourn here is not quite as bad as I had feared, without comparison better than in Ober Ohe, from which I worked on Hauselberg and Breithorn. A family lived there, whose head writes his name "Peter Hinrich von der Ohe zur Ohe" (in case he can write), whose property embraces perhaps one square mile, but whose children tend the pigs. Many conveniences are not known at all there, for example, a mirror, a toilet, and such. Thank God, that with the agreeable cool weather I have survived rather well the ten day sojourn there.

The line Hauselberg-Breithorn had to be opened. This involved cutting through a forest, but by strenuous work and with favorable weather all the measurements on Hauselberg were completed in six days, and a stone base was set on Breithorn. Stones were very scarce on the heath, and sometimes tombstones had to be brought a distance of several miles. At all triangle points a stone base about three and a half to four feet high was used to set up the heliotrope and the theodolite. The clearing from Breithorn to Scharnhorst through Haassel was largely completed. Timpenberg was chosen for a further continuation to the north; it could be connected with Wulfsode, Wilsede, and Hamburg. However, it had to be dropped, as Gauss found out there on September 22. Measurements in Wulfsode were completed on September 23 and 24. Numerous measurements in Wilsede were completed from September 26 to October 7. On one day in Wilsede with clear atmosphere Gauss measured 150 angles without the heliotrope, a record which he equaled on only two other occasions during the entire survey. Scharnhorst was the ninth and last station of 1822; he remained there three days and started home on October 13. The angle between Deister and Lichtenberg was to be measured the following year. Actually, this angle was not measured later.

During the winter 1822–1823 Gauss restricted his astronomical observations so as to conserve his health. Early in December he sent Schumacher a paper on the transformation of surfaces, a subject which he had attacked the previous winter as the solution of a prize question. In consequence of a misunderstanding of a notice in the *Leipziger Literaturzeitung,* according to which no papers handed in to the Royal Society of Copenhagen were deemed worthy of the prize, Gauss early in 1823 requested the immediate return of his paper. He had not kept a copy of it. Schumacher explained to him that the news item referred to an entirely different prize question. After a lapse of two years Gauss published in February, 1823, the second part of his *Theoria combinationis.* Two problems in it (Articles 35 and 26) are probably the direct result of geodetic work.

The measurements of 1822 had progressed more rapidly than Gauss had anticipated. He attributed this to the use of the heliotrope and the additional aid rendered by his son Joseph, who enjoyed the work so much that he dropped his plan to study law and enrolled as a cadet in the Hanoverian artillery corps. His father felt that he had aptitude for practical work, but insufficient inclination to abstract speculation demanded of a professional mathematician.

On March 22, 1823, Gauss was thrown to the pavement by a horse which had not been properly broken in, and had a narrow escape. The injury was limited to a black eye, cuts on the arm and the nose, and a bruise under the right eye, which fortunately was not permanently affected. He reported a week later to Olbers that rainbow colors under the eye were the only reminder of the accident.

The authorities of Bremen became interested in Gauss' survey and gave him as an assistant young Klüver, who had previously studied under him in Göttingen. An inspector of the waterworks named Blohm, and his brother, accompanied Klüver on a reconnaissance of the triangle Haverloh-Wilsede-Falkenberg. Gauss decided to go to Bremen on May 15, 1823. On the way he learned in Hanover that the ministry had approved his plans for an extension of the survey. On May 19 he went on to Bremen and Rothenburg near Bullerberg, where he remained until May 28. He observed the Bremen church towers while Müller made a reconnaissance between Bremen and Wilsede.

On May 30, 1823, Gauss began the real work of the summer on the Timpenberg. Niendorf, the next station, could be reached from his lodging at Bäzendorf only by a difficult walk. A clearing in the woods was required to connect it with Timpenberg. On June 11 Gauss went to Lüneburg for further measurements and on June 24 he met Schumacher in Altona. On June 27 he went to Blankenese and took readings on twenty-seven points. On July 21 Frau Minna's serious illness called Gauss home, but he stopped one day in the city of Hanover to measure one hundred positions of church towers and other points in the Hildesheim area.

On September 13, 1823, Gauss went to Brocken a second time, in order to repeat the defective measurements of 1821. In late 1823 connection was made with the Hessian triangulation, which was being carried out by C. L. Gerling, former student and close friend of Gauss. Gerling went to Inselsberg. The weather was very favorable, although at the beginning Gauss sat for three days in fog on the Brocken. On September 27 he left Brocken and from October 5 to 16 observed on the Hohenhagen, where he closed his work, which in 1823 had kept him away from Göttingen two and a half months.

Gauss' plan of connecting Göttingen with Hamburg by a chain of large triangles had not been realized. On the contrary, the network of triangles was rather complicated. His work was thereby increased; at the stations he was often busy until midnight at his desk. There were twenty-six triangles, all of whose angles he had observed himself. In the late summer of 1824 he proposed the determination of the latitudes of the observatories in Altona and Göttingen. He wrote Schumacher that he did not like to observe in the early morning hours because his eyes seemed much weaker then; that he preferred to stay up until 2 A.M. rather than get up at 5 or 6 A.M. Gauss was unable to undertake this project at once owing to the illness of his wife and the necessity of teaching two courses. He did not succeed in carrying it out until January and February, 1824.

In July, 1823, Gauss received official notice that the prize of the Royal Society of Copenhagen had been awarded him. Frau Minna's illness, with the resultant changes in his home, as well as some financial losses, caused him to request Schumacher to exchange the medal for cash.

On February 15, 1824, Count Münster's rescript went in, and on March 8 Gauss was directed by the cabinet ministry to continue the survey. Since Müller and Hartmann could not leave the city of Hanover before the close of the military-school course, the beginning of the measurements was postponed until April 18. Klüver, who had been under consideration for the preparation of the new land registry in Bremen, was appointed to aid; he proved very useful to Gauss.

On May 18 Gauss went to Visselhövede in order to resume the observations of 1822 on the Falkenberg. Four heliotropes were used this year, and thus the work progressed more rapidly. By May 23 he had finished at Falkenberg and had begun on the new point Elmhorst. Moor haze delayed the completion of these observations until June 5. Tall trees made a forest path necessary in all directions. From June 7 to 18 Gauss observed on the Bullerberg.

Gauss retained his living quarters in Rothenburg when he had completed his measurements on the Bullerberg and began to observe on the Bottel, a mile and a half from Rothenburg; this work occupied his time from June 19 to 24. He went to Zeven on June 27 and found the forest path already open. In Zeven Gauss enjoyed the beautiful natural surroundings, the rather cool weather, the pure air and good living quarters in the "Posthaus" at first, and he often walked the four kilometers to his observation station in Brüttendorf. Soon however he was complaining about the sultry heat, and on July 10 when he had finished his observations in Brüttendorf and had gone to Bremen, the fatigue of the field work and mode of life in the city affected him to such an extent that he doubted his ability to continue his work.

Gauss found his friend Olbers, now sixty-six years old, in good health; he visited in Bremen for six weeks. Next he went to the Weser river near Vegesack. He took lodging in Osterholz, a mile southeast of his triangle point. On Sunday, August 22, 1824, he had procured an open coach in which he left at 1 P.M for Garlste. The sun was shining, but suddenly rain set in when he was half way to his destination, and lasted the rest of the day. Gauss was drenched to the skin, saw nothing, and returned at half-past eight in the evening. Later he went on foot several times from Gnarrenburg to Brillit.

On August 26 he went to Brillit and finished his observations there on August 30, although the atmosphere was very unfavorable. His assistant Baumann was using the heliotrope at a church tower in Bremen, but moor fires prevented success. A heat wave came and, as usual, affected Gauss, although the stay in Osterholz had benefited his health.

Olbers, Schumacher, and Repsold visited Gauss in Zeven; he reported that he never saw Olbers in a happier mood. On September 5, 1824, these measurements were finished; from September 17 to 24 measurements were made on the Steinberg. On Septem-

The birthplace of C. F. Gauss in Brunswick (picture taken 1884), which was destroyed in World War II

(*Top*) The Gauss coat-of-arms; (*bottom, left & right*) silhouette of Gauss in his youth bust of Gauss probably by Friedrich Künkler (1810)

The Collegium Carolinum in Brunswick

The Schwarz portrait of Gauss (1803)

Minna Waldeck, second wife of Gauss

Portrait of Gauss by S. Bendixen (1828)

A sketch of Gauss by his pupil J. B. Listing

The observatory of the University of Göttingen

The courtyard of the Göttingen observatory as it appeared in Gauss' time

Gauss' personal laboratory in the Göttingen observatory as he left it

The Gauss-Weber telegraph (Easter, 1833)

Gauss' principal instrument, the Repsold meridian circle

Biermiller's copy (1887) of the Jensen portrait of Gauss (1840)

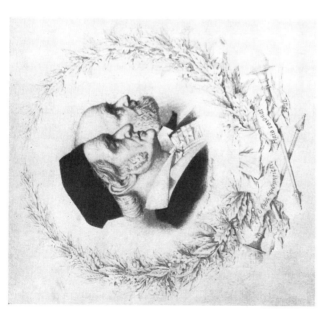

Gauss and Weber

Ritmüller's portrait of Gauss on the terrace of the observatory

Gauss in 1854

Gauss about 1850

Portrait of Wolfgang Bolyai by János Szabó

Johann Friedrich Pfaff

ber 25 Gauss again took measurements on the Bottel. The next day Gauss went to Apensen, a mile southwest of Buxtehude (where he boarded with a merchant named Köster). He was now 5.3 kilometers from his observation point Litberg, and wrote to Schumacher:

> If your assistant prefers a worse but closer lodging to a somewhat better but more distant one, Sauensiek is only about 15 minutes (by foot) from Litberg. I myself have not yet been in Sauensiek, but Müller has lodged there, as well as Klüver and Baumann. I do not know whether the name comes from the fact that sows get sick there. But two breweries are there which also supply my lodging with their brew.

In Apensen Gauss could see the houses of Altona, from whose church tower the tent on the Litberg could be seen. On October 3 these measurements on the Litberg were completed, and he had to complete the measurements of 1822 at Wilsede. He boarded in Barl and enjoyed the place. Schumacher provided him with beer and wine, the latter being especially welcome. On October 17, 1824, he wrote Schumacher that he had not felt so well in years, as in the last three or four weeks—in spite of ten days of rainy weather. Schumacher had sent Gauss two sergeants to use the heliotrope; moreover, Gauss mentioned three artillery men who were aiding him. The names of two petty officers, Biester and Querfeld, are mentioned. In Barl, Gauss became acquainted with Thomas Clausen, who brought him a letter from Schumacher. Clausen seems to have aided him in proofreading the prize essay, which was now in press.

When Gauss returned home the last of October, he found out how serious Frau Minna's illness in the summer months had been. Soon his home became the scene of disorder and worry when three of his children took the measles; his wife, weakened after two years of illness, also contracted measles and for a time hovered between life and death.

At this time strong efforts were being made to call Gauss to the University of Berlin. Soon after his return a letter came from the ministry promising him a substantial raise in salary if he would reject the Berlin offer. He did not immediately agree to remain. However, the matter dragged on and he found out by the grapevine that a proposal for a liberal increment in his salary had been sent to the King in London. Soon the King's approval arrived, and Gauss decided to remain in Göttingen although many of his friends urged him to accept the Berlin offer. His salary was raised to 2,500 thalers, and this at least in part can be attributed to satisfaction with his survey—which seemed important to Hanover.

Gauss soon overcame his disinclination to continue the survey. In March, 1825, he bought a new coach in the city of Hanover and decided to renew the triangulation early that year. He invited Schumacher and Bessel to visit him at his station Zeven, where he was also expecting Olbers. On April 18 or 19 he traveled via Hanover, Walsrode, and Rothenburg, reaching Zeven at noon April 25. In Hanover, where he stopped several days on business, he met, in the Hotel Hasenschenke, Encke, who was returning from Hamburg to Seeberg. At this time Schumacher, his guest Thune, professor of astronomy in Copenhagen, Bessel, Hansen, and Repsold went to Bremen. Schumacher, Bessel, and Hansen made some determinations of the latitude and longitude of Bremen. On Sunday, April 24, they accidentally met Gauss in Rothenburg. Upon his arrival in Ellermann's Inn Gauss was more surprised than pleased when he first saw Thune and then the entire group. Rumpf in Göttingen was sick; this meant that during his last days at home Gauss had been forced to take apart and clean the instruments himself. The rigors of the trip and the running of errands in Hanover had left him tired. He had wanted to talk over privately with Bessel the call to Berlin. Bessel had first met Gauss in 1807 at Lilienthal. He also regretted this unsatisfactory visit with Gauss after such a long time. He thought that Gauss looked quite well.

On April 25 Gauss was informed that the heliotrope light of the Bremen tower had appeared on the Brüttendorf mountain. Although still quite exhausted, he traveled there at once and measured some good angles. On the way back on the road from Ottersberg to Zeven, five hundred paces from the spot where his Brüttendorf base stone stood, his coach turned over as the result of a deep rut. The box with the theodolite fell on his thigh, and in his side he felt pain. The coach was not moving fast, hence there were no serious consequences, and the pains began to disappear the next day. Even the instruments were not damaged, but Gauss felt an unprecedented discouragement. From May 6 to 8 Olbers visited Gauss in Zeven and gave his spirits a lift. Gauss left Zeven and went to Bremen on May 10; measurements kept him busy there until May 22, when he went again to Osterholz in order to observe in Garlste. Thick moor haze and one thundershower after another prevented results up until May 28; hence he spent the first days in calculating station observations.

On June 6, 1825, he journeyed to Bremerlehe, where from June 7 to 13 he completed these measurements. Poor lodging at a Frau Muhl's again affected Gauss' health, and he longed for Varel, where the lodging was said to be good. But in Varel the weather was not very favorable; his health improved only slowly in spite of cool weather and the use of warm sea baths in Dangart, which was one hour distant.

Gauss arrived at noon on June 27 in Langwarden and remained there until July 12, 1825. Here the connection was to be made with the Danish triangulation. These measurements tortured Gauss indescribably for a long time and almost drove him to distraction; they were such as had annoyed him nowhere else. There were troubles with the instrument, refraction, the wind, cramped space in a tower, and so forth. In Langwarden Gauss took zenith distances of the ocean's surface at ebb tide and from various points on the dikes, which he determined by a small triangulation. He did this to deduce the Göttingen observatory's elevation above sea level.

In Jever a platform was built on the tower in order to get a view in all directions. Gauss spent two days in Wangeroog, but dropped further plans in this area, particularly since he again felt weakened by the heat. He finished on July 19 and spent the night of July 23-24 with Olbers in Bremen. He now went to Gnarrenburg near Brillit, where he stayed from July 29 to August 2. He had left this place in the spring on account of moor haze, but now had the same thing again to contend with. There was another heat wave, and he felt unable to make further measurements, except for several in Zeven, which he completed on August 4 and 5; he then returned directly to Göttingen. Gauss had planned a continuation of his survey, but actually he was now at the end of it.

Frau Minna's physician had prescribed mineral baths for her. In the early fall of 1825 she and Gauss took a trip to South Germany via Marburg and Mannheim to Baden-Baden and back via the Black Forest, the Murgtal, Tübingen, Stuttgart, Würzburg, and Gotha. Gauss again saw Gerling, Nicolai, and Lindenau; he made the acquaintance of Eckhardt, Bohnenberger, and Wurm. He was pleased to see that many heliotropes were in use in the Darmstadt, Baden, and Württemberg surveys.

The journey did not have a good effect on Gauss' health, since it was a very hot summer, and he was ailing most of the following winter. In October he began again to carry on some of his research on curved surfaces, which was to form the basis of his projected work on higher geodesy. This extensive and difficult subject kept him away from other work, and in the mornings he felt strained in preparing for his lectures and then returning to his meditations. At night he was getting only one or two hours' sleep, and Olbers compared Gauss' complaints with the pangs attending the birth of a stately hero. He doubted whether he ought to separate the purely geometrical parts from the work and publish them in the proceedings of the Royal Society, so he decided to put everything on paper. Gauss confessed that during no other period of his life did he work so strenuously and reap so little.

In the summer of 1826 Gauss busied himself with preparations for observations on the zenith sector simultaneously with Schumacher. He began in the spring of 1827 the calculation of his triangle network, which contained 32 points, 51 triangles and 146 bearings. Next he undertook a check on Krayenhoff's measurements in the interior of Holland. This work stimulated him to apply the method of least squares, taking as examples one of his own triangulations and one of Krayenhoff's.

Early in 1827 Gauss did not know whether he ought to regard his field work as complete or not. He regarded the determination of the difference in latitude between Altona and Göttingen as an urgent necessity. He was glad to learn that the Duke of Sussex had presented to the observatory a fine clock by Hardy. In October, 1827, Schumacher visited Gauss on the way to Munich, and also on the return trip. Gauss used the occasion to give Schumacher some instruction on the improved model of the heliotrope. Shortly thereafter Schumacher published in his journal an article on this subject by Gauss, who now began actual work on his third geodetic memoir.

In the spring of 1828 observations for determination of the difference in latitude between Altona and Göttingen began. A lieutenant of the engineers corps named von Nehus was sent at the expense of the Danish government as assistant to Gauss, who secured living quarters for him near the observatory. He would have preferred Schumacher himself. In the beginning the weather was very unfavorable, but by July 20 the measurements were finished in Altona, where Gauss had enjoyed the hospitality of Schumacher. In Göttingen and Altona together Gauss had made about nine hundred observations. The mean error of an observation was approximately the same as on the meridian circle. The observations were very strenuous and often lasted until 4 or 5 A.M. All of this had a bad effect on Gauss' health; he was also disturbed about his son Joseph, who because of nearsightedness was having difficulty in entering an officer's career.

Gauss had finished his memoir on curved surfaces in March, but did not present it to the Royal Society because no issue appeared at Easter. He presented it in October and published his own notice of it on November 5.

During the survey Gauss had a daily allowance of five thalers, Captain Müller four thalers, Lieutenant Hartmann and Joseph Gauss three thalers each; four assistants received only sixteen groschen each per day. After completion of the survey Gauss received a personal bonus of a thousand thalers in gold.

On March 25, 1828, King George IV decreed the extension of the triangulation over the whole Kingdom of Hanover under the direction of Gauss. The use of general staff officers was recommended, and costs were not to exceed five thousand thalers per year. Each spring Gauss presented a working plan for the summer and in the fall reported on progress made. The direction of these operations entailed extensive daily correspondence. After completion of the field work Gauss spent several weeks calculating the results, a task in which he had no assistance. The determinations of three thousand coordinates are contained in sixteen volumes, which are preserved. These coordinates formed the basis for maps in Papen's *Atlas*. Completion of the calculation connected with the survey lasted as late as 1848.

The triangulation work lay in the hands of Hartmann, Müller, and Joseph Gauss. Only once did Gauss visit a station—September 7, 1828, on the Hohenhagen. He dated a historical report on the survey February 8, 1838, prepared as result of the separation of England and Hanover after the death of George IV in 1837.

In the summer of 1828 Schumacher, on his way to Königsberg to receive Bessel's chronometer, visited Gauss for several days. On August 14, 1828, Gauss requested a leave of absence for reasons of health. Immediately after the visit to Hohenhagen just mentioned, he went to Berlin, where he attended a scientists' convention and for three weeks was a guest of Alexander von Humboldt. There he became acquainted with Wilhelm Weber, who at that time was on the faculty at Halle and in 1831 was called to Göttingen as professor of physics. From the time of this visit to Berlin, Gauss did not spend a night away from Göttingen during the remainder of his life, except in 1854, when the railroad was opened. This journey marked a turning point in Gauss' life and introduced the third epoch, when physics was dominant.

Gauss took on another burden when he was appointed member of a commission on weights and measures for Hanover. In March, 1829, he participated in a conference of this commission in the city of Hanover, and in May of the same year he journeyed there again to discuss that year's triangulations. His son Joseph was his only field assistant that year, and Gauss was highly pleased with his work. Gauss devoted much time to processing the measurements of Hartmann and Joseph, amounting to two hundred pages each.

Toward the end of 1830 Gauss secured for his son Joseph

a furlough of six weeks ending December 20, so that he could help his father in processing the measurements. At the time he was experiencing domestic trouble and grief, which will be touched on at another point in the story. He sought briefly some diversion in crystallography. In the following years the work was delayed by lack of assistants and also by the cholera epidemic.

Gauss did not publish the geodetic formulas he used in calculating coordinates. In later years his assistant Goldschmidt indicated, but did not develop, them. Goldschmidt became observer at Göttingen in 1835 after the death of Harding; in 1831 he solved the prize question of the philosophical faculty, dealing with the catenoid, which Gauss had set up in 1830.

At the instigation of Gauss, or at least with his cooperation, a trigonometric survey of the Duchy of Brunswick was begun in 1833 under the direction of Professor Friedrich Wilhelm Spehr (1799–1833), a pupil of Gauss and a professor at the Collegium Carolinum. After the sudden death of Spehr on April 24, 1833, the project bogged down. Gauss was happy to know that his native land was to be surveyed. Joseph Gauss did not have a high opinion of Spehr's ability in field work.

In 1834 Gauss devoted four months of strenuous work to calculating measurements made in the Harz area. He did the same for measurements in the Lüneburg area, Westphalia, and the Weser region in 1836, but not with the same vigor. Captain Müller devoted himself in 1836 to the triangulation of the area on the Upper Weser between Uslar, Göttingen, and Münden and in 1837 made preparations for gathering of data in the Osnabrück area. He also reconnoitered the Aller area for future trigonometric measurements. Later Müller succeeded in connecting his points with those of the Hessian survey.

Gerling asked Gauss whether he considered a determination of longitude between Göttingen and Mannheim important. Gauss gladly promised his cooperation in observing powder signals. This determination of longitude materialized in 1837, after Gerling had ended the Hessian survey. The rooms of the Göttingen observatory had to be cleaned for the centennial jubilee, September 17–19, 1837, and this task was completed beforehand.

The signals lasted from August 22 to September 9. Besides powder signals, the heliotrope was used for sending 179 signals from Meissner and 58 from the Feldberg; at these points there were 92 and 83 powder signals, respectively. The receiving points were Göttingen, Frauenberg, and Mannheim.

In late 1838 Gauss was busy with the reduction of Müller's measurements on the Aller. Müller then covered the western part of the area around Bremen in 1839 and in 1841 set up the network in eastern Frisia, which connected the North Sea islands with the mainland. After Müller's death in 1843, the completion of the work fell to Joseph Gauss.

Gauss published in 1843 a memoir on topics in higher geodesy; it has been reprinted in Volume IV of his *Collected Works* (pp. 259–300). A second memoir in geodesy appeared in 1846 and is to be found on pages 301–340 of the same volume. He planned a major work in geodesy, but never got far with it. An outline for it, the introduction, and the early part of a first chapter were found in his papers after his death and published on page 401 of Volume IX of the *Collected Works*.

The last reference to a geodetic question is found in a letter addressed to General Baeyer date June 22, 1853, and printed on page 99 of Volume IX of the *Collected Works*. Gauss' activity in geodesy was especially important for his measurement of a degree, which was to have been part of a measurement of the meridian. The value of his survey was lessened by the fact that except for several church towers and the like, a generation after his death the trigonometric points were no longer extant in nature. It is regrettable that so much of Gauss' valuable time was taken up by the survey. He himself estimated that he used a million numbers in it. The great value of his work here lies in the revolutionizing of methods, especially the invention of the heliotrope and the process of angle measurement. The method of least squares found one of its richest applications in geodesy. The importance of this geodetic survey for Gauss' work in surface theory must be discussed in another chapter. He was one of the first to show geodesy its real goal by defining the form of the earth and explaining the causes of its irregularities.

CHAPTER ELEVEN

Alliance With Weber: Strenuous Years

On November 30, 1830, the chair of physics at Göttingen became vacant through the death of Tobias Mayer. The Hanoverian cabinet minister in charge of university affairs[1] promptly asked Gauss for his views on filling the place. These were fully expressed in a confidential memorandum to the minister dated February 27, 1831. Gauss called attention to the fact that the University of Göttingen had always adhered to the viewpoint that it was not merely a school for the instruction of students and a preserver of scientific knowledge, but also a center for participation in the development and extension of the sciences as common property of humanity. He also called attention to the fact that the Royal Society of Sciences in Göttingen was established with the last-named purpose in view. In that way Göttingen assumed its proper rank as an institution of world importance, and as long as it remained loyal to this viewpoint, it could withstand the fluctuations to which its enrollment was subject in stormy times, but must always assume its former rank with the return of more peaceful times. Such was Gauss' opinion.

He went on to state that the new physicist must lecture to students of varying background and preparation, that he must be able to illustrate his lectures by means of well-executed experiments, that he must be well versed in all parts of physics and that he must be well trained mathematically. Gauss felt that the degree of mathematical knowledge should not be a measure of

---

[1] Georg Ernst Friedrich Hoppenstedt (1779–1858).

one's value as a physicist. On the contrary, he wrote that a physicist who was a mathematician of top rank might not be the best professor of physics, at least he might not meet the needs of his students. Gauss stressed the fact, however, that a thorough physicist in the full sense of the word must be well grounded in the higher branches of mathematics. He regretted to admit that few professors of physics at German universities in his day met these high requirements.

Further on in the memorandum attention was directed to the fact that some professors of physics did not claim to be mathematically or theoretically trained, but limited themselves to applied topics in the area of chemistry, electricity, and the like. Gauss then reminded the cabinet minister that certain others had superficial and defective mathematical knowledge, in which instances ridiculous things frequently happened. He expressed the hope that the Royal Society of Göttingen might be spared such danger.

Gauss then proceeded to name five physicists who in his opinion met these requirements: Bohnenberger in Tübingen, Brandes in Leipzig, Gerling in Marburg, Seeber in Freiburg, and Weber in Halle. The background, training, experience, and abilities of these men were very carefully compared and discussed. Gauss wrote that he was not on such safe ground in discussing the abilities of these men as *teachers;* he knew their scientific ability, but as to teaching he was reduced to a knowledge of their personalities, and was attempting to be as conscientious and objective as possible. He called attention to the great difference in their ages, especially to the fact that Weber was only twenty-seven years old, and that he was the only one of the five Gauss had actually heard lecture. Weber's lecture at the convention of scientists in Berlin in 1828 impressed Gauss as being well organized and excellent in quality. In addition, Weber's publications showed great promise, a fine research spirit, and talent in experimenting, according to Gauss.

The question of administrative ability in connection with filling the place was next touched on, as well as membership in the academic senate. Gauss felt that all five were men of high character in every respect. We are especially interested in what he wrote of Weber: "Weber made on me the impression of a modest amiable character, who lives more in science than in the external world. In such characters the development for real life is wont to gain direction only through events which guide it to them."

Due to Weber's age and previous position as assistant professor in Halle, Gauss felt that he could not immediately enter the academic senate, and that perhaps he should not be given administrative duties. He wrote that if he had made any error in the memorandum it was *sine ira et studio*.

As a résumé *in nuce,* Gauss thus expressed his preference: He voted for Bohnenberger if the officials desired a sixty-year-old man and did not object to the high salary necessary to get him. He voted for Gerling if the authorities placed the main emphasis on strengthening the faculty and the senate, implying administrative ability. He voted for Weber if the officials especially took into account genius and future productive research. In closing, Gauss drew attention to the fact that Seeber had already applied for the position. On January 27, 1831, Gauss had already told Weber that it would be proper for him to call the attention of the authorities to his qualifications.

On April 29, 1831, Weber was offered the position as full professor of physics at Göttingen, which he accepted on May 14. On July 5, Gauss wrote a letter to Weber congratulating him and expressing great happiness at his coming to Göttingen. The first order of business after his arrival was the purchase of instruments and equipment for the physics laboratory, which stood where part of the university library is located today. He was also busy with the preparation of his lectures.

From the very first, an intimate friendship, both scientific and personal, sprang up between Gauss and Weber. He was a frequent dinner guest in Gauss' home, who was also frequently a guest in Weber's home.[2] The only other person with whom

---

[2] He lived at Prinzenstrasse 3 in the years 1831–1837, and at Jüdenstrasse 40 in the years 1848–1891.

Gauss enjoyed so deep a friendship was Bolyai. In a letter to their brother Ernst Heinrich dated June 2, 1832, Wilhelm's sister, Lina, who kept house for him, complained about the frequent and in part unexpected dinner invitations he extended to Gauss. She wrote:

> Wilhelm can enjoy Gauss every day as long as he desires. Gauss lives a very lonely life, and Wilhelm is welcome at any hour. Gauss is such a socially trained man that in my presence he never talks of learned things and demands that I shall be present; he has talked with us from 12 to 5 o'clock on all kinds of matters. Recently Wilhelm had Hofrat Gauss for dinner three days in succession (his daughters are away on a trip).

Friends of Weber repeatedly expressed the view that he stood too much in the shadow of Gauss. It is incorrect to believe that Gauss in any way belittled Weber's work. His character was too high above such petty weaknesses of human nature. He always accepted Weber as an equal in research, a distinguished scientist, and a dear personal comrade. His letters show that he was unhappy in the years Weber was away from Göttingen. They made a perfect working team and complemented each other. Weber gave Gauss the stimulus for work in physics and was more of an experimenter, whereas Gauss leaned more to the theoretical development or mathematical side. Gauss was always making efforts to gain more recognition for Weber. The relationship between the two was at all times exceedingly harmonious. Gauss usually referred to Weber as "Friend Weber." In view of the great difference in age, it would not be amiss to think of Gauss acting as a father to Weber.

At the time of Weber's arrival in Göttingen, political disturbances were still in the air, a consequence of the Paris Revolution of July, 1830. The professors did not seem much concerned, even though the university's enrollment had decreased by six hundred students. The Hanoverian government provided the university with good finances, but at the time was politically reactionary. A main source of agitation centered around the preference the nobility enjoyed in the filling of top government jobs. Events in Paris accentuated this. The chief target of liberals was Count Münster, who in London exercised a despotic control over the Hanoverian territories. Matters came to a head when a man named von der Knesebeck published a song of praise of the nobility and called it the only support of the throne. His windows were broken and he was chased out of Göttingen. A political pamphlet by a young professor named Ahrens was denied the permission necessary for publication. Furthermore, the city administration was in incompetent hands, and the citizens had waited a long time for a new city charter.

On January 18, 1831, armed citizens and students moved on the city hall. The rebels occupied the city hall, dissolved the regular council, and set up an emergency common council. More students and citizens joined the movement, but the university merely looked on and did nothing. Some students stormed two university buildings, searching for arms. The students looked on the whole affair as a fine prank; finally classes were temporarily closed and the students sent home.

The new government sent a delegation to Hanover, in order to ally itself with the kingdom. The Duke of Cambridge demanded unconditional subjection. In mid-January seven or eight thousand men under the command of General von der Busche approached the city, and on the fifteenth he sent an ultimatum to the city. On the evening of that day the regular authorities were in full control.

This little drama lasted exactly eight days, yet early in February Count Münster had to resign. As a pronounced Tory he did not want to support the new laws of 1831 and 1833. King William IV appointed the Duke of Cambridge, Adolf Friedrich, a Göttingen alumnus, vice-regent. Preparations were now under way for the separation of Hanover from England. A great blunder was committed in not insisting on the legally necessary agreement of Duke Ernst August of Cumberland, successor to the throne, to the new constitution of 1833. We shall soon see how serious this oversight was.

Gauss' devoted pupil and friend C. L. Gerling, now professor of physics in Marburg, wrote him a letter of inquiry and concern, as soon as he heard of the rebellion in Göttingen. Other friends of Gauss wrote in the same vein. His reply to Gerling, dated January 29, 1831, is illuminating:

I thank you very heartily, my dear Gerling, for your friendly sympathetic letter. Basically I am little touched directly by the local events; indirectly up until now mainly only by the difficult communication with the physician, since in the last three or four days no carriages could come out or go in; communication for unsuspected pedestrians has never been interrupted. Unfortunately my domestic suffering has been very much increased in the last three or four weeks by the complete bedfastness and increasing weakness of my wife, as well as by several added circumstances, namely also through the difficulty of care of the poor sick one. My oldest daughter and my mother-in-law, for reasons which are partly obvious, can do little or nothing in this matter. My youngest daughter accomplishes uncommonly much, but of course besides our maids is no longer sufficient and the sick one could not allow strangers around her. For a long time I have urged her to try to get again the young woman whom you remember having seen in my house as our cook, and who a year ago was forced to leave it owing to illness; since then she has been keeping house for a widowed relative in Osterode, but has always kept a great attachment for my house, in which she lived almost seven years, but all my urging was in vain. On the morning of the twenty-fourth my wife awakened me before dawn; one of our maids had committed a crude bit of carelessness in the night, whereby the former had come into direct danger to life, and had thereby ripened the wish (long cherished but not confessed) to have our Hauerschildt here as quickly as possible. I at once took post horses and in the evening she was here. She herself had been quite willing immediately, but her relative was somewhat surprised and was probably left in the opinion that we were talking only of a short leave of absence. Here we soon agreed, however, if she was to be truly useful, she should arrange her affairs in Osterode for an indefinite absence, which might not be easy. She therefore returned there on the 27th; I gave her a letter to her relative, in which I offered all my eloquence to move him to agreement, and we expect her back on the 30th, if it is successful, but, as you can imagine, with great anxiety, since the weakness of the poor sick one has increased.

I wish to remark that, in so far as I found out in Osterode, the disturbances there were basically very unimportant, or rather, before they

ALLIANCE WITH WEBER

became important, they were immediately suppressed. As soon as the two rabble-rousing lawyers were led away, it was over, and when the local disturbances caused the calling away of troops from Osterode, it was found sufficient to leave 60 men there. In Göttingen the great number of troops (perhaps 4,000 to 5,000 men) remained only several days; afterward the greater part was again moved out. Now, 1,600 to 1,700 men may be here; in Osterode when I was there, perhaps nearly 300. May heaven preserve peace in Frankfurt and peace in the complicated relations with Belgium, then everything will straighten out quietly among us, I hope.

Miss Hauerschildt did, after some delay, return, but Gauss was continually afraid that her relative would insist that she resume her place in Osterode. It meant much to Gauss to have her in the household. Unfortunately she was not entirely well herself and could not do as much as formerly, which was desired.

Gauss' daughter Minna at an early age had candidates for her hand. He often said that she was the image of her mother, yet in mentality she was similar to Gauss, and writers feel that she was his favorite child. Alexander von Humboldt considered her beautiful, and we know that she had a pleasant personality. She helped in the nursing of Frau Minna, her stepmother, corresponded with her father during his absence from home, did part of the housework, and helped care for her grandmother Gauss and her younger sister, Therese.

The last of February, 1830, Minna became engaged to Georg Heinrich August von Ewald (1803–1875), young theologian and professor of Oriental languages at Göttingen. Gauss was greatly pleased by the match; he esteemed Ewald very highly both as man and scholar, and their relations were always most cordial and intimate. The wedding occurred at Grone near Göttingen on September 15, 1830. Ewald was born in Göttingen as the son of a linen weaver who had migrated from Brunswick. In spite of his youth he was already a scholar of reputation, and known as a man of strong moral volition and a pronounced sense of justice. Ewald revered Gauss, and used to say that his early married years in Göttingen were the happiest of his life. There was one sad feature of the marriage, however; almost from the start Minna began to have lung trouble, perhaps contracted in the care of her stepmother. There were no children of the marriage.

Ewald had many interests and liked to converse with Gauss about them. He had a hobby of Oriental numismatics, and the university still has a collection which was in his care. The history of ancient Persian religion and the origin of the Afghans attracted his attention. His interest extended from Hebrew to Chinese, and he was the first person to teach Sanscrit in Göttingen. His monumental *History of the People of Israel* was the result of thirty years' labor. His most enduring work was in Old Testament exegesis and Hebrew grammar, as well, of course, as the history of Israel just mentioned. Ewald turned out a large number of distinguished pupils in his field, including many in Britain. He visited England in 1838 and 1862, France and Italy in 1829 and 1836. In 1837, aided by other Orientalists, he founded the valuable periodical *Zeitschrift für die Kunde des Morgenlandes,* which prepared the way for the formation in 1845 of the German Oriental Society.

CHAPTER TWELVE

The Electromagnetic Telegraph

The first mention of the telegraph occurs in a letter which Gauss wrote to his friend the astronomer Olbers in Bremen, dated November 20, 1833:

I don't remember my having made any previous mention to you of an astonishing piece of mechanism that we have devised. It consists of a galvanic circuit conducted through wires stretched through the air over the houses up to the steeple of St. John's Church and down again, and connecting the observatory with the physics laboratory, which is under the direction of Weber. The entire length of wire may be computed at about eight thousand feet; both ends of the wire are connected with a multiplicator, the one at my end consisting of 170, that in Weber's laboratory of 50, coils of wire each wound around a one-pound magnet suspended according to a method which I have devised. By a simple contrivance—which I have named a commutator—I can reverse the current instantaneously. Carefully operating my voltaic pile, I can cause so violent a motion of the needle in the laboratory to take place that it strikes a bell, the sound of which is audible in the adjoining room. This serves merely as an amusement. Our aim is to display the movements with the utmost accuracy. We have already made use of this apparatus for telegraphic experiments, which have resulted successfully in the transmission of entire words and small phrases. This method of telegraphing has the advantage of being quite independent of either daytime or weather; the ones who receive it remain in their rooms, and if they desire it, with the shutters drawn. The employment of sufficiently stout wires, I feel convinced, would enable us to telegraph with but a single tap from Göttingen to Hanover, or from Hanover to Bremen.

Gauss gave first public notice of the telegraph in the *Göttingische gelehrte Anzeigen,* issue of August 9, 1834. He gives detailed information about the "great galvanic circuit" between the physics laboratory and the observatory, to which the new magnetic observatory was connected, and emphasizes that this "unique setup" is due to Weber. The two had begun their electrical measurements on October 21, 1832. Gauss ordered magnetometers of varying size, from which he proceeded to larger magnets, and thus a number of galvanometers of varying size and construction came into being, although we do not know which ones were finally used in telegraphing. They were all to be used for this purpose, but the two scientists realized that the apparatus with the smallest magnets was best adapted for telegraphing on account of the magnets' small period of oscillation.

It is reported that the first words sent on the telegraph were: *Michelmann kommt*. Michelmann was a servant who ran errands for Gauss and Weber. At first individual words were sent, and then complete sentences. The telegraph was operated once in the presence of the Duke of Cambridge, who seemed to take special interest in it.

Weber's correspondence with the city council in April and May, 1833, gave exact information about the purpose and date of origin of the telegraph. At first Weber had used thin copper wire for the lines, which, however, did not stand up very well and had to be replaced by stronger wire. Even the latter did not resist weathering and was replaced by soft steel wire of one-millimeter strength. The lines existed until 1845, when they were destroyed by lightning on December 16. Gauss describes this incident thus, in a letter to Schumacher dated December 22:

The brief newspaper account (sent in by Listing) about the local thunderstorm of December 16 you have probably read. It is one of the strangest incidents which has ever occurred. You know that since 1833 the wire connection had existed between the observatory and the former physics laboratory (via the magnetic observatory, the obstetrical clinic, and St. John's steeple). It was broken off six weeks ago between the magnetic observatory and the clinic by a windstorm, and the two ends of the wire on the other side, otherwise in the physics laboratory, ended in front of the window, since this laboratory has moved. Therefore four wires ran from St. John's steeple, two to the clinic, and two to the window of the former physics laboratory. The very strong stroke of lightning on St. John's steeple was probably distributed entirely on these wires, destroyed them all, partly into rather large, partly into rather small pieces, pieces of four to five inches in length and numerous little balls like poppy seed, all of which formed a brilliant rain of fire. I myself, at home, room darkened by shutters, did not see any of it, but merely heard the quite unexpected strong clap of thunder, which is in the immediate vicinity, as usual, quite simple, but this one I heard for some duration (about two seconds). No damage occurred except that a lady's hat had two holes burnt in it by falling incandescent pieces of wire, but very probably the wires protected the steeple, which offers no lightning conductor, and, ignited, would perhaps have brought great danger to the city and library. Finally the electric matter in the subject lightning rods (library and obstetrical clinic) probably reached the ground, aside from a partial leaping off on a gutter of the pharmacy where the outer wall is somewhat torn open.

In a lecture before the Royal Society of Sciences in Göttingen on February 15, 1835, Gauss gave more detailed information on changes and improvements in the telegraphic equipment. Especially he mentioned his new apparatus for the generation of induced currents, the "inductor," which he used not only for his scientific experiments but also as a substitute for the insufficiently constant voltaic elements in telegraphing.

In a letter to his friend and former pupil H. C. Schumacher in Altona, dated August 6, 1835, and in a memoir on terrestrial magnetism and the magnetometer published in Schumacher's *Jahrbuch* (1836), Gauss expressed himself very optimistically about the future of telegraphy. Twenty years later these "fantasies" were in process of realization, and Gauss lived long enough to experience the beginning of it. In the letter he wrote:

In more propitious circumstances than mine, important applications of this method could no doubt be made, redounding to the advantage of society and exciting the wonder of the multitude. With an annual budget of 150 thalers for observatory and magnetic laboratory (I make this statement to you in strictest confidence), no grand experiments can be made. Could thousands of dollars be expended upon it, I believe electromagnetic telegraphy could be brought to a state of perfection and made to assume such proportions as almost to startle the imagination. The Emperor of Russia could transmit his orders in a minute, without intermediate stations, from Petersburg to Odessa, even peradventure, to Kiachta, if a copper wire of sufficient strength were conducted safely across and attached at both ends to powerful batteries, and with well-trained managers at both stations. I deem it not impossible to design an apparatus that would render a dispatch almost as mechanically as a carillon plays a tune that has been arranged for it. One hundred millions worth of copper wire would amply suffice for a continuous chain to reach the antipodes; for half the distance, a quarter as much, and so on, in proportion to the square of the distance. . . . That at least the first alphabet is easy to learn you can conclude from the fact that recently my daughter at once read correctly several letters without any instruction.

The success of Gauss and Weber with their telegraph aroused great attention at that time, at least in Germany. Steps were under way to use it on the railroad. Gauss and Weber both wrote memoranda on the subject to the directorate of the Leipzig-Dresden railroad then under construction. Weber wanted to use one rail for conducting the current and the other for its return, while Gauss proposed a copper wire of 1.6-millimeters or an iron wire of 3.8-millimeters strength for conducting and the rails for return. The railroad sent its expert to Göttingen, and after conference with them, he decided the lines would have to be underground. The plan was dropped because of its high cost. Thus Germany lost the honor of being the first to produce a practical telegraph. In a memorandum dated March, 1836, Weber recommended to the railroad Gauss' principle of the needle telegraph, which however did not find practical application until five years later. Nevertheless, we know that Gauss and Weber were the first scientists to put electrical current in the service of communication. Lord Kelvin's marine galvanometer of 1858 was nothing more than a Gauss-Weber needle telegraph.

The fact remains that in later years the invention of Gauss and Weber was almost forgotten, so that others claimed it. The reason for this is to be sought not merely in the personal vanity of competitors, in business interests, or in considerations of national prestige. Communication of reports on scientific work was poorly developed, and achievements of the first order were often made public in a foreign country only incompletely or at a very late date. Also, a widespread knowledge of the German language was lacking. Latin was still commonly used for scientific memoirs. Gauss and Weber had only 181 regular subscriptions to their journal on magnetism. Of these 30 went to the Prussian Academy of Sciences, 20 to the Bavarian, and 15 to the Russian.

It is almost incredible that Sir David Brewster wrote thus to Gauss on December 4, 1854:

DEAR MR. GAUSS,
I had lately a visit from our distinguished friend Mr. Robert Brown, who mentioned to me, when talking of the electric telegraph, that you had, many years ago, constructed and used one. As I am, at present, writing on the subject I would esteem it a particular favour if you would oblige me by a notice of what you have done, and of the time when you used it publicly.

The answer to this letter is probably one of the most interesting documents on the invention of the electric telegraph. It was the last letter written by Gauss, and several persons have made efforts to find it, but without success. There are two possible explanations for the loss of the letter. In leaving St. Andrews in February, 1859, and moving to Strathavon Lodge near Edinburgh, Sir David suffered an annoying and irreparable loss. He packed his carriage with valuable silverware, papers, and personal, treasured souvenirs. Through the carelessness of officials, his baggage was allowed to drop into the Firth of Forth, in the process of being transferred from the landing to the steamer. Some of the papers were destroyed and others badly defaced. Sir David married Juliet, younger daughter of James "Ossian" Macpherson, and in years after Brewster's death most of his effects were kept at the Mansion House of Balavil estate, Kingussie, Scotland. On Christmas eve, 1903, a fire there destroyed many of Sir David's personal papers, instruments, and other possessions. It is believed that one of these two events accounts for Gauss' missing last letter.[1]

---

[1] The Gauss Archive in Göttingen possesses five letters from Sir David to Gauss, dated 1816–1854.

If the Gauss-Weber telegraph had been set up in a large city, it probably would have had more recognition. In those days Göttingen was a town of 9,968 inhabitants and 843 students; the railroad did not come through until 1854. Also, Gauss and Weber both shied away from priority disputes, placing their invention at the disposal of all, rather than trying to protect their rights. At the World's Fair in Vienna (1873) and Chicago (1893) their telegraph was properly exhibited. In the Deutsches Museum at Munich and elsewhere it has "come into its own" in more recent years.

CHAPTER THIRTEEN

Magnetism: *Physics Dominant*

Weber gave Gauss a great stimulus by searching for new phenomena, but Gauss was usually attempting to systematize and express in exact mathematical terms the experimental results obtained by others. A letter to Olbers shows that Gauss was deeply interested in the study of terrestrial magnetism as early as 1803. He urged Olbers to do something in this field. After his return from his American journey in 1804, Alexander von Humboldt began to encourage Gauss and his friends to take up research in magnetism. Olbers again tried to persuade Gauss to do some work in magnetism in 1820. The truth is that he was too busy with other subjects; it was not a lack of interest.

In 1820 Oerstedt discovered electromagnetism, and as a result the greatest scientific minds were occupied with this study in the following years. One need mention only the work of Biot and Savart 1820–1821, and the work of Ampère. Ohm's law was enunciated in 1827. The memoir of George Green (1793–1841) on the theory of the potential appeared at Nottingham in 1828; it is uncertain whether Gauss was familiar with it. By far the most important contribution was Faraday's discovery of induced currents in 1831. Fortunately it coincided with Weber's arrival in Göttingen.

Humboldt had a little collection of magnetic instruments in his home, which he showed to Gauss during his visit there in 1828 and strongly urged him to give some time to magnetism. All these factors were enough to force Gauss definitely into the study of magnetism. He wrote Olbers in a letter dated October

12, 1829, about the visit of the Belgian physicist Quetelet:[1] "The acquaintance of M. Quetelet has been very pleasant to me; in my yard we put on with his splendid apparatus various series of experiments on the intensity of magnetic force which granted an agreement scarcely expected by me."

Quetelet's visit stimulated Gauss on a point which was discussed in a paper of 1837, namely, the determination of the period of oscillation of a magnetic needle. He proposed to count the beginning of an oscillation at the point of greatest velocity instead of at the point of greatest elongation, as had been customary. Declinations of the magnetic needle were regularly made at the Göttingen Observatory as early as January, 1831. Sartorius von Waltershausen wrote that in the winter of 1832 he accidentally entered the observatory. Gauss picked up a small magnet and began to teach his friend; he showed that all the iron bars on the windows had become magnets through the effect of terrestrial magnetism. By January, 1832, he had thrown himself with all force into the investigation of magnetism, and by February of that year had succeeded in reducing the intensity of terrestrial magnetism to absolute units. The conception and theory were complete, and it was now a matter of making accurate measurements and refining the methods.

At this time Gauss had the idea of writing a major work on magnetism, but decided to publish separately a small part of it dealing with the intensity of terrestrial magnetism. In August, 1832, he began writing this memoir and on December 15 read it before the Royal Society of Göttingen. It was a memoir filling thirty-six pages, published in 1833 under the title *Intensitas vis magneticae terrestris ad mensuram absolutam revocata*, and may be found in Gauss' *Collected Works* (V, 79). The German version appeared in Poggendorff's *Annalen* (1833) and later as Number 53 of Ostwald's *Klassiker der exakten Wissenschaften*. R. S. Woodward called this work "one of the most important papers of the

---

[1] Quetelet was on journey making measurements of terrestrial magnetism in Holland, Germany, Italy, and Switzerland.

century." It reduces all magnetic measurements to three fundamental magnitudes: M or mass, l or length, and t or time. Mathematical and astronomical accuracy were thus brought to this part of physics: Gauss showed how absolute results could be obtained and not merely relative data based on observations with some particular needle. Coulomb's law stated that the force of attraction or repulsion between two poles varies inversely as the square of the distance between them. Paragraph 21 of the *Intensitas* confirmed this law. Measurements of the intensity of magnetic force had been somewhat crude before the time of Gauss. He postulated that every magnetic body contains equal quantities of the two magnetic "fluids," north, or positive, and south, or negative. He was the first person who recognized that such a determination is necessary if one is to arrive at a rational measurement of magnetic quantities.

Gauss noted that magnetic forces depend on temperature and that it is necessary to determine experimentally the temperature influence in order to be able to reduce all observations to the same temperature. He then turned to the fact that the earth exerts magnetic force, discussed briefly the variations of declination, and emphasized that practically nothing was known on this subject. Then he proceeded to formulate this theory mathematically.

In Paragraph 25 of the *Intensitas* Gauss collected a series of numerical results on the absolute values of the horizontal intensity of terrestrial magnetism. He used as units the milligram, the millimeter, and the second; it is customary today to use the gram, the centimeter, and the second. He felt that more accurate results could be obtained by using heavier needles whose weight would be as high as two thousand or three thousand grams, and that this would help to reduce the influence of air currents.

But Gauss had more extensive plans; he desired to measure all magnetic elements (declination, inclination, variation) with the same accuracy as that achieved in measuring the horizontal intensity. In addition, he wanted to investigate the influence of temperature. He felt that terrestrial magnetism is the result of all polarized pieces of iron contained in the earth, both at the center and near the surface. Toward the end of 1832 he began to apply his methods to galvanism.

On January 29, 1833, Gauss sent an official memorandum to the university board, proposing the erection of a magnetic observatory. This proposal was immediately approved, and the building was ready for use in the fall of 1833. Gauss gave a description of it in the *Göttingische gelehrte Anzeigen* of August 9, 1834. Weber published a description of the equipment in 1836. Except for the side rooms, the building was a long rectangle oriented exactly in the geographical meridian, 32 feet long and 15 feet wide; all iron usually found in buildings was replaced by copper. The ceiling was 10 feet high; double doors and double windows eliminated all air currents. The observatory cost 797 thalers, a large proportion of which was occasioned by the replacement of iron by copper.

The observatory was equipped with a theodolite mounted on a special base, an astronomical clock, and a magnetometer with a cabinet. The observations in the observatory were the determination of the declination and its variation at various hours, months, and years. Readings were taken daily at 8 A.M. and 1 P.M., because the greatest variations occurred in Göttingen at this time. Gauss and Weber had seven assistants in observing, including Gauss' son Wilhelm. On certain days in the year forty-four hours of uninterrupted observation of the variation in declination was carried out, in order to study the regular course of the variation and its frequent anomalies, such as the influence of the aurora borealis. The first such observations occurred in Göttingen March 20–21, May 4–5, June 21–22, 1834. Intervals of observation ranged from five to twenty minutes. Friends of Gauss now undertook observations in Berlin, Frankfurt, and Bavaria. In September, 1834, Leipzig, Brunswick, and Copenhagen were added to the list.

On March 21, 1834, Gauss urged Encke to come and inspect the observatory in Göttingen before setting up any equipment in Berlin. In another letter he reminded Encke that one of the main difficulties was the procurement of good mirrors. The magnetic research in Göttingen attracted wide attention, and many German as well as foreign scientists now visited Gauss in order to inspect the equipment. Among them were Oerstedt and Hansteen; Göttingen now became the center of magnetic research. Meyerstein in Bonn and Airy in Greenwich began to make observations, and in Russia this activity occurred as far away as Nertschinck. In 1836 Gauss published in Schumacher's *Jahrbuch* a popular essay "Erdmagnetismus und Magnetometer" in which he reported that operations were beginning in Freiberg, The Hague, Halle, Munich, Upsala, Vienna, Dublin, Breslau, Cracow, Naples, and Kasan. Gaussian instruments were set up at more than twenty points.

A Magnetic Association was formed and Gauss and Weber decided to publish the research of its members in a special periodical. Humboldt used his great influence to persuade the British government to erect magnetic observatories at as many places as possible in the colonies. The first number of the periodical appeared in 1837 under the title *Resultate aus den Beobachtungen des magnetischen Vereins im Jahre* . . . edited by Carl Friedrich Gauss and Wilhelm Weber. Its six volumes covered reports of the years 1836–1841. A commercial publisher handled the venture, but from the start there were difficulties. There were only 181 subscribers, and 110 copies went to individuals. In the six volumes, Gauss published fifteen articles and Weber twenty-three, which together amounts to two-thirds of the entire content. Nevertheless, the periodical was important as a center for publication of research on terrestrial magnetism and related areas.

Airy and Christie in a report to the Royal Society of London in 1836 proposed the erection of magnetic stations at Newfoundland, Halifax, Gibraltar, the Ionian Isles, St. Helena, Paranatta, Mauritius, Madras, Ceylon, and Jamaica. Later (1839) Montreal, the Cape of Good Hope, Van Diemen's Land, Bombay, and a point in the Himalayas were added. The fact that the apparatus and time of observations were in agreement with the German Magnetic Association shows the high esteem in which Gauss and Weber were held.

In 1837 Gauss and Weber collaborated in the invention of the so-called bifilar magnetometer. They used the idea of measuring the magnetic force of direction by means of the direction force which a fixed body suspended by two cords experiences when it is diverted from its condition of equilibrium. In the same year they published in the *Resultate* a paper on the invention and use of the new instrument. It seems that Snow Harris in Britain had invented the instrument independently of Gauss and Weber, but that his work was unknown to them. Lloyd in Dublin used it independently of Gauss for magnetic measurements, but Gauss has generally been given credit for the instrument.

By 1832 Gauss possessed all the essential elements of a general theory of terrestrial magnetism, and was prevented from working out and publishing it only by the lack of experimental material. He did not carry out this plan until the winter of 1838. Finally his "Allgemeine Theorie des Erdmagnetismus" appeared in Volume III of the *Resultate* (April, 1839).[2] He proceeds from the assumption that the cause of normal terrestrial magnetism is in the interior of the earth, while doubting this assumption for disturbances. He postulates further that terrestrial magnetism is the resultant of the action of all magnetic parts of the earth, regardless of whether one assumes two magnetic "fluids" or Ampère currents. Here the magnetic potential of the earth is developed mathematically. The two points at which the horizontal magnetic force is equal to zero (the whole magnetic force being directed vertically) he calls the magnetic poles of the earth.

Gauss now applied his theory to observations. The agreement between calculation and observation encouraged him to believe that he was near the truth. He was overjoyed in 1841 when an American, Captain Charles Wilkes, found the magnetic south pole at a point which deviated only slightly from his calculations. Captain Ross had found the magnetic north pole three degrees and thirty minutes to the south of the point indicated by Gauss' calculation.

---

[2] It covered forty-seven pages.

MAGNETISM

In closing the "Allgemeine Theorie" Gauss calculated the direction of the magnetic axis, that is, the direction of the magnetic moment of the earth, the magnitude of the moment, and the strength of magnetization of the earth. He found that the earth is very weakly magnetized in comparison with steel. His calculation of the moment was within 2 per cent of the correct value.

Of special importance at the close of the "Allgemeine Theorie" was a theorem which Gauss did not prove, but merely enunciated. It stated that instead of any desired distribution of magnetic fluids inside the space of a body there can be substituted a distribution on the surface of this body, so that the effect at every point of the outer space remains exactly the same, from which one easily concludes that one and the same effect in the whole external space is to be deduced from infinitely many different distributions of magnetic fluids in the interior. He raised the question whether one may assume that in every element of volume there is an equal quantity of north and south magnetism, and thought this old assumption should be checked; also whether the seat of all normal terrestrial magnetism is in the earth's interior. Modern research has shown that 94 per cent of the earth's magnetic field comes from the interior and the remainder from external causes. This shows us that Gauss was correct in his theory, even though he did not have experimental facts to prove it. The cause of terrestrial magnetism and the fact that the direction of the earth's magnetic moment almost coincides with the axis of rotation are not touched on. Gauss probably had private views on the subject, but never felt ready to give them to the public. It was the methods more than the results of earlier magnetic theories which Gauss criticized. He possessed the basis for his theory of terrestrial magnetism in 1806, but had to wait thirty years for the experimental data!

Gauss and Weber, assisted by C. W. B. Goldschmidt, published at Leipzig in 1840 an extensive *Atlas des Erdmagnetismus,* based on their theory. In later years it was out of print, and therefore was reprinted in Volume XII of Gauss' *Collected Works* (1929).

In a sense the close of Gauss' work in magnetism was marked

by the publication in the *Resultate* (1840) of his "Lehrsätze in Beziehung auf die im verkehrten Verhältnisse des Quadrats der Entfernungen wirkenden Anziehungs- und Abstossungskräfte." He was stimulated to write this paper by his work in magnetism. It is true that his measurements of inclination came after this time. In this paper the word "potential" as the name for a certain function occurs for the first time. He had used it previously in a note of October, 1839, which, however, was not published until after his death. George Green had used the name "potential function" for this function in 1828. Green's paper remained almost unknown even in England, and there is no evidence that Gauss was acquainted with it. Probably both of them got the term from the vocabulary of medieval scholastic philosophy. As a matter of fact it is known today that Daniel Bernoulli used the name "potential" about a century earlier than Gauss or Green, although they certainly were unaware of it.

Gauss' basic theorem in this memoir had already been set forth in his 1813 memoir on the attraction of homogeneous spheroids, and he probably knew it even before 1810. In Paragraph 24 of this paper he set up another important theorem of the potential: "If a closed surface is a surface of equilibrium for the attracting and repelling forces of masses which are totally in the outer space, then the resultant of forces at every point of that surface as well as at every point of all the inner space is equal to zero." (*Collected Works,* V, 307.) From a letter to Bessel dated December 31, 1831, we know that Gauss discovered this theorem several months earlier that year.

Another theorem concerns the case where the attracting and repelling masses are inside space bounded by a closed surface. At every point of the surface, if it is a surface of equilibrium, the resulting force will be directed to one and the same side; according to whether the aggregate of the former or the latter is larger, the resultant at all points will be directed inward or outward. If, however, the aggregate of the attracting masses is equal to the repelling, if there is a closed and inclusive equilibrium, then the resultant of forces at every point of the same and simultaneously in all outer space will be equal to zero. (*Collected Works,* V, 307.)

The theorem which Gauss regarded as the most important part of his memoir is a conclusion from the previous theorem that there is always one and only one distribution of given mass over a surface, so that the potential of this mass at all points of that surface assumes prescribed values (*Collected Works,* V, 240).

Gauss made electromagnetic measurements for the first time on October 22, 1832, and continued them until 1836. At first he was testing Ohm's law under the most varied conditions. In 1833 he discovered Kirchhoff's laws of branched circuits, discovered by the latter in 1845. He established the principle of minimum heat, later set forth by Kirchhoff in 1848, according to which the heat produced by the current is a minimum for the actual current distribution. In addition, Gauss gave exact proof of the identity of frictional electricity and that produced by galvanic elements and thermoelectric forces, a fact of which all physicists at that time were not convinced. These were preliminary studies.

Gauss' main interest lay in the study of the laws of induction. In 1834 he constructed an induction coil, and was enabled to recognize the damping of a magnet vibrating in a coil as a result of induced current; this led him to the construction of a copper damper for his magnetic apparatus, and to the discovery of "sympathetic vibrations." January 23, 1835, marked the acme of this work, when he formulated[3] the law of induction known now as the "Franz Neumann Law of Potential," although Neumann did not enunciate it until 1845.

By means of certain transformations on Ampère's law, Gauss arrived at a law formulated by Grassmann in 1845, which differs quite substantially from the fundamental law of Ampère. Grassmann did not realize that his law was equivalent to that of Ampère, and he proposed experiments to decide between the two. Gauss realized that this was impossible if only closed currents are available, and that Grassmann's law contradicts the law of action and reaction for current elements, whereas Ampère's law obeys the last-named law for the elements.

The peak of Gauss' work in physics was concerned with the

---

[3] Gauss' entry states: "7 A.M. before getting up."

magnitude of electrodynamic forces. He discovered the first part of Neumann's Law of Potential and then made use of the known fact that an induced current has a direction such that the electrodynamic forces resist the motion, so that in consequence positive work is performed during this motion. Thus he deduced from the law of the electrodynamic forces the law of the phenomena of induction. It is the same train of thought followed later by Franz Neumann and then much later by Helmholtz in his celebrated memoir on the conservation of force, and ties up with Clerk Maxwell's equations.

In those days the general fundamental law was sought in a generalization of Coulomb's law, which agreed in form with Newton's law of gravitation. Newton's work enjoyed unique authority. Gauss was not satisfied with what he had done. The problem was not regarded as having been solved until it was possible to trace it back to the position and motion of electrical charges. In 1835 Gauss attempted to formulate such fundamental laws, and one of them, called "Gauss' Fundamental Law," is examined in detail in Clerk Maxwell's *Treatise*. Gauss had proved that the basic law represents correctly the electromagnetic phenomena, but Maxwell showed that it fails for the phenomena of induction. Gauss probably saw this failure, for he did not publish this law.

In later years Gauss approached Faraday's view on field action. In 1845 he stated in a letter to Weber that the generalizations of Coulomb's law had not satisfied him, because they assumed an instantaneous propagation, whereas *his* real aim had been the derivation of the forces from an action which is not instantaneous, but propagated in time in a manner similar to light. He stated frankly that, in view of the fact he had not solved this problem, he saw no justification for publishing anything about these electrodynamic investigations. In 1858 Gauss' pupil Bernhard Riemann was the first to attempt to replace Poisson's equation for the potential by one which results in propagation of the potential with the velocity of light, c.

CHAPTER FOURTEEN

Surface Theory, Crystallography, and Optics

The general theory of curved surfaces, developed in the nineteenth century, owed its origin to geodesy. When Gauss visited von Zach at the Seeberg observatory near Gotha in 1812, he found his solution to the problem of finding the attraction of an elliptic spheroid, which he published in 1813. (*Collected Works*, V, 1.) In the time from 1812 to 1816 he was busy with the theory of the shortest lines on the elliptic spheroid, and conceived of the cartesian coordinates of a point of a curved surface as functions of two auxiliary magnitudes.

In the spring of 1816 Gauss had proposed as a prize problem for a new astronomical journal, founded by von Lindenau and Bohnenberger, the projecting of two curved surfaces on each other, yet preserving similarity in the smallest parts.[1] At the time he had the solution and had thought out a projection by means of parallel normals on the sphere of radius unity. In his concept the development or bending of curved surfaces was a special case of projection. At this time he had worked out the concept of the total curvature of a portion of a surface, and the concept of the measure of curvature at one point of a surface, as well as preservation of the measure of curvature in spite of bendings.

Schumacher stimulated the Copenhagen Society of Sciences in 1820 to set up as a prize question for 1821 the above-mentioned problem, which Gauss had proposed to Lindenau and Bohnenberger. There were no applicants, hence the problem was renewed

---

[1] It was not chosen as the prize question.

for 1822. Gauss sent in his solution and received the prize. The work did not appear until 1825 in the third and last number of Schumacher's *Astronomische Abhandlungen*.

The concept of projection is at the center of Gauss' theory of curved surfaces. For projections in which complete similarity is preserved, he coined the word "conform" in 1843. This type of projection had a long history, particularly in the field of cartography; it reaches back to the time of the Greeks. It is thought that Gauss was led to this work by the study of shortest lines on curved surfaces. For years he planned to write a major work on the theory and practice of higher geodesy.

In generalizing for geodetic triangles of any curved surface, the first step was to find out the sum of the angles of such a triangle. Gauss proved in his Copenhagen prize essay that for every point of a shortest line the osculating plane contains within itself the subject surface normal. He then showed that the sum of the angles of a geodetic triangle of two right angles deviates by an amount which is given by the area of the corresponding triangle on the sphere of radius unity, if one equates its surface to eight right angles.

After a long, hard struggle Gauss succeeded by the end of 1825 in generalizing his theory of curved surfaces. The *Disquisitiones generales circa superficies curvas*, one of Gauss' major works, was presented to the Royal Society of Sciences in Göttingen on October 8, 1827, and published in the *Commentationes recentiores* of the Society in 1828. It was reprinted in Gauss' *Collected Works* (IV, 217), and has been translated into four languages. In it he established the famous theorem that in whatever way a flexible and inextensible surface may be deformed, the sum of the principle curvatures at each point will always be the same. The so-called Gauss Theorem states that the measure of curvature of a surface depends only on the expression of the square of a linear element in terms of two parameters and their differential coefficients. Modern progress in the theory of surfaces begins with this work of Gauss. Two things in it profoundly affected subsequent development in the theory. The first was the systematic employment of curvilinear coordinates, and therewith a demonstration of the great advantages which could be derived from their use; the second was the conception of a surface as a two-way extension, not rigid but flexible, which could be made to assume new shapes by bending without stretching. All surfaces derived from a given surface by bending are said to be *applicable* or *developable* upon each other. The analytical criteria of whether two given surfaces are applicable upon each other constitute one of the interesting chapters in the general theory.

Gauss worked out the general formula for measure of curvature in 1826. It is not known how he arrived at the general concept of bending of curved surfaces. In December, 1822, he worked out the question of surfaces developable on a plane.

In Gauss' generalization of Legendre's theorem of 1789 on the reduction of a small spherical triangle to a plane triangle with sides of the same length, the theory of the shortest lines is connected with the theory of measure of curvature. Thus Gauss' theorem on the reduction of small geodetic triangles to plane triangles appears as the pinnacle of the structure of Gauss' general theory of curved surfaces.

Only a portion of his results in the field of curved surfaces was presented in the *Disquisitiones generales,* and a second memoir on the subject was planned. Soon after completing this work, however, Gauss was led to research on the foundations of geometry, and at the same time was studying surfaces of constant negative curvature. At the instigation of Gauss, the philosophical faculty of the University of Göttingen in 1830 set up the prize question: the determination of a minimum surface by rotation of a curve joining two given points around a given axis. It was solved by his fellow countryman from Brunswick, later his pupil and assistant, C. W. B. Goldschmidt, who received the prize.

Euler was the only geometer mentioned by name in the *Disquisitiones generales;* Gauss evidently knew Euler's *Recherches sur la courbure des surfaces* (1763), but it is uncertain to what extent he knew Euler's other work on surface theory. Gauss' remark that the partial differential equation of the second order for the surfaces developable on the plane "was not yet proved with requisite rigor" was a reference to Monge. The French geometer's investigations of special classes of surfaces had no influence on Gauss; the same is true of his descriptive geometry, which Gauss praised in a review (1813).

Gauss' work in surface theory was pioneer research for the later nineteenth century in two respects. First, Gauss moved on to the use of an infinite group, in the sense of Sophus Lie, while up until his time in geometry only finite groups of transformations had been considered. Second, he treated the theory of curved surfaces as the geometry of a two-fold extended manifold in a manner which paved the way for the general theory of the multiply extended manifolds, or n-dimensional space.

In 1831 Gauss suddenly manifested a great predilection for crystallography, probably as a hobby or diversion. After a few weeks he had fully mastered the subject as it was then known. He measured crystals with a twelve-inch Reichenbach theodolite, and calculated and sketched their most difficult forms. On June 30, 1831, he wrote to his friend Gerling:

Recently I have begun to busy myself with the study of crystallography, which was formerly quite foreign to me. At first I found it very difficult to orient myself somewhat in it; it seemed to me as though the memoir of your colleague Hessel[2] in the physics dictionary is the model of a confused lecture, to follow which a greater patience than mine is necessary. The reflecting goniometers, as they are equipped up to now according to Wollaston, seem to me to be rather incomplete instruments; I have thought out a very simple apparatus and had one made, by means of which the crystal is fastened at the telescope of a theodolite and thereby can keep its correct position with the greatest sharpness, and I am quite curious as to how experiments begun with it will turn out. I thus hope to be able to determine with ease and without repetition the angles between two surfaces as sharply as only the plane quality of the surfaces allows, and shall then go through a series of crystals.

---

[2] J. F. C. Hessel (1796–1872), professor of geology and mineralogy in Marburg. In his next letter Gerling explained to Gauss that Hessel's manuscript had been brought into disorder at the printer's but the proofreader had not noticed it.

Gauss was interested in the question of the rationality or irrationality of the ratios of the crystallographic coefficients. His system of crystallographic notation was essentially the one later devised by Professor William Hallows Miller (1808–1880) of Cambridge University. As a matter of fact, this system's use of the indices was first devised in 1825 by Whewell. In Germany crystallography had been developed at that time by Franz Ernst Neumann and Grassmann. Miller was the pupil and successor of Whewell at Cambridge, and did not publish his system until 1838. His works have a certain laconicism which is reminiscent of Gauss' style. Miller's crystal-notation system represented the face by a symbol composed of three numerals, or indices, which are the denominators of three fractions with unity for their numerator and in the ratio of the multiples of the parameters. He asserted the principle that his axes must be parallel to possible edges of the crystal. This system brought the symbols of the crystallographer into a form similar to that employed in algebraic geometry, and obtained expressions suitable for logarithmic computation. Gauss and Miller followed Whewell, Neumann, and Grassmann in representing the faces of a crystal by normals to the faces, which are conceived as all passing through a common point. This point is taken as the center of an imaginary sphere, the sphere of projection. The points, or poles, in which the sphere is met by these normals, and which therefore give the relative directions in space of the faces of the crystal, can have their positions on the sphere determined by the methods of spherical trigonometry. A great circle (zone circle) traversing the poles of any two faces will traverse all the poles corresponding to faces in a zone with them. Miller, Gauss, and Neumann used the stereographic projection, and thus were able at once to project any of these great circles on a sheet of paper with ruler and compasses. Thus elaborate edge drawings of crystals became of comparatively little importance. Their system gave expressions for working all the problems that a crystal can present, and it gave them in a form that appealed at once to the sense of symmetry and appropriateness of the mathematician.

Gauss complimented Miller with having "exactly hit the nail on the head" in his crystallography; after a short time he laid aside all his papers, observations, calculations, and sketches. Strangely, Gauss never published anything on crystallography and never talked about it again.

Work in astronomy forced Gauss into the problems of optics in the early period of his life. More accurately expressed, it was the problems of dioptrics, which were caused by the insufficiency of telescopes of that day. The main problem was the calculation of achromatic and spherically correct telescope objectives. As early as 1807 J. G. Repsold, owner of the well-known Hamburg optical works, turned to Gauss with questions on the construction of an achromatic double objective. Gauss wrote to Repsold on September 2, 1809, asking for exact values of refraction and dispersion of two kinds of glass, and gave him some advice on carrying out exact measurement. Twice in 1810 Schumacher in the name of Repsold asked Gauss for the calculation of a new, double objective of eight-foot focal length.

The first two objectives which Repsold produced according to Gauss' formulas yielded poor results. The thickness of the glasses was so slight that the glass lost its spherical form in polishing. On October 6, 1810, Gauss wrote Schumacher that he was ready to repeat the calculation for somewhat greater thickness of lenses. Schumacher soon wrote Gauss that a new experiment by Repsold was crowned with full success. As a matter of fact, up until 1810 Gauss had not paid special attention to the theory of achromatic objectives. His calculations went back essentially to Euler's *Dioptrik*. The double objective calculated for Repsold was corrected spherically for the marginal rays and chromatically in the axis.

In 1817 Gauss published an article on achromatic double objectives, especially with reference to complete removal of chromatic dispersion. He came to the following result: Complete removal of chromatic dispersion among the marginal rays and rays next to the axis is of course possible, or more definitely, an objective can be calculated, which unites at one and the same point all rays of two definite colors, those which impinge at a definite distance from the axis, as well as those which impinge infinitely near it (and indeed, as is always presupposed here, parallel to it). In this result Gauss dropped the old condition that the convex lens would have to be biconvex.

Gaussian objectives were now frequently produced and calculated with other kinds of glass—apparently with varying success. According to Steinheil of Munich it was done for the first time in England, but the result was bad. In 1860 Steinheil himself successfully produced a Gauss objective. Later experts were not of one opinion as to the relative value of the Gauss and Fraunhofer objectives. The Gauss objective met its greatest success in microscopes. At the close of his paper on achromatic double objectives Gauss discussed the correction of spherical aberration.

About 1840 Gauss returned to the question of constructing achromatic objectives, when the Viennese mechanic and optician Simon Plössl began constructing his so-called dialytic telescopes. This type owes its origin to the difficulty of producing large clear pieces of flint glass. The French and English governments had offered large prizes for the solution of the problem.

Gauss stated in a letter to Encke, dated January 2, 1840, that the two conditions for complete achromatism are (a) that the red and the blue image fall on a plane normal to the telescope axis, and (b) that they be of the same size. It was his recognition of b that made him so skeptical of the dialytic telescope. He felt that in three lenses far apart the second condition of achromatism could be fulfilled, but not if two of them, as was the case in the dialytic telescopes, are cemented in and are close together.

In the fall of 1840 both Encke and Schumacher sent Gauss a dialytic telescope for study. Encke was so affected by Gauss' skepticism that he was ready to drop his original view about the excellence of Plössl's instruments. We can imagine Encke's surprise when he received Gauss' letter of December 23, 1840, which informed him that the chromatic error of the dialytic objective can be completely compensated by the eyepiece, so that the eye receives an image completely free of color. This principle is used today in the so-called compensation eyepieces of modern microscopes. Gauss was thus finally convinced of the advantages of the Plössl instruments and purchased one in the spring of 1841; he was pleased with its performance.

Scattered through Gauss' correspondence from 1813 to 1846 is the discussion of many matters in optics which are considered very elementary today. However, it is well to remember that such topics were not so well understood by physicists of his day. They deal with general properties of the course of rays in telescopes, such as magnification, brightness, modification for nearsighted eyes, and so forth.[3]

The so-called Gauss eyepiece is still used today for the purpose of autocollimation. In spectrometers and refractometers it sets the axis of a telescope accurately at right angles to a plane polished surface. The Gauss eyepiece tube has an aperture in the side through which light is admitted to a piece of plane unsilvered glass at an angle of 45 degrees to the axis of the telescope. The light is thus reflected past the cross wires and down the telescope tube to the plane polished surface. If the latter is exactly at right angles to the telescope axis, the light will be reflected back down the telescope, and an image of the cross wires will be formed exactly coincident with the cross wires themselves. The observer must adjust the position of the telescope until this coincidence is obtained. On October 31, 1846, Gauss wrote to Schumacher: "The point is, that no glass must be between the mirror inclined at 45° and the cross-wire system."

Of special importance was a prize problem of the Royal Society of Sciences in Göttingen for November, 1829, proposed by Gauss in the mathematical class of the society, and touching a method for photometry of the stars. The society did not grant the prize to any of the papers sent in, and renewed the problem at the insistence of Gauss, who placed at the disposal of Gerling his own ideas on such a photometer, so that the latter could work out his thoughts on the subject and compete for the prize. As a matter of

---

[3] Both Gauss and his son Joseph were nearsighted.

fact, Gauss was interested in the problem of heterochromatic photometry. Actually, the problem was not solved until 1920 by Schrödinger. Gerling constructed a photometer according to the principle indicated by Gauss, but he merely got honorable mention. The prize went to Steinheil of Munich who participated in the competition with a photometer based on other principles of construction.

By far Gauss' greatest achievement in the field of optics was his *Dioptrische Untersuchungen,* which appeared in 1840. According to his own claim he had possessed the results for forty or forty-five years, but had always hesitated to publish such elementary meditations. A work of Bessel on the determination of the focal distance of the Königsberg heliometer objective gave him the impetus necessary to publication. Bessel's method assumed mistakenly that the usual lens formula

$$\frac{1}{g} + \frac{1}{b} = \frac{1}{f}$$

is correct for lenses of finite thickness. As a consequence of this mistake, Bessel greatly underestimated the possible error of his measurement. *Dioptrische Untersuchungen* treats the problem of pursuing the course of a ray of light through a centered system of refracting spherical surfaces. The equations of the ray before the first refraction are to be set in relation to the equations of the ray after the last refraction, that is, the coefficients of the latter equations are to be deduced from those of the former.

In the *Dioptrische Untersuchungen* there are data on the construction of the image when the principal points and foci of the system are given, and finally formulas for a simple lens of nonvanishing thickness are given. Bessel's determination of the focal distance of the Königsberg heliometer objective was examined. While Bessel estimated the error of his result at 1/75,000, Gauss showed that it amounted to 1/1,300.

Gauss wrote in unpublished notes that reflections on spherical surfaces are to be incorporated into his theory by making the index of refraction negative. The light rays fall on a lens, are refracted the first time on the front surface, reflected at the rear surface, and refracted again at the front surface. His formulas indicate principal points and foci for this case.

The *Dioptrische Untersuchungen* emphasized that the position of the principal points and foci depends on the index of refraction of the lenses of the system, that is, it varies from wave length to wave length, and that in general chromatic aberration occurs. For achromatism he demanded that all parallel rays independently of color converge at one point, that is, not only such as are parallel to the axis, but such as are inclined to it. In the usual achromatic objectives, in which lenses are close to each other, these conditions are approximately fulfilled, but not in the case of dialytic objectives. This explains why Gauss had doubts about the dialytic principle.

Gaussian dioptrics represents the perfection of those investigations which relate to central rays (paraxial rays), that is, to the point by point projection of means of narrow pencils of rays. In the century which has passed, practically nothing has been added to the Gaussian theory. The importance of his work is slight for practical optics today, since one cannot be limited to narrow pencils of rays, but must project by means of wide open pencils of rays.

One can arrive at the fundamental concepts and properties of projection in a purely geometrical manner. Gauss' deductions produce the impression that all these concepts are limited to narrow pencils of rays, yet a geometric study shows that the relationship and these fundamental concepts are possible in wide open pencils of rays, while with Gauss only the physical production of such relationship is limited to narrow pencils.

Gauss stated that his interest in dioptrics reached back to about 1800. Nothing has been found in his manuscripts to offer proof of this statement. However, some notes on a subject closely related to his optical research date from the period 1814–1817. In them he treated fully the systems of principal rays and aperture rays. A note of 1811 dealt with the experiments of Malus on the polarization of light.

In 1836 Gauss purchased a Schwerd instrument, in order to make diffraction experiments. For a very brief time he was intensely interested in the subject, but had to give it up for lack of time and accomplished nothing in this field.

CHAPTER FIFTEEN

Germination: Non-Euclidean Geometry

No other phase of Gauss' career has aroused as much controversy as his work in non-Euclidean geometry. There are several reasons for this. Each of the three men (Lobachevsky, J. Bolyai, and Riemann) generally credited with being the founders of non-Euclidean geometry had either direct or indirect connections with Gauss. Felix Klein made some rather exaggerated statements concerning the influence of Gauss on these men, and since Gauss' manuscripts were not open to scholars[1] until very recent years, no final verdict could then be reached. Gauss himself published practically nothing on the subject, and scholars were reduced to the examination of certain passages in his letters and several book reviews where he alluded to the possibility of such systems of geometry. He was extremely sensitive to public criticism and unwilling to have anything so revolutionary appear over his name.

Critics, both contemporary with him and in our own time, have attacked Gauss' character because he so often stated that the result of some new discovery had been in his possession for many years. They did not realize that in his early years such a flood of new developments came to his mind that he could hardly control them, and that it took him long years of work before they fully matured. He felt that a publication should be a "completed work of art," and in fact was not too much concerned when (if ever) many of his results were published. To him, the main concern was his occupation with new truth. His similarity to Newton in this respect is striking.

---

[1] Other than the editors of his *Collected Works*.

Since he adhered to the above viewpoint, he was not particularly concerned when someone anticipated him by prior publication. His goal was primarily the extension and establishment of truth, not personal glory. Since his death, a careful study of his papers and letters has disclosed that his statements about early discovery were correct. In addition, one should remember that Gauss had an unusually accurate memory. He used notes of the years 1796–1815 in his scientific diary, which was not "discovered" until 1898. Later in life he entered in notebooks mathematical results which he had mentioned in letters. Thus it has been possible to compare his utterances with the evolution of his thoughts.

On November 28, 1846, Gauss wrote to Schumacher that in 1792 at the age of fifteen he had thought of a geometry "which would have to occur and would occur in a rigorously consistent manner, if Euclidean geometry is not the true one," that is, if the eleventh axiom (of parallels) is not valid. Of course, Gauss meant merely the first dawning of the thought. On October 2, 1846, he had told Gerling that the theorem that in every geometry independent of the parallel axiom the area of a polygon is proportional to the deviation of the sum of the external angles from 360 degrees is "the first theorem (likewise at the threshold) of the theory, which I recognized in 1794 as necessary."

We know that Gauss was giving much thought to the possibility of non-Euclidean geometry in the years 1797–1802. In his diary an entry in September, 1799, contains these words: "In principiis geometriae egregios progressus fecimus." On May 17, 1831, Gauss reported to Schumacher that he had begun to write up some of his meditations on parallel lines, which in part were about forty years old. Germination, therefore, occurred in the years 1792–1794.

When Gauss began his studies in Göttingen in October, 1795, he had recognized the weak position of the Euclidean system, and in the case of polygon areas had considered the consequences which arise from a rejection of the parallel axiom. This type of "meditation" was in the air at that time. There was a flood of publications on the question of parallels, especially in France in the years after the Revolution.

Legendre, who had a strong dislike for Gauss, tackled the parallel problem, but was unable to solve it. In a letter to Olbers dated July 30, 1806, Gauss remarked that it seemed to be his fate in almost all his theoretical works to be competing with Legendre, and mentioned the theory of numbers, transcendental functions, elliptic functions, the foundations of geometry, and the method of least squares. One might add to this list geodesy and the attraction of homogeneous ellipsoids.

At the University of Göttingen there was lively interest in the question of parallels. Kästner had eagerly collected literature on the subject and in 1763 had directed a worth-while dissertation of his pupil Klügel on previous attempts to prove the axiom. Kästner believed that no one in his right mind would attack Euclid's axiom. Gauss' friend Pfaff believed that the only thing to do was to replace the parallel axiom by a more simple one. When Gauss went to Göttingen, J. Wildt (1770–1844) gave a trial lecture on the theory of parallels (1795). In 1800 he published three "proofs" of the eleventh axiom, and in 1801 Seyffer, the professor of astronomy, published two reviews of attempts to prove the parallel axiom. Seyffer had come to the conclusion that it was more than doubtful, perhaps impossible, to prove the eleventh axiom without drawing aid from a new axiom. Gauss was not on intimate terms with Kästner and Wildt, but he was very close to Seyffer, and their correspondence continued until the latter's death. Their conversations frequently touched on the theory of parallels.

At Seyffer's home Gauss met Wolfgang Bolyai, the young Hungarian student who became the most intimate friend of his entire life. Needless to say, the theory of parallels was one of their chief mutual interests. After Gauss returned home to Brunswick in 1798, Bolyai made efforts to prove the axiom and in May, 1799, believed he had reached his goal. On May 24, 1799, the two said farewell forever at Clausthal in the Harz mountains; Bolyai was returning to his native Hungary. He told Gauss of his "Göttingen theory of parallels." In a letter to Bolyai dated December 16, 1799, Gauss regretted that he did not have time to find out more about Bolyai's work on the foundations of geometry. He felt that it would have saved him much toil, and that much remained to be done on the subject. Gauss stated in the letter that he was far advanced in this work, but that he did not have time to work it out properly. He felt that his work would make the truth of geometry doubtful. Gauss was not satisfied with what had been done; he felt that all so-called proofs of the parallel axiom were failures (1799). No notes exist of the work which Gauss had done up until that time. Certainly he recognized that the Euclidean was not the only possible system.

The works of Saccheri (1733) and Lambert (1766) on the theory of parallels were available to Gauss at the University of Göttingen library. Indeed, the record shows that he drew out volumes by Lambert in 1795 and 1797. These works were probably also known to Wolfgang Bolyai. In later years Lambert's theory of parallels was discussed among Gauss' pupils. Gauss owned the *Mathematische Abhandlungen* of J. W. H. Lehmann (Zerbst, 1829) in which Saccheri and Lambert are quoted. Marginalia and traces of use show that Gauss read it and paid special attention to the passages on parallel theory. Neither Lambert nor Saccheri was able to reach the level of non-Euclidean plane trigonometry.

Bolyai reached home in Transylvania in the summer of 1799, but had no time for mathematics until 1804, when he was appointed to a professorship of mathematics and physics at Maros-Vásárhely. He pulled out his "Göttingen theory of parallels," polished it up, and sent the sketch to Gauss on September 16, 1804, asking for criticism and forwarding of it to some reputable scientific society for judgment.

On November 25, 1804, Gauss replied that he was delighted by the genuine, fundamental ingenuity of the little work. He stated that the train of thought was very similar to his own, but that up until that time he had been unable to solve the problem completely. Bolyai's procedure did not satisfy him fully, but he wrote that he hoped to break through the difficulties before his death. He added that he was too busy with other matters at the time to give attention to it. Gauss promised that if Bolyai succeeded in surmounting all the hindrances, he would be glad to be anticipated by so intimate a friend and would do all in his power to make the work properly known to the public.

Bolyai was greatly encouraged by the letter, and on December 27, 1808, sent Gauss a supplement. To this he got no answer, and their correspondence was broken off until 1816. In those years Bolyai had put up a hard fight against the problem, and finally was so discouraged that he felt all his effort had been wasted.

When his son Johann Bolyai (1802–1860), who had entered the Academy of Military Engineers in Vienna in 1818, reported to his father in 1820 that he was attempting to prove the eleventh axiom, the elder Bolyai was horrified and in the most moving terms begged his son to leave the theory of parallels in peace:

> Do not lose one hour on that. It brings no reward, and it will poison your whole life. Even through the pondering of a hundred great geometers lasting for centuries it has been utterly impossible to prove the eleventh without a new axiom. I believe that I have exhausted all imaginable ideas. Furthermore, if Gauss too had spent his time with puzzling over the eleventh axiom, his theories of polygons, his *Theoria motus corporum coelestium,* and all his other works would not have appeared, and he would have lagged behind. I can prove in writing, that he racked his brains on parallels. He averred both orally and in writing that he had meditated fruitlessly about it.

Johann was not terrified by his father's warning; on the contrary, it spurred him on to action. Toward the end of 1823 he began to sense victory. His unexpected solution was that which Gauss had attained in 1816 after long hesitation and doubt.

In the letter of November 25, 1804, Gauss had indicated that his train of thought was quite similar to that of the elder Bolyai. The latter had considered the line which originates when one erects perpendiculars of the same length, on the same side, at points equally distant from a straight line, the end points being connected by a straight line. In Euclidean geometry one thus gets a parallel as a base line, but non-Euclidean geometry yields a broken line which consists of equally long segments approaching each other at equal angles. Wolfgang Bolyai had attempted to prove that a broken line of this kind, if one moves out on it far enough, would have to intersect the base line. That would prove that the assumption that the eleventh axiom is not valid leads to a contradiction. Gauss was making attempts in the same direction, as shown by the last page of his notebook "Mathematische Brouillons" (*Collected Works,* VIII, 163). These notes were begun in October, 1805.

Between 1799 and 1804 Gauss had been trying to move ahead on another path. Notes dated 1803 show several attempts by means of geometric constructions and functional equations derived from them (the same process he had applied to the area of a triangle) to develop the relations between the parts of a triangle. But these efforts were in vain. There was still some doubt in his mind. As late as 1808 he considered his work on the parallel axiom unfinished and was not yet fully convinced that it could not be proved.

The following remark by Gauss in 1813 throws some light on his views at the time: "In the theory of parallel lines we are now no further than Euclid was. This is the *partie honteuse* of mathematics, which sooner or later must get a very different form." (*Collected Works,* VIII, 166.)

In 1816 Gauss published a book review of two attempts at proof of the axiom. He expressed a different tone, and spoke of the "vain effort to conceal with an untenable tissue of pseudo proofs the gap which one cannot fill out." He wanted to indicate his conviction of the impossibility of proving the eleventh axiom.

By 1816 Gauss was in possession of non-Euclidean trigonometry. His pupil Friedrich Ludwig Wachter (1792–1817), who was a professor in Danzig, visited him in Göttingen. Incidentally, Gauss had a high opinion of Wachter's talent and was deeply distressed in 1817 when Wachter disappeared and was never found.[2] Wachter had reviewed in 1816 one of the two works on parallel theory which Gauss had reviewed. During his visit with Gauss in April, 1816, their conversation turned to the foundations of

---

[2] He was declared legally dead in 1827.

geometry. Gauss stimulated Wachter to do some research on what he called "anti-Euclidean" geometry. Gauss had talked to Wachter about his "transcendental" trigonometry, and Wachter had made vain efforts to penetrate it. In a letter to Gerling on March 16, 1819, Gauss wrote that he had developed non-Euclidean geometry to such an extent that he could completely solve all problems, as soon as the constant = C was given. On April 28, 1817, he wrote Olbers:

> I am coming more and more to the conviction that the necessity of our geometry cannot be proved, at least not *by human* intelligence nor *for* human intelligence. Perhaps we shall arrive in another existence at other insights into the essence of space, which are now unattainable for us. Until then one would have to rank geometry not with arithmetic, which stands a priori, but approximately with mechanics.

It is not known in what way Gauss arrived at non-Euclidean trigonometry. In his papers is the development of formulas which was probably written down in 1846. He used the process of geometric constructions and of functional equations derived from them, a process which he had used without success in 1803. It is probable that between 1813 and 1816 he advanced on this path.

His hints about the impossibility of proving the parallel axiom, contained in his book review of 1816, did not meet the success Gauss expected, in fact they were criticized, and he decided not to publish his views during his lifetime. He was soon surprised and made happy to find two others moving in the same direction in which he was traveling.

The jurist F. C. Schweikart (1780–1859) had published a work on parallel theory in 1807 in which he objected to the introduction of the infinite in the usual explanation of parallels as nonintersecting straight lines, and demanded that in building up geometry one should proceed from the existence of squares. Later, between 1812 and 1816, without the help of the eleventh Euclidean axiom, he had developed a geometry which he called "astral geometry." In 1816 he was called from Charkov to the University of Marburg, and in 1818 discussed his system with his colleague Gerling. On January 25, 1819, Gerling wrote thus to Gauss:

I told him how you had expressed yourself publicly several years ago (1816), that fundamentally one had not advanced here since Euclid's time; indeed, that you had told me several times how you through manifold occupation with this subject had not come to a proof of the absurdity of such an assumption [of a non-Euclidean geometry].

Schweikart asked Gerling to forward a short sketch of his "theory of astral magnitudes" to Gauss and request his opinion. In his reply Gauss stated that almost all of it was as though it had come from his own pen. He found his concept that space is a reality outside of us, whose laws we cannot completely prescribe and whose properties rather are to be completely ascertained only on the basis of experience. The word "astral" chosen by Schweikart was supposed to express the fact that in measurements of magnitude, as they occur in the celestial world, deviations from Euclidean geometry could be observed. It seems to have pleased Gauss, for in later notes he used the word.

A nephew of Schweikart, also a jurist, F. A. Taurinus (1794–1874), had as a young man busied himself with the parallel theory. He was definitely stimulated by his uncle's book. In October, 1824, he sent an attempted proof to Gauss, who knew since 1821 through the uncle that Taurinus was investigating the foundations of geometry, and recognized in the young man "a thinking mathematical head."

In a long letter dated November 8, 1824, Gauss replied to Taurinus. He fully explained his views on the parallel axiom, but enjoined the recipient of the letter to make no public use of this private communication or any use that could lead to publicity. Gauss wrote:

> The assumption that the sum of the three angles (of a triangle) is less than 180° leads to a special geometry, quite different from ours (Euclidean), which is absolutely consistent and which I have developed quite satisfactorily for myself, so that I can solve every problem in it, with the exception of the determination of a constant which cannot be found out a priori. The larger one assumes this constant, the closer one approaches Euclidean geometry and an infinitely large value makes the two coincide. If non-Euclidean geometry were the true one, and that constant in some relationship to such magnitudes as are in the

domain of our measurements on earth or in the heavens, then it could be found out a posteriori.

This letter stimulated young Taurinus to continue his research with increased zeal. In 1825 he published his *Theorie der Parallellinien* in which he was convinced of the unconditional validity of the parallel axiom, but he began to develop the results which are yielded by a rejection of it. Thus he arrived at that constant which would be peculiar to non-Euclidean geometry. In the simultaneous possibility of infinitely many such geometries, each of which is without internal contradiction, he saw sufficient reason to reject them all.

A copy of the book was sent to Gauss, who also received a copy of Taurinus' *Geometriae prima elementa* (1826). Gauss did not send the author an acknowledgment of either one; probably he was offended because Taurinus had mentioned him in the preface of each work. In the second book the author developed the formulas of non-Euclidean trigonometry with one stroke by making the radius of the sphere imaginary in the corresponding formulas of spherical trigonometry. He applied these formulas to the solution of a series of problems and calculated correctly the circumference and area of a circle and the surface and volume of a sphere.

The work of Taurinus was forgotten for many decades. In a letter to Gauss dated December 29, 1829, he wrote: "The success proved to me that your authority is needed to furnish recognition for them (his works), and this first literary attempt, instead of recommending me, as I had hoped, has become for me a rich source of dissatisfaction."

Letters of Gauss show that in late 1827 he began to do some intense research on the foundations of geometry—he called it "the metaphysics of space doctrine." In 1828 he wrote that he had first studied these foundations forty years previously and that he would probably not live long enough to work out for publication the full results. The main reason was that he feared what he called "the clamor of the Boeotians" if he were to speak out fully on so revolutionary a matter.

In a short note dated November, 1828, Gauss proved, independently of the eleventh axiom, that the sum of the angles of a triangle cannot be greater than two right angles. In April, 1831, he had begun to write down some of this research because he did not want it to perish with him. Some of these notes have been published in Volume VIII of the *Collected Works* (pp. 202-209). In them the fundamental properties of parallels or "asymptotic straight lines," as Bolyai called them, are deduced. Gauss arrived in the last note at the paracycle, the curve which results when the radius of a circle becomes infinite. He calls it "trope" (*cercle tropique*), a clear sign that he conceived the paracycle as the transition from actual circles to hypercycles. A letter from Gauss to Schumacher dated July 12, 1831, discussed the results of the abolition of similarity in non-Euclidean geometry and indicated the formulas valid there for the circumference of a circle.

On November 3, 1823, Johann Bolyai wrote his father that he had "created a new, different world out of nothing." In February, 1825, he presented to the elder Bolyai the first sketch of his work. The father was not in agreement with it; he was annoyed by the occurrence of the indefinite constants and the many hypothetical systems thereby rendered possible. Father and son could not agree, and finally Johann decided to compose the essence of the work in Latin, attach it to Wolfgang's planned *Tentamen,* and send it to Gauss for his opinion.

Reprints of the *Appendix scientiam spatii absolute veram exhibens* were ready in June, 1831, and one of them was sent to Gauss. Owing to the cholera epidemic, Gauss received only the covering letter, at the close of which Johann wrote a brief outline of his work. After a long time the reprint was returned to Wolfgang. It finally reached Gauss in February, 1832, through a friend of the Bolyais, Baron von Zeyk, who was studying in Göttingen.

The first impression on Gauss was favorable. He found in the book all his own ideas and results developed with great elegance, although written in a form difficult to follow for one who was not familiar with such research, because of the great concentration demanded. Gauss recognized that his own ideas of 1798 on the subject were less mature and developed that those of Johann. He considered young Bolyai "a genius of the first magnitude." Gauss wrote Wolfgang on March 6, 1832, expressing surprise at the close agreement of Johann's results with his own and sent the young geometer hearty greetings along with assurances of his especial high esteem. Wolfgang wrote his son: "Gauss' answer respecting your work is very fine and redounds to the honor of our fatherland and nation. A good friend says it would be a great satisfaction."

Johann was greatly disappointed, insulted, and embittered because Gauss gave no public recognition to the *Appendix* and claimed priority of discovery. In later years relations between him and his father were strained, probably due in large measure to this matter.

In the above-mentioned letter Gauss gave as a sample of his own research a simple proof of the theorem that in non-Euclidean geometry the area of a triangle is proportional to the deviation of the sum of the angles from 180 degrees. It is certain that this represents a part of his research of September, 1799, since the receipt of Johann's *Appendix* must have evoked in him memories of his friendship with the elder Bolyai at that time. In the same letter Gauss urged Johann to busy himself with the corresponding problem for space, namely, to determine the cubic content of the tetrahedron (space bounded by four planes). The papers of Johann contain several processes which can serve as a solution, among them the method which Gauss had in mind and which he indicated in one of his notebooks at the time of sending his letter to Wolfgang on March 6, 1832.

A second note by Gauss, dated about 1841, refers to the volume of the tetrahedron. It is on a piece of paper which was found between the leaves of a reprint of the memoir (1836) of the Russian mathematician Nikolai Ivanovitch Lobachevsky (1793–1856) on the application of imaginary geometry to several integrals. By the term "imaginary" geometry he meant non-Euclidean geometry.

As late as 1815–1816 in lectures he gave at the University of Kasan, Lobachevsky was still in the Euclidean fold and had made several attempts to prove the parallel axiom. In the years immediately following he dared to explore the consequences of assuming that the axiom does not exist, and gradually familiarized himself with the thought of the impossibility of proving it. He supported this viewpoint in an unpublished textbook on geometry (1823). Soon thereafter he arrived at the recognition that there is a noncontradictory geometry which does not need the axiom of parallels. He developed this geometry so far that he could treat all its problems purely analytically and gave general rules for the calculations of lengths of arc, areas of surfaces, and volumes. The results of these researches were presented to the Kasan scientific society on February 12, 1826, but were not published until 1829 and 1830 in the Kasan *Messenger*. A series of additional memoirs followed them in the years 1835–1838.

In order to disseminate his ideas in western Europe, Lobachevsky published in 1837 in *Crelle's Journal*[3] a short résumé of his imaginary geometry; unfortunately it was poorly adapted as an introduction to the subject. This article seems to have escaped the attention of Gauss, who probably never heard of Lobachevsky until 1840, when he read in Gersdorf's *Repertorium der gesammten deutschen Literatur*[4] an unfavorable review of the German version of Lobachevsky's *Geometrische Untersuchungen zur Theorie der Parallellinien*. Gauss considered the review rather silly.

About the same time (1840) Ernst Knorr (1805–1879), a physicist at the University of Kasan and a friend of Lobachevsky's, visited Gauss and gave him a copy of Lobachevsky's above-mentioned memoir of 1846. Later Gauss' friend Friedrich Georg Wilhelm Struve (1793–1864), director of the observatory at Pulkova, sent him the other memoirs by Lobachevsky which had appeared in the Kasan learned society journal. It is not known where Gauss got the memoir in the Kasan *Messenger* of 1829–1830.

---

[3] XVII, 295.
[4] Vol. 25, part 3 (1840), pp. 147 *et seq.*

Fortunately Gauss was able to read the works of Lobachevsky in the original. In a letter to Gerling on February 8, 1844, he compared the short memoirs to "a confused forest through which it is difficult to find a passage and perspective, without having first gotten acquainted with all the trees individually." On the other hand he praised the conciseness and precision of the *Geometrische Untersuchungen*. He repeated this praise in a letter to Schumacher dated November 28, 1846, and stated that Lobachevsky had taken a path different from his own, but in a masterly fashion and in genuinely geometric spirit. He got "exquisite enjoyment" from reading it.

On a piece of paper found in one of the two copies[5] of Lobachevsky's *Geometrische Untersuchungen* belonging to Gauss is the outline of a deduction of formulas of non-Euclidean trigonometry. Probably he wrote it in 1846 when he had occasion to look through the work again. The note gives Gauss' results of 1816, but a concept is added which is of a later date. As a final result formulas are derived which are identical with the equations of spherical trigonometry, referred to a sphere of radius $1/k$; the corresponding equations of non-Euclidean trigonometry follow from those if a purely imaginary value is given to the constant k. Lobachevsky had noted this relationship at the close of the *Geometrische Untersuchungen*. It occurs there as a special case. It is not known whether Gauss desired to indicate by the letter k that the two geometries can be subordinated to the more general concept of the geometry of a manifold of constant measure of curvature. In the last mentioned letter to Schumacher he wrote: "You know that I have had the same conviction for 54 years (since 1792), with a certain later extension which I do not want to mention here."

It is not known just what this "later extension" was. Possibly he granted full equality to the geometries yielded by the sign of the measure of curvature. His pupil Riemann later developed

---

[5] He had ordered one copy after reading the review in Gersdorf's *Repertorium;* Otto Struve, son of Wilhelm, had later given him the other one.

the thought that one need only conceive space as an unlimited not as an infinite manifold.

On November 23, 1842, Gauss proposed Lobachevsky as a corresponding member of the Royal Society of Sciences in Göttingen, with the citation that he was one of the most distinguished mathematicians of the Russian Empire, and he was immediately elected. Gauss wrote in his own hand a letter to accompany the diploma of membership. Lobachevsky was highly pleased to get this recognition from a foreign country and especially from the hands of a man he had learned to admire in his youth. His cordial letter of thanks to Gauss was dated June, 1843, in which he excused the tardiness of his reply by reference to the burning of the city, which had affected his health and personal affairs and overburdened him with official business.

Why did Gauss never mention Bolyai and Lobachevsky in his printed works? This question has often been asked, and Gauss has been criticized for it. The only answer is that he had firmly resolved never to publish anything on parallel theory during his lifetime. And he adhered to this decision. He would do anything short of publication to aid and encourage others, but he felt he had to avoid controversy. He did not mind if others anticipated him and published results he had had for many years.

Otto Struve visited Gauss for the last time in August, 1843, and found him reading one of the short works of Lobachevsky, which, as he said, interested him on account of their content as well as on account of the Russian language, which he was then eagerly studying. A result of the conversation was that by the end of 1843 Struve had sent Gauss as many of Lobachevsky's writings as he could find in St. Petersburg. It is interesting to note that Dirichlet as early as 1827 discussed non-Euclidean geometry with Gauss on the occasion of his visit in Göttingen.

In a little book published in 1851 Wolfgang Bolyai praised Lobachevsky's *Geometrische Untersuchungen,* which was probably the only work of his that he or his son Johann ever saw. Johann studied the relationship of this work to his *Appendix.* Lobachevsky had the priority of publication.

When Johann Bolyai reported to his father on November 3, 1823, about his new discoveries, the latter urged him to hasten publication because "many things have an epoch simultaneously when they are found in several places, just as in the spring violets come to light in several places." We know today how right he was. Gauss was undoubtedly the first man to free himself from the fetters of Euclidean tradition, but the evidence does not show that he had direct influence on Bolyai and Lobachevsky, either by stimulation or furnishing of basic ideas. There is no doubt that Taurinus was independent of Gauss, even though he had a certain stimulus from him through his uncle Schweikart. Wolfgang Bolyai tried to steer his son away from the subject of parallels with the warning that even Gauss had failed to solve the problem. In later years Johann accused his father of having given his ideas to Gauss, who then claimed them as his own. In addition, it is perfectly clear that Bolyai and Lobachevsky were entirely independent of each other.

Possible influence of Gauss on Lobachevsky needs to be examined more closely. Bartels, friend and teacher of Gauss, spent nine years in Switzerland, returned to Brunswick 1805–1807, and in the latter year became professor of mathematics in Kasan. Bartels was the teacher of Lobachevsky, and introduced him to the work of Gauss in the theory of numbers, but there is no evidence that he gave him Gauss' ideas on non-Euclidean geometry, simply because he did not know them. Gauss and Bartels were together only until 1807 and never saw each other again after that date. They exchanged letters of a personal nature in 1808 and again in 1821, but these letters contained nothing about mathematics. Of course, doubt of the correctness of the Euclidean axiom was in the air at that time, and we must assume that Bartels shared the common view—not due specifically to his friendship with Gauss. In later years Otto Struve, a grandson of Bartels, said that the latter regarded Lobachevsky as one of his foremost and most talented pupils in Kasan. The work on non-Euclidean geometry he regarded more as interesting intellectual speculations rather than as a work advancing science. Struve did not remember that Bartels ever spoke of similar ideas of Gauss'. The strange fact is that in the course of a little more than a decade four men independently liberated themselves from Euclidean tradition. First, Gauss and Schweikart, almost simultaneously, then, again almost simultaneously, Lobachevsky and Bolyai. Even more miraculous is the fact that all four found only one of the two possible solutions. At that time none of them seems to have thought of the possibility of a consistent system of geometry in which the sum of the angles of a triangle is greater than two right angles. Much later Gauss recognized the possibility, and the idea was fully developed by his pupil Riemann, who on June 10, 1854, read before the philosophical faculty of Göttingen his famous trial lecture *Ueber die Hypothesen, welche der Geometrie zu Grunde liegen*. Of the three themes submitted it was Gauss' choice which fixed upon this one, and he was in the audience.

There is a certain expression involving the equation of a surface and the coordinate of a point called the "Gauss curvature" which will be constant on all surfaces such as the sphere and the plane. It is also true, however, of certain other surfaces, at least if the dimensions of the figure are sufficiently small. Among the axioms of geometry which do not depend on the parallel postulate are those which secure the free mobility of a figure in space. This means that space is the same in nature throughout and that any figure can be carried along it without tearing or crumpling or stretching. Among the surfaces of constant Gauss curvature is that which is formed by revolving the tractrix about the X-axis. This surface is everywhere saddle shaped. The geometry of this surface is non-Euclidean and belongs to a type different from that of the sphere, being essentially Lobachevskian. The tractrix space is not a complete Lobachevskian plane, but a portion of it, with its edges joined to themselves by the same process by which a cylinder can be formed from a strip of Euclidean plane.

In a note dated about 1827 Gauss called the curved surface of constant negative measure of curvature generated by rotation of the tractrix (pseudo sphere) the "counterpart of the sphere." The formulas set up by him lead to the theorem that in the pseudo sphere (and only in it) surfaces of rotation are congruent to each other and that, preserving this property, one can move a geodetic triangle on the pseudo-sphere just as one can move a spherical triangle on a sphere. It is thus worthy of note that Gauss studied surfaces of constant negative curvature. It is not known whether he originated the expression "counterpart of the sphere."

CHAPTER SIXTEEN

Trials and Triumphs: Experiencing Conflict

The University of Göttingen very early began preparations to celebrate its centennial jubilee. It was decided to erect a new building and to dedicate it as part of the celebration. This large building in classical Greek style, known as the "Aula," is on the Wilhelmsplatz. In the small park in front of it stands a monument of King Wilhelm IV of Hanover, to whose memory the Aula was dedicated; it was unveiled as part of the jubilee. The Aula has always been used as the administration building of the university, and it contains a beautiful auditorium adorned with oil portraits of men important in the history of the university.

The jubilee was celebrated from September 12 to 20, 1837; large numbers of alumni, officials, and delegations from other universities were present. The curators of the university and the Ministers von Arnswaldt and von Stralenheim were escorted into the city by an armed guard. On September 17 all the church bells chimed in the festival. The prorector Bergmann at the salute of cannons turned over to the student body a special jubilee flag of the university.

A choir sang in the steeple of St. John's Church as King Ernst August, who had been officially received at the Weender Tor by the city council and a delegation of citizens, arrived. In the parade, which marched past the King, were students, the faculty, the pastors of the city, the curators, ambassadors and deputies of foreign courts, and representatives of many universities. The parade then moved to Wilhelmsplatz, where the Aula was dedicated and the monument unveiled. Karl Theodor Albert Liebner

(1803–1871), professor of theology, gave the jubilee sermon, and the service was closed by a great *Te Deum*. After a large banquet the day was closed by a concert in St. John's Church.

September 18 was the day of the actual academic celebration. Early in the morning Minister von Stralenheim handed to the prorector a golden key and thus turned over to the university the new Aula. At the same time he gave Bergmann a medallion and chain of pure gold, and decreed that each successor of the prorector should wear them. In the Aula, Carl Otfried Müller, Göttingen's great classical scholar, gave the principal address of the occasion in eloquent Latin.[1] After a gala dinner the citizens of Göttingen put on an imposing torchlight parade in the evening in honor of the curators of the university, followed by fireworks.

A large number of the leading scholars of Germany were present; the most important one was Alexander von Humboldt, who was Gauss' house guest during this visit to his alma mater. When Carl Bölsche saw Gauss walking out of the Aula arm in arm with Humboldt, he was conscious of the great impression made on the youth. He wrote that they said to each other:

"Such an age—what good fortune and what a call for everyone to use his talent in his own way as honestly and as restlessly." After I heard him [Gauss] mentioned and as often as I saw him in Göttingen, a feeling of edifying reverence and love overcame me. True greatness, without realizing it or desiring it, ennobling the youth!

On September 19, honorary doctorates were conferred in the Aula by the deans of the several faculties. Gauss regretted that Sophie Germain was no longer alive to receive one of them. In the afternoon he gave before an open session of the Royal Society of Sciences a lecture on his invention of the bifilar magnetometer. Gauss undertook the lecture at the special request of Humboldt, who was present at the session. He began his lecture with these words:

---

[1] Gauss wrote just before the jubilee that a deluge of poems had "broken out."

Our society approaches the celebration of the Georgia Augusta, as the daughter appears at the golden wedding anniversary of her mother, not in order to express her feelings in flowery language, but to share the joy of the house and to present a modest gift. According to native custom the daughter brings a simple work of her hands finished in the evening hours, or a fruit ripened in her own garden. But the feelings of the daughter on the day of honor of her dear mother, to whom she owes existence, care, and prosperity, the feelings of grateful, joyous emotion, are too much a part of her being to have need of words. Indeed, the day of honor of the mother is also the day of honor of the daughter.

While I have the honor, at this festive moment and before such a brilliant assemblage, of opening with a lecture the first session of our society in the new rooms in this sense, I am quite conscious of how very much I must count on a benevolent indulgence in more than one respect. A lecture from the area of the rigorous sciences, in itself slightly compatible, and in any case in my hands not adorned with eloquence, can in the most favorable instance arouse sympathy only among those who are more intimate with similar efforts. It will be more gratefully recognized if such persons as are more remote from these sciences do not deny their honoring attention to a lecture from which I cannot separate several developments appearing perhaps dry to them, without becoming superficial or even unintelligible.

In the evening of September 19 a great dance took place in the riding academy, which had been specially equipped for the occasion. Over two thousand people took part in the dance. On the following morning the students took their flags into the Aula and closed the jubilee by singing the traditional "Gaudeamus igitur" on Wilhelmsplatz.

Gauss was not a friend of official celebrations. How did he react to all this gaiety and ceremony? The answer is found in a letter which he wrote to his friend Olbers on September 26, 1837:

> After our jubilee ceremonies are surmounted—of which you get news through the press *ad nauseam usque*—I must give you a sign that I am still alive. Indeed, little would have been lacking for me to have met the same fate as Göschen,[2] who died day before yesterday most

---

[2] Johann Friedrich Ludwig Göschen (1778–1837), professor of law.

probably as a sacrifice of the same [ceremonies]. Another local professor, Dissen,[3] had died a few days earlier, but if in his case the jubilee had some effect, then it can only have been disturbances of mind, since he, confined to his room for many years, could not take any part in the ceremonies.

I myself had not been quite well previously and had decided to take no part in all the processions and the stay in church and Aula where the tasteless as well as oppressively heavy so-called official robe, a monk's cloak, had to be put on. Not until early Sunday, when we were informed that the aforesaid processions and sermon presentation would occur before the King, did I decide not to remain behind. But even at the meeting in the library, where 100 persons were crowded together in close quarters, I was really sick in the deoxygenized air; then came the procession in the heavy-as-lead robe; then the stay in the terribly overcrowded church, where one could scarcely breathe, and an insipid sermon which went on and on. I almost fainted. Then a new procession, then standing an hour in the open air at the unveiling of the statue, then presentation. When I finally came home, I was melting away in sweat, and had to change linens and clothes quickly, in order to attend the banquet.

I believe that I escaped Göschen's fate only by the fact that on the following day I avoided *all* participation and remained quietly at home, just as on Tuesday [I avoided] participation in the honorary degree ceremony; finally by the fact that just after it through the solicitude of my good Weber all windows were opened, in order to let in fresh air again, so that when I arrived after 12 o'clock for the session of the Royal Society, I found enjoyable air and could hold my lecture.

Gerling had taken quarters in my home during the week. I saw Humboldt every day at my home, mostly several times every day. He departed last evening. I admired his vigor. He traveled night and day from Berlin here; from here to Hanover he again traveled during the night.

It is now as quiet again in Göttingen as it had been previously noisy. During the jubilee many a goblet of wine was emptied to your health.

At the time Gauss was quite concerned about his youngest son, Wilhelm, who in the fall of 1837 migrated to America with his bride, a niece of Bessel. His oldest son was engaged at this time to Sophie Erythropel of Stade, daughter of a physician, who died in October, 1837. (A sketch of all Gauss' children is given in Appendix D.)

---

[3] Ludolf Dissen (1784–1837), professor of classical philology.

Humboldt on his return to Berlin spent two days in Hanover visiting ministers, ambassadors, and other members of the court. In a letter thanking Gauss for his recent hospitality he told of a meeting with the astronomer Caroline Herschel, who seemed to be especially glad to have some news of Gauss. Humboldt had an hour's audience with King Ernst August, who said that he was pleased with everything he had seen in Göttingen during the jubilee. The King remarked that he had never seen better behaved young people.

Yet the sounds of the jubilee with all its good feeling, pride, and cheer had scarcely died away when a great catastrophe hit the university. It was to affect Gauss deeply and directly the rest of his life. It affected the university adversely for many years to come. It affected many prominent persons over a long period of years. And in the long run little was accomplished by the act which caused the catastrophe.

On September 26, 1833, during a political uprising, King Wilhelm of Hanover (William IV of England) recognized and signed an amended constitution, by which certain rights were secured for the people. In June, 1837, he died, a weak but benevolent ruler who had enjoyed far-reaching sympathies among the populace of both countries. The English parliamentary reform of 1832 corresponded to the introduction of the Hanoverian constitution. William IV was succeeded by Ernst August, Duke of Cumberland (1771–1851), fifth son of George III of England. His first act[4] was to revoke the liberal constitution adopted by his predecessor, and to restore the constitution of 1819. One of the provisions of the new constitution declared that an heir to the throne who suffered from any physical or moral defect was thereby debarred from the throne. The new King's only son was blind, and probably his sole purpose in setting aside the constitution of 1833 was to make his son legal heir to the throne. Friedrich C. Dahlmann (1785–1860), professor of history and political science at Göttingen, made a motion in the academic

---

[4] November 1, 1837.

senate that the university should undertake steps to call the King's attention to the danger of his plan. This occurred just before the jubilee, and found no response. Few professors wanted to intervene in the matter. People in Göttingen thought they saw an evil omen in the fact that at the unveiling of the monument to Wilhelm IV during the jubilee, Ernst August turned his head the other way.

There were also financial reasons why Ernst August nullified the constitution of 1833. It secured the acknowledgment of the provincial diets, and the division of the chief diet or parliament into an upper and lower house; it gave the lower house the exclusive right of initiating legislation and handed over to the ministry the control of the finances. The new king never gave his official approval to the constitution of 1833, as prescribed. New elections were ordered according to the constitution of 1819. The king dismissed his cabinet ministers and immediately renamed them department ministers. His purpose was to release all his subjects from their oaths of loyalty to the constitution of 1833.

Leaders in both church and state, and the people at large, were vehemently opposed to the arbitrary and illegal course taken by the King. Wilhelm IV had granted concessions in answer to the demands emphasized and enforced by riots at Göttingen and elsewhere. Dahlmann had taught that a constitution can be changed only in a constitutional manner. Such a constitution, involving all national, moral, and intellectual forces, formed a symbol for the duty of the individual in the service of the fatherland. An oath of loyalty to the constitution was at least theoretically almost a religious matter. It was a matter of conscience. Such were the feelings of the high-minded men of the time.

The people protested all in vain. The King had a will of iron, and a furious temper withal. Opposition merely inflamed his passions and stiffened his resolution. The imperious old man swore he would leave the people no constitution at all if they refused to accept the one he offered. He proved himself to be bluff, brutal, and self-willed. He did not know the German language. He knew only "subjects," not full-fledged citizens, and hated constitutional government. Ernst August was also rather unfriendly to research and science.

One can imagine the King's indignation when he received a protest signed by seven Göttingen professors. They were Dahlmann, Ewald, Weber, the famous Grimm brothers, Wilhelm Eduard Albrecht (1800–1876), professor of law, and Georg Gottfried Gervinus (1805–1871), professor of history and literature. Dahlmann and Ewald were the prime movers in the protest, which was directed not to the board of curators of the university, but to the royal cabinet. The Seven declared that they felt bound by their oath of loyalty to the constitution of 1833. The King's anger greatly increased when he found out that the document had immediately been made public far beyond the borders of Hanover. Curator von Arnswaldt and Cabinet Councilor Hoppenstedt tried to get the Seven to take back their protest. Opinion in Göttingen was divided.

Prorector Bergmann and the academic senate secured an audience with the King at his castle Rotenkirchen in nearby Solling. Bergmann presented a document in which the dissemination of the protest was called "an unhappy event." The government used this in a release to the press, which attempted to isolate the Seven and make it appear that the whole university was against them. Six other leading Göttingen professors three weeks later released to the press a statement in which they said that they could never criticize the action of their seven colleagues.

On December 12, 1837, the famous Göttingen Seven were fired from their jobs. Jacob Grimm, Dahlmann, and Gervinus were ordered to leave the Kingdom of Hanover within three days, since they had "confessed" having a part in circulating the protest. The others were allowed to stay in Göttingen as long as they "behaved" and kept quiet. The government ignored the declaration of the other six professors.

In order to avoid disturbances, a strong military unit was sent to Göttingen and every livery owner was forbidden to rent coaches or carriages to anyone, to prevent a demonstration by the students who might want to accompany their popular teachers as they left the city. Dahlmann, Gervinus, and Jacob Grimm left the city as though they were traitors. But word got around, and a large number of students preceded the exiles to Witzenhausen and put on there an enthusiastic demonstration.

Ewald, Dahlmann, and Jacob Grimm published works defending themselves. They claimed that their motives were idealistic, involving honor and freedom, matters of conscience, not political agitation. The effects were far-reaching. After that time the professor in Germany was shoved more into the center of public and political life. Great sympathy for the Göttingen Seven was stirred throughout Germany; money was collected for the exiles. The University of Königsberg even gave Albrecht an honorary doctorate.

The events of 1837 threw Ewald off his accustomed way of life, that of a serene scholar, and he never returned to it completely. In fact, in later years he made himself rather ridiculous by his political action, and actually got himself into serious trouble when he opposed Prussia's annexation of Hanover. He believed in a patriarchal rather than a parliamentary form of government.

Many persons in Hanover and Göttingen thought that the goal of the Seven was too high, criticized them severely, and ascribed ulterior motives to them. Included were many Göttingen professors. It was difficult to fill the vacancies in the faculty. Other governments tried to attract the exiles to their university faculties. The Duke of Brunswick wanted to reopen the University of Helmstedt by employing all of the Seven, and there was a rumor that all would go to the University of Marburg. Gradually they found positions: Gervinus went to Heidelberg, Ewald to Tübingen, the Grimms to Berlin, Dahlmann to Bonn, Albrecht and Weber to Leipzig.

Gauss has been criticized for not having signed the protest. Dahlmann had counted on his signature. It is wrong to accuse Göttingen professors who did not sign of manifesting cowardice or weakness of character. Gauss' friends believed that he would leave Göttingen at least as a mild protest. A rumor without actual foundation circulated that he was going to Paris.[5] Why didn't Gauss sign the document? The reasons therefor are several, and his own letters indicate them. In the first place, Gauss was then sixty years of age and had been in Göttingen most of his life. Moreover, his mother, who was ninety-five years old and blind, lived with him. He felt it would be impossible to move her. Secondly, Gauss had a deep interest in politics and kept himself well informed in that field, but he was never active in any way politically. He abhorred anything that smacked of political radicalism and was thoroughly conservative. He felt that the step which his colleagues took was utterly useless and would hurt the university. In the last point he was correct. It required more than fifty years for Göttingen to recover from the blow, from a loss of prestige.

Gauss was especially hard hit by the fact that Ewald and Weber were among the signers. Both of them, soon after the affair, went to London for a visit. Ewald spent six months in England working mainly on Sanscrit and the works of Jewish linguists written in Arabic. He spent his time in the Bodleian Library, Oxford, and in the British Museum, London. Weber left Göttingen in March, 1838, on a tour of scientific study. He went via Leipzig to Berlin, where he was joined by Poggendorff; he did not return to Göttingen until August, 1838. Gauss did not know in advance that Weber was going to London.

Gauss exerted special effort on behalf of Weber during the period of Weber's absence abroad, requesting that he be restored. He turned to Count Friedrich von Laffert at Ilfeld, who was the government official in charge of the affairs of the University of Göttingen. People around the King regarded Weber as the least dangerous of the Seven, in fact they thought he had been "seduced" to sign. But Gauss' effort was in vain.

His next step was to attempt direct action. He knew that

---

[5] In a letter to Olbers he wrote that Paris was the *last* place to which he would go.

his friend Alexander von Humboldt would come in contact with the King during the latter's visit to Berlin the second half of May, 1838. Therefore, he asked Humboldt to use his great influence by pleading the cause of Weber before the King. Gauss wrote Humboldt that the continuance of his whole scientific activity depended on Weber's staying in Göttingen, and one should remember that he was not a person given to exaggeration. It is significant that Gauss did not take any steps in behalf of Ewald, merely because he was his son-in-law, feeling that it would be "improper." He hesitated to ask Weber to sign a statement that he had acted too hurriedly in signing the protest. In fact, he said that under the same circumstances he would not recant in this manner. Gauss felt sure it would place him in a bad light before colleagues and students if he tried to get Weber to recant in order to be restored. His last hope was Humboldt.

Meanwhile Gauss was further distressed, for he was now separated from his daughter Minna. In May, 1838, Ewald was appointed professor of Oriental philology at the University of Tübingen, where he labored with great intensity for ten years.

Humboldt did not have the opportunity[6] to discuss Weber's case with the King; however, he did take it up unsuccessfully with two members of the court and found that the King was willing to restore no member of Seven, except under conditions which were too humiliating to be accepted. When the Board of Curators requested Gauss to make proposals for filling Weber's professorship, he used it as a last attempt to get his friend restored. Weber's professorship was given to Benedikt Listing (1808–1882), a pupil of Gauss.

The Grimm brothers were on the polite terms of faculty colleagues with Gauss. They were suspicious of his efforts in behalf of Weber and thought he was concerned only about the fate of Ewald and Weber. They thought Weber was too much under the influence of Gauss. Letters of Gauss show that he was concerned

---

[6] At a banquet the King told Humboldt: "For my money I can have as many ballet dancers, whores, and professors as I want."

about the Grimm brothers, although naturally not so active in their behalf as in the case of Weber. At the time of the trouble there were rumors that letters were being opened, and Gauss was more careful about what he wrote than he would otherwise have been.

Gauss would have attempted to persuade Ewald and Weber not to sign the protest, but he did not know details of it until six or seven days later, after it had been made public. He felt that Weber, Ewald, and Albrecht signed the protest as a matter of conscience and that they regarded it as a private communication to the Board of Curators. On January 7, 1838, Gauss summarized his feelings about the entire matter in a letter to Schumacher in these words: "Just in the present affliction of our poor Georgia Augusta, I cannot free myself of a certain piety, and just now it would be harder for me to leave Göttingen than at any other time, at least as long as I don't have to give up all hope of saving something that is personally dear to me in it."

In a letter to his daughter Minna he stated that he wished to be ready to do what duty and honor demanded. He felt that his own position was insecure, but such was never the case.

In the summer of 1838 Gauss began to fear that he was losing his hearing. He became temporarily deaf, first in one ear, then in the other, sometimes in both. Then after several days he would suddenly have perfect hearing. Due to the suddenness, Gauss thought it was purely a local condition. He had little faith in doctors and was not convinced by the statement of Karl Friedrich Heinrich Marx (1796–1877), professor of medicine in Göttingen, that it was a general affection of the nervous system. Gauss felt sure it was merely a stoppage of the Eustachian tube and read several medical works on the subject. He thought of journeying to Dr. Kramer in Berlin for an operation, since he did not have sufficient confidence in any doctor in Göttingen, but finally decided against that course. The thought of "having a silver tube driven through the nose and the interior of the head" was abhorrent to him. Finally he decided it was trouble in the outer ear and began to use his own cure of almond oil. Gauss wrote to his friend Olbers that many physicians were inclined to seek unusual, more complicated, and rarer causes and to overlook the obvious ones. When the symptoms began, he took a small spoon and explored for earwax; at first the ear seemed dry and he found no wax, but later brought out hard wax. Olbers recommended spraying the ear with tepid, soapy water, keeping cantharides in water for a while behind the ear, and taking foot baths strengthened with ground mustard. Some months later the trouble disappeared, and Gauss attributed the cure to almond oil; he was sure the cause had been purely mechanical. The deafness never returned.

In March, 1839, Gauss sent to Ewald a copy of the volume on the university jubilee of 1837. The faculty had nothing to do with the editing of it, but he was pleased to see that his own lecture was printed exactly as he had turned it in. He was amused by the fact that the long sermon on the occasion had been greatly abbreviated. At the time, Therese had been in a grocery store and got some sacks made of paper on which the original sermon was printed. Gauss sent it to Ewald as a curiosity.

In November, 1838, the Royal Society of London conferred on Gauss the Copley Medal. At that time it was regarded as the highest honor of the society. He was pleased especially because he thought it would mean greater participation in magnetic research in England. The actual value of the metal in the medal was six louis d'or; Gauss wrote in a letter to his daughter Minna that if the value had been more he would have sold it and divided the proceeds between Joseph and Wilhelm.[7]

In the spring of 1839 Therese's health was poor, Gauss' aged mother was very feeble, and he himself suffered occasionally from catarrh and headaches, frequently from insomnia. When hot weather arrived, it always affected him noticeably. He could stand the winter without serious difficulty.

During his last illness in 1808, Gauss' father wrote his will, in which he gave his wife Dorothea a life interest in all his prop-

---

[7] After his death it was given to Joseph.

erty. If his two sons insisted on immediate division, she was to receive one third. Dorothea continued to live in Brunswick, but she was lonely and longed to be with her son. Yet she could not make up her mind to give up her home. This loneliness was accentuated by the fact that, although she could read printed matter, she could not read handwriting. All communication with the son of whom she was so proud was through the aid of others. Finally in 1817, at the age of seventy-four, she yielded to Gauss' pleading and moved with him into the new apartment at the observatory. The home in Brunswick was sold, and Gauss' brother received his share of the estate, which he would otherwise have gotten at the death of Frau Dorothea.

In her new home the aged mother of the mathematician lived on for many years, lovingly and tenderly cared for by Gauss, his daughters, and his wife. The old woman could never be persuaded to dispense with her peasant clothing or to take her meals at the table with the others. She never accustomed herself fully to her new home. Her room was a small one on the first floor, overlooking the charming garden of the observatory, and she could easily step out on the terrace. There was a large acacia tree in front of her window. She moved about unhampered in the family and among the intimate friends of her son.

To her grandchildren Dorothea dictated letters to her stepson, in which there was news about herself and the family in Göttingen, as well as questions about relatives and friends in Brunswick and her native village, Velpke. Whenever she heard that someone from Brunswick was in Göttingen, she looked the person up, asked questions, and gave some errand to be performed. One of her favorite occupations was to buy gingerbread at the Göttingen markets and make friends with the women she met there.

Dorothea's joy was great in 1830 when she heard that Gebhart Gauss (1811–1879), nephew of Carl Friedrich, was coming through Göttingen as a journeyman plumber. But Gauss' wife was on the sickbed from which she never got up, and Gebhart could not stay at the observatory. Dorothea hoped to see him on his return.

At the age of eighty-seven she thought of taking a trip to Brunswick; two years later she lost her eyesight and had to drop the plan. Her physical strength was remarkable up until the last months of her life, and her memory remained perfect. After the blindness she had to be led by someone when she visited Minna Ewald or friends in the city. In the last months she did not leave her room, and finally died on April 18, 1839, aged ninety-seven. Gauss was grief-stricken. He realized that the life of Dorothea Gauss had been a hard one. In one letter he spoke of it as "full of thorns." As early as 1810 he wrote that her marriage had been unhappy because she and his father were not compatible. The elder Gauss was often harsh and domineering in the home, although behind it all lay a good heart. In the same letter (1810) Gauss wrote that his mother had "many weaknesses," but that she was worthy of love.

Gauss was worried for some months about the safety of Wilhelm, his youngest son, and Wilhelm's bride. They went to Bremen and had to wait some time for favorable winds, but finally on October 29, 1837, sailed on the *Alexander,* whose captain, Mertens, became very fond of Wilhelm. The ship ran into a bad storm, but eventually reached New Orleans. Several months passed before Gauss got word of their safe landing. Meanwhile, he heard of the yellow fever epidemic in New Orleans at the time of their landing, and was quite concerned until much later when he heard of their safe arrival in Missouri. They had gone up the Mississippi on a river boat.

In the fall of 1838, Gauss was extremely happy to hear of the birth of his first grandchild on May 30, at St. Charles, Missouri. In the excitement of announcing this event to Olbers, he forgot to tell whether it was a boy or a girl. Olbers promptly inquired, and Gauss wrote him that it was a boy, who promised to become "a husky American." The baby was given Gauss' name in the Anglicized form, Charles Frederick. Later Gauss was surprised when he heard that his grandchildren were allowed to ride ponies, alone, on the open prairie. How impossible in Europe! When Charles Frederick was fourteen Gauss asked for a letter from him, even though it had to be written in English. Charles Frederick lived on until 1913 in St. Louis, Missouri, where he became a millionaire manufacturer of hats. In his last years Gauss had the great pleasure of hearing of the birth of eight grandsons and six granddaughters; four grandsons were born after his death, making a total of eighteen grandchildren.

The last letter from Olbers to Gauss was dated May 30, 1839; at the close of it Gauss wrote: died March 2, 1840. This was a heavy blow, for he had lost a most intimate friend, one who had stood by him since his youth. Olbers had a genial character, was financially well off, and was cared for by his son during the years as a widower. Yet he began to complain of sickness due to age, and of being lonely.

On the sunny side of this period must be chronicled the marriage of Joseph, Gauss' oldest son, on March 18, 1840, to Sophie Erythropel, daughter of a physician in Stade. He approved his son's choice and was very fond of his daughter-in-law. They frequently visited Gauss and Therese. Gauss began to fear that he would never see a grandchild, but finally a son, Carl August Adolph (1849–1927), was born to the couple. One Christmas, when Carl was a small boy about three years of age, they presented to Gauss an oil portrait of him. Carl lived most of his life in Hamlin, Germany, and was the only grandchild Gauss ever saw, since the others were in America. Of course, he had excellent daguerreotypes of the others. Carl remembered how his grandfather tried to show him a star through the great telescope; how he stood full of expectation near the eyepiece, while his grandfather, wearing a velvet cap, was turning the crank which moved the shutter on the dome of the observatory. Another time the child was playing in the garden of the observatory when his grandfather met him and asked: "What do you expect to make of yourself?" Whereupon young Carl replied: "Well, what do you expect to make of *yourself*?" Then the old man patted the child's shoulder and said smilingly: "My boy, I am already somebody."

Ewald and Minna were never entirely happy in Tübingen, and never felt quite at home in the South German atmosphere.

They shipped their furniture and belongings on leaving Göttingen, but there seems to have been a housing shortage in Tübingen at the time, and they had great difficulty in finding a home. The real cause of their unhappiness in Tübingen was Minna's poor health, which actually must have been tuberculosis. Finally she died at six o'clock on the evening of August 12, 1840, and was buried in Tübingen. On the morning after her death Ewald wrote a letter to Gauss conveying the sad news. This was probably the hardest blow that ever afflicted Gauss; there is some reason to believe that Minna was his favorite child. On August 22, 1840, Gauss replied to Ewald's announcement of Minna's death:

DEAR EWALD,
Several times since the receipt of your two letters, which arrived at the same time, I took up my pen, to mix my tears with yours, but my strength failed. Even now I cannot realize that my darling angelic child is lost to us on this earth. It was always my dearest, most consoling hope to be reunited with her here, and to see my last years thereby cheered. Now it is gone, this hope! God give us strength to bear the heavy grief. In need of consolation myself, dear Ewald, I can express to you no other [consolation] except that the splendid one is elevated above her earthly sufferings and has gone to the better home. You knew how to appreciate her value. The earth rarely sees such absolutely pure, noble creatures. She was the image of her mother.

The second paragraph of the letter expresses concern about Therese's health and voices the hope that heaven may strengthen her, then continues:

May heaven also strengthen you, dear Ewald, and may you give me from time to time report of your health. If we may measure the happiness of immortals by human standards, then the happiness of the deceased, who found her happiness in this world only in the happiness of those she loved, will be incomplete as long as the latter are bowed down by grief.
Always with hearty friendship,
GAUSS

Having been commissioned by the Czar of Russia to paint an oil portrait of Gauss for the Pulkova observatory, Christian

Albrecht Jensen (1792–1870) of Copenhagen arrived in Göttingen during the summer of 1840. Gauss himself and others, except Schumacher, were highly pleased with it. Three copies of it, one belonging to Sartorius, were kept in Göttingen. A later copy by Biermiller hangs in the observatory today, and the present biographer has an exact copy by J. H. Landry. It is the best-known portrait of Gauss, and has been frequently reproduced. Arrangements with the Czar were probably made by Struve. Gauss had these words of Edmund in Shakespeare's *King Lear* placed under it as his motto: "Thou, Nature, art my goddess; to thy laws my services are bound."[8]

When Jensen finished the portrait, Therese accompanied him on August 13 on the trip home as far as Hamburg, to visit her brother Joseph and his bride, who were living in Stade at that time.

---

[8] Act I, Scene II. The word "law" was altered to "laws."

CHAPTER SEVENTEEN

Milestones on the Highways and Byways

Writers on Gauss have frequently commented on the fact that he published relatively little in comparison with what he had discovered. The first reason for this phenomenon lay in the fact that the magnitude of the mathematical genius revealed in Gauss was rooted in the union of creative and critical power. This characteristic can be recognized in his doctoral dissertation as well as in his *Disquisitiones arithmeticae*. In his later publications criticism of the accomplishments of others recedes, but Gaussian rigor remains as a distinguishing characteristic.

This rigor is recognized externally in the form of presentation. Gauss always strove to give his research the form of completed works of art, and he never published a work until he had attained the desired form. He used to say that after a structure was completed one should no longer be able to see the scaffolding. This principle was exemplified in the letter seal which he used, a tree with very little fruit and the inscription: *Pauca sed matura*. In letters to Schumacher, Bessel, and Encke, Gauss explained why he would not and could not deviate from this mode of presentation.

When he again took up his theoretical work in the winter of 1825–1826, after completion of geodetic measurements in the field, he complained thus to Schumacher in a letter dated November 21, 1825: "The desire which I have always had in my works, to give them such a completion *ut nihil amplius desiderari possit*, makes them extraordinarily difficult for me."

Schumacher answered him on December 2, 1825:

In reference to your works and the principle *ut nihil amplius desiderari possit*, I would almost like to wish, and to wish for the good of science, that you did not hold to it so strictly. We would then have more of the infinite richness of your ideas than now, and to me the subject matter seems more important than the most complete form of which the matter is capable. But I write my opinion with trepidation, since you have certainly considered for a long time the pro and con.

On February 12, 1826, Gauss further elaborated his viewpoint:

I was somewhat astonished at your utterance, as though my mistake consists of neglecting the subject matter too much in favor of completed form. During my whole scientific life I have always had the feeling of exactly the opposite, that is, I feel that the form could have been more completed, and that bits of carelessness have remained behind in it. For you do not wish to interpret it, as though I would accomplish *more for science,* if I were satisfied with furnishing individual building stones, tiles, and so forth, instead of a structure, be it a temple or a hut, since the building to a certain extent is only the form of the bricks. But I do not like to erect a building in which main parts are missing, even if I devote little to the outer ornaments. In no case, however, if you were otherwise right in your objection, does it fit my complaints about the present works, where it is not a matter of what I call subject matter; and likewise I can definitely assure you, that, though I also like to give a pleasing form, *this* claims comparatively little time and strength or has taken little in earlier works.

When Gauss sent Schumacher a little paper on the heliotrope for the *Astronomische Nachrichten,* he added (November 28, 1826): "This time I have certainly not deserved the criticism, as though I had conceded *too much* to form at the cost of subject matter, but rather the opposite.

Schumacher now felt called upon to explain his opinion fully (December 2, 1826): "I believed that another can do this filing just as well, and in that I can have made a mistake; where I have not made a mistake is the claim that you cannot turn over the inventing to another. Every year of your life increases the hints of new ideas understandable only to you. Shall all this be lost?"

Gauss was cold to all such pleas. On August 18, 1832, he wrote to his pupil and friend Encke: "I know that some of my friends wish that I might work less in this spirit, but that will never happen; I can have no real joy in the incomplete, and a work in which I have no joy is only torture to me. May each one work in *that* spirit which promises him most."

On January 15, 1827, he wrote Schumacher that he was far advanced in working out his memoir on curved surfaces:

I find many difficulties in it, but that which one could rightfully call *filing* or *form* is in no wise what delays considerably (if I except the inflexibility of the Latin language), rather it is the inner concatenation of truths in their coherence, and such a work is not successful until the reader no longer recognizes the great effort which occurred in the execution. Therefore I cannot deny that I have no really clear idea of how I could carry out my work of such kind differently than I am accustomed, without, as I have already expressed myself, furnishing building stones instead of a building. Sometimes I have tried merely to bring to the public hints about this or that subject; either they are heeded by nobody or, as, for example, in some utterances in a book review, *G. G. Anz.*, 1816, p. 619,[1] they were afterward besmirched with mud. Therefore, in so far as the discussion is about important subjects, something essentially complete or nothing at all.

Such hints as those mentioned above are numerous in the works of Gauss' youth, and are not entirely missing in his later works. The letter from Gauss to Encke on August 18, 1832, already quoted, shows how careful he was with his hints:

It has always been my conscientiously followed principle not to make such hints, which attentive readers find in every one of my writings in large quantity, until I have mastered the subject for myself (for example, see my *Disquis. Arithmet.*, page 593, art. 335).

It was a noteworthy accident that Gauss, soon after he had complained to Schumacher about the failures of his hints, received through his friend on July 24 and August 14, 1827, the two letters of Jacobi with which his research on elliptic functions began, and that not long afterward he got acquainted with Abel's researches which anticipated "probably a third" of his

---

[1] *Göttingische gelehrte Anzeigen.*

own. The influence which the famous passage in Article 335 of the *Disquisitiones arithmeticae* had on Abel and Jacobi is well known. Here Gauss sowed the seed which bore much fruit, and other hints did not fall on stony soil.

Almost a quarter of a century later the same point of debate between the two friends turned up again when Schumacher printed in his *Astronomische Nachrichten* a work of Jacobi on Kepler's equation. On December 5, 1849, Schumacher replied to Gauss: "If I did not know how much time the final polishing of your work costs you, I would ask for your paper."

Gauss then replied that he was inclined to use a part of his leisure time for working out a paper on the subject; but it would demand considerable time to execute the whole theory in a form satisfactory to himself. He continued (February 5, 1850):

You are entirely in error if you believe I mean by that only the last polishing in relation to language and elegance of presentation. These cost comparatively only an unimportant expenditure of time; what I mean is the *inner* completeness. In many of my works are such points of incidence which have cost me years of meditation, and in whose later presentation concentrated in a small space nobody noted the difficulty which must first be conquered.

Similar utterances are in the letter to Bessel dated February 28, 1839; Bessel supported in a letter to Gauss dated June 28, 1839, the same viewpoint expressed by Schumacher.

The completed presentation, in which Newton and Archimedes were Gauss' models, was to be only the external sign of inner completeness. Here the word of Gaussian rigor gains its true meaning. In contrast to the habits of the eighteenth century, Gauss adopted what he called the *rigor apud veteres consuetus* or *rigor antiquus*. On September 1, 1850, he thus expressed his conviction to Schumacher:

It is the character of mathematics of modern times (in contrast to antiquity) that through our language of signs and nomenclature we possess a lever whereby the most complicated arguments are reduced to a certain mechanism. Science has thereby gained infinitely, but in beauty and solidity, as the business is usually carried on, has lost so much. How often that lever is applied only mechanically, although the authorization for it in most cases implied certain silent hypotheses. I demand that in all use of calculation, in all uses of concepts, one is to remain always conscious of the original conditions, and never regard as property *all* products of the mechanism beyond the clear authorization. The usual course is, however, that one claims for analysis the character of generality and expects of the other person who does not recognize the results produced as proved that he should prove the opposite. But one must expect this of him who for his part maintains that a result is wrong, but not of him who recognizes a result as not proved, a result which rests on a mechanism whose original, essential conditions are not pertinent in the case under consideration.

A second reason for the discrepancy between Gauss' richness of new thoughts and his relatively small number of publications during his lifetime lay in inhibitions which, owing to his mode of working, stood in the way of preparing to go to press.

In the above-mentioned letter of February 28, 1839, to Bessel Gauss stressed with unusual vigor of tone that he needed for working out his research "time, more time, much more time than you can imagine. And my time is often limited, very limited." Such complaints about lack of time for theoretical research are continually repeated in his letters. Probably the happiest years of his life were those from 1799 to 1807 when he was living as a private scholar on the bounty of his duke. In his old age he thought of these years with emotion and gratitude. Thus on February 15, 1845, he wrote to Encke about young Gotthold Eisenstein, who was pursuing mathematical research supported by the King of Prussia:

He is still living in the happy time when he can yield completely to his talent, without finding it necessary to be disturbed by something foreign. I am vividly reminded of the—long past—years when I lived in similar circumstances. On the other hand just the purely mathematical speculations demand an unencroached upon and undivided time.

The duties of a professorship weighed heavily on Gauss, especially his position as director of the Göttingen observatory. This is reflected by his words in a letter of June 28, 1820, to Bessel: "As much as I love astronomy, I feel nevertheless the burdens of the life of a practical astronomer, often too much, but most painfully in the fact that I can hardly come to any coherent major theoretical work."

In 1821 were added the geodetic measurements; the tedious and time-consuming field work ended in 1825, but for twenty more years much of Gauss' time was occupied by purely mechanical calculating which should have been done by a person of lesser creative ability.

Then came the duty of holding lectures for his students. His real desire was to be an astronomer at some observatory and to give much of his time to pure research. The optimum arrangement, according to Gauss, would be for a professor to lecture to his students on the research in which he was engaged at the moment. He disliked administrative duties, red tape, and the "business" of being a professor.

Gauss performed his duties with his usual conscientiousness, but in a letter to Bessel on January 27, 1816, he called the giving of courses "a very burdensome, ungrateful business." His complaints became very bitter when the burden of geodetic work was added. In 1824 when he was considering a call to Berlin, he cried out in a letter to Bessel (March 14, 1824):

> I am so far removed here from being master of my time. I must divide it between giving courses (against which I have always had an antipathy, which, although it did not originate in, is increased by the ever-present feeling that I am throwing away my time) and practical astronomical work. What remains to me for such works on which I could place a higher value, except fleeting leisure hours? A character different from mine, less sensitive to unpleasant impressions, or I myself, if many things were different from what they are, would perhaps gain more from such leisure hours than I generally can.

Such passages in letters to intimate friends could be multiplied considerably. One more, found in a letter to Olbers on February 19, 1826, will suffice:

> *Independence,* that is the great watchword for deep intellectual work. But when I have my head full of mental images hovering in the air, when the hour approaches that I must teach courses, then I cannot

describe to you how exhausting for me the digression, the animation of heterogeneous ideas, is, and how hard for me things often become, which under other circumstances I would consider a miserable ABC work. Meanwhile, dear Olbers, I do not want to tire you out with complaints about things which are not to be changed; my whole position in life would have to be a different one if such adversities did not occur rather often.

From the above statements it is perfectly clear that the progress of Gauss' theoretical work was hindered not merely by lack of leisure time, but also by his constitution of mind. Of course, this is to be tied up with the tragedies which struck Gauss several times in his domestic life.

The sensitiveness to unpleasant impressions, of which he spoke in his letter to Bessel on March 14, 1824, must have contributed to the fact that in his publications he avoided touching on subjects which could have led to debate. In his doctoral dissertation he was cautious about the use of imaginaries[2] and their geometric interpretation, which he possessed before 1799 but suppressed and did not publish until 1831. The same delay occurred with respect to his work in non-Euclidean geometry.

This shyness was strengthened by his unpleasant experiences in 1816 when he reviewed the parallel theories of Schwab and Metternich and hinted at the impossibility of proving Euclid's eleventh axiom. He had such attacks in mind when he wrote to Gerling on August 25, 1818: "I am glad that you have the courage to express yourself (in your textbook), as though you recognized the possibility that our theory of parallels, consequently all our geometry, might be false. But the wasps, whose nest you stir up, will fly around your head."

In addition there was the low opinion which Gauss had of the great majority of mathematicians. As early as December 16, 1799, he wrote to Wolfgang Bolyai, who had sent him an attempt to prove the axiom of parallels:

---

[2] He stated his full views on the imaginary in a letter to Bessel dated December 18, 1811.

Publish your work soon; certainly you will not reap for it the thanks of the general public (to which belongs many a one who is considered an able mathematician), for I am convinced more and more that the number of true geometers is extremely small and most of them can neither judge the difficulties in such works nor even understand them—but certainly the thanks of all those whose judgment can be really valuable to you alone.

When Bolyai sent Gauss in 1832 his son Johann's work on non-Euclidean geometry, in which the problem of parallels was solved, he got this reply from Gauss (March 6, 1832): "Most people do not have the right sense of what is involved, and I have found only few people who received with special interest that which I communicated to them. In order to do that one must have felt quite vividly what is really missing, and most people are in the dark about that."

In a letter to Gerling on June 25, 1815, Gauss expressed himself even more severely: "It seems to me that it is important in more than one respect to keep awake in pupils the sense of rigor, since most people are too inclined to pass over to a lax observance. Even our greatest mathematicians have mostly somewhat dull feelers in this respect."

On September 29, 1837, Gauss expressed to his young friend Möbius his low opinion of the intelligence of the general public: "One must always consider that, where the readers for whom one is writing take no offense, it would perhaps not be beneficial to penetrate more deeply than is profitable for them."

In 1831 even Gauss' pupil and intimate friend Schumacher believed he had proved the axiom of parallels, and Gauss had great difficulty in convincing him of the weakness of his process. Bessel could not find the flaw in Schumacher's work, but merely bowed to the authority of Gauss. Schumacher took no offense and recognized the importance of Gauss' motto and method of working when he wrote him on May 24, 1839: "The *pauca sed matura* are here,[3] golden apples of the Hesperides, ripened under

---

[3] Refers to Gauss' general theory of terrestrial magnetism.

the sun of genius, one of which has more value than a shipload of Borstorf [apples]."

Bessel urged Gauss to hasten the publication of his results in these words (January 4, 1839): "Mr. von Boguslawski[4] has told me that you have reached a point satisfactory to yourself in your research on terrestrial magnetism. I know the meaning of this word and therefore wish you good luck for the most complete success, also cherish the hope that you will no longer leave it in your exclusive possession."

Concerning his writing on terrestrial magnetism and the magnetometer, Gauss wrote to Schumacher on July 9, 1835:

As you know, I work slowly, most slowly in such (semipopular) subjects; I am almost ashamed to say how long I have been writing on these few pages.

I would think that now a nonmathematical scholar (for they are often difficult as to ideas), no less than, for example, an intelligent carpenter who has had only a little schooling, or to whom one has explained abbreviations like square and such, would have to be able to comprehend the subject.

In 1825 Olbers wrote to Bessel and praised him for prompt publication of his results. At the same time he indulged in some criticism of Gauss, who, he thought, always wanted to pluck the finest fruits to which his path had led him, before showing them to others. Olbers considered this a weakness of character, since with his wealth of ideas Gauss had so much to give away.

Bessel could not fully agree with Olbers on this point, for he wrote to Gauss on May 28, 1837: "You have never recognized the duty of advancing the *present* knowledge of subjects through prompt publication of one part of your researches adapted to the whole; you are living for posterity. Where would the mathematical sciences be now, not only in your dwelling, but in all Europe, if you had expressed everything that you could express!"

Gauss always wished his presentation to be as compact as possible and found that such writing demanded much more time

---

[4] P. H. L. Boguslawski (1789–1851), professor of astronomy and director of the observatory at Breslau.

than the opposite type of writing. Schumacher refused to give up in his attempt to change Gauss' viewpoint and in 1833 proposed a new seal with the words *multa nec immatura*. In 1836 he desired that Gauss use the words *nec pluribus impar*, but admitted he was afraid his friend would not use it.

A generation later, Kronecker, who was a leading mathematician of his day, had this to say about Gauss' style:

> The manner of presentation in the *Disquisitiones* [*arithmeticae*], as in Gaussian works generally, is the Euclidean. He sets up theorems and proves them, in which he industriously erases every trace of the train of thought which led him to his results. In this dogmatic form is certainly to be sought the reason for the fact that his work was not understood for so long and that the efforts and researches of Lejeune Dirichlet were needed to bring it to full effect and appreciation among posterity.[5]

Gauss was very slow in passing judgment on others. He expressed this viewpoint on February 12, 1837, in a letter to Schumacher: "Only when I have time to process something for myself and to transform *in succum et sanguinem*, can I permit a judgment."

He felt that he would have to carry much of his work to the grave, because it would be incomplete and he was unwilling to turn over such results to another. And that is exactly what happened. The editing of his *Collected Works* extended from 1863 to 1934. He insisted on purity in determining mathematical concepts and missed it among most of his contemporaries. Gauss admitted in a letter to Bessel in 1826 that he had sometimes spent months of effort in vain on a problem.

Gauss never placed a high value on his method of least squares because he did not appraise things from the utilitarian viewpoint. He felt that people who had to do a great deal of calculating would have hit upon the same artifice, and made a bet that Tobias Mayer must have used the same method. Later, in looking through the papers of Tobias Mayer, he realized he would have lost that bet.

---

[5] L. Kronecker, *Vorlesungen*, II, ed. Hensel (1901), p. 42.

He was pleased by Abel's work on transcendental functions because he felt that it relieved him of about one-third of the labor he had planned on that subject. Gauss praised the elegance and conciseness of Abel's work, and stated that many of his formulas were like copies of his own. He recognized Abel's independence by declaring that he had never communicated anything on the subject to anyone.

When Arago's works were translated into German a change was made where he discussed the number of men who were qualified to pass final judgment on exact researches. The original stated "eight or ten," but the translation by Hankel gave the figure at "seven or eight." Gauss was quite pleased and attributed the change to Alexander von Humboldt. Actually it was done on the advice of Dirichlet.

Gauss gave an interesting sidelight on his methods of writing in a letter to Schumacher on October 12, 1840: "What is to be printed is usually written more than once. On the other hand I very rarely make a rough draft of letters, I believe that in *my whole life* I have made in advance a rough draft of scarcely a dozen letters."

Archimedes was the man of antiquity whom Gauss esteemed most highly. He imagined him as a worthy old man of noble appearance. The only thing for which he could not pardon him was the fact that in his "sand calculating" he did not discover the arithmetic of position or decimal system of numbers. Gauss said, "How could he overlook that, and at what a pinnacle science would now be if Archimedes had made that discovery."

On February 12, 1841, Gauss wrote to Schumacher the secret of how his memory worked: "My memory has the weakness (and has always had it) that everything read, unless at the moment of reading it is connected with something directly interesting, soon disappears without trace."

Not long after this, Schumacher tried to pry out of Gauss the secret of his unique power in numerical calculating. After three weeks he got an answer from Gauss (January 6, 1842):

My occupation with higher arithmetic for almost fifty years now participates in the aptitude in numerical calculating ascribed to me, to the extent that of themselves many types of number relations have involuntarily stuck in my memory, relations which often occur in calculating. For example, such products as 13 x 29 = 377, 19 x 53 = 1,007 and the like, I look at directly, without taking thought; and in others which can be deduced immediately from such ones there is so little thought that I am scarcely conscious of it. Moreover, I have never purposely cultivated in any way skill in calculating, otherwise it could have been carried much further without doubt; I place no value on it, except in so far as it is the *means* but not the purpose.

One day a small circle of friends sat in Gauss' home discussing the intelligence of animals. Georg Heinrich Bode (1802–1846), young professor of classical philology, had many wonderful things to tell on this subject, in particular about his travels in America and about the wisdom of a parrot that he had brought along from Northampton, Massachusetts. It was so wise that Bode had named it "Socrates." At first Gauss listened silently to Bode's praise. But when the young scholar maintained that his "Socrates" could even answer questions asked in Greek, Gauss remarked smilingly that he had not taught "Hansi," his chaffinch from the Harz mountains, Greek, but that he had succeeded in teaching him several bits of his native Brunswick dialect, which the little animal knew how to use cleverly. A short time previously he had held out to him a cigar and his pipe and asked: "What shall I smoke, Hansi?" Whereupon the sly little bird after a short meditation promptly answered with *"Piep."*[6]

Georg Julius Ribbentrop (1798–1874), who first served in the University of Göttingen library and later for many years as professor of law, was a confirmed bachelor and an eccentric; he was widely known as the perfect example of the absent-minded professor. One evening he was invited to Gauss' home for dinner. After the meal a severe electrical storm came up, and it turned into a long-lasting cloudburst. In those days the observatory was some distance beyond the gates of the city. Therese Gauss, at that

---

[6] Low German for "pipe."

time her father's hostess, invited the guest to stay overnight on account of the long walk home. Ribbentrop agreed, but soon disappeared before the others missed him. After some time the doorbell rang. There stood Ribbentrop in front of his astonished hosts, drenched. He had hurried home to get his necessary sleeping articles for the overnight invitation.

Gauss had promised to show Ribbentrop the eclipse of the moon visible on November 24, 1836. On the appointed evening the rain was pouring down, so that Gauss assumed his astronomical guest would not appear. He was so much the more surprised when the latter, drenched, suddenly stood before him.

"But, my dear colleague," Gauss greeted him, "in this weather our planned observations of the sky might come to naught."[7]

"Not at all," opined Ribbentrop remonstrating, as he triumphantly waved his big umbrella. "My landlady has seen to it that this time I did not forget my umbrella."

Gauss' factotum, the optician J. H. Teipel, was an industrious and able artisan, who along with his upright, honest character possessed fine, native wit. He was proud of the fact that after he had learned his trade assisting Gauss at the Göttingen observatory, he was allowed to call himself "university optician." He had to endure many malicious references to this title, but in his dry, clever way always had a ready comeback. One day a wag asked Teipel the difference between *opticus* (optician) and *optimist*. He got this answer: "Approximately the same as between *Gustav* and *Gasthof*."

Teipel was not only the artisan assistant of Gauss; he also undertook to show visitors through the observatory, who as laymen wanted "to look at the stars" and were satisfied with a popular explanation of the starry heavens. The following comic scene occurred once: In observing planets Teipel was once asked by a lady about the distance of Venus from the earth, whereupon he explained: "That I cannot tell you, Madame. For numbers in

---

[7] German: *ins Wasser fallen.*

the world of stars Hofrat Gauss is there. I am here merely calling attention to the beauties in the sky."

Eugene Gauss used to tell that his father first thought of the heliotrope while walking with him and noticing the light of the setting sun reflected from a window of a distant house. Gauss' own version was slightly different. The heliotrope was his favorite invention. He emphasized that he was led to it not by an accident, but by mature meditation. It was true that from the steeple of St. Michael's in Lüneburg he had seen the windowpane of a Hamburg steeple flashing in the sun, an incident which merely strengthened his conviction of its practicality. Gauss liked to tell how, when the heliotrope was first used, a crowd of curious spectators gathered and let out a shout of joy at the first appearance of the distant light.

Of the works Gauss published, none refers directly to the *geometria situs,* although this subject occupied his attention most of his life. In the last years of his life, about 1847 to 1855, he gave special attention to it and hoped for great things in this field, because he regarded it as almost untouched. Fifty years earlier he had written to Olbers that he was interested in Carnot's *Géométrie de position,* which was then about to appear, and referred to the work of Euler and Vandermonde. In 1810 his friend Schumacher translated Carnot's work.

It should be remembered that Carnot meant by the term "geometry of position" something different from *geometria situs;* he had in mind the application of negative numbers to geometry. Later, projective geometry was frequently designated as geometry of position as opposed to metric geometry. But these are not what Gauss had in mind by the term *geometria situs.* He referred to that branch of mathematics which today is called topology.

On October 30, 1825, Gauss reported to his friend Schumacher that he had made great progress in his work on the general theory of curved surfaces, and then continued: "One must pursue the tree to all its root fibers, and much of it costs me weeks of strenuous meditation. Much of it even belongs in the *geometria situs,* an almost unworked field."

In a notebook under date of January 22, 1833, Gauss wrote: "Of the *geometria situs*, which Leibniz foresaw, and which two geometers[8] have been permitted to glimpse weakly, after a century and a half we still know and have not much more than nothing. A principal problem of the border field of *geometria situs* and the *geometria magnitudinis* will be to count the entwinings of two closed or infinite lines."

In 1834 the Göttingen *Gelehrte Anzeigen* contained a review by Moritz A. Stern (1807–1894), who was for many years professor of mathematics in Göttingen, of a book by the Dutch mathematician Uylenbrock dealing with the work of Huygens and other important seventeenth-century mathematicians. In the review, Stern mentioned having heard Gauss discuss his research in topology. It is regrettable that Gauss never found time to publish something on the subject. In a letter to his pupil Möbius (1790–1868) on August 13, 1849, Gauss thanked him for a copy of his paper on the forms of curves of the third order and urged him to investigate in a similar manner the form relationships of algebraic curves which occur in his own dissertation (1799).

One of the oldest notes by Gauss to be found among his papers is a sheet of paper with the date 1794. It bears the heading "A Collection of Knots" and contains thirteen neatly sketched views of knots with English names written beside them. This is probably an excerpt he made from an English book on knots. With it are two additional pieces of paper with sketches of knots. One is dated 1819; the other is much later, for it bears the notation: "Riedl, *Beiträge zur Theorie des Sehnenwinkels,* Wien 1827." Notes of Gauss referring to the knotting together of closed curves are printed in his *Collected Works* (VIII, 271-285). In particular he found in a note dated December, 1844, the numerous forms which closed curves with four knots can exhibit. He was conscious of continual semantic difficulty in topology.

The above-mentioned remark of January 22, 1833, concerns the concatenation of two curves in space, in which at the close the definite integral formula for the number of knots is given. It has

---

[8] Euler and Vandermonde.

been alleged that Schnürlein, a pupil of Gauss, carried on intensive research on the application of higher analysis to topology, with his help, but no one has been able to verify this.

The determination of the mutual position of curves in the plane is the means which Gauss used in his dissertation (1799) for the deduction of the fundamental theorem of algebra. This viewpoint is even stronger in his last proof of the theorem (1849).

Under the heading of topology should be mentioned the research which Gauss did on the possible types of distribution of geocentric points of a planet on the zodiac (*Collected Works*, VI, 106), and the case mentioned here of a chainlike overlapping of two planetary orbits, as is often the case among the asteroids.

There can be no doubt that Gauss had some influence on the later work in topology carried on by his pupils Möbius, Listing, and Riemann. His influence probably lay in giving stimulation; and one need not detract from their work by saying that Gauss stimulated them to work in topology.

In the last decade of his life Gauss continued his observations at the instruments of the observatory, and the numbers of the *Astronomische Nachrichten* of that period contain frequent communications from him on various observations of eclipses and locations of planets and comets. His last observation communicated to this journal, which he himself carried out, was that of the solar eclipse of July 28, 1851, when he was seventy-four years old; his assistants Klinkerfues and Westphal participated in this. Gauss' last communication to the periodical occurred in 1854. His diary of the Reichenbach circle from March 6 to June 21, 1846, shows twenty-one observation days. The diary closes with the latter date. From July 4, 1846, to June 27, 1851, he observed most of the previously observed fundamental stars—several of them ten to twenty times—and several other stars. Moreover he included the planets Venus and Mercury and observed the newly discovered planets Metis (Graham's planet), Parthenope, Victoria, Iris, Flora, and Neptune. He published the latter observations; they are printed in Volume VI of the *Collected Works*. He also observed the solar eclipse of October 8, 1848.

Gauss' love for astronomy was demonstrated by the fact that

he gave four of his six children the first names of the discoverers of the asteroids. On May 6, 1807, Bessel wrote to Gauss: "I saw with pleasure that you have calculated the orbit of Vesta; also the name chosen by you is splendid, and therefore certainly also pleasant to all your friends because it shows them to which goddess you sacrifice."

CHAPTER EIGHTEEN

Senex Mirabilis

In 1848 the so-called "literary museum" was founded in Göttingen as a protection against the Revolution. It was in part a social organization, designed to bring students and professors closer together. This museum, later called the "union," was set up at Hospitalstrasse 1 in the former home of Karl Otfried Müller, the professor of Greek who had died in 1840 at Athens while on a journey. As a means of educating the students many newspapers and large quantities of "unobjectionable" literature in every form were offered. When Göttingen got a warning from Hanover about the *Jung Deutschland* movement, officials proudly pointed to the literary museum. Only two professors in Göttingen were suspected of belonging to the movement. One was a former pupil of Gauss, Moritz A. Stern in mathematics, and the other was Theodor Benfey (1809–1881) in Oriental languages. This suspicion had arisen merely because they were Jews.

Gauss joined the club and in his last years almost the only physical recreation he took was a daily walk to the reading room between eleven and one o'clock. He rapidly scanned political, financial, literary, and scientific news, making a mental note or writing down things which especially interested him. Some of this was for statistical studies he made. It was his custom to collect recent issues of the newspapers he wanted to read, and to prevent their being taken away before he had read them, he arranged them chronologically on the seat of his chair and sat on them. Then he cautiously pulled out one after the other and passed them on after having read them. Students generally stood in awe of Gauss, and when one of them happened to be reading a newspaper which Gauss desired, he cast a questioning look at the student who then hurried to give him the paper.[1]

Personal and scientific correspondence with friends, relatives, and colleagues in all parts of the world took up much of Gauss' time in his later years. He received occasional letters begging for financial aid, several from persons who claimed to be his kin. Yet, in spite of new correspondents, he was both relieved of letter writing and grieved at the death of his intimate friends: Olbers died in 1840, Bessel in 1846, and Schumacher in 1850. At one period he had written weekly to Schumacher. Bolyai, Encke, Gerling, and Humboldt survived him; hence his correspondence with them continued to the last. He noted on his calendar, several months in advance, the day in 1853 when Humboldt would reach the age attained by Newton and sent him special congratulations. On February 15, 1851, Gauss' assistant Benjamin Goldschmidt died very suddenly at the age of forty-four. He had been observing the night before and had shown some visitors the Pleiades through the telescope. He was found dead in bed early the next morning. Goldschmidt was a man of benevolent character, and well informed in his field. His sudden death had a serious effect on the aged Gauss, who was sincerely attached to the young man and honored him for his ability. Goldschmidt was succeeded by E. F. W. Klinkerfues, a picturesque character of considerable talent, who remained at the observatory until his suicide there in 1884. Klinkerfues had early become an orphan, was a member of a large family, and had had a difficult life. Gauss had great sympathy for him.

One of the happy events of Gauss' old age was the visit paid him on July 26, 1849, by Bernhard August von Lindenau, astronomer and minister of state of the Kingdom of Saxony. He was the last of the old friends from the beginning of the century whom Gauss saw.

After Lindenau's visit Gauss seemed to rest more on his

---

[1] Students referred to him as a *Zeitungstiger*.

laurels. He explained to intimate friends that he did not like to be driven in his scientific work and that his working time was noticeably shorter in comparison with earlier years. At other times he complained of the burden of teaching his courses, which prevented him from carrying out some major research. His work dealt with the theory of the convergence of series, actuarial studies, and mechanical problems connected with the rotation of the earth, resulting from Foucault's experiment and the theoretical studies of Lagrange, Plana, Hansen, and Clausen. He had the Reichenbach meridian circle equipped with new microscopes, ordered optical instruments from Oertling, a well-known mechanic in Berlin, and also had a large Foucault pendulum set up, in order to show the rotation of the earth.

Although at times he grew weary of long arithmetical calculations, Gauss derived a certain pleasure from them as a rule. He used many different artifices and used to tell his students that there was a certain poetry in the calculation of logarithmic tables. It was difficult for him to forget that he was a man of science, even during his recreations. For years he played whist regularly with the same friends. It was his custom to write up how many aces each one had had in his hand in every game, in order to get an empirical corroboration of certain laws of the calculus of probabilities. He made very few errors in calculating and used many means of checking results. In extended numerical calculating he observed perfect order; every number was written in the neatest manner possible. Each number was in exactly the right place. Row after row manifested the same accuracy.

He always strove to carry out the work as accurately as the auxiliary means allowed. The last decimal in seven- or ten-place logarithms had to be verified as much as possible, and he carried out full-scale research to determine to what extent the last decimal in various tables was accurate. He got special pleasure from calculating with incorrect tables, because he then had the attractive job of correcting misprints or errors in calculating. His greatest joy was in simplifying long calculations of analytical or numerical nature and in compressing the result of a week's work on one octavo page, thus making it intuitively evident to the connoisseur. Even where he had to make extracts from the works of others, the content of a volume or the excerpt of an entire official document was brought together in a very small space in an unusually clear manner.

Gauss was always attempting to find some new application for mathematics. He kept numerous little notebooks with punctual and neat entries. There was an index of the length of life of many famous men and of his deceased friends, calculated in days. At nine o'clock the third evening before his death he calculated the number of days that he had lived and made this entry in an actuarial book he had possessed for many years. Another covered the monthly income of the Hanoverian railroads. He kept another on the walking distance (number of paces) from the observatory to the various places he often visited. There was one with the day and number of electrical storms in various years. One of the most interesting such registers gave the dates of birth of his children, date of vaccination, dates of cutting the first eight teeth, and date of beginning to walk. Each date was accompanied by the number of days old the child was at the time. He kept a chart of all the keys to the observatory and his home and entered an exact sketch of each key.

In his family and his home nothing was unimportant. Although he lived modestly, he enjoyed social intercourse. This enjoyment was somewhat limited in his last years for reasons of health. He did not like to travel, and never went farther than Austria.

Gauss was a great friend of music, especially of singing. When he heard a beautiful song, or, for that matter, any song that appealed to him, he wrote it down. His notes are as small and neat as though printed. One of his favorite songs was the well-known "Als ich ein Junggeselle war"; another was the English song that begins, "Tell me the tales that to me were so dear, long long ago." He wrote down Mignon's "Kennst Du das Land" and copied the following, which he designated as Jean Paul's favorite song:

> Namen nennen Dich nicht
> Dich bilden Griffel und Pinsel,
> sterbliche Künstler nicht nach
>
> Lieder singen Dich nicht
> Sie alle reden wie Nachhall
> Ferneste Zeiten von Dir.
>
> Wie Du lebest und bist
> So trag ich einzig im Herzen
> Holde Geliebte Dein Bild.

In 1850 Gauss copied down the revolutionary weavers' song as published in the *Volksblatt* of Nathusius, a song of which he definitely did not approve. Two additional songs that made a strong appeal to him were the patriotic "Schleswig-Holstein Meer umschlungen," by Chemnitz and Bellmann, and "Unser Leben gleicht der Reise eines Wanderers in der Nacht." As a university student Gauss read Lohlein's *Klavierschule,* Euler's *Nova theoria musicae,* Bach's *Ueber die beste Art, Klavier zu spielen,* and Margary's (?) *Anleitung zum Klavierspielen,* although this biographer knows of no evidence that he ever played a musical instrument. He enjoyed the following concerts in Göttingen: Carl Maria von Weber (August, 1820), Paganini (May 28, 1830), Liszt (November 24, 1841), and Jenny Lind (February 2 and 4, 1850). It is doubtful whether he knew Johannes Brahms and Joseph Joachim during their student period in Göttingen.

Gauss' physical appearance was impressive, according to his students. He was of medium height, perhaps slightly below average. His hair was blond and in later years a beautiful silvery white. His penetrating, clear blue eyes constituted a striking feature. His hands and feet were well formed and of normal size. He walked gracefully and with a measured gait. His black satin cap was a favorite article of clothing. Gauss was a typical Nordic and Nether-Saxon. He enjoyed conversation but never used more words than were necessary. He was heavy set, but not corpulent, and gave the appearance of being sturdy. His countenance conveyed an impression of affability and mildness. His eyebrows were rather heavy and, characteristic for an astronomer, the right one was noticeably higher than the left one. A high forehead symbolized great intelligence. His voice was pleasing. He had a rather prominent Roman nose. Gauss was nearsighted and had to use spectacles part of the time, yet his eyes and ears were keen, accurate, and well trained both in observation and experiment. He practiced moderation in all habits, enjoyed his glass of wine,[2] and smoked a pipe as well as an occasional cigar.

Personal or domestic distraction could and did occasionally exert a paralyzing effect on his scientific work, in spite of his great power of concentration. He has never been pictured as the absent-minded-professor type. The writer knows of only one such anecdote, but regards it as purely apocryphal. It is found in Carpenter's *Mental Physiology,* where the author gives it as an example of the remarkable power of concentration and attention. The story is that Gauss, while engaged in one of his most profound investigations, was interrupted by a servant who told him that his wife (to whom he was known to be deeply attached, and who was suffering from a severe illness) was worse. He seemed to *hear* what was said, but either did not comprehend it or immediately forgot it, and went on with his work. After some little time, the servant came again to say that his mistress was much worse, and to beg that he would come to her at once; to which he replied: "I will come presently." Again he lapsed into his previous train of thought, entirely forgetting the intention he had expressed, most probably without having distinctly realized to himself the import either of the communication or of his answer to it. Not long afterward when the servant came again and assured him that his mistress was dying and that if he did not come immediately he would probably not find her alive, he lifted up his head and calmly replied, "Tell her to wait until I come," a message he had doubtless often sent when pressed by his wife's request for his presence while he was similarly engaged.

As the son of poor parents Gauss was not accustomed to the

---

[2] For years he used a special brand of French wine.

luxury and refinements of more modern times. The limited means of his early years were sufficient for his simple needs. He practiced economy and laid aside a "nest egg" for a rainy day. Gauss was slow to accept financial aid from others. Throughout his life he remained true to his feelings of honor and intellectual independence. In the Napoleonic period he refused offers of financial aid from his friend Olbers and Laplace to pay the French war contribution. On the other hand he did not hesitate to accept support from his protector, the Duke of Brunswick. Although conditions of poverty in his youth were severe, they did not leave any scar on his later life. At the same time it must be admitted that Gauss was not the serene Olympian as he has occasionally been pictured. His wants were simple, and material possessions were sometimes regarded by him as exerting a disturbing influence on scientific work.

Gauss used to say that he was entirely a mathematician, and he rejected the desire to be anything different at the cost of mathematics. It is true that the research in physical science offered him a type of recreation. He called mathematics the queen of the sciences, and the theory of numbers the queen of mathematics, saying that she often condescended to serve astronomy and other sciences, but that under all circumstances top rank belonged to her. Gauss regarded mathematics as the principal means of educating the human mind. He recognized the value of studying classical literature, and said that although he chose mathematics as a career he had not neglected the latter. Gauss recommended to his students the study of the ancient mathematicians, in particular Euclid and Archimedes.

It was his custom to tell his friends that if others would meditate as long and as deeply as he did on mathematical truths, they would be able to make his discoveries. He said that often he meditated for days on a piece of research without finding a solution, which finally became clear to him after a sleepless night. His conversation with friends sometimes was interrupted for meditation; sometimes the conversation would be continued after a gap of several days.

Newton was probably the mathematician to whom Gauss was most closely related by discovery and temperament. To Newton alone he applied the adjective *summus*. He occasionally compared Newton and Leibniz, and, although recognizing the great talent of Leibniz and his merit in discovering the calculus, Gauss regretted that Leibniz had dissipated his energy by being too much the jack of all trades. Gauss was indignant about the legend of the apple in connection with the discovery of the law of gravitation. He thought it too simple an explanation. Gauss' version ran thus: "Some dumb upstart came to Newton and asked him how he had arrived at his great discoveries. Since Newton realized what kind of a creature confronted him and wanted to get rid of him, he answered that an apple fell on his nose, whereupon the fellow went away, completely satisfied."

In the manifold conditions of human life Gauss saw an extended field for the application of mathematical theories. The answering of economic, financial, and statistical questions furnished him rich material for such research. He placed special value on tables of mortality and the investigation of laws governing the length of human life, partly for further application in calculating life insurance, annuities, and widow's funds. Gauss took special interest in mortality rates of infancy and old age. He felt that there were too many extraneous influences to derive laws between those two periods of life. He used to tell his friends that he had carried on research on the mean life expectancy of children up to age one and a half years which exhibited such admirable regularity that it was almost like astronomical observations. Likewise it was his view that at an advanced age the mean length of human life follows a definite law, although we do not possess sufficient data for a satisfactory answer to this question. He said that we would be able to perfect these data by rewarding with premiums people who could prove that they were ninety or a hundred years old, adding that if he were a rich man he would offer large capital for this purpose.

Beginning in 1845 the academic senate at Göttingen entrusted Gauss with a gigantic task, a study and reorganization of the

SENEX MIRABILIS

fund for professors' widows. He devoted himself to it with his usual vigor; here his mathematical ability and knowledge of financial operations linked up with his practical talent in organizing. This job took much of his time as late as 1851, and his work saved the fund from ruin. His long memoir discussed the principles which govern the administration of such a fund; it is printed in Volume IV of the *Collected Works*. Gauss received recognition for this achievement, and widows as well as orphans were grateful to him. In a letter to Gerling on July 26, 1845, he explained what the situation was:

Now several words about the work mentioned at the beginning of this letter. It concerns the local fund for professors' widows, where a year ago the number of widows had increased so much (to 22), that the income was no longer sufficient to pay the pensions, although other adversities were connected with it—long-drawn-out law suits, in which all interest is absent and on the other hand considerable court costs are to be borne in cash—and that worries were therefore stirred up. I did not share these worries *to the extent* that I would have attached a preponderant importance to the present large number of widows (which since then has decreased by 3), but rather I see a *much* greater danger in the present inordinately large number of participants (50 or 51, while several decades ago there were only 30 or a few more). The fund subsists almost entirely on *its property,* while the contributions are quite insignificant (10 thalers annual contribution, 250 thalers annual pension, both gold). In addition there is an *obscure version* of a highly important part of the statutes, to which, in my judgment, a highly unintelligible interpretation has been given in practice. Several months ago I was commissioned to investigate the situation basically, which I am convinced can occur only by selling for *cash* and balancing with the property the three types of obligations of the fund, namely: (1) to the present 19 widows, (2) to the possible survivors of the present 51 members, (3) to the survivors of all future members. Especially difficult (that is, protracted) is the calculation of (2), and it cannot be completed except by means of an *auxiliary table* for an obligatory income, in which annual differences and years are the basis. I have now begun and almost finished the construction of such an auxiliary table, difference of age, $-1 \ldots +20$ and age from the maximum up to twenty years back, all by individual years and according to two rates of interest, 4 per cent and 3.5 per cent, and according to Brune's mortality table, after I used almost all my time for a month on it. In itself such a work is demanding, over

100,000 figures (according to my method of writing, where half or more is calculated in the head), a work very protracted for the mind, but I submitted in view of the usefulness of that *work,* for which it is a preparation, for the university to which I owe my position in life, although I have to count not on thanks for it, but only on vexation.

In Quetelet's *Annuaire* Gauss found the ratio of widows to existing marriages as one to four. He felt that this was inaccurate and that many widows' funds were ruined by the use of such figures. In his opinion, the ratio which is valid for a whole land is not valid for individual provinces, and even less for individual classes of society. He thought for professors' widows the ratio 1:2 would yield too few. In the case of the Göttingen widows' fund, even if the ratio were known to be 7:12, no use could have been made of it. Gauss found that, owing to laxity of administration, the Göttingen fund had accurate information on existing marriages for only two years, 1794 and 1845, even though the fund had existed for over one hundred years when he attacked the problem.

On January 31, 1846, Gauss gave Gerling some further details on his work for the fund:

Our widows' fund affair has cost me an enormous amount of time. Studying, with the help of more than 100 years of accounts, for the construction of detailed auxiliary tables (using Brune's mortality tables, certainly the only ones which are based on correct principles), and the actual application of them to 42 married couples and 20 widows, according to two different rates of interest, have demanded a work of about 5 months, later the working out of a memoir on the condition as well as a critical revision of earlier negotiations, on which the existing regulation is based—again more than 6 weeks. Now I have finally worked out a new memoir with proposals for remedy of the threatening evils, which is accepted by the whole commission under me, but which, as I have great cause to fear, will be received by a large number of the academic senate in an extremely Abderian manner. In order to be able to bear this as calmly as possible, I (foreseeing such success) withdrew six months ago.

Gauss' special hobbies included public finance, its sources and duties, the administration of banks and railroads, the relationship between paper money and value of the metal, amortizations, and the like. Daily he rapidly scanned the press for financial news, particularly market reports and the course of foreign bonds. He considered paper money very dangerous for the credit of nations, because governments in time of trouble could too easily be led to overestimate their financial strength. Gauss was glad that the Kingdom of Hanover had not introduced paper money in his day. He was a definite enemy of all petty finance operations, if they only burden the public, without leading to any important result, and he used to call them "penny-pinching" (*Pfennigfuchsereien*), ascribing to their authors little intelligence and feeling of fairness. Gauss was known to his friends as a wise investor, although actually none of them knew how wealthy he really was. Perhaps it is well that his talents in this field were not more widely known, for he would surely have been continually annoyed by inquiries and thus hindered in his scientific work. The opinion has been expressed that he would have made an excellent minister of finance. On one occasion he expressed himself about the matter; it occurs in a letter to Schumacher, June 27, 1846, just after the death of Bessel:

> I believe that under similar circumstances I would have ranked as high [as Bessel] in mercantile knowledge. For my part I possessed in myself up until an advanced age nothing, as the world is, which could have given me a safe treasure even against starvation, except schoolteaching, which has always been repugnant to me.

During the time of disturbance over the Göttingen Seven, on March 4, 1838, Gauss wrote to his friend Olbers:

> Of course I have taken initial steps to be able to lift off the many lead weights which hang on me here. To that belonged the preparations to be able to mobilize the small estate which I have partly to administer, partly my own property. If everything is successful, as I more desire than hope, then in the next 4-5 months I shall be able to manage something more than 15,000 thalers, and perhaps you can give me good advice in this connection, since you probably often have the opportunity for important transactions. For my part I am in general not in favor of hanging everything on one nail. I thought of Austrian bank stock and Russian papers. Do you perhaps have detailed information about the profit and safety of Belgian property such as Brussels bank stocks?

Olbers was unable to give Gauss any definite advice. He wrote that he did not know of a good opportunity for investment of money and that he had to accept the reduction by Hanover and the Württemberg Credit Union of the interest rate from 4 to 3.5 per cent on the bonds of his native city Bremen, because he did not know any way of reinvesting them at a higher rate.

In the above letter Gauss referred to the fact that he administered the estate of his father-in-law, J. P. Waldeck, who had died in 1815. Eventually half of the estate went to Gauss' wife and half to her sister, Luise Christiane Sophie, who had married a municipal court director and syndic, Dr. Carl Andreas Hoefer, and lived in Greifswald. The Hoefers had three sons and two daughters. At her death on September 12, 1831, the estate of Frau Minna Gauss was appraised at 18,263 thalers. She left it to her three children, but attached certain special conditions to the inheritance of Eugene, who had caused his parents trouble. The codicil provided that if by 1838 (two years after becoming of age), Eugene showed fully valid proofs of improvement, he would then enjoy the interest from his share. If by 1843 he proved himself to be financially responsible, he would receive the capital. If he should try to fight the provision in court, he would get the legal minimum. Gauss was administrator of his wife's estate, but he did not know of this condition in her will, which she had written a year before, until after her death.

Gauss turned over to his son Wilhelm his share in Frau Minna's estate on August 4, 1837, shortly before Wilhelm left for America; it amounted to 6,837 thalers. Wilhelm and Therese were to receive 4,500 thalers each, while the residual estate of 13,763 thalers was divided into three equal parts. Eugene's share amounted to 4,587 thalers. Beginning with his birthday in 1836 the interest at 3.5 per cent was added to his capital. Eventually Eugene received his full share of his mother's estate, with interest.

Frau Osthoff, the mother of Gauss' first wife, died in 1821, and made a friend of the family her administrator. She named Gauss' two children of his first marriage her only heirs, but stipulated that they were to receive the inheritance after coming of age. They were not allowed to touch the interest until that time. This will caused Gauss great vexation, for he wanted to use some of the money in educating the children. Frau Waldeck, his second mother-in-law, died in 1848 and named Joseph Gauss her administrator; the Gauss children received their proper share of her estate.

Joseph was also the administrator of his father's estate. To their great surprise his children ascertained that Gauss' estate in stocks and bonds was valued at 152,892 thalers. Each of the children inherited 38,215 thalers. A balance due Therese from her mother's estate brought her share to 44,975 thalers. An even greater surprise was in store—on the two days between his death and burial the children found cash amounting to 17,965 thalers hidden away in Gauss' home. It lay in his desk, his wardrobe, in cabinets, and dresser drawers.

Gauss received a basic salary of only 1,000 thalers per year, plus student fees. The number of his students was always small. During the years that he served as director of the Royal Society of Sciences he was paid 110 thalers annually. It is fabulous how he was able to build up so large an estate. He invested in Vienna bank stock, in the Austrian metal league, and in the Hamburg Fire Insurance Company. Other investments included Norwegian and Swedish bonds, particularly Swedish mining stock. There were Russian, Hanoverian, Prussian, Belgian, Baden, Württemberg, and Mecklenburg Credit Union bonds, and some Bückeberg municipal bonds. Rates of interest ranged from 2.5 to 5 per cent and dates of maturity extended as late as 1893. He knew when to sell at a profit. The last transaction was made only three weeks before his death.

As a youth Gauss had been undecided whether to dedicate his life to philology or mathematics. Outside mathematics and science his greatest talent lay in the learning of foreign languages. He had a reading knowledge of the modern European languages and spoke and wrote the principal ones quite correctly. From boyhood he was familiar with the ancient languages and well versed in their literature. He published his major works in Latin. On more than one occasion his Latin style has been highly praised by competent authorities. His use of his mother tongue was wellnigh perfect.

At the age of sixty-two Gauss decided that he ought to learn a new language or a new science, in order to keep his mental faculties fresh and to be receptive to new impressions. For a brief time he thought of taking up botany, but for physical reasons decided against it. Then he tried Sanscrit, but did not like it, and soon dropped the subject. Finally, Gauss turned to Russian.

On August 17, 1839, he wrote to Schumacher: "At the beginning of last spring, regarding the acquiring of some new aptitude as a type of rejuvenation, I had begun to busy myself with the Russian language and found much interest in it." He added that since May, 1839, his study of Russian had been almost completely interrupted, but requested Schumacher to procure some Russian books for him, since he wanted to begin again. On August 22, 1839, Schumacher sent him a Russian calendar, and in a letter of August 8, 1840, Gauss wrote his friend he wanted to read some literary prose. Schumacher was planning a trip to St. Petersburg, and Gauss asked him to bring back some novels. He bought for Gauss the works of Betúscheff and made him a gift of five volumes of memoirs of the Russian topographical bureau. On December 29, 1841, Gauss wrote Schumacher that he was finding little time for Russian. As late as June 13, 1845, Schumacher recommended Bolotoff to him for explanation of Russian pronunciation. He used C. P. Reiff's Russian-French dictionary.[3] The Gauss library contained seventy-five volumes of Russian literature, including eight volumes of the works of Pushkin, Russia's greatest poet.

It seems that a desire to read Lobachevsky's works in the original was not the first stimulus for Gauss' study of Russian, as has been sometimes assumed. Gauss devoted incredible energy to the study of Russian, and in two years he had mastered it by himself to such an extent that he not only read fluently all books

---

[3] He also used Schmidt's *Russian-German Dictionary* (Leipzig, 1842).

in prose and poetry, but also carried on his Russian correspondence in that language. One day when he was visited by a Russian minister of state he conversed in Russian with the visitor, who declared that Gauss' pronunciation was perfectly correct.

In general, Gauss evaluated languages according to their logical sharpness and the wealth of ideas which they could express. Often he complained of insufficiency, especially when it was a matter of expressing some scientific subject matter exactly. He then attempted cautiously to introduce new nomenclature for new concepts, and these terms were usually accepted, although he resorted to this procedure only when there was an urgent need.

One of the few recreations in which he indulged as a change from his mathematical studies was extended reading in the most varied branches of human knowledge. German and English literature particularly attracted him, and, as was noted above, Russian literature furnished him many pleasant hours in his final years.

His favorite German author was Jean Paul (1763–1825). Some people have been surprised at this fact, but upon closer examination the reasons stand out clearly. Gauss appreciated Jean Paul's great wealth of similes, his depth of intellect, and his inexhaustible humor. He was *the* best-seller of his day—more widely read than Goethe and Schiller. Jean Paul and Gauss both manifested the same polarity between rationalism and romanticism. As a young man he reveled in the beautiful descriptions found in Jean Paul's works. He enjoyed the sentimental and patriotic elements of Jean Paul's writing; he was firmly attached to his fatherland and his people, not the aristocracy, royalty, or nobility, but the small-time, everyday folks. There were also religious reasons why Gauss was attracted to Jean Paul, but these will be mentioned in a later chapter.

Gauss often complained because, as he said, Jean Paul was misguided by a belief in animal magnetism, and this weakened his enjoyment of those pleasant elements mentioned above. He called *Doctor Katzenbergers Badereise* a masterpiece and always laughed about the struggle of the doctor and the druggist over the eight-legged rabbit, and the art of making the ducats full weight by means of earwax. Jean Paul's character and style reminded Gauss of his friend Bolyai, which is another explanation of the attraction. Gauss and Jean Paul respected each other highly, although they never saw each other. There is no evidence that they corresponded.

On January 2, 1813, Charles F. D. de Villers (1765–1815), a Frenchman who from 1811 to 1814 served as professor of philosophy in Göttingen, wrote thus to his friend Jean Paul:

> Among your warmest admirers here is to be counted the sky, star, and number man, Prof. Gauss. The quiet, gentle and intellectual Gauss reads and loves you almost as passionately as I—this common inclination has established mutual attraction between us, and I have you to thank for the friend with whom I otherwise would have perhaps had few points of contact.

Occasionally Gauss used an appropriate quotation from Jean Paul in his letters and works. On December 31, 1839, he wrote to Minna that his younger daughter Therese was enjoying better health than usual that winter and had been reading *Dr. Katzenbergers Badereise;* she was actually enjoying it, a fact which surprised her father. Gauss read Tobias Smollet's *Peregrine Pickle* and *The Expedition of Humphrey Clinker,* in each of which there is a doctor similar to Katzenberger, who represented the medical cynic.

In his home Gauss had a number of hand billets of Jean Paul, his complete works, the biography by Spazier, and a silver medal of the poet. After his death the medal was given to one of the sons in America. Gauss had a close friend, A. W. Eschenburg in Detmold, the son of his old teacher who had translated Shakespeare. The following words which he wrote in Eschenburg's album are strongly reminiscent of Jean Paul's *Streckverse*:

> For the darling of heaven all his paths are strewn with roses; happily he looks back and with happy confidence into the cloudless future. Joy, the daughter of heaven, is his inseparable companion, when he looks at

grief it turns into a gentle smile; all hearts beat for him and everybody vies for his love. Happy is he whom heaven loves, but more happy is he whom heaven's darling loves.

Gauss' tastes in his recreational and general reading were well-nigh universal. He studied subjects far removed from his own field, such as bookkeeping and shorthand, occasionally using the latter in his notes. In the field of ancient classics he read Aristotle, Plato, Theophrastus, Cicero, Virgil, Tacitus, Livy, Thucydides, Hesiod, Euripides, Pindar, Anacreon, Xenophon, Julius Caesar, Diogenes Laertius, Aulus Gellius, Herodotus, Suetonius, Cornelius Nepos, Terence, Phaedrus, Ovid, Curtius Rufus, Seneca, Pliny, Horace, Sallust, Lucian, Juvenal, Plautus, Martial, and Tibullus.

His acquaintance with French literature was gained mainly through the works of Lesage, Montaigne, Rousseau, Voltaire, Maupertuis, Montesquieu, Condorcet, de Luc, Saussure, Brigard, and Boileau. In reading Boileau, he entered copious marginalia.

Gauss read all of Holberg in the original Danish. His knowledge of Swedish, Italian, and Spanish was rather superficial.

English was the foreign language in which Gauss was most fluent. He owned a copy of Dr. Samuel Johnson's *Dictionary of the English Language* (ed. 1778), and used George Crabb's *English Grammar* and Flügel's *Practical Dictionary of the English and German Languages* (Leipzig, 1847). The Gauss library contained twenty English novels by G. P. R. James. Other important English writers Gauss enjoyed were Pope (*Rape of the Lock*), Sheridan (*School for Scandal*), Smollet, Moore, Milton, James Thomson, Swift (*Gulliver's Travels*), Goldsmith (*Vicar of Wakefield*), Richardson (*Clarissa Harlowe*), and Robertson (*History of Charles V*).

After his two sons migrated, Gauss paid special attention to American literature. He owned and read all the works of James Fenimore Cooper; Harriet Beecher Stowe's *Uncle Tom's Cabin* interested him especially, since one of his sons had become a slave owner and suffered considerable financial loss from runaway slaves. In the *New York Review,* October, 1840, he was pleased to read a favorable account of some of his work. Some of his friends sent him a copy of the Harvard catalogue for 1845 (printed in Latin) and copies of the *American Almanac* (1847) and the *Boston Almanac* (1849).

The Gauss library contains a French-Polish Dictionary, but it is extremely doubtful whether he ever gave attention to Polish. Through reading Jean Paul he was attracted to the study of Pestalozzi. His library contained the *Gentleman's Magazine*, 1754–1771; it is mere conjecture to seek the reason why he acquired this set.

No attempt is made here to discuss Gauss' reading and knowledge of German literature, so extensive that it would carry us far afield. Goethe could not understand or appreciate mathematics, hence it is not surprising that the greatest mathematician of modern times did not fully appreciate Goethe. They never saw each other or corresponded. Goethe probably heard much about Gauss through mutual friends, the Sartorius family. The style and mode of thought in Goethe did not appeal to Gauss and could not satisfy him, although he knew all of the poet's works. He considered Goethe too poor in thought content, but recognized the value and perfect form of his lyric poetry.[4] Schiller's philosophical views were totally repugnant to him, and he cared less for him than he did for Goethe. He called the *Resignation* a blasphemous, morally ruined poem and in his edition wrote on the margin in Gothic script the word "Mephistopheles!" Of Schiller's dramas Gauss valued highly *Wallensteins Lager;* the *Piccolomini* and *Wallensteins Tod* left him quite cold, since the hero did not interest him at all. He was very fond of Schiller's little poem "Archimedes und der Schüler" (1795), although he considered the treatment of the distichs a failure. In 1807 Gauss used this poem in his inaugural lecture on astronomy.

The tragic was in general not the element in which Gauss liked to move about. Misanthropic, cynical, melancholic, or pessimistic tendencies, as found in Lord Byron, or through his

---

[4] Gauss made a thorough study of Goethe's *Farbenlehre*.

influence on German literature, were repugnant to Gauss. He found Byron's mode of thought too unpleasant and demonic. Even some of the tragic aspects in Shakespeare's works were too much for Gauss. There is so much tragedy in daily life that he did not want it in his literature.

Gauss knew all the works of Sir Walter Scott very thoroughly and he passionately admired them. The tragic ending in *Kenilworth* made a painful impression on him and he would have preferred not to read it. He read Scott's *Life of Napoleon* with great interest and felt quite satisfied, being in full agreement with the author. One day he found a passage in Scott which set him to laughing. It was just too much for an astronomer. Gauss compared all the editions he could get his hands on to make sure it was not a misprint. The words were: "The moon rises broad in the northwest." He made a note on the margin beside this passage. In his last years he enjoyed reading Gibbon's *Decline and Fall of the Roman Empire* and Macaulay's *History of England*.

A nephew of Gauss' wife became well known in German literature, Edmund Franz Andreas Hoefer (1819–1882), who lived most of his life in Stuttgart and Cannstadt. He was editor of the *Hausblätter* from 1854 to 1868. Hoefer's novels and short stories were widely read in their day; he also published a history of German literature. His work was considered a splendid example of regionalism (*Heimatkunst*). Hoefer was a close friend of Wilhelm Raabe, one of Germany's most important writers in the nineteenth century.

Gauss always paid close attention to political events, particularly those in his own country. He made a long list of what he regarded as the most important political events, beginning with the French Revolution in 1789 and ending with an event on July 1, 1810. It makes interesting reading today. His strong character and consistency manifested itself in the field of politics. Gauss was an absolutely conservative and aristocratic person. In his later years students used to call him a reactionary. He preferred government by a strong leader of high intelligence to any other form of government. Mob rule with its acts of violence, and especially the Revolution of 1848, aroused in him indescribable horror. On May 17, 1849, he wrote to Schumacher:

Our public affairs are getting more and more dismal. I do not know what philosopher[5] it was who set up the doctrine that one should neither mourn nor laugh at bad times, but *understand* them. I confess that the first prohibition is *very* difficult to fulfill, but even more difficult is the carrying out of the third command. Sometimes those seem to me to be right who believe that not merely St. Paul's Church,[6] but almost all Germany has become a madhouse.

In 1848 the government of Hanover tried to overcome the bad effects of the events of 1837 by calling back all of the Göttingen Seven. Ewald and Weber were the only two who returned. Their return brought joy to Gauss in his last years. Actually the Revolution of 1848 did not have a bad effect on the university. Göttingen was reputed to be almost the most radical city in the Kingdom of Hanover, outstripped only by Hildesheim. A military unit had to be sent to Göttingen. The students paraded out of the city and there were some minor disturbances between the students and police. Students on the conservative side formed companies and patrolled the streets of the city at night. Rumors were rife. Most students were on the conservative side and were more interested in internal reform of the university. Johann Miquel, a student leader on the revolutionary side, was a close associate of Karl Marx. He later became minister of finance and vice-president of the ministry of state in Prussia.

In 1927 Max Schneidewin, then an old man of eighty-four years, the son of a Göttingen professor of classical philology, at his home in Hamlin gave his reminiscences of Gauss in the Revolution of 1848. In his parents' home he and his brothers and sisters always heard the name "Gauss" mentioned with a certain amount of awe. During the Revolution, professors who were physically fit were chosen by the state or municipal authorities to patrol and preserve order. As a small boy of six, he was astonished to see his father wearing a light-weight black coat, similar

---

[5] It was Spinoza. See *Tractatus politicus,* Chapter I, paragraph 4.
[6] The National Assembly, 1848–1849, met in this church at Frankfurt.

to a uniform, with a large belt. He saw a lance leaning against the bookcase. One day at noon his father related that he had to stand guard with other professors at the Geismar Tor. Suddenly it was announced that Gauss, who lived only five minutes' walk from the Geismar Tor, was coming. At once the whole guard of professors lined up in order to "present arms." Schneidewin said that could have happened only for Gauss. After that, the boy always felt a special admiration for Gauss whenever he saw him.

In a letter to Encke, dated December 14, 1848, Gauss expressed himself thus concerning the Revolution:

> I read your speech of October 18 with so much more pleasure, since in the sad period from March 18 to December 6 one so rarely got to hear the voice of calm reason from Berlin. In this time I was able to think of your king[7] only with a painful feeling. I do not know whether there was ever another prince who in all his steps so sincerely meant well, and in return was always rewarded with such shameful ingratitude. May his newest experiment now be more successful. It made the same impression on me which is reflected between the lines in an utterance attributed to a French diplomat in the newspapers, that this constitution gives *more* than the need of an enlightened nation demands, and that one will have to admire the Prussian people if they prove able to bear this one.
> The direction of Jacobi's political views had long been known to me from the newspapers; even Steiner and Erman are mentioned in them, sometimes in a similar though not quite so blunt manner, but that even Dirichlet joins this tendency was to me quite unexpected. In Göttingen we are happier to the extent that in the faculty (now) only a very small number who worship such idols can be pointed out.

Gauss had a very low opinion of the intelligence and ethics of "the people"; he frequently expressed this view with reference to political, religious, and scientific matters. He used to say *mundus vult decipi* and pursued agitators and rabble rousers with the eye of mistrust, with the steadfast gaze of a falcon. He had a low opinion of constitutional systems of government and was unremitting in his efforts to demonstrate either logical errors or lack of expert knowledge on the part of politicians. His

---

[7] The King of Prussia.

friends, although differing with him occasionally, felt that he often succeeded. As an old man he loved above all quiet and peace in the land, and to him the thought of a civil war's breaking out in Germany was equivalent to thinking of his own death. Yet he did not cling to the traditional merely because it was traditional. If there was some really demonstrable progress involved, whether in the intellectual or the material area, he was just as much in favor of reform as his contemporaries. However, he did not like change in his home, preferring the simplicity to which he was accustomed in his youth.

Gauss' home was his castle, and he demanded the same independence for the state. He deplored the political behavior of Germany in his day, her lack of harmony, and longed for national unity.[8] Like so many Germans, he would gladly have entrusted the fate of his nation to the hand of a strong ruler, and did not wish to adhere to a reed swayed by every passing wind or a ship without a pilot. He was definitely opposed to foreign rule or occupation of one nation by another.

Apparently Gauss attached peculiar significance to the fact that he discovered the division of the circle into seventeen equal parts on the same day that Napoleon left Paris to journey to the Italian army. He used to mention the fact to his friends. Napoleon reached the Italian army at Nice on March 26, 1796. In the *Hamburger Correspondent* for March 29, 1796, there is a note dated March 17, "General Bonaparte has left here (Paris) for the Italian army." Of course his actual departure could have occurred one or two days earlier.

On April 20, 1848, Gauss wrote to Bolyai, the intimate friend of his youth:

The violent political and social earthquake, which in ever increasing extent is upsetting almost all European conditions, has not yet touched your fatherland in the narrow sense (I mean Transylvania). Indeed I cherish the confidence that *in the end* enjoyable fruits will proceed from it; but the transition period will first bring manifold oppressions, and

---

[8] He studied Böhmer's *Kaiser Friedrichs III Entwurf einer Magna Charta für Deutschland.*

(*quod tamen deus avertat*) can last a long time. At our age it is always very doubtful whether we shall experience the golden age just ahead.

Through the present devaluation of Austrian bonds, in which the greatest part[9] of my 40 years' savings is invested, I am threatened with being able to leave little or nothing to my children at my death.

A wave of "table rapping" swept over Europe and America at the mid-century. German newspapers were full of it, and Gauss' friend Rudolf Wagner became interested and published on the subject. This table movement or table turning was supposed to be a form of psychic phenomena. Many believed that spirits from the other world were "rapping for admission to our world." A group of people would sit around a table with arched hands resting upon it, and wait to see what happened. Gerling became interested and performed many experiments in this "field." He reported at great length to Gauss about it, and asked his advice. The reply, dated April 21, 1853, is last letter Gauss wrote to Gerling:

I had intended to answer you this afternoon and to explain my own view of the matter. Since this could not occur without some prolixity, I was very glad this morning among so much nonsense which the newspapers offer us on the subject to find an essay in a journal, whereby I am spared that effort. The journal is the one that arrived here this morning, I have forgotten whether it was the Didaskalia or Konversationsblatt, and the essay had two signatures, the first one I have forgotten, the other was Poppe. I would have had to write you exactly the same thing that is in this essay. I have made an experiment on our round, rather massive (perhaps 50 pounds) dining-room table[10] with Therese, in order to check how much pressure is necessary for the four hands lying flat (not touching one another) to set the table in rotating motion, of course with a pressure purposely not perpendicular, but acting simultaneously tangentially, and I found to my own amazement, that only a *very slight* pressure was necessary. Moreover, the three-foot table was on a rug. If I now consider that of that pressure I myself exerted by far the greatest part, then I believe that in the case of the tiny, light weight table described by you that each of the eight hands has

---

[9] This statement is inaccurate. His investments in Austrian bonds amounted to 26,604 thalers, out of an estate of 152,892 thalers.

[10] It was a circular table of 47 inches diameter; the feet formed an equilateral triangle, each side of which was 23 inches.

such a slight share that after a half hour's wait the hands become insensitive to it.

I myself shall certainly make no experiment after the manner of Andreas. My patience would not suffice to hold my hands one half hour or longer in an unchanged position on the table top: but if I did it, then I am certain in advance that my hands, even if they did not get cramped, would get into such a condition, that I would no longer be master of them as in the normal condition, I mean, that I would no longer remain certain whether I was not pressing or whether I was pressing perpendicularly or at an angle. Therefore, in so far as only movements are produced by the experiment, to my mind it proves nothing. It would be different if the objective effects claimed in the newspaper articles, and similar to the galvanic, were corroborated: sparks, decomposition of water, paralysis, or great augmentations of a horseshoe magnet. But all these claims are without a guarantor, and *in dubio* I would give them the signature Münchhausen.

In almost all experiments I find something of which I must disapprove. As soon as a movement is noted, the people jump up, push back their chairs with their feet and run after the moving table. I would demand most strictly that each one remain seated on his chair. Since by regulation the hands are to lie only very loosely on it, why doesn't one allow the table to glide away under the hands, without lifting the contact? Tables whose oblong form do not permit that, should be excluded. If you say that it will then stand still, then this proves that just a not-too-slight pressure is essentially necessary, and this [pressure] in the running after the table will certainly become a pushing after it, even though the people forced by the long martyrdom are not definitely conscious of it. In this connection I was especially amused by the pittoresque description of the Heidelberg experiment, where the whole law faculty, including two young women, ran after the table like mad, so that poor Zöpflein, who, as I hear, is said to be a true Falstaff in figure, finally couldn't keep up with the others.

Table rapping was alleged to be connected with animal magnetism in some occult manner. The following lines in a letter of May 10, 1853, show that Gauss was unimpressed by all that he heard and read on the subject. The letter is addressed to Humboldt:

I have been able to observe the present-day tomfooleries rather calmly, indeed to laugh heartily about several genre pictures like the experiments of the Heidelberg law faculty with table rapping. I have long been accustomed to have a low opinion of the sterling quality of the higher culture which the so-called upper classes believe they can acquire by reading popular writings or attending popular lectures. I am rather of the opinion that in scientific areas genuine insight can be gotten only by use of a certain measure of one's own effort and of one's own processing of that offered by others.

This was a mild criticism of Humboldt, who spent considerable time in delivering popular lectures and publishing popular works. At least, Gauss let him know that he doubted the efficacy of the results.

CHAPTER NINETEEN

Monarch of Mathematics in Europe

Augustus De Morgan in his *Budget of Paradoxes* (p. 187) relates the following story. Francis Baily[1] wrote a singular book, *Account of the Rev. John Flamsteed, the first Astronomer-Royal*. It was published by the Admiralty for distribution, and the author drew up the distribution list. Certain rumors led to a run upon the Admiralty for copies. The Lords were in a difficulty; but on looking at the list they saw names, as they thought, which were so obscure that they had a right to assume Mr. Baily had included persons who had no claim to such a compliment as presentation from the Admiralty. The secretary requested Mr. Baily to call upon him.

"Mr. Baily, my Lords are inclined to think that some of the persons in this list are perhaps not of that note which would justify their lordships in presenting this work."

"To whom does your observation apply, Mr. Secretary?"

"Well, now let us examine the list; let me see; now—now—now—come! Here's Gauss—who's Gauss?"

"Gauss, Mr. Secretary, is the oldest mathematician now living, and is generally thought to be the greatest."

Their lordships ultimately expressed themselves perfectly satisfied with the list.

When Alexander von Humboldt returned in 1804 from his

---

[1] When David Forbes (1828–1876), the mineralogist, chemist, and metallurgist of Edinburgh, visited Göttingen in the summer of 1837, Francis Baily gave him a letter of introduction to Gauss.

American tour and remained in Paris for a time, he found the name of Gauss mentioned with the greatest respect. Before his tour he had never heard the name. In 1807 he wrote his first letter to Gauss and told him of Laplace's high esteem for him. No other name in nineteenth-century mathematics has received recognition equal to that of Gauss. His standing in physics and astronomy was perhaps slightly less. Contemporary scholars recognized his unconditional intellectual superiority in his own field. Gauss enjoyed an almost superhuman respect and admiration at the hands of those competent to judge him. There were a half-dozen exceptions to this, persons motivated by purely personal feelings.[2]

As early as 1813, his friend Olbers while serving as a deputy of Bremen in Paris during the Napoleonic period tried to persuade Gauss to take a trip to Paris, assuring him he would find such a reception as no other scholar had ever received. Laplace is said to have urged Napoleon to spare Göttingen because "the foremost mathematician of his time lives there." Even Legendre, who strongly disliked Gauss, in the foreword to the second edition of his *Théorie des nombres* (1808) spoke of Gauss' *Disquisitiones arithmeticae,* praised its high value, rich content, and recognized its complete originality. Gauss' career does not show a gradual rise, as in the case of most scientists; it begins at a high point and continues at that level.

In 1805, when the King of Prussia requested Humboldt to enter the Berlin Academy of Sciences in order to lend it the splendor of his name, acquired on the American tour, Humboldt informed the King that his appearance would not be of importance; the *only man* who could give the Berlin academy new splendor, he wrote the King, was Carl Friedrich Gauss.

Even Jacobi, who perhaps ranked second to Gauss in mathematical attainments among the contemporaries and who was inclined to be critical of Gauss' character, admitted his intellectual supremacy. After spending a week with Gauss in 1840, Jacobi

---

[2] Legendre, Abel, Jacobi, Dühring, Tait, and Halstead.

wrote his brother, the physicist: "Mathematics would be on an entirely different spot, if practical astronomy had not diverted this *colossal genius* from his glorious career."

Was Gauss conscious of his high position in science? Yes, without question and at an early age. His demands of himself, however, were as great as those he made of others. Yet he was modest and never showed this consciousness of greatness unless there was some special stimulus. He never wore the various medals and orders which had been showered on him, except that when the King came to Göttingen he would wear the Guelph Order. As a youth, Gauss is said to have gloomily sensed the fact that he did not have a friend of equal genius with whom he could discuss his scientific problems.

In 1808 an excusable show of pride was manifested by Gauss. Breitkopf and Härtel were printing his *Theoria motus corporum coelestium* for the publisher Perthes. The printer proposed that the final reading of proof be dropped in order to save time and postage. Gauss boldly wrote to the publisher: "I place such a value on the correctness of this work, which has caused me much work for several years, and if I am not mistaken, will still be studied even after centuries, that I am satisfied if you charge to me the outlays made by Messrs. Breitkopf and Härtel in this respect, if I myself can first check all folios before and afterward."

This assurance of superiority had the advantage of lifting him above petty polemics. When Schumacher called his attention to the fact that an Italian scientist in a periodical had made a "highly insolent" attack on Gauss' theory of magnetism, and Schumacher wanted to know whether Gauss thought it desirable that someone else "straighten out the impudent fellow," Gauss answered that he had already heard of the criticism from another source and added that the deductions of the critic were "dumb stuff," a judgment which agreed with his own impression gained by reading another essay of the same author in the field of magnetism. Gauss added: "Supposedly that journal is in our library, but I do not consider it worth a library catalogue card, and since you report that it is rudely written, I shall in no case read it."

Unfortunately the Italian critic did not live long enough to see the publication of this letter. Gauss not only ignored such criticism, but he prevented others from replying in his name. Polemic natures were an abomination to him. Justus von Liebig, the great chemist, was once candidate for a professorship in Göttingen. He later told that Gauss and the geologist J. F. L. Hausmann (1782–1859) were against him because he was always involved in some polemic and they did not want to have a cantankerous colleague.

Modesty and self-criticism characterized Gauss' behavior when he had to dissent from the views of those scientists he had learned to esteem highly. He stated that he distrusted his own views if they touched on a subject he had not studied thoroughly. On the other hand if some mathematical matter was involved for which he had rigorous proof he spoke out openly and freely. It must be added that non-Euclidean geometry was an exception to this behavior. He was timid if subjective opinions were concerned, or if the issue depended on individual estimate of probability.

In purely scientific matters Gauss could be very sharp in private, where he came upon stupidity, arrogance, or pretense. Of a friend he once wrote: "Mr. Benzenberg seems to write his letters as well as his books in his sleep." The astronomer Franz von Paula Gruithuisen (1774–1852), a professor in Munich,[3] frequently aroused the ill will of Gauss, who accused him of "mad chatter" and said that whenever he picked up anything by Gruithuisen he expected to find nonsense. Gruithuisen claimed to have discovered a city and highways on the moon. Actually Gauss was greatly amused at the polemic between Gruithuisen and the philosopher Schelling. After Herbart had given him some of Schelling's writings, he said: "The two opponents seem to me to be completely worthy of each other."

On May 14, 1826, Gauss discussed in a letter to Olbers a *cause célèbre* in the history of astronomy. Encke had accused[4] the

---

[3] He visited Gauss in August and September, 1825.

[4] "Imposture astronomique grossière du Chevalier d'Angos," *Correspondance Astronomique*, IV (1820), 456.

Chevalier d'Angos (d. 1836), knight of the Maltese Cross, of deception in the matter of a comet he had allegedly discovered in 1784. The astronomical world regarded the case as proved after Encke's publication. Gauss alone was more cautious and humane. He called d'Angos a "wind bag" and considered the deception very probable, but hesitated to pronounce him guilty until there was proof positive. He told Olbers in the letter: "I take the expression proof here not in the sense of the lawyers, who set two half proofs equal to a whole one, but in the sense of the geometer, where $\frac{1}{2}$ proof $= 0$, and it is demanded for proof that every doubt becomes impossible."

Felix Klein stated[5] that some Gaussian manuscripts reveal a knowledge of the fundamental ideas of quaternions, the discovery (in 1843) of the great Irish mathematician, Sir William Rowan Hamilton. P. G. Tait declared[6] that Klein was mistaken and that the Gaussian restricted forms of linear and vector operators do not constitute an invention of quaternions.

As a matter of fact, Hamilton had expressed himself on the subject in a letter to Augustus De Morgan on January 6, 1852:

In fact, with all my very high admiration . . . for Gauss, I have some *private* reasons for believing, I might say *knowing,* that he did not anticipate the quaternions. In fact, if I don't forget the year, I met a particular friend, and (as I was told) pupil of *Gauss, Baron von Waltershausen,* . . . at the second Cambridge meeting of the British Association in 1845, just after *Herschel* had spoken of my quaternions and your triple algebra, in his speech from the throne. The said Baron soon afterwards called on me here (Dublin), . . . he informed me that his friend and (in one sense) master, *Gauss,* had long *wished* to frame a sort of *triple algebra;* but that his notion had been, that the third dimension of space was to be symbolically denoted by some *new transcendental,* as *imaginary,* with respect to -1, as that was with respect to 1. Now you see, as I saw then, that this was in *fundamental contradiction* to my plan of treating *all dimensions of space with absolute impartiality, no one more real than another.*[7]

---

[5] *Mathematische Annalen,* LI (1898).
[6] *Proceedings* of the Royal Society of Edinburgh, December 18, 1899.
[7] Graves, R. P., *Life of Sir William Rowan Hamilton,* III (1889), 311–312.

A tradition states that Laplace, upon being asked who was the foremost mathematician in Germany, replied, "Bartels,"[8] whereupon the questioner wanted to know why he didn't name Gauss. Then Laplace said: "Oh, Gauss is the greatest mathematician in the world."

When Niels Henrik Abel (1802–1829) of Norway, one of the most important mathematicians of the nineteenth century, went to Germany in 1825 he had originally intended to visit Gauss. He was not well known at the time. A copy of his proof of the impossibility of solving the general equation of the fifth degree had been sent to Gauss, and Abel thought Gauss did not do enough to put him before the public. After that he had no use for Gauss and was extremely critical of him. He did not go to Göttingen; it is regrettable that the two did not meet. Abel had wanted to use the splendid university library in Göttingen. Too late, Gauss realized what he had missed, for he wrote to Schumacher on May 19, 1829: "Abel's death, which I have not seen announced in any newspaper, is a very great loss for science. Should anything about the life circumstances of this highly distinguished mind be printed, and come into your hands, I beg you to communicate it to me. I would also like to have his portrait if it were to be had anywhere."

Aurel Edmund Voss (1845–1931), professor of mathematics in Munich, commented, in a lecture he delivered for the Gauss Centenary in 1877, on the isolated position which Gauss seemed to his contemporaries to assume. Abel and Jacobi were acutely conscious of this and ascribed it to pride. But they were wrong. It was merely the isolation of a pioneer. Posterity knows more about such matters than do contemporaries. On at least one occasion Gauss expressed himself clearly on this point. His personal view is found in a letter to Bolyai on April 20, 1848:

It is true, my life has been adorned with much that the world considers worthy of envy. But believe me, dear Bolyai, the *austere* sides of life, at least of mine, which move through it like a red thread, and

---

[8] Ernst Schering's version of the anecdote mentions Pfaff instead of Bartels.

before which one is more defenseless in old age, are not balanced to the hundredth part by the joyous. I will gladly admit that the same fate, which has become so difficult for me to bear, and still is, would have been much easier for many another person, but the mental make-up belongs to our ego, the creator of our existence has given it to us and we can do little to change it. On the other hand I find in this consciousness of the nothingness of life, which the greater part of humanity must in any case express on approaching the goal (death), that it offers the strongest guarantee for the succession of a more beautiful metamorphosis. My dear friend, let us console outselves with this and thereby seek to gain the necessary equanimity, in order to tarry with it until the end. *Fortem facit vicina libertas sanem,* says Seneca.

In reviewing an essay Laplace had published in the *Connaissance des temps* (1816), Gauss wrote Olbers on December 31, 1831: "The essay . . . is in my judgment quite unworthy of this great geometer. I find two different, very gross blunders in it. I had always imagined that among geometers of the first rank the calculation was *always* only the dress in which they present that which they created not by calculation, but by meditation about the subject itself."

The monarch expressed himself on Lagrange in a letter to Schumacher on January 29, 1829: "The reproof hits Lagrange, like almost all analysts of modern times, of not *always* keeping the subject actively in mind during the game of signs."

Gauss expressed himself in general terms about mathematicians in France and England in a letter to Schumacher on August 17, 1836: "For my part . . . I joyously accept all *true scientific* progress which is made beyond the Rhine and beyond the channel, but when they commit stupidities over there . . . then there is nothing further to be done, except to take *no* notice *at all* of it."

In 1837 Humboldt had to decide whether to attend a convention of scientists in Prague or the Göttingen University Jubilee, since they met at the same time. As indicated in Chapter XVI, he spent about a week as Gauss' house guest. The reason for his decision is found in a letter he wrote to Gauss on July 27, 1837: "Several hours with you, dear friend, are dearer than all the sectional meetings of the so-called scientists, who move in such great masses and so gastronomically that there has never been enough scientific intercourse for me. At the end I have always asked myself like the mathematician at the end of the opera, *en dites-moi franchement ce que cela prouve.*"

When the publication of the correspondence between Gauss and Bessel[9] and the correspondence between Bessel and Olbers was being planned in 1848, Gauss wrote Schumacher on December 23, 1848, that he knew well how many of the compliments in almost all Bessel's letters ought to be omitted. This occurred because Bessel liked to say something pleasant to people, or something he presumed they would like to hear. Gauss felt that if the letters were published during his own lifetime, such passages should be cut out. When the Bessel-Olbers correspondence appeared, Gauss felt that there were certain passages in Olbers' letters that should have been omitted, since they indicated a lack of mature judgment and were more of a private chat on matters of science, contemporary scientists, and related subjects.

Moritz Cantor (1829–1920), later world-famous as a historian of mathematics and professor at the University of Heidelberg, delivered a lecture there on November 14, 1899, almost exactly a half century after his arrival in Göttingen late one October evening in 1849. Gauss was the subject of the lecture, at the close of which Cantor gave intimate and charming reminiscences of the great master:

I intended to enroll for the winter courses under Stern, Weber, Listing, and Gauss. I heard the former, but Gauss was not lecturing. It was the same in the summer of 1850, Gauss was not lecturing. In the winter 1850–1851 Gauss taught the announced course on the method of least squares, and I attended it. As far as I know, it was the last course which he taught.[10] Later, mostly just as previously, he always had an excuse for not conducting his announced courses. It seems to me as though I were looking into the office in which the course was held. We listeners were sitting around a large table covered with books. Gauss

---

[9] It did not appear until 1880.
[10] This statement is not correct.

was sitting in an armchair at one narrow side of the table and beside him on an easel there was a moderately large wooden blackboard on which he calculated with chalk. Gauss wore the black velvet house-cap ... and when he got up, he continually had his left hand in his trouser pocket. ... The use of ink was excluded by the small amount of space available for each listener at the table, but Gauss disliked even the taking of notes with pencil. Once when we wanted to take notes, he said: "Dispense with writing here, and pay better attention." It is easy to comprehend that even with the strictest attention and the best memory one was not able later to finish any accurate notebook. One was able to reproduce only especially ingenious, separate deductions. Moreover such turns of speech and interpolated remarks as had nothing to do with science, stuck in one's memory, but they characterized the speaker. ... On another occasion the table was covered with logarithm tables. Gauss explained their differences according to color of paper, formation of numbers, whether of equal size or whether they projected above or below the line; he spoke of the number of decimal places in the tables, of their calculation and rose to the remark, uttered very seriously: "You have no idea, how much poetry is contained in the calculation of a logarithm table." In other utterances one noted the wag. . . . Laplace also wrote besides his great scientific work on the calculus of probabilities an *Essai philosophique sur les probabilités,* which found the quickest sale. Gauss had laid the three first editions on the table and showed us in the first edition a statement that the conqueror only harms his own country instead of helping it, which is missing in the second edition and returns in the following ones. The first edition appeared while Napoleon was on Elba, the second during the hundred days, further editions followed in measured intervals. Gauss even knew how to put in linguistic remarks, one of which I will not hold back. He declaimed against the phrase *möglichst gut.* Not the possibility is increased, but the goodness, therefore one would have to say *bestmöglich.*

This course on the method of least squares just described by Cantor had nine students in it. Of the nine, only five paid the required fee, and Gauss received the handsome sum of twenty-five thalers for giving the course. Later on, two additional students paid their fees. The names of the students were: A. Ritter, M. Cantor, R. Dedekind, Lieutenant von Uslar, Chr. Menges, A. Valett, L. Hildebrand, G. Wagener, J. C. Lion. Most of these names are forgotten today.

August Ritter (1826–1908) was born in Lüneburg and worked

for the Leipzig-Dresden railroad as a draftsman until 1850. In that year he went to Göttingen where he studied until 1853 and took his doctor's degree. Later he spent some time as an engineer in Rome and Naples, and in 1856 was called to teach engineering and mechanics at the Technical Institute in Hanover. From 1869 until his retirement in 1899 he was a professor in Aachen. His last years were spent in Lüneburg.

It would be interesting to know more about the others. Cantor made an international reputation for himself. By far the most important person among the students was Richard Dedekind (1831–1916), one of the creators of the modern theory of algebraic numbers. In 1901 Dedekind set down his own reminiscences of the course in great detail. Part of them have special biographical interest for us:

As a native of Brunswick I early heard Gauss mentioned, and I gladly believed in his greatness without knowing of what it consisted. It made a deeper impression on me when I first heard of his geometric representation of imaginary or, as one still said at that time, impossible quantities. At that time as a student at the Collegium Carolineum (the present-day Institute of Technology) I had penetrated a little into higher mathematics, and soon afterward, when Gauss celebrated the golden jubilee of his doctorate in 1849, our faculty sent to him a congratulation composed by the brilliant philologist Petri, in which the passage, he has made possible the impossible, especially attracted my attention. At Easter, 1850, I went to Göttingen, and there my understanding increased somewhat when I was introduced to the elements of number theory in the seminar by a short, but very interesting course by Stern.[11] On my way to or from the observatory, where I took a course of the excellent Professor Goldschmidt on popular astronomy, I occasionally met Gauss and rejoiced at the sight of his stately, awe-inspiring appearance, and very often I saw him at close range at his usual place in the Literary Museum, which he visited regularly in order to read newspapers.

At the beginning of the following winter semester I considered myself mature enough to hear his lectures on the method of least squares, and so, armed with the lecture attendance book and not without heart palpitation, I stepped into his living room, where I found him sitting at a desk. My announcement seemed to gladden him very little, I had

---

[11] Moritz A. Stern (1807–1894).

also heard that he did not like to decide to conduct courses; after he had entered his name in the book, he said after a short silence: "Perhaps you know that it is always very doubtful whether my lectures materialize; where do you live? at the barber Vogel's? Well, that's a piece of good luck, for he is also my barber, I shall notify you through him."

Several days later Vogel, a character known throughout the city, quite filled with the importance of his mission, entered my room in order to tell that several other students had announced themselves and that Privy Court Councillor Gauss would conduct the course.

There were nine of us students, of whom I gradually became more closely acquainted with A. Ritter and Moritz Cantor; we all came very regularly, rarely was one of us absent, although the way to the observatory was sometimes unpleasant in winter.[12] The auditorium, separated from Gauss' office by an anteroom, was rather small. We sat at a table, whose long sides offered a comfortable place for three, but not for four, persons. Opposite the door at the upper end sat Gauss at a moderate distance from the table, and when we were all present, then the two of us who came last had to move up quite close to him and take their notebooks on their lap. Gauss wore a lightweight black cap, a rather long brown coat, grey trousers; usually he sat in a comfortable attitude, looking down, slightly stooped, with hands folded above his lap. He spoke quite freely, very clearly, simply and plainly; but when he wanted to emphasize a new viewpoint, in which he used an especially characteristic word, then he suddenly lifted his head, turned to one of his neighbors and gazed at him with his beautiful, penetrating blue eyes during the emphatic speech. That was unforgettable. His language was almost free of dialect, only sometimes came sounds like our Brunswick dialect; in counting, for example, in which he was not ashamed to use his fingers, he did not say *eins, zwei, drei,* but *eine, zweie, dreie,* and so forth, as one can even now hear among us at the market place. If he proceeded from an explanation of principles to the development of mathematical formulas, then he got up, and in stately, very upright posture he wrote on a blackboard beside him in his peculiarly beautiful handwriting, in which he always succeeded through economy and purposeful arrangement in making do with the rather small space. For numerical examples, on whose careful completion he placed special value, he brought along the requisite data on little slips of paper.

On January 24, 1851, Gauss closed his presentation of the first part of his course, through which he familiarized us with the essence of the method of least squares. There followed now an extremely clear development of the fundamental concepts and principal theorems of the

---

[12] The course met three hours a week.

calculus of probabilities, explained by original examples, which served as an introduction to the second and third mode of establishing the method, which I must not go into here. I can only say that we followed with ever increasing interest this distinguished lecture, in which several examples from the theory of definite integrals were also treated. But it also seemed to us, as if Gauss himself, who previously had shown little inclination to conduct the course, sensed some joy in his teaching activity during it. Thus the close came on March 13, Gauss got up, all of us with him, and he dismissed us with the friendly farewell words: "It only remains for me to thank you for the great regularity and attention with which you have followed my lecture, probably to be called rather dry." A half century has now passed since then, but this allegedly dry lecture is unforgettable in memory as one of the finest which I have ever heard.

Throughout his later years Gauss was continually sought out by German and foreign scientists for conferences and personal meetings. A sample of this is found in a letter he wrote to his daughter Minna on July 16, 1839:

This summer I am still expecting many visits here. One or two Englishmen will come probably in the next weeks, in order to discuss magnetic observations with me, which are to be instituted on the expedition of two English ships to the south polar regions. At about the same time Kupffer[13] of Petersburg (a former student of mine) will come, also principally on account of observations to be made in the Russian Empire. In a similar connection Hansteen of Norway has announced himself for the end of August. In September comes Dirichlet of Berlin, a man whom I esteem very highly, and who is especially friendly with Dirichlet [sic!].[14]

As I hear, Listing will also arrive very shortly, although at first only for several days, in order to journey then to his native city of Frankfurt, until he begins his professorship here.

Although all these visits from outside are in themselves dear to me, yet it will be strenuous for me if they fall on hot days.

On July 15, 1838, Gauss mentions in a letter to Olbers a visit of Sir John Herschel: "Mr. Herschel arrived here yesterday eve-

---

[13] Adolf Theodor Kupffer (1792–1865) studied in 1820 in Göttingen, and went to St. Petersburg in 1828 as professor of mineralogy, and after 1840 physics. In 1843 he was made director of Russia's Central Magnetic-Meteorological Institute.

[14] Probably he meant to write "Humboldt."

ning. He had journeyed by stage-coach directly from Harburg to Hanover, and came here after he had been there 2 or 3 days. Perhaps I can persuade him to return via Bremen."

Christopher Hansteen (1784–1873), mentioned in the above letter to Minna, was professor of astronomy and director of the observatory in Oslo. He early became interested in terrestrial magnetism and published several works in the field, having been one of the first who participated in the observations of the Magnetic Association. Hansteen stayed two weeks with Gauss and practiced the use of the two magnetic instruments which Gauss had invented. He ordered reproductions of them for the magnetic observatory which was being built in Oslo. Kupffer also ordered both instruments for St. Petersburg and three or four places in eastern Russia.

Kupffer returned to Göttingen in mid-October to attend a Magnetic Congress at which Sabine [15] of London, Lloyd [16] of Dublin, and Steinheil [17] of Munich were present. Mrs. Sabine, who had translated Gauss' *Allgemeine Theorie des Erdmagnetismus,* accompanied her husband. The translation was made primarily for the use of army officers who were going on the Antarctic magnetic expedition.

In the last year of his life Gauss was annoyed several times by a religious crank, a Quedlinburg schoolteacher and doctor of theology named Carl Schöpffer. He delighted in attacking the Copernican system of astronomy and had delusions of persecution. On several occasions he reported miraculous escapes, claiming to have been in physical danger from his imagined foes. Schöpffer had received rather short shrift at the hands of Karl von Raumer, Encke, and Johann von Lamont.

---

[15] Edward Sabine (1783–1883), a British general, was the director of various scientific expeditions and active in magnetic research.

[16] Humphrey Lloyd (1800-1881), professor of physics in the University of Dublin, was widely known for his work on conical refraction. He erected a **magnetic observatory in Dublin.**

[17] Karl August Steinheil (1801–1870) after 1832 was professor of mathematics and physics at the University of Munich.

Schöpffer went to Göttingen in the spring of 1854, having resolved to use the university library in his astronomical "research." He founded a monthly magazine called *Blätter der Wahrheit*, which died after the ninth issue. Soon after his arrival in Göttingen Schöpffer got acquainted with Gauss, who treated him in a clever, yet friendly manner. Gauss gave him some books to read and told him to seek his advice whenever he believed he needed it. Alexander von Humboldt had also accorded him a friendly hearing.

Schöpffer recounted at great length the course of his previous rebuffs in opposing the Copernican system, at the same time mentioning as authority the names of several earlier astronomers and philosophers who had been on his side of the controversy. Gauss listened to the long harangue in silence, which Schöpffer fortunately interpreted as approval. Gauss was in poor health during his last year and did not feel able to argue with this fanatic who actually believed that Gauss had some doubts about the Copernican system. Humboldt had given him a shrewd answer by saying that if he could find some astronomer of repute to come out against the system, he (Humboldt) would immediately declare himself against it. When Gauss heard this alleged statement of his old friend, he merely answered: "If I were only twenty years younger!"

Schöpffer reported that he was subjected to arsenic poisoning in Göttingen and that one day on the street there he escaped injury from a nearby explosion.

The thirty-first annual convention of German scientists and physicians met in Göttingen September 18–23, 1854. Schöpffer attempted to get on the program in order to make another attack on the Copernican system. Three Göttingen professors, Baum, Listing, and Weber, were instrumental in keeping him off the program. Finally he planned to walk into the meeting and start speaking, mistakenly believing that he could count on the support of Gauss and Hausmann. On the way to the meeting he learned that Gauss was ill and could not attend and that Hausmann was out of the city on an unexpected journey. Thus Schöpffer had to drop his bold plan.

The entire episode ended when Schöpffer returned to Quedlinburg from Göttingen at Easter, 1855, soon after Gauss' death.

CHAPTER TWENTY

The Doyen of German Science, 1832-1855

Heinrich Ewald married as his second wife on December 12, 1845, Auguste Schleiermacher (1822–1897), daughter of a prominent finance councilor and head librarian in Darmstadt.[1] Gauss was quite pleased when his son-in-law wrote him of his engagement, and sent him a cordial letter of congratulation on October 30, in which he stated that he had long hoped Ewald would marry again, and realized how lonesome he must have been, for under similar circumstances he married in less than a year, whereas Ewald had been a widower five years. He reminded Ewald in the letter that he had corresponded with the bride's uncle Ludwig Schleiermacher, author of a book on analytical optics, and had a high opinion of his scientific training.

In 1847 the King of Hanover restored the liberal constitution of 1833, and, at the earnest solicitation of the academic senate, an offer was made in 1848 to Ewald to return, on most generous terms, to the position from which he had been discharged ten years before. The offer was gladly accepted, for Ewald had never been quite happy in Tübingen, and he was a 100 per cent Hanoverian.

A daughter was born to Ewald on May 5, 1850, and she was given the name Caroline Therese Wilhelmine, Therese Gauss serving as godmother. At the christening Gauss, Ewald, Weber, and the others in the group played croquet on the lawn after

---

[1] Schleiermacher held the title of *Geheimrat* and was on intimate terms with the Grand Duke.

they had drunk chocolate. Miss Ewald continued to live at her father's old home in Göttingen until her death on May 5, 1917; she never married. She and her mother in later years often visited Joseph Gauss and family at his home on Wilhelmstrasse in Hanover. In July, 1859, Miss Ewald and her mother visited Therese Gauss at her home in Dresden. From 1848 to 1866 Ewald worked happily and assiduously at Göttingen, his death occurring there on May 4, 1875.

Gauss was again made happy in 1849 when Weber was called back to his professorship in Göttingen. But Gauss was too old to collaborate with him again on the brilliant research in which they had engaged more than a decade before. It cheered him to have his intimate friends visit him in the late afternoon and early evening hours. Just to have Ewald and Weber near meant much to him. At least two members of the Göttingen Seven had returned to the fold. Weber lived on to attain the ripe old age of eighty-seven, and died June 23, 1891. He was busy with his work in physics during most of his life, but so retired was his life in his last years that many scientists had forgotten he was still alive. Near the close of his life Weber's memory failed. The guest book of the students' mathematical club in Göttingen shows his name entered in January, 1891. Hermann Amandus Schwarz (1843–1921), who was professor of mathematics in Göttingen until 1892, had taken him to a meeting of the club. After his memory failed, Weber used to express frequently the desire to return to Göttingen, and when he was told that he *was* actually in Göttingen, there came the answer: "No, that is not the Göttingen of Gauss."

Gauss was often sought out by young scientists recommended to him by Alexander von Humboldt. As he grew older he became more and more convinced that mathematics could be studied from books without a teacher. In his day many students were poorly prepared, and he complained repeatedly about this condition in his letters. He felt that his lectures must be worthy of himself and his science. This explains why he usually had only a handful of students at any time while several blocks away his colleague B. F. Thibaut (1775–1832) was lecturing to a hundred students on elementary mathematics. It was Gauss' custom to announce the same courses year after year; there was very little variety. The result is that it is hardly possible to speak of a Gaussian School in mathematics. Distinguished mathematicians profited by his writings, but perhaps only one of them enjoyed the intimate relationship of pupil to teacher. That was Bernhard Riemann, and as a matter of fact he was already at the level of a master. Probably one should include Eisenstein, who will be discussed shortly, and Dedekind on this list with Riemann.

On the other hand, it is correct to speak of a Gaussian School in astronomy, at least for a certain time. Schumacher led off in 1808, then in 1810, and soon thereafter came Encke, Gerling Möbius, and Nicolai, who sought and found theoretical and practical training at the Göttingen observatory.

Gauss always took a personal interest in his students. He was grieved that two of his gifted students died young. One of them, Johannes Friedrich Posselt, was born in 1794 as the son of a clergyman on the Danish island of Föhr. He served from 1819 until his death in 1823 as professor of mathematics at the University of Jena. Posselt was acquainted with Goethe through the latter's official connections with the university.

Friedrich Ludwig Wachter was born in Cleve in 1792 as the son of a high school principal who was later transferred to Hamm in Westphalia. Wachter in his "vita" for the Göttingen faculty expressed thus his purpose in study: "not to burden the memory with many things, but to recognize the true reason of scholarship and to sharpen the intellect and to train judgment." He specialized in astronomy under Gauss in 1810 and 1811 and even as a student published some astronomical calculations at the instigation of his teacher. For acquiring the doctor's degree in 1813 he wanted to use a purely mathematical topic from the theory of curved surfaces. Meanwhile Wachter had been called to teach at the high school in Altenburg, on the basis of the splendid report of his examiners. Gauss and Thibaut were not among the examiners; the examination in mathematics was handled by Tobias Mayer, the physicist. Wachter had been admitted to the oral examination after he had promised to make up the dissertation. He had refused to defend theses because of inexperience in speaking Latin! The case of Gerling was mentioned as a precedent. This request was granted; it shows how progressive Göttingen was at that time in contrast to many other German universities, even in later decades. Wachter owed his job in Altenburg to Gauss, whose strong recommendation was enough to get it for him. His teaching in Altenburg was interrupted from November, 1813, to July, 1814, by army service.

Wachter's dissertation, written after his return from the army, was on an astronomical subject; to it was attached a five-page note on the parallelogram of forces. The mysterious and tragic death of Wachter was discussed in Chapter XV.

Another pupil of Gauss, who name is still well known in mathematics for his work in synthetic geometry and the geometric interpretation of imaginary elements, was Karl Georg Christian von Staudt, who was born in 1798 in Rothenburg ob der Tauber and served as professor of mathematics at the University of Erlangen until his death there in 1867. A remark in a letter to Bessel on December 15, 1826, shows the high opinion Gauss had of von Staudt. During his stay of several years in Göttingen von Staudt felt that he was not only benefited by the master's teaching, but also gladdened by his recognition and praise. Whenever von Staudt handed Gauss his solution of an assigned problem, Gauss would give the student his own solution, with the joking remark that he counted on mutual satisfaction.

In a letter on April 14, 1822, Bessel recommended to Gauss his pupil Heinrich Ferdinand Scherk, who was born in 1798 in Posen and went to Göttingen with financial support from the Prussian government. Gauss was highly pleased with Scherk and considered him one of the best pupils he ever had. Scherk's later success justified Gauss' praise. In 1826 he became professor of mathematics at the University of Halle, where he numbered among his students E. E. Kummer, who ranked as one of the leading mathematicians of the nineteenth century. In a sense, then, Gauss may be thought of as the mathematical grandfather of

Kummer. Scherk was called in 1834 to Kiel, which was then Danish; his teaching there was highly successful, but he was too much of a German patriot and agitated for the founding of a German navy. In 1852 he lost his Kiel professorship through politics and took refuge in Bremen as a teacher in a public school, dying there in 1885.

H. B. Lübsen (1801–1864), a former Oldenburg petty officer, worked himself up by his own efforts and was very successful as a private teacher of mathematics in Hamburg. His textbooks for self-instruction in algebra and arithmetic went through many editions and were popular until recent times; they are said to have exerted great influence on young students. In the foreword to the first edition of a textbook[2] on higher geometry, Lübsen wrote: "Theory, said my revered teacher Gauss, attracts practice as the magnet attracts iron."

Another student of Gauss deserves mention, even though his name has almost been forgotten. Ludwig Christoph Schnürlein was born on April 14, 1792, in Ansbach as the son of an innkeeper. He was sickly from birth and could not be sent to school until the age of nine. After confirmation he thought of preparing for his father's business. One Buzengeiger, who was then teaching at the high school in Ansbach and frequented the Schnürlein tavern, discovered the boy's special talent for mathematics. He encouraged him to devote himself to this study, gave him private instruction and all manner of aid. On his recommendation Schnürlein was able to begin studies at the University of Tübingen at the age of twenty-seven (Easter, 1819), without having graduated from high school. In his fourth semester Schnürlein transferred to the University of Erlangen. Soon afterward he passed an examination in Munich with such brilliant success that he received from the Academy of Sciences for three years a stipend of five hundred florins, and was advised to study astronomy in Göttingen.

Schnürlein joyfully seized the opportunity and used it in the

---

[2] This book was dedicated to the spirit of Descartes.

best possible manner. For him Gauss was a model of the teacher and the object of his highest reverence. In the Gauss Archive there is a letter from a high official to Gauss, in which Schnürlein is discussed and these words occur: *Für Schnürlein sind Sie der liebe Gott.* In the published Gauss correspondence occasional mention is made of Schnürlein's comet calculations. In 1824 Schnürlein became assistant at the observatory in Bogenhausen. When this position was abolished after two years, he participated in Munich in a competitive examination for a teaching job in mathematics and physics; he got the grade of "excellent ability." The same year he was appointed to the high school faculty in Erlangen, and in 1830 to the same job in Hof. There he was active for about twenty years and enjoyed great success, until his retirement, when he moved to Bamberg; his death occurred on November 7, 1859. In August, 1850, the University of Erlangen conferred the honorary doctorate on Schnürlein, at the instigation of von Staudt.

Schnürlein published an elementary calculation of the length of arc of the ellipse and hyperbola, relations among surfaces of the second order, extension and generalizations which are connected with Bernoulli's numbers. He also published interpretations of Gauss' methods for calculating the elements of the orbit of a comet.

One of Gauss' prominent students was Justus Georg Westphal, who was born on March 18, 1824, in Colborn near Lüchow in Hanover, studied in Göttingen, and was assistant at the observatory 1851–1855, having been appointed lecturer in 1854. His dissertation was *Evolutio radicum aequationum algebraicarum e ternis terminis constantium in series infinitas* (1850). He discovered the Comet 1852 III on July 24, observed solar and lunar eclipses, star occultations, comets, and asteroids. Westphal left Göttingen in the fall of 1855 and died on November 9, 1859, in Lüneburg.

No discussion of Gauss' students would be complete without mention of Moritz A. Stern, who was born in Frankfurt am Main on June 28, 1807, took his doctorate in Göttingen in 1829, and served on the mathematics faculty there from 1829 until his retirement in 1884. Stern died on January 30, 1894, in Zurich. He published extensively both in mathematics and astronomy. His address at the Gauss centenary in 1877 contained much valuable information and was well received.

A name frequently connected with Gauss in the late nineteenth century was that of Ernst Christian Julius Schering, who was born on July 13, 1833, in Sandbergen, northeast of Lüneburg. In 1852 he went to Göttingen, where he studied under Gauss, Weber, and later Dirichlet. In 1857 he took his doctorate with the dissertation *Zur mathematischen Theorie elektrischer Ströme* and in 1858 was appointed lecturer with a paper *Ueber die conforme Abbildung des Ellipsoids auf der Ebene*. Schering married Maria Malmsten (1848–1920), daughter of a mathematician and diplomat in Stockholm. He edited the first six volumes of the *Collected Works* of Gauss, and made them models of perfection; he had the advantage of personal association as a young man with Gauss and was very versatile. His widow placed his papers at the disposal of his successors in the editorship of the Gauss *Collected Works*. Schering served on the Göttingen faculty from 1858 until his death on November 2, 1897. He published works in mathematics and astronomy, as well as two important monographs on Gauss.

The last of Gauss' students who achieved fame in astronomy was Friedrich August Theodor Winnecke (1835–1897), who studied in Göttingen from 1853 to 1856. He served as director of the observatory in Strassburg from 1872 to 1886. Winnecke investigated the paths of double stars and comets and made determinations of the solar parallax. He published a splendid pamphlet on Gauss at the time of the centenary, in 1877, in which he gave many interesting glimpses of the master in his last years. The pamphlet was in the popular style.

Ernst Wilhelm Gustav von Quintus Icilius (1824–1885) studied under Gauss and served from 1849 to 1853 on the Göttingen faculty. He then served as professor of physics at the Polytechnic Institute in Hanover. Gauss considered his dissertation *Die Atom-*

gewichte vom Palladium, Thallium, Chlor, Silber, Kohlenstoff, und Wasserstoff nach der Methode der Kleinsten Quadrate berechnet (1847) only an *exercice de collège* and forced him to cut out of it some material which was contrary to the method of least squares. The reader may wonder how the poor fellow got the name of Quintus Icilius. Guichard was originally the family name. Colonel Karl Gottlieb Guichard had dared to correct Frederick the Great when he once erroneously called the Roman centurion Icilius by the name "Ilicius." Old Fritz forced the colonel to attach the two Latin names to his own.

Two other students of Gauss should be mentioned in passing, men who have not been entirely forgotten. The first was Theodor Wittstein,[3] who later served on the faculty at the Polytechnic Institute in Hanover. The other was Alfred Enneper (1830–1885), who served on the Göttingen faculty from 1859 until his death. He was regarded as a capable mathematician and published extensively.

Among the mechanics who furnished instruments to Gauss was Moritz Meyerstein (1808–1882), who came from the neighboring town of Einbeck. He was the owner of a large mechanical workshop, and Göttingen conferred an honorary doctorate on him. His successor was August Becker, who was born October 23, 1838, in Göttingen.

Gauss' own records show that he took in 2,267 thalers in student fees from 1808 to 1821. In 1845 they amounted to 422 thalers. During his last years of teaching, 1846–1853, Gauss had a total of 790 thalers; 67 paid promptly, 4 were free, and 15 paid some time later.

In his lectures Gauss frequently liked to have examples calculated, and assigned problems whose results he sometimes published. He had many calculations carried out by Gerling, Nicolai, Encke, and others, so that in a letter to Encke Nicolai once congratulated him on the completion of a little example in calcu-

---

[3] Wittstein published an important Gauss lecture which he delivered for the centenary in 1877.

lating (the preliminary calculation of the orbit of Pallas)—a problem which Gauss had assigned him as a problem.

Humboldt, in writing to Schumacher about Jacobi on July 3, 1844, made the following statement: "He is serene and miles gloriosus, recognizing in the triumvirate only two besides himself, *Gauss* and *Cauchy*—tout le reste lui parait de la vermine. I don't like these exclusions."

When Hansteen sent Abel's memoir on elliptic transcendentals to Schumacher with a request that he print them in his journal as soon as possible, he told Schumacher that when he handed Abel the last number of the *Astronomische Nachrichten*, the latter became pale and had to run to a bar and take some strong liquor to overcome his excitment. Abel felt that his method, which he had discovered several years previously, was general and more comprehensive than Jacobi's theorems. Abel feared that Jacobi would publish before he did. In recounting this incident by letter to Gauss on June 6, 1828, Schumacher added: "When you publish your researches it will probably cost him even more liquor."

One of the most peculiar attacks ever made on Gauss, indeed the most bitter among the few attacks, was made by Eugen Dühring (1833–1921). It is an interesting *curiosum* showing how far a brilliant, but pathologically deranged, mind can go, and should be preserved in the archives of science. Dühring had written a splendid essay on the principles of mechanics[4] and won a coveted prize. He worshiped Robert Mayer and bitterly attacked anyone who he thought had in any way wronged Mayer. His principal target was Helmholtz, whom he used to call "Helmklotz." Dühring made a study of personalities ranging from Archimedes to Lagrange. He was strongly anti-Semitic. The offenses against Mayer of most of those he attacked are obscure. In later years he published a magazine which he called the *Personalist and Emancipator*, a semi-monthly against "corrupt"

---

[4] *Kritische Geschichte der allgemeinen Prinzipien der Mechanik* (2d ed., 1877).

science. In his eyes Abel was a plagiarizer. When Justus von Liebig and Clausius defended Gauss and Riemann, he launched out at them too. Dühring used sharp words against Cauchy. Dühring wrote that university advertising had lifted Gauss up to heaven and stamped him a god. He accused Gauss of being religiously narrow, of having accomplished nothing worth while in mathematics, and of being too proud of his title *Hofrat*. Dühring scoffed at the peasant origin of Gauss, as well as his support by the Duke of Brunswick. He wrote that Gauss represented no type in the sense of the eighteenth century. Dühring liked to think of himself as an iconoclast in the temple of science; in reality, he was an arsonist, for he went beyond the mere besmirching of many great names. In the field of non-Euclidean geometry he really let himself go, for we read in the chapter on Gauss and "Gauss worship":

His megalomania rendered it impossible for him to take exception to any tricks that the deficient parts of his own brain played on him, particularly in the realm of geometry. Thus he arrived at a pretentiously mystical denial of Euclid's axioms and theorems, and proceeded to set up the foundations of an apocalyptic geometry not only of nonsense but of absolute stupidity. . . . They are abortive products of the deranged mind of a mathematical professor, whose mania for greatness proclaims them as new and superhuman truths! The mathematical delusions and deranged ideas in question are the fruits of a veritable *paranoia geometrica*.

It is astounding how history has turned the tables on Dühring and now applies the term "paranoia" to his distorted aberrations.

In June, 1844, Gauss received a letter from Humboldt recommending highly a most brilliant young mathematician, Ferdinand Gotthold Maximilian Eisenstein (1823–1852), the last survivor of six children of a Jewish mercantile family in Berlin. During the vacation in 1844 he visited Gauss, who was literally carried away by the genius of young Eisenstein. Humboldt had given him a hundred thalers for the trip. Eisenstein's papers began to appear in 1843 and came out rapidly; in a short lifetime he published fifty papers, principally in the theory of numbers and elliptic functions. He worshiped Gauss and Gauss looked on him almost as a son; he was his favorite mathematician. Eisenstein was never very healthy and was hypochondriac. He longed for companionship and did not get along too well with his family. Dirichlet, Jacobi, Stern, Gauss, Encke, and others tried to cheer him up. The University of Breslau conferred the doctorate on him in 1845. Gauss wrote the foreword to his *Mathematische Abhandlungen* (1847). In rapid succession he became a member of the Academies of Sciences in Breslau and Berlin, and of the Royal Society in Göttingen. His advancement was due largely to Humboldt and Gauss. Eisenstein was also talented in music. It is tempting to speculate what he would have accomplished had he attained an advanced age. Gauss deeply mourned his early death. Moritz Cantor quotes Gauss as having said there had been only three epoch-making mathematicians in all history: Archimedes, Newton, and Eisenstein. History has reversed the decision and given Gauss himself the place he gave Eisenstein.

Eisenstein, according to his own account, merely visited the so-called "democratic clubs" several times, but did not join them during the Revolution of 1848. He was only mildly involved, but the conservative Gauss was greatly concerned when he heard about Eisenstein's activity, and wrote thus to Encke on August 11, 1849:

From Mr. Dirichlet I believe I must conclude that our young friend, to whom I ask you kindly to give the enclosure, was exposed to the crudest insults on the unhappy night of March 18–19 (1848). I almost suppose that this is a misunderstanding (namely a misunderstanding of Dirichlet's story on my part). For that Eisenstein would have stood on the barricades is quite impossible, and if by an unhappy accident he had landed among the rioters, been captured, beaten with a club, and led off to Spandau, then certainly the democratic newspaper reporters would have made the most of something which fitted into their trash (even if it was the consequence of an unhappy error). In my enclosure I didn't want to mention the matter to Eisenstein himself.

On July 16, 1849, occurred the greatest outward triumph[5] of Gauss' life—the celebration of his golden jubilee. It was exactly

---

[5] Accounts of the jubilee appeared in the London *Athenaeum,* the *Astronomische Nachrichten,* and in the *Deutsche Reichszeitung,* No. 162 (July 21, 1849), under the title *Die Jubelfeier des Dr. Gauss in Göttingen.*

fifty years since he had attained his doctorate at the University of Helmstedt and offered in his dissertation the first rigorous proof of the fundamental theorem of algebra. Gauss was deeply moved on his day of reminiscence, honor, and retrospection. A large circle of friends, admirers, and grateful pupils had gathered around him to do him honor. Among those present were Jacobi, Dirichlet, Gerling, the astronomer Hansen, and Professor W. H. Miller, the mineralogist and crystallographer from Cambridge University. He received renewal of diplomas, medals, orders, congratulatory documents, and—what he prized most highly— honorary citizenship from the cities of Brunswick and Göttingen.

The Aula was decorated with flowers for a session of the Royal Society of Sciences. Gauss delivered a lecture on "Contributions to the Theory of Algebraic Equations," returning to a subject he had embraced in his dissertation and treating it from a more general viewpoint.

At a banquet in his honor he spoke of the ever active, serious scientific effort which had always blessed the university, and stressed the thought: "Trivial words have never rated in Göttingen." Then he gratefully emphasized the benevolence of the university's Board of Curators, under whose discerning direction the representatives of science, protected against the adversities of life, had been almost undisturbed in their research and therefore in the enjoyment of such peace frequently attained a noteworthy old age.

An interesting sidelight on the jubilee occurred when Gauss was about to light his pipe with a piece of the manuscript of his *Disquisitiones arithmeticae*. Dirichlet was horrified at what seemed to him sacrilege, rescued the paper from Gauss' hands, and treasured it for the rest of his life. The editors of Dirichlet's works found the manuscript among his papers.

Jacobi wrote a letter to his brother on September 21, 1849, in which he gave a glimpse of the banquet at the jubilee:

You probably know that I was with Dirichlet at Gauss' jubilee. I had the seat of honor there beside him and delivered a great speech. You know in 20 years he has quoted neither me nor Dirichlet; this time however after several glasses of sweet wine he was so carried away that he said to Dirichlet, who boasted to him of having studied his writings more than anyone else, that he had not merely studied, he had gone far beyond them. It is no longer easy to enter a scientific conversation with Gauss; he seeks to avoid it, while he discusses the most uninteresting things in a continuous stream. Except for Hansen and Gerling of Marburg nobody was there; our journey was therefore important in order to support to some extent a manifestation in honor of mathematics.

Therese Gauss wrote to her brother Eugene in St. Charles, Missouri, on December 5, 1850, and gave in her letter some interesting details of the jubilee:

A year and half ago in July '49, he [Gauss] celebrated his fiftieth anniversary jubilee of the doctorate—or rather the university and the city celebrated it for him with general love and sympathy. He himself was very much opposed to having this day noticed, but, without his knowledge, everything had been prepared for it. From near and far the university had invited strangers; father's friends and eminent scholars came, many delegations from other cities, who brought him congratulations, honorary doctor's diplomas, and three new orders. From Brunswick and Göttingen he received honorary citizenship; from the King, congratulations in his own handwriting and a higher order. There was no end of letters and communications. In the morning festive processions began to congratulate him, all the authorities of the city, of the university, of the public school, strangers, acquaintances—probably about fifty persons. Then father himself delivered a lecture in the Aula of the university, which was overcrowded with spectators and listeners and had been decorated with garlands and flowers like a fairy hall. Even the houses in the streets were decorated with flowers; in the city there were waves of people in festive attire, as on a holiday. When, at last, in the evening at seven, father came home from the great banquet, he was indeed quite exhausted, and it was well that the torchlight procession that the students had intended for him was abandoned upon his wish, but the love and sympathy which had been shown him from all sides had, in spite of all fatigue, pleased him indescribably. How sad it was though, that where so many strangers had congregated on his day of honor, not one of his beloved sons could be with him!! Even Joseph had been compelled to decline, as his position as railway director did not then make his absence from Hanover possible.

Since the University of Helmstedt, which had conferred the doctorate on Gauss, no longer existed in 1849, it was the University of Göttingen which renewed the diploma for the jubilee. On March 23, 1849, the University of Kasan had conferred an honorary doctorate on him, probably at the instigation of Lobachevsky, who felt grateful that Gauss had gotten him into the Royal Society of Göttingen. Of all the honors received, Gauss appreciated most the honorary citizenship conferred by Brunswick and Göttingen. The latter document is dated July 14, 1849. The city council of Brunswick had been reminded of the approaching jubilee by Friedrich Wilhelm Schneidewin (1810–1856), who was professor of classical philology in Göttingen from 1836 to 1856. He was a native of Helmstedt and had served as a high school teacher in Brunswick from 1833 to 1836. The text of the letter of honorary citizenship of Brunswick, dated July 8, 1849, is preserved, although the document itself is apparently lost. At the ceremony it was Schneidewin who handed Gauss the document. His letter of thanks to the city council and officials of Brunswick was dated August 5, 1849. It required several weeks for Gauss to thank in writing all those who had honored him at the jubilee. Duke Wilhelm of Brunswick had conferred on him the Commander's Cross of the Order of Henry the Lion. His letter of thanks to the Duke, dated August 2, was full of deep emotion and special gratitude, for in it he reviewed his early life and referred to the aid he had received from the Duke's ancestor.

Another honor which pleased Gauss immensely at the time of the jubilee was the congratulatory letter from his Alma Mater, the Collegium Carolinum, forerunner of today's Institute of Technology. Humboldt had written him a hearty letter of congratulation on July 12, regretting his inability to be present. Schumacher was also unable to attend.

On August 20, 1849, Gauss wrote to his old friend Wilhelm Arnold Eschenburg (1778–1861), a government official in Detmold, a touching letter of thanks for his congratulations, in which the following passages occur:

Through your letter of greeting at my doctoral jubilee, dear Eschenburg, you have caused me a very great joy. While most of the other letters received on this occasion had their last roots more or less in some scientific relationship or other, your letter is intended not for the astronomer or geometer, but for the unforgettable boyhood friend. Reminiscences of boyhood and youth vividly came before me. From the first time when I became acquainted with you as a fellow pupil (October, 1789), I felt attracted to you. There are renewed in me the images of our boyhood games, when we, the worthy Drude[6] in our midst, shouting with joy went to the Wends' tower[7] or the *grunen Jäger*.[8] There is renewed in me the image of your deceased father[9] in later years, who always appeared to me as a model of the καλὸς κάγαθός, and his family like a temple of purest earthly happiness under special protection of a kind guardian angel.

I always imagine your own home, too, in similar glory, according to all that I have found out about your wife and children from your letters and otherwise. There has lingered with me a very pleasant impression of your son,[10] who once visited me here several years ago, and who supposedly is the same one whom you mention as an officer of justice attached to a regiment of Lippe troops. I heartily wish you happiness on your retirement [1848]. Public conditions are so unpleasant everywhere in Germany that nobody who is involved in them is to be envied. Even in Lippe Detmold—which I had always thought of as a patriarchal little land, in which genuine purity of morals sits on the throne, so that I had once even thought of moving there for my last days—a part of the Pandora's box seems to have been shaken out of the so-called March achievements. May the storms in Germany soon wear themselves out and may you enjoy a peaceful happiness to an advanced age.

The plan of moving to Lippe-Detmold was evidently conceived by Gauss in 1837 or 1838, at the time of disturbance over the Göttingen Seven.

---

[6] Friedrich Ludwig Heimbert Drude (1752–1840), director and teacher at St. Katharine's School, Brunswick.

[7] Beer garden at the location of an old defense tower near Brunswick.

[8] Forest tavern at Riddagshausen east of Brunswick.

[9] J. J. Eschenburg (1743–1820), professor of literature and philosophy at the Collegium Carolinum; translator of Shakespeare.

[10] August Eschenburg (1823–1904), later a government official in Detmold.

CHAPTER TWENTY-ONE

Gathering Up the Threads: A Broad Horizon

Among his friends in Great Britain Gauss counted the astronomer Airy, Sir David Brewster, Humphrey Lloyd, General Edward Sabine, and the family so well known in the annals of astronomy—William Herschel, his sister Caroline, and his son Sir John Herschel.[1] With all these people Gauss corresponded regularly; in 1825 he visited Caroline Herschel at Hanover. Occasionally Sir John Herschel sent him some young Englishman with a letter of introduction. The friendship with Lloyd and Sabine came about through common interest in terrestrial magnetism. Another Englishman who was on intimate terms with Gauss was Thomas Archer Hirst (1830–1892). He took his doctorate at Marburg in 1852 and visited Gauss the same year, presenting to him a copy of his dissertation, *Ueber conjugirte Diameter im dreiaxigen Ellipsoid*. Gerling of Marburg arranged the visit. Hirst served for many years at the Royal Naval College in Greenwich.

One of the first Americans to visit Göttingen was Benjamin Franklin; at the time he was making plans for the University of Pennsylvania, but this occurred some years before the time of Gauss. Aaron Burr visited Germany in 1809–1810, and spent Christmas Day, 1809, in Göttingen. Through letters of introduction he met Gauss and Arnold Heeren, taking tea with the latter. He was greatly impressed by Gauss, who showed him about the

---

[1] Gauss met Charles Babbage (1792–1871), the noted British mathematician, at the Berlin meeting in 1828, which 463 members attended.

observatory and conversed at length with him. His letter of introduction to Gauss was from E. A. W. Zimmermann, the early patron of the scientist.[2]

During the second half of his life Gauss had a lively interest in America, for a group of Harvard students began to enroll in Göttingen: J. G. Cogswell, Edward Everett, Ticknor, Bancroft, Longfellow, Motley, and William Emerson, the brother of Ralph Waldo. Of these Harvard men Gauss was on especially intimate terms with Cogswell and Everett. Everett studied in Göttingen from 1815 to 1817; on January 30, 1822, he caused Gauss and Olbers to be elected Fellows of the American Academy of Arts and Sciences in Boston. His letter to Gauss, accompanying the diploma, was written in excellent German and dated March 31, 1822. In it he referred to Gauss' kindness during his student days in Göttingen and reminded him that his scientific work was properly esteemed in America. Everett made mention of Bowditch's review (1820) of the *Theoria motus* in a journal he edited. The diploma was signed by John Quincy Adams, John Thornton Kirkland, John Farrar, and Josiah Quincy. On his American tour in 1836 Joseph Gauss visited Edward Everett during his term as governor of Massachusetts.

Joseph Green Cogswell (1786–1871), the well-known librarian and bibliographer, was one of the early American students in Göttingen. He served as librarian of Harvard College, 1820–1823. In 1840 he was highly recommended to Gauss by Humboldt and visited him in Göttingen, seeking advice in spending several hundred thousand dollars for the Astor Library in New York, which he served as bibliographer and superintendent from 1848 to 1861. He acquired from Gauss a copy of the *Theoria motus* and his manuscript of 321 pages in folio of explanations and commentary upon it. Cogswell also got for the Astor Library a copy of Gauss' *Determinatio attractionis* (1818) with a twenty-eight-page autographed quarto manuscript by Gauss of this memoir and a fifty-eight-page quarto manuscript of illustrations and

---

[2] *The Private Journal of Aaron Burr* (Rochester, 1838; new ed. 1903).

remarks on the memoir. Other items acquired for the library included the 1817 memoir on quadratic residues and a twenty-nine-page quarto autographed manuscript by Gauss, as well as a manuscript of three hundred pages in folio of astronomical calculations illustrating the orbits of Juno, Pallas, Ceres, and Vesta, and one of thirteen pages of formulas.

Rear Admiral Charles Henry Davis, U.S.N., later renowned for his victories in the Civil War, was one of Gauss' frequent correspondents; he translated and published in 1857 the English edition of his *Theoria motus*.

Frequent mention of Alexander Dallas Bache (1806–1867) is found in the published correspondence of Gauss. He was a great-grandson of Benjamin Franklin and carried a letter of introduction from Humboldt when he visited Gauss in January, 1838. Bache served as president of Girard College from 1836 to 1842. At the time he was seeking Gauss' advice on plans for the college. Bache served as director of the U.S. Coast and Geodetic Survey, 1843–1867.

Two budding American classical scholars became friends of Gauss during their student days in Göttingen. They were attracted to him because of his greatness and his vast classical learning. The first one was George Martin Lane (1823–1897), Latinist of Harvard University, 1851–1894. Lane was also widely known for his ballad "The Lay of the Lone Fishball." The other was Basil Lanneau Gildersleeve (1831–1924), who studied in Göttingen about 1850. He became the leading Greek scholar in America, serving the University of Virginia and later The Johns Hopkins University.

Gauss was on friendly terms with Nathaniel Bowditch (1773–1838), American astronomer and translator of Laplace's *Mécanique céleste*. During his American tour Joseph Gauss visited Bowditch in July, 1836.

There were several other Americans with whom Gauss was in close contact and whose work he followed with interest: Elias Loomis (1811–1889), the mathematician, who served Western Reserve University, City College of New York, and Yale University; James Pollard Espy (1785–1860), the meteorologist, who was known as the "storm king" for his theory of storms enunciated in 1835; the astronomer James Melville Gilliss (1811–1865); Benjamin Silliman (1779–1864), chemist and geologist of Yale University, whose *American Journal of Science and Arts* (better known as *Silliman's Journal*) Gauss used to read; Major General Ormsby MacKnight Mitchel (1809–1862), who during his professorship at Cincinnati College, 1836–1859, did much to popularize astronomy; Commodore Matthew Fontaine Maury (1806–1873), oceanographer known as the "pathfinder of the seas"; Maria Mitchell (1818–1889), famed astronomer of Vassar College; and Benjamin Peirce (1809-1880), the Harvard mathematician.

The American who became one of Gauss' favorite disciples was Benjamin Apthorp Gould (1824–1896), who graduated from Harvard in 1844. Gould went to Europe for study soon thereafter, arriving in Berlin in May, 1846; he worked there for a while under Encke, but was not satisfied. His fervent desire was to study under Gauss. In Berlin he became an intimate friend of Eisenstein. On March 23, 1847, he sent Gauss a highly favorable letter of recommendation written for him by Humboldt. Gauss immediately agreed to accept him as a student, and he arrived in Göttingen the first week of April, 1847. Gould received his doctorate in astronomy under Gauss in 1848 and then visited many of the observatories of Europe, serving for a time as assistant at Altona under Schumacher. Gauss wrote for Gould a warm letter of recommendation, which he used when he returned to America in 1849 and was seeking a position. That same year Gould established the *Astronomical Journal* at Cambridge, Massachusetts, maintaining it until it was suspended on account of the Civil War. In 1851 Gould entered the service of the U.S. Coast and Geodetic Survey, where his accomplishments were outstanding. When the transatlantic cable had been completed in 1866, he established an observatory at Valentia, Ireland, and made the first determinations of transatlantic longitude by telegraphic cable. From 1856 until 1859 Gould was director of the Dudley Observatory at

Albany, New York. In 1870 he went to South America and established a national observatory for Argentina at Cordova, where he remained until 1885 when he returned to Cambridge, where he re-established his *Astronomical Journal*. Gauss was highly pleased by the first copies of this journal, which Gould sent him in 1849 and 1850. Gould wrote hearty congratulations to Gauss on the occasion of his golden jubilee in 1849, as well as condolences on the death of Schumacher (1850) and Goldschmidt (1851).

Gould revisited Europe in late 1851 and had the pleasure of seeing Gauss again. He was disturbed because his journal was not doing well in his absence. On September 9, 1853, Gould wrote Gauss of his extensive travels in America, covering 11,000 miles. On his return he found eighty unanswered letters on his desk. He told Gauss that he kept informed of his health through Americans returning from Germany. In the Coast Survey he took over the work of the late Sears Walker. The work did not exactly please him, but the pay enabled him to keep the journal going. In the preceding two years he had spent $1,100 for the journal. His salary was $1,500 with traveling expenses, so that he could give $600 per year to the journal. On February 13, 1855, Gould wrote a touching letter to Gauss concerning his last illness and stating that he had no hope of seeing him again in this world. The letter arrived in Göttingen after Gauss' death.

William Cranch Bond (1789–1859), the eminent American astronomer, was a friend of Gauss. In 1839, Bond supervised the construction of the Harvard observatory and became its director, serving in that capacity until his death in 1859. He was the inventor of a method of measuring time to a very small fraction of a second and among the first to employ photography in stellar observations.

Bond's son, George Phillips Bond (1825–1865), succeeded his father as director of the Harvard observatory. Young Bond and his father discovered the satellite of Neptune and the eighth ring of Saturn. He published a work on the rings of Saturn and the orbit of Hyperion, and won a gold medal for work on Donati's comet. George P. Bond kept a diary of his European tour in 1851, and the following entry is dated Leipzig, September 2: "... Encke receives but thirteen hundred thalers, Gauss but one thousand thalers. Doctor Gould, he [D'Arrest][3] told me, was probably to take Doctor Goldschmidt's place as second Professor of Astronomy at Göttingen, Gauss being the first."

On September 4, 1851, George Bond was in Göttingen and made this entry in his diary:

> At ten I went to the observatory to see Professor Gauss. It is a little singular that the landlord should not have known where he lived. He knew Doctor Goldschmidt, or rather had known him (died Feb. 1851, aged 44). Gauss had just left the house, to be home at eleven. I walked half a mile or more into the country to spend the time, returning at eleven. He had not come back. I returned to my room,[4] and presently ... some one knocked at my door, and in walked Doctor Gould, who, singularly enough, had arrived from Altona at two in the morning, and I at four. Our rooms were almost opposite to each other. He appears well, and must have improved since leaving home. At dinner there were six Americans, four students, besides Doctor Gould and myself. After dinner I repaired once more to the observatory, and spent half an hour in conversation with Gauss, and gave him the last daguerreotype of the moon I had left. It was the best of the first series. He showed me daguerreotypes of his son and grandson now living in America—in St. Louis; he has another son there also. He showed me his library, which can scarcely contain above 700 or 1,000 volumes.[5] He had Cooper's works; *Merry Mount,* a history of the country about Boston before its settlement, published in 1849; also Frothingham's *Siege of Boston,* with which he seemed much taken. He brought out also Dr. Bowditch's translation of the *Mécanique Céleste.*
>
> Shortly before leaving [sic!] he spoke of Mr. Peirce's position with respect to what he said about the discovery [of Neptune] being accidental. He thought that the calculations of both Adams and Leverrier rested on an "infirm" basis, inasmuch as the assumed distances were so wide of the truth. That the discovery was accidental, and might have failed because the planet could have been 30° from the predicted place. ...

---

[3] Heinrich Louis d'Arrest (1823–1875), astronomer in Berlin, later in Leipzig, and finally in Copenhagen.
[4] Probably in the Hotel zur Krone on Weenderstrasse.
[5] Actually his personal library contained 5,000 volumes.

Staid [sic!] up in Doctor Gould's room until near one o'clock in the morning, talking of various subjects. He had just seen Gauss, who had intimated to him the probability of his taking his professorship after his death. This is certainly no small honor, offered as it is to an American, and by such a man as Gauss, while there are so many in Germany who would be glad of the situation, D'Arrest among others, whom Doctor Gould mentioned as equally fitted for it.

Gould had been embarrassed because Peirce published on the distribution of cometary orbits and included material on the unequal division of planes which Gould had received from Gauss. Gould communicated orally to Peirce, who had not given credit.

For many years Gauss corresponded with Ferdinand Rudolph Hassler (1770–1843), a Swiss who migrated to America in 1805 and in 1807 became founder and first superintendent of the U. S. Coast and Geodetic Survey, during the administration of Thomas Jefferson. Hassler later held professorships at West Point and Union College. From 1811 to 1815 Hassler was on an official mission in Europe, procuring instruments, books, and equipment for the survey. In later years he also served as superintendent of the United States bureau of weights and measures. On August 31, 1829, Hassler wrote Gauss a letter in which he reminisced about his student days in Göttingen, and at the same time sent some of his works by one Thomas Cooper of New York City, who was then traveling in Germany. Joseph Gauss visited Hassler in 1836 in Washington, but unfortunately Hassler was ill at the time. He presented to Hassler two of his father's publications. The same year Hassler gave one of his assistants, Edward Blunt, a letter of introduction to Gauss. A. D. Bache also carried such a letter when he visited Göttingen in 1838. By the fall of 1837 Hassler was using seven Gaussian heliotropes, produced by Apel. Gauss had written a favorable review of Hassler's logarithmic tables in 1831. In 1836 Gauss wrote to Hassler and invited him to cooperate in magnetic observations, but technical difficulties and occupation with other tasks prevented Hassler's participation.

Gauss' interest in America was highly intensified during the last two and a half decades of his life because it was the home of his two younger sons and their families. He was quite pleased at their material progress and financial security. Writing to Gerling on June 23, 1846, and discussing Eugene's business interests, he stated: "Life in America is capable of waking talents which would never emerge in Europe."

The reports which his son Joseph wrote Gauss during his tour of America in 1836 greatly astonished him as to the fabulous salaries people were earning, particularly the railroad builders. At least they seemed fabulous to him. Joseph wrote his father that he had gained the impression that almost every American was thinking only of making money and was greedily chasing after profits.

In 1837 Gauss and Olbers began to read rather amazing, sensational reports from America concerning the use of electricity for medical and locomotive purposes. Both were skeptical; their published correspondence contains much discussion of it. It furnished a topic of conversation when A. D. Bache visited Gauss in 1838. There was a report that electricity was being used in America to run a printing press. Another visitor to Gauss in 1837 enlarged upon the entire matter. This was Anthony Dumond Stanley (1810–1853), professor of mathematics at Yale University and intimate colleague of Benjamin Silliman, whose name had been involved in the reports. Stanley accompanied Gauss' son Wilhelm on the journey to America and visited Olbers in Bremen on the way out. Gauss, in attempting to assess the truthfulness of the reports, quoted to Olbers a remark in Fenimore Cooper's *Notions on the Americans* to the effect that in America there is not so much lying as elsewhere; he added, for Olbers' benefit, that "one can lie much less than in Europe and *still* put out terribly strong lies, and in contrast to Cooper many maintain that Americans like to brag and exaggerate."

An examination of calling cards kept by Gauss reveals that his visitors included Friedrich Ludolph Karl Leue, the university judge; C. P. Metropulos; Nikolaus von Fuss (1755–1825); Anders Jonas Angstrom (1814–1874), from 1839 until his death astronomer and physicist at the University of Uppsala; Philipp Schoenlein; Ad. Torstrick, *cand, phil.*[6] Gauss always played a prominent role in the affairs of his university. The minutes of the meetings of the philosophical faculty disclose noteworthy contributions he made to its deliberations; he served as its dean from July 3, 1833 to July 2, 1834; from July 3, 1841 to July 2, 1842; and from July 3, 1845, to July 2, 1846. There is a tradition that he refused to serve as rector of the university. He was a corresponding member, mathematical class, Royal Society of Sciences in Göttingen, from 1802 to 1807; from 1807 to 1855 he was *ordentliches Mitglied*. Gauss served as director of the Royal Society from May to Michaelmas 1831, and for the following academic years, 1833–1834, 1836–1837, 1839–1840, 1842–1843, 1845–1846, 1848–1849, 1851–1852, 1854 to his death. Weber succeeded Gauss as director. The position yielded special compensation. In 1852 Gauss had the honor of presiding at the sessions of the Royal Society celebrating the centennial of its founding. He used impeccable Latin in conferring degrees as dean.

On December 28, 1815, the Prince Regent at Hanover conferred on Gauss the Knight's Cross of the Guelph Order. Gauss prized this order more highly than the many others he received. As a rule he did not wear them, but whenever the king or a member of the royal family visited Göttingen, he wore this one. On November 29, 1816, King George at Carlton House issued to Gauss the letters patent which granted him the title of *Hofrat* (court councilor).[7]

Among the many names Gauss considered when filling the vacancy left by Goldschmidt's death in 1851 was that of Franz Friedrich Ernst Brünnow (1821–1891), who had become director of the observatory at Bilk in 1847. In 1854 he accepted the same position at the University of Michigan, then went to the Univer-

---

[6] August Wilhelm Schlegel, the great Romantic writer, visited Gauss in 1813.

[7] On July 1, 1845, King Ernst August gave Gauss the title of *Geheimer Hofrat*.

sity of Dublin in 1865 and became astronomer royal for Ireland. The position was given to Klinkerfues.

In 1802 the post of astronomer to the Academy of Sciences and of director of the observatory in St. Petersburg was to be filled. Pfaff drew the attention of Nikolaus von Fuss,[8] for many years secretary of the St. Petersburg academy, to Gauss. Early in 1802 Fuss asked Pfaff to sound out Gauss on a possible call. In a long letter of September 5, 1802, Fuss issued to Gauss the formal offer of a position in St. Petersburg. At this, his protector, the Duke of Brunswick, laid plans to build him an observatory in his native city. But this plan never materialized. E. A. W. Zimmermann wrote to Fuss on January 20, 1803, that Gauss had decided to remain in Brunswick. After the political disasters of 1806 Gauss wrote to Fuss (October 20, 1806), inquiring if he still might come to St. Petersburg. Before the formal offer could be made, Brandes in Hanover sent him a call to Göttingen on July 25, 1807, and he accepted. Gauss' election as a corresponding member of the academy in St. Petersburg on January 31, 1802, was on the basis of his work in astronomy. On March 24, 1824, he was unanimously elected a foreign honorary member of the academy, mathematical class. The fame of the *Disquisitiones arithmeticae* was specifically mentioned as a basis for this action.

In the spring of 1809 the chair of mathematics and astronomy at the new University of Dorpat became vacant because its holder, Professor Pfaff,[9] was returning to Germany. On May 31, 1809, the council of the university elected Gauss to the place and commissioned Professor Parrot[10] to write him and get his acceptance. Gauss wrote to Parrot on August 20, 1809, rejecting the offer, although leaving the door open for a future offer. He stated that the government did not wish him to leave Göttingen, but that it

---

[8] Fuss was a son-in-law of Euler.

[9] Johann Wilhelm Andreas Pfaff (1775–1835), younger brother of the better known J. F. Pfaff. From 1809 to 1816 he was professor of mathematics at the Real-Institut in Nuremberg, later in Würzburg and Erlangen.

[10] Georg Friedrich Parrot (1767–1853) after 1826 was in St. Petersburg as *ordentliches Mitglied* of the Academy of Sciences. He died in Helsinki.

could make only vague promises of betterment and promotion in the future. At the time the university was more than five months in arrears with payment of salary. His real reasons for declining were plainly stated in the letter. He felt that the financial terms were not good enough, particularly as to widows' pensions and travel allowance. He stated that he would have to buy new household furniture, since he could not ship his that far. Also, Gauss had some fear of the climate in Russia. However, his main reason for refusal of the offer was based on the fact that he desired more time for research.[11] In Dorpat mathematics and astronomy were represented by one professor, and Gauss thought they should be separated. He wrote Parrot that he did not want to teach the ABC of mathematics, admitting at the same time that he was attracted by the large amount of money available for instruments and equipment. At the close of the letter to Parrot he recommended Schumacher for the position. However, the position was finally given to Johann Sigismund Huth, who had been a professor in Frankfurt on the Oder and in Kharkov. The separation of mathematics from astronomy at Dorpat, as suggested by Gauss, did not occur until 1820, but he never received another call from that institution. Gauss' old friend and teacher J. M. C. Bartels became professor of mathematics there in 1820 and continued in that capacity until his death in 1836.

When the time came to fill the position left vacant by the death of Christian Friedrich Rüdiger (1760–1809), who had been astronomer at the University of Leipzig Observatory in the tower of the Pleissenburg since 1791, Gauss became interested in the place. He asked Olbers, who had visited there in 1806, for a report on it. The latter wrote Gauss that the observatory was poorly located and that not much could be done to improve it. Rüdiger had been unable to get sufficient appropriations. Olbers told Gauss that the few instruments set up were shaky and difficult to use. Some fine instruments donated by Count Brühl at London

---

[11] Olbers had informed Gauss that the university curator von Klinger was a difficult person for the Dorpat professors to deal with.

were still in crates. The observatory had a good library, and there were two secretaries.

In the summer of 1809 Gauss entered into correspondence with the Reverend Dr. Franz Volkmar Reinhard (1753–1812) of Dresden, court pastor and church councilor, who was in charge of filling the position. He offered Gauss a full professorship at a salary of 880 thalers plus 120 thalers allowance as rent for the residence attached to the observatory. He reminded Gauss that one could not get such a house in Leipzig at an annual rental of 120 thalers. Reinhard promised a liberal allowance annually for increasing the number of books in the observatory library and also promised to have the Brühl instruments mounted in a new building. He also reminded Gauss that the King of Saxony had already done much for the observatory, but that the matter could not be presented to him until it became clear what turn the Napoleonic war would take. Reinhard asked Gauss to declare himself ready to accept the call; he regretted that the salary was not as large as he had desired. According to him, the philosophical faculty had designated for Gauss *primo loco* as *professor ordinarius novae fundationis*. Reinhard also promised Gauss a couple of hundred thalers for moving and travel expenses. But Gauss did not accept the Leipzig offer; the only reason seems to have been that he merely desired to postpone a decision in order to see what turn the war would take.

On April 18, 1810, Gauss was elected a member of the Berlin Academy of Sciences, and on April 25, 1810, Wilhelm von Humboldt issued to Gauss in the name of the section for public instruction in the ministry of the interior a call to Berlin, offering him 1,500 thalers and a position as *ordentliches Mitglied* of the academy. He wrote: "You are in no wise obligated to teach courses, you are only requested to lend your name as a full professor to the new university and, as much as your leisure and health allow, to teach a course from time to time."

In an accompanying private letter he added: "At the university I release you, as you desire, from every obligation, and there is therefore nothing that could impede you on the path of quiet, secluded, and peaceful research."

Alexander von Humboldt and others attempted to persuade Gauss to accept the Berlin offer. It is not known definitely why he refused. Perhaps he did not want to leave the Göttingen observatory, which was under construction. Some have thought that he did not trust the Prussian government. Gauss was probably reluctant to leave the fresh graves of his wife and son; the call occurred just at the time when he was courting his second wife and making preparations for marriage.

In 1821 negotiations were renewed to get Gauss to Berlin.[12] Frau Waldeck, his mother-in-law, wrote a letter to Olbers on May 14, 1821, trying to get him to intervene. She begged him to keep secret the fact that she had written, and stated that Gauss was very unhappy in Göttingen. There were difficulties between him and Harding, the other astronomer. In addition, he did not have sufficient time for research. She also stated that Gauss was quite worried about the future.

Olbers communicated the secret that Gauss was more inclined than ever to leave Göttingen to Councilor von Lindenau, at that time minister of Sachsen-Gotha, formerly director of the observatory on the Seeberg. Lindenau turned to General von Müffling, the influential chief of the general staff in Berlin, a man who had done much for the geodetic surveys. Negotiations now began. Gauss demanded a free residence and an annual salary of 2,400 thalers. After almost four years of bickering and letter writing without end Gauss received the following counterproposal in November 1824: He was to get 1,700 thalers as *ordentliches Mitglied* of the academy, 300 thalers as secretary of the mathematical class, 600 to 700 thalers from the ministry as a consultant on all questions referring to mathematical study. The offer was not a bad one. Why did not Gauss accept? The most logical explanation is that he felt insulted by the long-drawn-out procedure. He for-

---

[12] Humboldt wanted Gauss to be director of a Polytechnic Institute to be founded in Berlin.

mally gave two reasons for rejection: The Hanoverian government had given him an increase in pay and had approved the entrance of his son Joseph into the artillery corps. In the years 1828–1836 Alexander von Humboldt tried unsuccessfully to reopen negotiations to get Gauss to Berlin.

During the last negotiations in 1825 Leopold von Buch, the famous Berlin geologist, wrote to Gauss: "From the *first* day on you would assume the dominating place in the academy, which is due you . . . what beneficial results for the whole land, for all Germany your mere presence, the rule of such a mind would produce. Tear us out of the barbarism into which we are in danger of sinking. The chairs of Euler, Lagrange, and Lambert call loudly."

In 1821 plans were under way to build a new observatory in Hamburg. Gauss was very much interested in the position of director; he asked Olbers to get information about it and to recommend him for the place, stating that he would accept it if the financial terms were satisfactory. Olbers in replying tried to persuade Gauss not to consider the place. He stated that such positions in the free states and small republics were poorly paid and that there was usually unpleasantness in dealing with the senate and other officials. Olbers told Gauss he would have no trouble in securing a new position as soon as it became known he wanted to make a change.

Gauss was offered a free residence and six thousand marks salary in Hamburg, about nine-fifths of what he got in Göttingen. He felt that living costs in Hamburg would not be greater than in Göttingen. However, Gauss feared that a large part of his wife's estate would be lost if he made a change. Olbers admitted that the offer of six thousand marks was a good one, but he advised Gauss to accept the Berlin offer as soon as possible.

As late as 1842 an effort was made to call Gauss to the University of Vienna.[13] Schumacher visited there that year and evidently acted as intermediary.

---

[13] Cf. the published Gauss-Schumacher correspondence, IV, 85–89.

Gauss occasionally denied or strongly doubted prevailing views on topics in astronomy, without fully communicating his own opinion. Among other things he considered an organization and mental life on the sun and on the planets very probable, and occasionally remarked how in this question gravitation acting on the surface of celestial bodies was of predominant interest. He said that with the general constitution of matter, therefore, only beings as small as June bugs could exist on the sun with a twenty-eight-fold greater gravitation; on the other hand our bodies would be pressed together and all our members crushed. Then he continued in humorous fashion: "Yes, on the sun there is room for us all, but each one of us will need his servant."

Gauss considered it possible to set up a telegraphic communication between moon and earth and in reference to this question had calculated the size of the necessary mirrors, which yielded the result that such communication could be instituted without great cost. He used to say that it would be a greater discovery than that of America if we could communicate with our moon neighbors, but did not consider it probable that the moon was inhabited by a population of higher intelligence.

Richard Adams Locke (1800–1871) published in August, 1835, his gigantic moon hoax in the New York *Sun*. Locke went down in journalistic history as the first feature-story writer and the perpetrator of the greatest hoax of all time. He purported to describe miraculous discoveries of life on the moon made by Sir John Herschel at the Cape of Good Hope. The report was ascribed to a "supplement to the Edinburgh *Journal of Science*," a journal which had ceased to exist several years earlier. Churchwomen in Springfield, Massachusetts, started a fund to send missionaries to the moon. There were French, British, and German versions of the story. Gauss and Augustus De Morgan erroneously attributed the story to Jean-Nicolas Nicollet (1756–1843), an able French astronomer who fled to America and who became acquainted with Eugene Gauss while exploring the Upper Mississippi Valley. Gauss regarded the hoax as very crude and merely an example of how gullible the public is. When Sir John heard of the affair he was "overcome." Paris newspapers published pictures of the moon's inhabitants going through the streets singing "Au clair de la lune." The New York *Daily Advertiser* thought Herschel had immortalized his name and placed it "high on the page of science." Edgar Allan Poe, who knew and admired Locke and wrote he had the finest forehead he had ever seen, tore up the second installment of his own fiction story, "Hans Pfaall," because he felt Locke had left nothing for a fiction writer to offer about the moon. Finally Locke admitted he had written the story and that the whole thing was a fake. Incidentally, it may be added that the French scientist Arago was entrapped by the story. The story was published in book form in 1852 in New York under the title *The Moon Hoax*. Locke also wrote another hoax, *The Lost Manuscript of Mungo Park*, but it attracted little attention.

In a letter to Olbers dated September 3, 1805, Gauss gave an interesting sidelight on his methods of work and difficulties he frequently encountered:

Perhaps you remember . . . my complaints about a theorem which . . . had defied all my attempts to find a satisfactory proof. This theorem is referred to in my *Disquisitiones arithmeticae*, page 636[14] . . . it is just the determination of a sign of a root that has always tortured me. This lack has spoiled for me everything else that I found; and for four years a week has seldom passed when I would not have made one or another vain attempt to solve this problem—recently very lively now again. But all brooding, all searching has been in vain, sadly I have had to put down the pen again every time. Finally I succeeded several days ago—not as a result of my tedious searching but by the grace of God, I might say. The puzzle was solved as lightning strides; I myself would not be able to show the guiding thread between that which I previously knew, that with which I had made the last attempts—and that by which I succeeded. Strangely enough the solution of the puzzle now appears easier than many other things that have not held me up as many days as this has years, and certainly nobody will have any idea of the long squeeze in which it placed me, when I someday lecture on this topic.

Gauss stated in several places that valuable insight on problems came to him upon waking up. His discovery of the inscription of the regular polygon of seventeen sides falls in this category;

---

[14] *Collected Works*, I, 442–443.

and in his *Collected Works* (V, 609) he wrote that he discovered the laws of induction on January 23, 1835, at 7 A.M. just before getting up. Descartes and Helmholtz also testified to the great value of such early morning thoughts.

As a student in Göttingen Gauss used to maintain jokingly that his teacher Kästner was the foremost mathematician among the poets and the foremost poet among the mathematicians. As late as December 22, 1845, in writing to Schumacher he said about Kästner: "Kästner had a very eminent native wit, but strangely enough, he had it in all subjects *outside* of mathematics, he even had it when he talked *about* mathematics (in general), but was often completely deserted by it *inside* mathematics."

In an address before the Berlin Academy of Sciences in 1874 Emil du Bois-Reymond called Gauss a grand master of thought; he had in mind especially the extreme care with which Gauss prepared his works. On July 21, 1836, Schumacher wrote Gauss: "I don't know anyone who possesses to such a high degree as you the talent of expressing himself sharply and briefly."

Three days later Gauss modestly mentioned his friend's praise in this manner: "The compliment which you . . . make me on account of my talent of expressing myself briefly and sharply, I must reject, at least in so far as it refers to a language[15] other than German or, at most, Latin."

Many years earlier (1821) Gauss had explained this difficulty to Schumacher: "In a foreign language attention to the direction, so that I myself am quite satisfied with the finish and naturalness, always costs me as much time as the topic itself."

Gauss had great difficulty in finding a publisher for his *Theoria motus* (1809). When Perthes finally did accept it, he insisted that it appear in Latin. This meant a tremendous amount of extra work for Gauss, but he understood the publisher's reasons and did not criticize him. In those disturbed Napoleonic years Gauss was glad enough to find a publisher.

In 1845 on the day before his sixty-eighth birthday Gauss

---

[15] Schumacher had English in mind.

perhaps somewhat wearily wrote to Schumacher: "I shall never again write Latin about scientific things."

Then on December 4, 1849, he explained more fully the difficulty of using Latin, again in a letter to Schumacher:

In times when I had to write most of my works in Latin, very often I had to turn back and forth for a long time the thought hovering before me until I had found a somewhat satisfactory turn and yet one which often in no wise quite satisfied me. Yet such never occurs as long as one moves in the purely mathematical field (I might say in the technical-mathematical), but principally where one considers the subject and the characteristic of its nature from a higher, equally philosophical—as Lagrange used to say, metaphysical—standpoint.

CHAPTER TWENTY-TWO

Religio Scientiae: *A Profession of Belief From the Philosopher and Lover of Truth*

Gauss' mature philosophy of life was closely connected with his strongly religious nature, which was characterized by tranquillity, peace, and confidence. All pretense was especially repugnant to him, and he treated all charlatanism, especially on the scientific side, with disdain and often with bitter irony. He once said that the most despicable human being is the one who persists in his errors after he has recognized them. A thirst for truth connected with an urge for justice was the leading element in his character. The principle of least compulsion was the mathematical embodiment of that basic ethical thought which he recognized as binding on the universe.

All philosophical studies possessed a powerful charm for Gauss' mind, although he often disliked the ways by which scholars arrived at certain viewpoints. He once said to a friend: "There are questions on whose answers I would place an infinitely higher value than on the mathematical, for example, concerning ethics, concerning our relationship to God, concerning our destiny and our future; but their solution lies quite unattainable above us and quite outside the area of science."

By science he understood that strictly defined logical structure whose foundation rests on certain truths generally recognized by the human mind, truths which once admitted furnish an immeasurable field of the most complicated researches connected with each other by a strong chain of thought. Therefore he placed arithmetic[1] at the top and in reference to questions which we cannot fathom he was wont to use the words: Ὁ θεὸς ἀριθμετίζει, by which he recognized logic as valid for the whole universe, even for those areas where our human mind cannot penetrate.

Gauss was continually meditating on religious and philosophical subjects, but unfortunately he seldom spoke or wrote on them. He was always attempting to harmonize his scientific experience with his philosophy of life. He considered all philosophical ideas subjective and he kept them quite separate from genuine science since he knew that they lacked a rigorous basis.

In Gauss' character was a notable tendency to strew out before him like leaves in the wind his thoughts (often accompanied by deep emotions) on these eternally unsolved problems. But often they were blown away just as suddenly as they had come, either by some humorous turn or by a quick shift of conversation to the most unimportant matters of daily life, or they were covered by the impenetrable veil of secrecy. One day he said to a friend that it was immaterial to him whether Saturn has five or seven satellites—there is something higher in the world. Having said that, he immediately became silent, but his eye sparkled and a stream of thoughts passed through his mind.

The most popular long poem published in England in the eighteenth century was James Thomson's *The Seasons*. In a sense it covers the history of ideas and the relationship of the fields of human knowledge. The poem reflects the trends of the time in these fields. Thomson was deeply indebted to the scientific work and theories of Newton, whom Gauss revered so highly. Gauss possessed a copy of the poem, read it repeatedly, and enjoyed it thoroughly. He went to his notes and copied down the following passage which made a special appeal to him and seemed to epitomize so completely his own views and feelings:

> Father of light and life! thou Good Supreme!
> O teach me what is good! teach me Thyself!
> Save me from folly, vanity, and vice,

---

[1] Theory of numbers.

From every low pursuit; and feed my soul
With knowledge, conscious peace, and virtue pure—
Sacred, substantial, never-fading bliss![2]

It is not known just what Gauss believed on most doctrinal and confessional questions. He did not believe literally in all Christian dogmas. Officially he was a member of St. Albans Church (Evangelical Lutheran) in Göttingen. All baptisms, burials, and weddings in his family occurred there. It is also not known whether he attended church regularly or contributed financially. A faculty colleague called Gauss a deist, but there is good reason to believe that this label did not fit well.

Gauss possessed strong religious tolerance which he carried over to every belief originating in the depths of the human heart. This tolerance is not to be confused with religious indifference. He took special interest in the religious development of the human race, especially in his own century. With reference to the manifold denominations, which frequently did not agree with his views, he always emphasized that one is not justified in disturbing the faith of others in which they find consolation for earthly sufferings and a safe refuge in days of misfortune. He demanded the same tolerance from others respecting his own views.

Gauss' religious consciousness was based on an insatiable thirst for truth and a deep feeling of justice extending to intellectual as well as material goods. He conceived spiritual life in the whole universe as a great system of law penetrated by eternal truth, and from this source he gained the firm confidence that death does not end all.

One day he said: "For the soul there is a satisfaction of a higher type; the material is not at all necessary. Whether I apply mathematics to a couple of clods of dirt, which we call planets, or to purely arithmetical problems, it's just the same; the latter have only a higher charm for me."

---

[2] "Winter," lines 217-222.

For him science was the means of exposing the immortal nucleus of the human soul. In the days of his full strength it furnished him recreation and, by the prospects which it opened up to him, gave consolation. Toward the end of his life it brought him confidence. Gauss' God was not a cold and distant figment of metaphysics, nor a distorted caricature of embittered theology. To man is not vouchsafed that fullness of knowledge which would warrant his arrogantly holding that his blurred vision is the full light and that there can be none other which might report truth as does his. For Gauss, not he who mumbles his creed, but he who lives it, is accepted. He believed that a life worthily spent here on earth is the best, the only, preparation for heaven. Religion is not a question of literature, but of life. God's revelation is continuous, not contained in tablets of stone or sacred parchment. A book is inspired when it inspires. The unshakeable idea of personal continuance after death, the firm belief in a last regulator of things, in an eternal, just, omniscient, omnipotent God, formed the basis of his religious life, which harmonized completely with his scientific research.

One day he said:

In this world there is a pleasure of the intellect, which is satisfied in science, and a pleasure of the heart, which consists principally of the fact that human beings mutually ease the troubles and burdens of life. But if it is the job of the highest being to shape creatures on special spheres and to let them exist 80 or 90 years in order to prepare such a pleasure for them, then that would be a miserable plan.[3] Whether the soul lives 80 years or 80 million years, if it perishes once, then this space of time is only a reprieve. One is therefore forced to the view, for which there is so much evidence even though without rigorous scientific basis, that besides this material world another, second, purely spiritual world order exists, with just as many diversities as that in which we live—we are to participate in it.

One of Gauss' intimate friends was Rudolf Wagner (1805–1864), professor of comparative physiology and zoology in Göttingen from 1840 to 1864 and head of the department of physi-

---

[3] Another time he said that the problem would be shabbily solved.

ology after 1842. Wagner was inclined to the spiritual and had become involved with Karl Vogt in a public controversy on religious matters. He wished to use the great name and authority of Gauss to support his views. During the last months of Gauss' life Wagner would visit him and discuss religion as well as related matters. He then hurried home and wrote down all that he could recall. One should remember that Gauss was then very old and not in good health; he probably said things which he would not have said at other periods of his life. Also, one should remember that Wagner had an ax to grind and should discount his report accordingly. And yet much of what he wrote agreed with material found in other sources. One cannot help being reminded of Eckermann's conversations with Goethe. Wagner composed an essay on the views of Newton, Haller, and Gauss on religion—an essay which he read to various circles but never published because Weber, Sartorius von Waltershausen, and Therese Gauss forced him to suppress it.

On December 19, 1854, Wagner went to Gauss at noon. Gauss was sitting in a large easy chair, was breathing with difficulty, and thanked Wagner for sympathy about his health, as he pressed his hand. Very soon in the course of the conversation he regained his usual freshness and vivacity of speech.

His first question was whether Wagner had read the fifth volume of Radowitz' writings. It lay open in front of him on his desk. Gauss took up the remarks of the famous statesman[4] about Goethe, immortality, and Catholicism. He discussed Radowitz' Catholic views. In order to avoid this delicate point, Wagner said that Radowitz had also been busy with mathematics. Gauss did not know his works but said that Dirichlet had told him they were unimportant and asked whether Wagner (as he had heard) was close to Radowitz.

Wagner soon turned the conversation to Whewell's *History of the Inductive Sciences*. As a matter of fact, Wagner had given much attention to the book because he wanted to ask Gauss

---

[4] Joseph Maria von Radowitz, Prussian general and secretary for foreign affairs.

several questions about Newton. This led Gauss to Whewell's work *Of the Plurality of Worlds,* which the author had recently sent him. Gauss arose from his chair without apparent difficulty and fetched the book from his library in an adjacent cold room. He then explained briefly his opinions of it and gave its main results in the following closing words: ". . . only the earth is inhabited by intelligent beings, because the Savior had appeared only for these." Wagner said he could not imagine that the other cosmic bodies were created on the side just for fun, and that he could not support the arguments of many theologians from the Bible because from brief references which Holy Writ gives about the astronomical world they seemed to conclude too much. Gauss seemed to agree with this view without expressing himself definitely. He thought that the earth was only the third planet and in contrast to many other cosmic bodies was "only a clod of dirt."

The conversation then led them to Kepler, Newton, and Leibniz, their scientific accomplishments, and their views on supernatural things. Gauss said: "Kepler often expressed himself very inappropriately and strangely in his astronomical works about religious things. Newton also treated such matters, but always quite differently, always moderate, dignified, never losing himself in peculiar and excessive speculations."

Gauss rated Newton very high, but Leibniż much lower. On this occasion he said:

In a certain respect one can rather say of Leibniz that he spoiled mathematics. His invention of the differential calculus is not anything so great; it was actually already prepared by Archimedes. Leibniz' universality is indeed admirable, but it would have been better if he had limited himself more, then he would have accomplished more in mathematics. But the greatest thing is purely mathematical thinking; this is worth much more than the application of mathematics. Here I can give no applause to my friend and pupil Encke in view of certain remarks in his last rector's address (in Berlin).

Gauss now made fun of a circle of young mathematicians in Berlin, who in their arrogance had formed a sort of alliance to turn their powers only to pure mathematics, while they designated the application to physics as something degrading. Gauss said further that he rated his *Disquisitiones arithmeticae* and a few of his shorter memoirs much higher than his *Theoria motus* and the works on terrestrial magnetism, even though he had gladly busied himself with practical questions. But the above researches were pure works of observational material. Wagner reports that Gauss then spoke in his "grand and amiable" manner of the wonderful feelings and the great inner pleasure of such thought movements. He also remembered Newton, who in various passages expressed himself about these feelings.

He then spoke of his relationship to Kästner, Lichtenberg, and Olbers. When Wagner left, Gauss gave him Whewell's book to read. As Wagner was walking across the courtyard of the observatory, he noticed on the white paper jacket of Whewell's book a number of Bible references in Gauss' neat handwriting.[5] He saw at once to what they referred and resolved to ask Gauss about them next time.

On December 23 Wagner again visited Gauss at eleven o'clock. Before he sat down Gauss asked, "Well, have you read Whewell's book?"

"Yes, at least thumbed through."

"What do you say about his arguments which are purely theological?"

"I am in general not very much satisfied with such treatment of that kind of materials; it doesn't accomplish much. But meanwhile I find that Whewell clings to the correct relationship between scientific research and faith. But first of all permit me a question: On the jacket you noted a number of Bible passages; I would like to know from what source and for what purpose?"

Gauss answered, "Oh, you mean these?" He pointed to them. "They are passages which refer to immortality. At the moment I can't say where the collection came from. But I find these passages are all not so striking and coherent. In general, dear col-

---

[5] Daniel 12:1–3, Job 19:25, Psalms 17:15, 49:15–16, 16:9–11, Ecclesiastes 12:7, Isaiah 26:19, Ezekiel 37:5 *et seq.*, Wisdom of Solomon, 3:1 *et seq.*, II Maccabees.

league, I believe you are more believing in the Bible than I. I am not, and," he added, with the expression of great inner emotion, "you are much happier than I. I must say that so often in earlier times when I saw people of the lower classes, simple manual laborers who could believe so rightly with their hearts, I always envied them, and now," he continued, with soft voice and that naïve childlike manner peculiar to him, while a tear came into his eye, "tell me how does one begin this?"

Wagner was overwhelmed by great inner emotion and was in some embarrassment as to how he was to answer. He reports that he perceived the whole seriousness and greatness of the moment. The manner of Gauss' question reminded him of the ancient, often mentioned question: "What shall I do to inherit eternal life?" Again reverence for the entire personality of the man gripped him. He was reminded that Gauss' colleagues treated him not as an equal but as someone on a higher plane, as someone for whom others step aside and make way. Wagner felt that there was something imposing in his sharp gaze overshadowed by highly arched eyebrows.

After Wagner had regained his composure, he meditated and began to tell the story of his life. Gauss listened very quietly and interrupted him only once with words: "Do you perhaps have the good fortune to have had a believing father or a believing mother?"

Wagner closed his short narrative with the remark, "Faith is a gift."

Gauss commented pensively, "You say that faith is a gift; this is perhaps the most correct thing that can be said about it."

"Of course, faith is a gift; but as a rule it is conferred only when one seriously asks for it. Then this gift is never denied. For there exists concerning it a promise sublime above all doubts: To him who asks, shall be given."

Gauss said, "Even in my life, experiences occurred which often disconcerted me and which led me to a Providence in the individual case, which you assume. Such, for example, is the dispensation which made me an astronomer, which also led me to

Göttingen. I was to go to St. Petersburg. There I would have become a pure mathematician. Now Zimmermann, professor at the Carolineum in Brunswick, at the moment of his departure for Weimar gave me the numbers of Zach's *Monatliche Correspondenz* in which the discovery of Ceres by Piazzi was reported."

In the further course of the conversation Gauss asked whether Wagner assumed a continued development of the soul in the direction of earthly occupations. The invention of the method of least squares, for instance, which Gauss had already made as a young man, but at that time kept secret, had enabled him to send in calculations of the Ceres orbit with speed and precision which astonished all contemporary mathematicians and astronomers.

Wagner expressed the opinion that such a development from the standpoint of faith was indeed not favored directly by Scripture on the basis of revelation, but was also not directly excluded, and in a limited way appealed to him personally. He added that he was talking only of the advanced individual in the field of higher intellectual recognition.

Wagner then began to develop his views on the harmony of worlds, on the necessity of a mutual action between physical and moral world order, whose laws of causality we could not now penetrate.[6] These laws would of necessity exist, if moral freedom and the course of world events were not to appear senseless. A history of humanity without continuation and perfection at another scene is inconceivable and a contradiction of infinite harmony and fixed legitimacy of the world of physical phenomena. Behind the scene of *this* world there must be another, terrible power to punish evil, and also in the field of freedom there must exist a connection resting finally only in "divine mathematics."

---

[6] This remark refers to an earlier conversation between Gauss and Wagner on the applicability of mathematics to psychology, where Gauss said: "I scarcely believe that in psychology data are present which can be mathematically evaluated. But one cannot know this with certainty, without having made the experiment. God alone is in possession of the mathematical bases of psychic phenomena." Gauss viewed unfavorably the work of his colleague Herbart, Fechner, and several other philosophers in this area.

Wagner mentioned that our own life teaches us that the smallest things, all events of the individual, point to this. He said that it was impossible for him to mistake a pre-established harmony even in the course of human science. As an example, Wagner cited here the rescue of the famous physicist Fraunhofer, who in 1801 in Munich was buried in rubble as the result of the collapse of a house, but after several hours came out alive. This directed the attention of King Max Joseph to him. After the healing of his wounds the King gave him eighteen ducats, which the fifteen-year-old boy used to purchase a glass cutting machine. This and the attention of other patrons directed to him led him to the profession in which he accomplished such great things through a rare union of theoretical and practical training.

Gauss accepted the above assumption and spoke repeatedly of Radowitz, whose writings Wagner had sent him. He showed Wagner passages which he considered suitable to stimulate thinking, yet he repeatedly explained that the arguments of the author did not correspond to his wishes. He said that Radowitz came forward again and again with ecclesiastical and dogmatic proofs and had only a few words for other deductions.

Then he asked Wagner whether he had read Jean Paul's views on immortality. Besides the *Kampaner Tal* he quoted the other essays. All these presentations had always been very attractive to him, except where Jean Paul referred to animal magnetism, those were the weakest supports of immortality.

Wagner agreed completely and after he had referred to the mistaken interpretations of alleged magnetic phenomena, for example, in the seeress von Prevorst, he quoted the well-known Bible passage: "If you do not believe Moses and the prophets, you will not believe that one rises from the dead."

Gauss then returned to the continued development of the soul after death in the sense of earthly occupations; only thus could he explain many things in the course of the fate of individual human beings, for example, the early death of Eisenstein, who barely attained the age of thirty and gave promise of the greatest accomplishments.

Wagner asked, "Do you really consider Eisenstein so important?"

"Yes! One of the greatest talents of *all* times. He did some things which give testimony of the most refined and rarest concepts."

As Gauss now returned to Jean Paul and the quoted Bible passage about Moses and the prophets, he said: "I must confess that such old theologians and song writers as Paul Gerhard have always made a great impression on me; a song by Paul Gerhard always exerted a wonderful power on me, much more than, for example, Moses, against whom as a man of God I have all sorts of qualms."

After Wagner had again taken up his remarks on pre-established harmony and reciprocal action of physical and moral world order, Gauss became very silent and listened with the greatest attentiveness without interrupting him. Then he said very seriously: "It always appealed strongly to me when Jean Paul calls out to one who doubts and asks about the solution of the secrets of earthly things in another world: 'Look over the cemetery wall at the graves, there lies the answer!' "

His voice had become weak and trembling, and he broke out in a stream of tears. Wagner reports that this emotional outburst in a man usually so resolute made the strongest impression on him. He remained silent before him and grasped his hand. Thus they sat quietly for a long while. To Wagner it was one of the most solemn moments of his life. He thought of Gauss' words: "How does one begin to believe?"

Wagner comments thus on his own feelings:

I perceived what was going on in his soul and how to this vigorous thinker with all foreboding of future things the world which lies behind the gates of eternity was shown less certain in its outlines than to the consciousness of a believing child. But the very childlikeness of his own question awakened in me a happy hope for him. I thought of the wonderful beauty and sublimity of feeling in the occupation with the theory of numbers, of which he had spoken to me. But perhaps I felt more strongly in this unforgettable moment that there is a far more splendid feeling, namely, that which the evangelist had when he could say: "I believe, Lord, help thou mine unbelief" (Mark 9:24).

Finally Wagner got up slowly and pressed Gauss' hand. Then Gauss said in serious serenity: "Yes! The world would be nonsense, the whole creation an absurdity without immortality."

Several days after this incident Wagner again went to Gauss, but only in order to inquire about his health, and did not think of evoking conversation from him. However, he resolved not to avoid conversation. In order to prevent at the very start any fear of urging the issue, he had taken along the *Life of Sir Humphry Davy*,[7] which he had edited, to remind Gauss of an earlier epoch of his life, when he met Davy at Olbers' home in Bremen. Gauss sent out word that he was feeling very ill that day, but that he would see Wagner for a short time. The conversation lasted less than half an hour. Gauss visibly avoided emotional matters such as had occurred the previous time. Wagner, too, avoided them.

Down to the smallest details Gauss remembered the dinner with Davy and Schumacher at Olbers' home in 1824, without knowing the passages in the book which Wagner was reading to him. Gauss said: "I do not believe that Davy made false personal observations, he appeared harmlessly happy to me. That must refer to a man who understood the natural history of catarrh and whom Olbers summoned because Davy had great disciple interest in this matter. I still remember the astonishment of this man; his name was Böse."[8]

Wagner began with that episode from Davy's life, which Gauss retold with the smallest details. But Gauss, who was very short of breath and who was sitting on the sofa, said that he would not have received Wagner if he had not had something urgent to say to him, after he had read his pamphlets *Menschenschöpfung und Seelensubstanz* and *Ueber Wissen und Glauben* with great interest. They lay open beside him on the sofa. Wagner felt pleased because he assumed that Gauss agreed with the contents. He said: "I see that you assume the possibility of a trans-

---

[7] John Davy, Sir Humphry's brother, actually edited the book; the German translation (4 vols.; Leipzig, 1840) was by Dr. C. Neubert. Wagner wrote the introduction. See Vol. 3, pp. 308, 311, for Gauss references.

[8] His name was not mentioned in Davy's book.

position of the soul after death to another cosmic body. I too am of this opinion about soul substance, but would prefer a comparison of the speed of locomotion with a galvanic current to your simile of the movement of light, since in that instance one can think more easily of the advancement of something, even though imponderable but substantial, than in the case of light."

Then Gauss explained his thoughts on how and under what conditions, magnitudes, and forms a new "change of clothing" of the soul (which Wagner would also have to assume) might appear on the sun, or a small planet or asteroid. He even took up Wagner's thoughts about the assembling of souls of deceased human beings after the end of generations, in a great cosmic space. He had calculated that on the sun in accordance with the law of gravitation we would have to have a much smaller body "about like a big bug," and added, "If we were as big on the sun as we now are, then we would believe that we had lead in our arms and legs and would not be able to stretch out our arms, which would be awful. On Ceres or Pallas our body could be much larger than now. On the sun there would be room enough for all of us."

He added that he wanted to tell Wagner this because sometimes he had read it wrongly indicated, as though human beings on the smaller cosmic bodies would also have to be smaller. Gauss elucidated all this with great seriousness. When Wagner could not suppress a smile, Gauss himself joked in a mildly serious manner about these plays of fancy.

Wagner was again impressed by the greatness of the man. He realized how much Gauss had meditated about the future of the human soul, and how he was always striving to harmonize such views with the principles of mathematics. He closed these reminiscences with the words of Leibniz: "The mathematical sciences, which treat of the eternal truths rooted in Divine Mind, prepare us for the recognition of substances."

Two religious works which Gauss read frequently were Braubach's *Seelenlehre* (Giessen, 1843) and Süssmilch's *Göttliche Ordnung gerettet* (1756); he also devoted considerable time to the New Testament in the original Greek. On November 11, 1835, Gauss wrote to Olbers that he had had a recent letter from his son Eugene, who at the time was attached to Company F of the First Regiment, U.S. Infantry, at Fort Crawford, Minnesota. During part of that time he performed the duties of librarian to the post. Captain S. Loomis, the commanding officer, wrote Gauss about the exemplary conduct of his son and went on to testify that in a religious meeting Eugene had publicly declared his decision "henceforth to serve God." Loomis testified to Eugene's "conscientious behavior not only as a soldier of the Government of the U.S., but also as a soldier and a follower of Our Lord and Saviour Jesus Christ." Eugene wanted to enter a theological seminary and prepare to be a missionary. Actually he never carried out this wish, but he remained very pious the rest of his life. In the same letter to Olbers Gauss commented thus on Eugene's attitude: "Even though much error and hypocrisy may *often* be mixed in such pietistic tendencies, nevertheless I recognize with all my heart the business of a missionary as a highly honorable one in so far as it leads to civilization the still semisavage part of earth's inhabitants. May my son try it for several years."

Although a serene man of science, Gauss attained such objectivity and calmness not without struggle. There was a streak of the mystic and Romantic in him, which sometimes penetrated the hard outer shell of logical reserve. Newton, Descartes, Pascal, Gauss, Helmholtz, and certain others were "pious," but their faith varied in degree. In his savage attack on Gauss (1877), Eugen Dühring accused him of a narrowness in general thinking that originated in religious superstition. Today no responsible person accepts such a charge. Dühring launched out against Justus von Liebig and Clausius because they had dared to defend (with "brazen tongue") Gauss and Riemann against his assault.

On September 3, 1805, Gauss wrote thus to Olbers: "Whomsoever the comely goddess of truth does not always shun, he who has a bride, as I have, and a friend, as you are, can overlook trivialities."

Some writers have stated that Gauss' life moved along like a majestic river and that he was so favored by nature as to have satisfaction from his unusual genius. But such was not the case, even though he did manifest great kindness and rectitude of character. Such a view is contradicted by the following passage in a letter to Bolyai, dated April 20, 1848:

It is true, my life is adorned with much that the world considers worthy of envy. But believe me, dear Bolyai, the austere sides of life, at least of mine, which move through it like a red thread, and which one faces more and more defenselessly in old age, are not balanced to the hundredth part by the pleasurable. I will gladly admit that the same fates which have been so hard for me to bear, and still are, would have been much easier for many another person, but the mental constitution belongs to our ego, which the creator of our existence has given us, and we can change little in it. On the other hand I find that this consciousness of the nothingness of life, which in any case the greater part of humanity must express on approaching the goal, offers me the strongest security for the following of a more beautiful metamorphosis.

On March 14, 1824, Gauss wrote Bessel that "all the measurements in the world do not balance *one* theorem by which the science of eternal truths is actually advanced."

At the time of his first wife's death in 1809 Gauss discussed his feelings on fate in a letter to Schumacher: "Of all the reasons for consolation, which I have tried, none has been stronger for me than that if fate had laid before me the alternative of choosing my present misfortune or dying myself and leaving the deceased [his first wife] disconsolate, I would have had to approve what fate has now decided."

Again he expressed himself to Schumacher on September 24, 1831, concerning the death of his second wife:

A week ago the mortal remains, which were a main source of the indescribable pains of the poor sufferer [his second wife], were given back to the earth, and I still cannot think of these pains for one moment without the most intense shock. I could not write to you sooner. In time the promise of reason will finally replace the feeling, namely, that she, like all others, is to be congratulated on having departed from a scene where joys are fleeting and vain, where suffering, failures, and painful disappointments are the basic color. How much I would long to depart from it if so many bonds did not fetter me.

His mood was much more cheerful in a letter to Schumacher on February 9, 1823: "Dark are the paths which a higher hand allows us to traverse here . . . let us hold fast to the faith that a finer, more sublime solution of the enigmas of earthly life will be present, will become part of us."

At the time of negotiations to get him to the University of Berlin, Gauss expressed a similar thought in a letter to Bessel on April 25, 1825: "In such apparent accidents which finally produce such a decisive influence on one's whole life, one is inclined to recognize the tools of a higher hand. The great enigma of life never becomes clear to us here below."

At the time Schumacher's mother died, Gauss wrote him on November 10, 1822: "I do not attempt to console you, in such events there is no consolation except the strongest conviction that we are sitting here *in ultima* and hereafter are promoted in turn to a higher school."

In another letter to Schumacher dated September 1, 1846, he expressed a similar thought: "It is the sad lot of old age gradually to see depart from us so much that was near and dear to us and to see ourselves more and more isolated, and there is no consolation in this, except the prospect of a higher world order which will some day balance everything."

Gauss was not a professional philosopher, and although he attempted to apply the rigid standards of mathematics to philosophy for his own satisfaction, he was very reluctant to express himself in this field. He possessed and studied carefully the works of Hume, Francis Bacon, Kant, Christian Wolff, Fries, Descartes, John Locke, and Malebranche. On one occasion he criticized some of Schelling's work. But even among the greatest philosophers, occasionally in Kant, he found certain confusion in concepts and definitions. His special target was always Hegel, and in this connection Schumacher once wrote him: "Among Noah's sons was one who covered the shame of his father, but the Hegelians are still tearing away the cloak which time and oblivion had sympathetically thrown over the shame of their Master."

To this Gauss replied that he was uncertain whether the comparison with Noah was entirely suitable:

> Holy Writ only tells that he was circumcised once, while he otherwise passes for a sensible man to whom we (since *abusus non tollit usum*) may be thankful for the fact that he saved the sprigs of the vine from the flood, even though he would have done better to turn over many other things to the flood. Hegel's *insania* in the dissertation under question seems to be wisdom compared with his later ones.

In 1844 Schumacher called the attention of Gauss to a definition in one of the works of Christian Wolff which states that the center of gravity is that point through which the body is divided into two equally heavy parts.[9]

Immediately he got a reply from Gauss, dated November 1, 1844:

> I am almost amazed that you consider a professional philosopher capable of no confusion in concepts and definitions. Such things are nowhere more at home than among philosophers who are not mathematicians, and Wolff was no mathematician, even though he made cheap compendiums. Look around among the philosophers of today, among Schelling, Hegel, Nees von Esenbeck, and their like; doesn't your hair stand on end at their definitions? Read in the history of ancient philosophy what kinds of definitions the men of that day, Plato and others, gave (I except Aristotle). But even in Kant it is often not much better; in my opinion his distinction between analytic and synthetic theorems is such a one that either peters out in a triviality or is false.

He then proceeded to give a correct definition of the center of gravity. In a letter to Moritz Wilhelm Drobisch dated August 14, 1834, Gauss wrote that "most so-called professional philosophers, when they venture into mathematics, sell us only *aegri somnia* (sick man's dreams) for philosophy." Most of Gauss' references to philosophers refer to Kant's theory of space.

On May 11, 1841, Gauss wrote to Jacob Friedrich Fries and thanked him for the copy of *Geschichte der Philosophie* which the author had sent him. In the letter he made this remark:

---

[9] Wolff used the word *gleichwichtige*.

I have always had great love for philosophical speculation, and rejoice so much more to have in you a reliable leader in the study of the fate of the science from the most ancient to the most modern times, since in my own reading of the writings of many philosophers I have not always found the desired satisfaction. Namely the writings of several frequently mentioned (perhaps better, so-called) philosophers, who have appeared since Kant, reminded me sometimes of the sieve of the goat milker,[10] or, instead of the ancient, to use a modern image, they reminded me of Münchhausen's pigtail by which he pulled himself out of the water. The dilettante would not *dare* to give such a confession before the master if it had not occurred to him that the latter was almost of the same opinion about those merits. I have often regretted that I don't live in the same town with you, in order to be able to gain pleasure as well as instruction from conversation with you on philosophical subjects.

Many years previously Gauss had thanked Fries for a copy of his *Mathematische Naturlehre* (Heidelberg, 1822). Concerning this book M. J. Schleiden, professor of botany in the University of Jena, told this story:

When I was studying in Göttingen (1830-1834), one of the more advanced students came to Gauss, saw on his desk the work mentioned, and said: "But Professor, do you devote yourself to that confused philosophical stuff?" Whereupon Gauss turned very seriously to his questioner with the words: "Young man, if in your three-year course you get to the point that you can appreciate and understand this book, then you have used your time far better than most of your fellow students."

Schleiden told the philosopher Rudolf Eucken that Gauss read Kant's *Critique of Pure Reason* five times. The fifth time he is reported to have said: "Now it's dawning on me." A young scholar (whose name is now unknown) had made slighting remarks to Gauss about Kant, and this led Gauss to the remark that he had devoted great effort to the *Critique of Pure Reason,* and he believed he was just gradually getting to understand it fully. Incidentally, Schleiden was a follower of Kant and Fries. Apparently the views of Gauss and Kant on the distinction between right and left were not so contradictory as might appear at first glance.

---

[10] What the Romans called *hircum mulcere*.

Gauss was interested in the manner in which Kant founded his concept of space and pioneered in the field of non-Euclidean geometry. His work led to n-dimensional space, and the whole subject has been revolutionized in our time by the work of Einstein. Gauss' basic concept of the essence of space is contained only in separate, scattered utterances which are collected in Volume VIII of his *Collected Works*. It all goes back to his work in non-Euclidean geometry, as we saw in Chapter XV. In Kant the basic concept is that of absolute space, namely, that absolute space is independent of the existence of all matter and itself has its own reality as the first reason of the possibility of its composition. Reality here means empirical reality and not merely transcendental ideality ascribed to space. Kant mentions the fact that two bodies which are exactly alike, do not therefore have to be congruent, that is, through a mere movement they cannot be necessarily moved over into one another. Of such kind is a body and its reflection, or the left and right hand.

Gauss disagreed with Kant, as shown in the following passage of a letter to Schumacher on February 8, 1846:

The distinction between right and left cannot be defined, but only shown, so that it is thereby a case similar to sweet and bitter. *Omne simile claudicat*, but the latter is valid only for beings which have organs of taste, while the former is valid for minds to whom the material world is perceptible, two such minds, however, cannot make themselves directly understood concerning right and left unless one and the same individual thing forms a bridge between them.

In a letter to Gerling on June 23, 1846, Gauss repeats the same thought:

One cannot reduce to concepts the distinction between two systems of three straight lines each (directed lines, of which the one system points forward, upward to the right, the other forward, upward to the left), but one can only demonstrate by holding to actually present spatial things. Two minds cannot reach agreement about it unless their views connect up with one and the same system present in the real world.

Gauss and Kant thus differed in the manner of founding a theory of space, but there is scarcely any difference in their views about space. Gauss did not deny the a priori character of space. Kant once predicted the geometry of n-dimensions and was thus in agreement with Gauss on this point also. Gauss denied that space is merely the form of our external perception, but he did not want to maintain the existence of a real space independent of our perception. Since Kant one can regard space neither as a thing nor as a property of things, but as a kind of presentation of actually present things. Gauss was undoubtedly right when he said that the distinction between right and left can be transmitted only from one person to another. Physicians say that it is frequently impossible to teach this distinction to the feeble-minded.

When people read the lives of men like Goethe, Schiller, Humboldt, and even Kant, those of ordinary intelligence may form some idea of what existence meant for those great men, and what lessons their lives convey to humanity. What has been presented in this chapter gives an answer to the inquiry embodied in the following quotation from Gauss' Göttingen colleague, the philosopher Lotze, in his *Microcosmus:*

If the object of all human investigation were but to produce in cognition a reflection of the world as it exists, of what value would be all its labor and pains, which could result only in vain repetition, in an imitation within the soul of that which exists without it? What significance could there be in this barren rehearsal—what should oblige thinking minds to be mere mirrors of that which does not think, unless the discovery of truth were in all cases likewise the production of some good, valuable enough to justify the pains expended in attaining it? The individual, ensnared by that division of intellectual labor that inevitably results from the widening compass of knowledge, may at times forget the connection of his narrow sphere of work with the great ends of human life; it may at times seem to him as though the furtherance of knowledge for the sake of knowledge were an intelligible and worthy aim of human effort. But all his endeavors have in the last resort but this meaning, that they, in connection with those of countless others, should combine to trace an image of the world from which we may learn what we have to reverence as the true significance of existence, what we have to do and what to hope.

CHAPTER TWENTY-THREE

Sunset and Eventide: Renunciation

Almost the only physical recreation which Gauss permitted himself in his last years was a daily walk to the literary museum, several blocks away from the observatory, where he spent most of the time between eleven and one o'clock. During the last twenty years of his life he took little part in social activity and during the last ten scarcely any. At the literary museum he scanned rapidly the various scientific, literary, and political periodicals, noting on paper or storing in his remarkable memory little interesting titbits of new information. He seemed to need more rest, and his friends noted that at home he relaxed more than formerly by engaging in light reading. His correspondence with friends and colleagues in all parts of the world claimed a considerable portion of his time. Up until the last he corresponded regularly with Alexander von Humboldt. Death interrupted his correspondence with Olbers in 1840, with Bessel in 1846, and with Schumacher in 1850. With Schumacher he corresponded weekly.

In the winters of 1852 and 1853 Gauss complained repeatedly about his health. Throughout most of his life he had been very healthy and had enjoyed a strong constitution. He suffered from congestion due to phlegm and mucus, which he considered his principal complaint. As a consequence of this condition he had become accustomed to get up at 3 A.M. and to drink Seltzer water and warm milk. It was a simple remedy but seemed to relieve him. Forty years earlier Olbers had given him two prescriptions, but all use of medicine had been otherwise foreign to him. Gauss had little confidence in medical science and for a long time could not make up his mind to call in a doctor.

Finally after the repeated persuasion of his family and closest friends he decided to seek the aid of his university colleague and friend Dr. Wilhelm Baum (1799–1883), professor of surgery and director of the surgical clinic from 1849 until his death. Baum made his first professional call on January 21, 1854, and after a thorough examination continued through several days he gave enlargement of the heart as his diagnosis. From the very beginning he held out little hope of recovery or survival for very long. The condition, which became more pronounced in advanced age, seemed to be an old one, for Olbers had recognized or supposed it years before and had advised Gauss on certain precautionary measures. The application of practical remedies and the coming of spring had a favorable effect on Gauss' condition, so that in the course of the spring and summer of 1854 he was able to put in his regular appearance at the literary museum, as well as to take little walks in the neighborhood.

Gauss had a lively interest in the building and operation of all railroads, although for more than twenty years he had not spent a night away from Göttingen and thus did not know the new improvements from direct experience. On June 16, 1854, he visited the construction of the railroad line between Göttingen and Kassel. Unfortunately, the horses were frightened by a passing locomotive, the coach was turned over, and the coachman was seriously injured. Gauss and Therese were not hurt and returned to the observatory shaken up but otherwise uninjured.

Through the newspapers Gauss learned about the death on May 21, 1854, of his old friend Bernhard von Lindenau. He was painfully affected by the report and his friends avoided talking about it, but he kept returning to the subject and reminisced about the long years of their friendship. He considered Lindenau's character to be noble and unselfish, and thought that the success of his diplomatic activity was to be ascribed to this. In these conversations he did not seem to think of the nearness of his own death. Lindenau had last visited Gauss on July 26, 1849.

On July 31, 1854, occurred the official opening of the railroad line between Göttingen and Hanover. It was a fine summer day and Gauss felt well enough to go into the city and observe the festivities from various vantage points. In 1936 Baron August Sartorius von Waltershausen of Gauting near Munich, then eighty-four years old, a son of Gauss' intimate friend and colleague Baron Wolfgang Sartorius von Waltershausen, still remembered how Gauss held him in his arms so that he could see the passing train. At the time he was nearly three years old. This was the last day on which Gauss' friends saw him in fairly good health.

Gauss survived the intimate friends of his youth in Brunswick. During his last years his only remaining close tie with the city of his birth was his brother Johann Georg Heinrich and family. The brother died on August 7, 1854, aged eighty-six, and Gauss wrote to his nephew immediately:

I have received with heartfelt sympathy the sad news of your letter of the 8th. It was grievous for me that for several years I had remained without any news of my brother. As long as Professor Goldschmidt lived, I always had some connection with Brunswick, since he was accustomed to journey there a couple of times a year to his father, who was then still living, and then always made inquiry about the health of my brother and communicated with me. Professor Goldschmidt has been dead now for several years, just as all the friends of my youth there. It is the fate of man when one gets old. I am already in my 78th year, but I shall not equal my brother, since for a year I have been feeling the loss of strength.

I am unusually glad that I must conclude from your letter that the last years of my brother's life were eased as much as the course of affairs allowed, by the true care of your mother, to whom I ask you to convey my hearty sympathy and greetings.

I myself have not seen my native city again for 33 years and even then only for one day. Now the journey is considerably shortened by the train, because one can go there from here via Hildesheim or Hanover in six or seven hours, and I suppose in one or two years, when the side line is opened, in half the time. Whether I will survive until that moment, or whether my strength will permit me to make use of the train, in order to see my native city just once again, is questionable. But it always remains my sincere wish that everything may be well with you and yours.

In the fall of 1854 Gauss' illness got worse from week to week. An earlier symptom, swelling of the feet, became aggravated, but Gauss did not seem to regard it as dangerous. Finally he had to stay in the house and could not take his short daily walk to the literary museum. Owing to increasing asthma he could walk about his home only with great effort. On December 7, 1854, alarming symptoms were manifested, and Dr. Baum feared that he would not live through the night, but the will to live was strong, and after a good night's rest Gauss expressed the hope that he could soon return to his usual order. He seemed to be much better.

Dropsy now developed, and conversation became so difficult that he had to decline visits from his intimate friends. Baum and his daughter Therese were the only ones who saw him during the last weeks. Although Gauss was not able to carry on any of his usual work, he was still mentally active, still read extensively, and did some writing at his desk. His handwriting, usually so fine and neat, had become shaky. In December, 1854, he wrote his will. On account of asthma he spent the last days and nights in his big easy chair. During the last thirty hours he was at times delirious and at other times in a coma.

His friend Sartorius, as was his wont, visited Gauss on New Year's Eve, 1854. He found him in good spirits but left with the feeling that the end was near. During the first week of January, 1855, he was suffering greatly but still hoped for the restoration of his health. On January 5 he wrote to Otto Prael, the university's superintendent of buildings and grounds, the following note, probably the last thing he ever wrote, concerning repairs in his home to be made in the spring:

To your honored letter of December 21 last year I have the honor of replying that, of repairs about which I am competent to judge, at the moment I can name only one, that in the wainscoting of my living room through drying of the wood openings have originated through which one can stick his hand. Probably new paint for the woodwork in the room and new paint for the walls will then also have to be connected with the repair of this defect. Most devotedly,

GAUSS

After the illness had vacillated back and forth several times Sartorius saw Gauss again on January 14. Christian Heinrich Hesemann (1815–1856), pupil of Rauch and official court sculptor in Hanover, had just arrived on orders of the King, in order to make a medallion of Gauss. He was told that he could start work the following day. Hesemann died suddenly on May 29, 1856, and C. Dopmeyer, another sculptor in Hanover, completed this medallion, which was later placed on Gauss' tombstone. This medallion or plaque was the basis of the medal which the King ordered just after the death of Gauss. The seventy-millimeter medal was produced at Hanover by Friedrich Brehmer (1815–1889), a well-known German sculptor and medalist. Sartorius found Gauss in a serious condition but in good spirits. His blue eyes sparkled.

On February 21, 1855, Sartorius again saw Gauss soon after noon, but only for a few moments. His consciousness was clear, but a great change had occurred, and his friend realized that death was hovering near. Sartorius pressed his hand for the last time and left the room. Soon after noon on February 22, Gauss suffered the last heart attack, then toward evening he seemed to be better, and consciousness had not left him although his eyes were closed. He heard everything that was going on around him, asked who was present in the room, and requested a drink of water. His intimate friends sat in an adjoining room and hoped for a better night. His heart was still beating, but the intervals between heartbeats became longer and longer; his breathing became more and more quiet. At 1:05 A.M. on February 23, 1855, he peacefully breathed his last. Sartorius reports the almost incredible incident that Gauss' pocket watch, which he had carried most of his life and which an astronomer does not easily forget to wind, stopped at a few minutes after one. During his illness it had been carefully kept going.

When King George V of Hanover visited the observatory on April 27, 1865, he ordered a large copper tablet placed over the door of the room in which Gauss died. The inscription reads thus:

Heinrich Wilhelm Matthias Olbers

Heinrich Christian Schumacher

Alexander von Humboldt

Friedrich Wilhelm Bessel

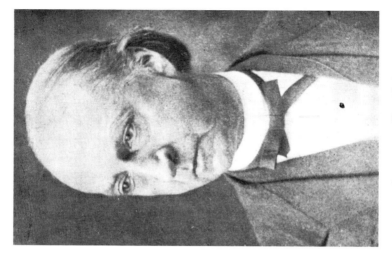
Johann Franz Encke

Johann Benedikt Listing

The Philipp Petri daguerreotype of Gauss on his deathbed

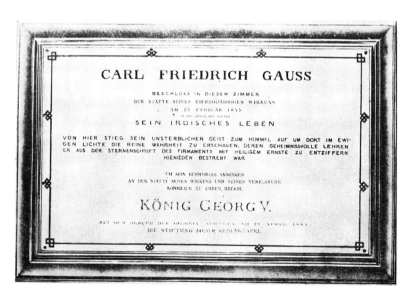

The copper memorial tablet in Gauss' death chamber, given by King George V of Hanover (1865)

The grave of Gauss in St. Albans Cemetery in Göttingen

The bust of Gauss by C. H. Hesemann (1855) in
the University of Göttingen library

Joseph Gauss

Heinrich Ewald

Minna Gauss Ewald, a watercolor by L. Becker (1834)

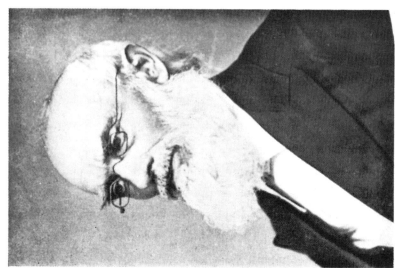

Eugene Gauss

Wilhelm Gauss

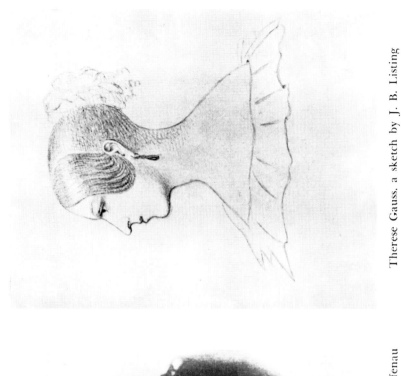

Therese Gauss and her husband, Constantin Staufenau

Therese Gauss, a sketch by J. B. Listing

Schaper's monument to Gauss in Brunswick (1880)

Bust of Gauss by W. Kindler (in the Dunnington Collection)

Bust of Gauss by Friedrich Küsthardt in Hildesheim

The Gauss-Weber monument in Göttingen, by Hartzer (1899)

The Gauss monument by Janensch, in Berlin

The Gauss Tower on Mount Hohenhagen, near Dransfeld, dedicated 1911

The Brehmer medal of Gauss (1877)

The Eberlein bust of Gauss in the Hohenhagen Tower

Commemorative postage stamp issued for the Gauss Centennial, 1955

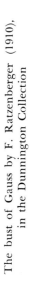

The bust of Gauss by F. Ratzenberger (1910), in the Dunnington Collection

Carl Friedrich Gauss closed his earthly life in this room, the scene of his forty years of activity, on February 23, 1855, in the arms of his own. From here his immortal spirit ascended to heaven, in order to contemplate pure truth there in eternal light, whose mysterious doctrines he strove with holy seriousness to decipher here below from the starry writing of the firmament. In order royally to honor his famous memory at the scene of his activity and his death King George V ordered the erection of this tablet during his visit at the Georgia Augusta on April 27, 1865.

At the news of his death his intimate friends and close acquaintances hurried to the observatory. They found him in his big easy chair, both hands resting on his knees, his feet extended, and his imposing head of silver-gray hair sunk to his chest. The feeling of peace was dominant and impressed them. They were touched to see Therese at his feet; she was parting his silvery locks, weeping, kissing and caressing his forehead as though she wanted to call him back to life. Absolute silence prevailed, tears unheeded bathing their faces.

With permission of the family a careful autopsy was performed the day after his death. Those participating were Dr. Baum, Listing, Wagner, and Professors Förster,[1] Fuchs,[2] and Henle.[3] The skull and brain were carefully weighed and measured. Wagner reported that the brains of Gauss and Dirichlet (d. 1859) were equipped with very rich and deep convolutions, the most remarkable he had observed. The frontal convolutions were especially noteworthy. Specific forms and arrangements did not occur. Wagner regretted that he could not compare the brain of Gauss with that of Laplace.[4] Gauss' brain showed no unusually great mass development. With the meninges it weighed 1,492 grams; after the meninges, the water infiltration, and certain blood vessels had been removed, it weighed 1,410 grams. The weight was considerable if we consider that Gauss was a man of medium stature, that he was seventy-eight years old, and that

---

[1] August Förster (1822–1865), professor of pathological anatomy.
[2] Konrad Heinrich Fuchs (1803–1855), professor of pathology.
[3] Jakob Henle (1809–1885), professor of anatomy and physiology.
[4] It was in the possession of his physician, Magendies.

atrophia senilis had already set in. The brains of Byron and Cuvier were heavier, 63.8 ounces and 64 ounces respectively. Dante's brain weighed 50.2 ounces. Robert Gauss, brilliant Denver, Colorado, newspaper editor and grandson of Carl Friedrich, at his death in 1913 left instructions for his brain to be weighed and studied. This was carried out, and it was found to weigh 55.7 ounces, three ounces heavier than that of his grandfather, and exactly equal to the weight of Schiller's brain. The brains of Gauss and Dirichlet are preserved in the department of physiology at Göttingen.

The sculptor Hesemann was summoned by telegram from Hanover and made a death mask and an accurate cast of the whole head and the inner surface of the skull. He later made the heroic-size white marble bust of Gauss for the university library. Alexander von Humboldt considered it the best likeness of Gauss. Several sketches and measurements of the entire body were made with the idea of using them in future monuments. A local photographer, Phillipp Petri, was called in and made four daguerreotypes, two showing the head and shoulders, and two the entire figure in death. Three of these pictures have disappeared.

On February 25 Gauss spent the last night in his room. His intimate friends laid him out in his academic robe with its purple velvet trimming in a simple black coffin costing fifty-eight thalers. A laurel crown was bound around his head, and early spring flowers surrounded his silent figure. Early on the morning of February 26 his body lay in state in the rotunda of the observatory under the dome. Among the visitors were many who had never seen him in life. Church bells tolled and reminded the citizens of Göttingen that their greatest son was about to be laid to rest. The coffin was surrounded by cypress, and two branches of palms were inclined toward the body. Twelve candelabra poured out a solemn light over him.

At nine o'clock twelve students of mathematics and science (including Dedekind) carried the coffin out on the open stone terrace of the observatory where a large and distinguished group had assembled. Among those present were his daughter Therese, his son Joseph and family, his son-in-law Ewald, Regierungsrat Adolf von Warnstedt (1813–1897) of Hanover, curator of the university, most professors and students, members of the city council, and a vast multitude of friends, neighbors, and admirers.

The choir now sang Luther's great hymn "Ein' feste Burg ist unser Gott," with its rude tone of combative faith:

> A mighty fortress is our God,
> A bulwark never failing;
> Our Helper He,
> Amid the flood of mortal ills prevailing.
> For still our ancient foe
> Doth seek to work his woe:
> His craft and power are great,
> And armed with cruel hate
> On earth is not his equal.
>
> Did we in our own strength confide,
> Our striving would be losing,
> Were not the right man on our side,
> The men of God's own choosing.
> Dost ask who that may be?
> Christ Jesus it is he,
> Lord Sabaoth his name,
> From age to age the same,
> And he must win the battle.
>
> And though this world, with devils filled,
> Should threaten to undo us;
> We will not fear, for God hath willed
> His truth to triumph through us.
> The prince of darkness grim,
> We tremble not for him;
> His rage we can endure,
> For lo! his doom is sure,
> One little word shall fell him.
>
> That word above all earthly powers,
> No thanks to them abideth;
> The Spirit and the gifts are ours
> Through him who with us sideth;

Let goods and kindred go,
This mortal life also;
The body they may kill:
God's truth abideth still,
His kingdom is forever.

Ewald then delivered the first funeral sermon, and Sartorius gave the second sermon (see Chapter XXIV). The coffin was closed and carried down to the hearse, and a long procession slowly moved to a beautiful, shaded spot in St. Albans cemetery a few blocks away. There the Reverend Dr. Sarnighausen, associate pastor of St. Albans Church, pronounced the benediction. After the coffin had been lowered into the grave, it was covered with palms and laurel. Nearby is a picturesque pond with well-shaded banks; on it graceful white swans glide slowly about. An impressive monument costing 750 thalers was erected in 1859 by Gauss' children. Ivy soon covered the lot. Beside him are buried both wives, his son Louis, and his mother, although no separate tombstones mark their graves.

The person most deeply affected by Gauss' death was his daughter Therese, to whom he was so tenderly attached. It was she who had been his constant companion for three decades, had kept house for him, and nursed him. On May 16, 1855, she wrote to her brother Eugene in Missouri:

MY DEAR EUGENE,

Your letter in which you kindly express your feelings for me so sympathetically on the loss of our father, before I was able to write you myself, has touched me deeply and I thank you with all my heart for the words of brotherly sentiment. The benefit of all love shown me, especially by the few who through the dear departed still have a closer connection on earth with me, I feel so very much in my inner impoverishment, even though no consolation and no substitute can be brought for that which I have lost; for with the earthly separation from father everything is indeed extinguished, in which and for which I have lived! It is not so much the external emptiness and desolation now surrounding me, by which the feeling of loneliness is forced so inexpressibly painfully into my soul, as rather the pain for a torn *inner* life relationship which existed so richly, so holy, so fervently between father and me, as perhaps exists only seldom between two human beings on earth. The secluded, quiet life with him and for him was my world, in which every passing year moved me more ardently and closely to him, made me richer in all that his warm, tender heart, his intimate confidence, his splendid, eternally fresh spirit poured out to me! My whole existence was tied up in father and as I believed I could not live without him, so he believed he could not live without me. He so often told me and I have found in it my happiest consciousness, that according to his feeling it was something deeply intimate and more than mere habit which bound us so closely and firmly together in life, that the thought of a voluntary separation appeared inconceivable to him as well as to me. Since grandmother's death I did not leave him even for one day; I was homesick even if I was away from him only several hours! The last year of suffering, full of sickness demanding constant attendance, bound me still more closely to him; the last weeks had scarcely a moment when he permitted me to be away from him, and he expressed the desire that we might not be separated even by death, for only a few days before he died he said to me: "The best and greatest that God could grant both of us would be the one thing: that we might die *together* on *one* day." At first I believed that my own longing would also quickly pull me after him; but the corporeal bond holds too tightly to be torn asunder without physical necessity, and thus years and years may pass in which I must learn to bear the inner isolation. For I shall remain inwardly lonesome, wherever I may turn; half of my existence rests in the grave and although I am resigned and very quiet and peaceful in the suffering, yet I feel more deeply with every passing day that father's death has made a cleft in my innermost life, which nothing, not even time, can heal again.

My dear Eugene, the fact that I did not write you for a long time, that I left your letter of April, 1854, unanswered, had its reason in my painful mood, in the continuous change between hope and fear in his sickness. I didn't know *what* I should write you, for I would have reproached myself for expressed hopelessness at the first appearance of improvement, and yet that (appearance) never gave me enough hopeful courage to write.

My last letter to you, dated, I believe, April 30, 1853, is two years old, and if at that time I wrote that father's health was no longer quite robust, it nevertheless did not cause any unusual anxiety. But in the course of the summer following he began to complain to such an extent as to cause alarm. Part of the time he suffered much, and, his strength failing rapidly, I, full of apprehension, besought him in vain to call in a physician. Not till January, 1854, as the disease in a few weeks had made rapid progress, did he consent. The physician, who has since with unremitting love, care, and sympathy attended him, lessened his suffering where cure was impossible, and doubtless somewhat prolonged his life, declared to me positively, after his first visit, that his condition was dangerous and hopeless. He recognized the disease at once as a heart condition, which probably had been coming on for years, in course of which there had been an accumulation of water around the heart, which in a few weeks also extended to other parts of the body. At that time the disease advanced rapidly and left little hope, but under the careful treatment of our loving physician, Dr. Baum, some improvement followed like a miracle. Some symptoms of the disease disappeared entirely, and father was able to go out for short distances, though only slowly and with immediate exhaustion. Mr. Angelrodt's erroneous reports of his health, which he gathered perhaps superficially at the hotel and communicated to you, may be based on these facts. But suddenly in November the old trouble returned in more decisive form, increased from day to day, and at the beginning of the present year the physician said to me that the life of our beloved one would be of only short duration. The last weeks of suffering were terrible, as the disease of dropsy in general is terrible, because it visibly approaches death inch by inch. But father bore all his suffering to the end with unvarying, touching serenity, friendliness, and patience. Entirely hopeless he never was; he always believed in the possibility of recovery as long as one spoke encouragingly to him. Ah! how difficult this has often been, when I, hopeless, knew the nearness of death! He never lost complete consciousness. Four hours before his death he still knew me, when, for the last time, he took a drink from my hand, drew my hand toward him and, kissing it, looked lovingly at me. He then closed his eyes and seemed to sleep, but I believe he did not sleep, but that his spirit, clear and conscious as ever, had freed itself from its earthly shell and had gone to its heavenly home.

You ask me about plans for my future, good Eugene; you attach to it the thought of seeing your lonely sister in America and the dearest offers to accompany me to you and home again. But I have no plans. I have at the moment only a heart filled with gratitude for the good deed—since a filling of the position is not now intended—of being allowed to remain in possession of the beloved rooms for an indefinite time, rooms which on account of father are so inexpressibly dear to me, rooms in which he spent half of his life, rooms where everything bears traces of him, the room in which he died! Every spot in them bears holy memories for me, and memories and everything that nourishes them are now my only consolation! Father's living room and bedroom are still unchanged, everything in them is in the same place as it was during his illness and his death. I go in so often, so often, and always believe that I feel closer there to the dear departed! In this sense it is a melancholy joy for me that father's wish leaves me in possession of the household furniture and that the dear thought of my brothers finds in it no injustice against any of us. Certainly, dear Eugene, it is not on account of the financial value of the things, but partly because every article *used by father* is for me a holy relic for which I would gladly give away everything else and partly because the order of my home is not threatened by division and sale.

God has given me a great miraculous strength of endurance in nursing father, for my otherwise not strong health did not break down during weeks of night vigils and never ending struggle with grief and worry; but after father's death the unnatural tension of my nerves relaxed, I felt completely exhausted and still can't properly recover again. Therefore it is also a benefit for me in this respect, namely, that I may remain in these quiet, peaceful rooms, without having to busy my thoughts now with plans for the future. In spite of their painful desolation I feel less lonesome in them than I would in any other place, for I still always have here the feeling of father's spiritual nearness and of home. Joseph has visited me rather often, although always only for one day.

Actually one can never judge definitely for oneself concerning the future, but according to my present feeling I do not believe that I shall leave Göttingen in order to live in another place. At least years must first pass—and who knows whether I shall still be living after these—until my emotions again assume a less exclusive direction to the only thing that now fills me; I love father's grave and all the places where he lived, too much, not to want to remain near them. Here in Göttingen I do not have very many, but several very intimate, friends who have always given me in difficult and sad times proofs of ardent and true sympathy and love. They moreover share with me so many reminiscences of father, and for me, momentarily at least, all that is one more tie to Göttingen.

Far-reaching plans so seldom lead to the goal, life always treats us differently from what we expected; but, dear Eugene, if the desire that we see each other once again on earth is fulfilled, if in the course of the next years you desire to visit your first homeland once more, in which very little is left for you except graves—than we shall probably see each other in Göttingen again. It would be so inexpressibly good and fine if I could embrace you and my beloved Wilhelm once more in life! Under your protection the long journey over the sea in itself would probably not be terrifying to me—but I have now too little vital energy to be able to answer for myself the question whether I would like to step out again from the narrow circle of my quiet life.

A week ago I also had a letter from Wilhelm, in which he with sincere love made me the same offers that you make: to come over in order to fetch the lonely sister; and like you he *had* the hopeful plan of soon coming to Germany in order to see our father once more in life. Today I am sending simultaneously with this letter my answer to him and hope that both of you will soon receive the letters. I shall rejoice so heartily too, dear Eugene, to hear from you more often than formerly, and shall gladly write to you again. From Joseph you yourself now probably get direct news rather often; four weeks ago he, his wife, and child visited me for one day. The latter two are very well, but unfortunately he complains rather often about his poor health due to overwork, for the improvement of which he will soon visit a spa.

And now farewell, my good brother, greet your wife and children for me and keep me dear.

Truly your sister
THERESE GAUSS

CHAPTER TWENTY-FOUR

Epilogue

1. Apotheosis: Orations of Ewald and Sartorius

On the morning of February 26, 1855, Heinrich Ewald, son-in-law of Gauss and professor of theology and Oriental languages at the University of Göttingen, delivered the following funeral oration before a large audience on the terrace of the university observatory:

Dear departed one! How shall I eulogize you at the last moment when we, deeply mourning and standing even closer to you, surround your soulless remains? It has not been six years since we surrounded you on a rare festive day[1] in the embellishment of all your power; at that time not only our city but also strangers hurrying in from near and far shared our general joy; and the happy glance of your eye, the fresh green laurel of your silvery hair promised us the possession of you for the longest span of human life. In two more years and a few months a long-desired and even loftier festive joy would have again united us with you, since then a full half century would have passed since you were called to the bosom of our university in a somewhat unusual manner. The Lord of Life and Death has decreed otherwise: these halls, once dedicated by you to their permanent purpose, long the scene of your works and the rich as well as mature, noble fruits of the same, are now to be bereft of you; we are to renounce the sweet habit of not knowing them without you, indeed of scarcely being able to think of them without you; we are bowed down, we mourn.

But, my friends, let us far more thank Him, Who gave him to us, Who preserved him for so long in the finest and fullest vigor, Who allowed him to act among us just as he did act. How should I attempt to describe and to appraise his services to so many and such various sciences all closely connected by a closer bond. Others, more expert and

---

[1] A reference to Gauss' golden jubilee of the doctorate, July 16, 1849.

more skillful, will do this now after the end of his earthly career even more than they have long since done at this university, as in all places of the earth, where a considerable number of the loftiest and most serious sciences blossom in fine league with their happy application and successful activity on an extended area of lofty and necessary efforts of our time. When he measured the infinite space of the heavens and the distant surfaces of the earth as nobody before him, when he taught how to find and esteem correctly with aids of science and recognition of work and art that which is attainable by the human eye, aids most of which he handled most happily, in the boldness and sharpness of research, in the certainty of the most important results, and in the completed presentation involuntarily reminding many persons of the final and loftiest efforts of all antiquity and towering above the pinnacles of the equally lofty and even loftier efforts of our last three or four centuries in many places. Oh, how little did he forget the Infinite, which is even higher than those infinite spaces and which ever surrounds and presses us. Or, when he called Nature his goddess whom he served, whose gentle signals or even powerful words and commands he like a genuine priest considered it his life duty to note (but that is a poetic word, and in his eyes the genuine poet's word possessed something holy): Oh, how little did he forget Him through whom this work miraculous to us has all its miracles as its limits and its eternal laws. And did he not love to show such thoughts as lift the enigmatic veil of all earthly existence; and herein belonged, as otherwise everywhere, wise reserve, caution, and thoughtfulness not to the least of his many virtues, especially in the face of the many types of supertensions and injurious errors of our times; is the total effect of a mind fathoming and explaining the deepest laws of all creation anything other than an ever stronger elevation to Him Who gave them and preserves them with His hand? May I be permitted to mention briefly here how unforgettable to me are the hours when our friend and teacher now removed from the earth, still in the full power and vigor of his years, at some accidental stimulus, talked in an intimate group about immortality and the whole relationship of the visible to the invisible, with a clarity and certainty, a brightness and assurance which astonished nobody more than those who had expressed their doubts and had introduced such conversations.

But, my friends, if all sciences, however difficult they may be to count in detail, finally fall into two quite different and not to be intermingled areas, with all mutual contact, into that of nature and that of human-divine things, and if one must always rightfully expect only of those who cultivate the latter area the most personal intervening sympathy in all changes and fates, all sufferings and injuries of all humanity, of peoples and societies, of the small and large realms of the earth, then we can also admire here how firm and loyally in his own area, under all changes of time, he always kept unmoved in his mind the highest task of his own life, apparently untouched by all such changes in the outer world. And again though—what soulful sympathy for all sorrows and joys of the changing world glowed through his heart, what inexhaustibly deep benevolence for all humanity always lived in him, what joyful recognition and esteem of every foreign merit, what pure, unshakable, fruitful friendship with so many intimates, lifelong with some of the greatest minds of his time, what warm zeal in service and what indefatigable readiness to serve, perhaps not according to the moods or the whims of the time, but everywhere that he could help through advice or deed, through the treasures of his knowledge or otherwise.

And what noble modesty and self-contained serenity dwelt beside such greatness loudly recognized by the world! On his jubilee when the signs of high honor and recognition poured in from all sides, from universities and academies, from princes, from friends and colleagues, hardly anything pleased him as much as the honorary citizenship conferred on him by this city.

And now you are torn away from us and we are to take one last look at your transfigured features, you our joy and our jewel, our model and our illumination! But thanks, imperishable thanks to you for all that you were to the university for almost a half century, for all that you were to your friends and pupils, the intimate and the casual, the near and the far ones, thanks, sincere thanks for all that you were to your own family, most intimately attached to you, what you were to me also! Oh, my friends, we are looking into a nearby grave, but nothing pulls up our thoughts more strongly to the certainty of immortality than the open consideration of human infirmity even of such ones whom we desired to keep eternally among us. Yea, you will remain eternally among us, honored and admired by those in the most distant future. And in addition there remain your hope and our hope. The never withering laurel, the never vanishing fragrance of eternal thanks and of eternally elevating memory, of eternal love and eternal hope we now lay on this your coffin adorned in another manner by loving hands. Thus farewell and may your earthly remains rest in peace, beloved friend, teacher, father!

At the conclusion of Ewald's oration Baron Wolfgang Sartorius von Waltershausen, professor of geology at the University of Göttingen and intimate friend of Gauss, delivered the second funeral oration:

With deeply moved hearts and mightily affected minds, my honored friends, we have just stepped up to this coffin, which contains the earthly remains of a man full of peace. Gauss' sublime mind is no longer among us! His glittering famous career is completed, his clear, blue, investigating, intelligently sparkling eye is closed forever and the peace of death surrounds his lofty, noble forehead; the laurel wreath winds around it! If I dare to begin speaking in this serious ceremonial hour, it is not for the purpose of piling up words of praise on the great deceased one, above which he was highly sublime even in life, nor to present to you, honored listeners, a true, well-sketched picture of his life and his work in science; I feel it deeply, infinitely deeply, that I am not equal to this task. A proper solution of it belongs alone to the judge of our intellectual development, the history of science, which with the illuminating fiery strokes of truth has listed in her holy books her unpartisan judgment concerning the deceased and concerning his unexcelled accomplishments from the end of the past century up to the middle of the present one.

It is only my purpose in this ceremonial moment to collect our spirits and to elevate our thoughts to God; but also at the same time to fulfill a pious childlike duty in the name of several friends and pupils present, as well as in my own name, to express once again publicly the feelings of our sincerest gratitude, of our sincerest love, of our sincerest honor.

Carl Friedrich Gauss was born April 30, 1777, at Brunswick; his illustrious genius forged ahead in his first years. We must eternally thank the benevolence and insight of a noble prince, who unfortunately closed his unhappy life with his mortal wound and a lost battle, that he furnished his loyal fatherly assistance to a young, vigorously striving mind on its scarcely opened career; he thereby erected for himself a permanent monument to which future centuries will not deny their full recognition.

In the most difficult, most fateful years which have ever broken in upon our German fatherland, Gauss by a higher dispensation became ours as a compensation for so many sufferings and he remained ours for forty-eight years, for the fame of our land and for the glory of our university, until a few days ago the cold hand of death touched his temples.

In that time of humiliation, of wretchedness, in which Germany lay under a foreign yoke, I must emphasize it loudly and clearly, Gauss proved himself a solid German man, who, loyally adhering to the fatherland, with his giant mind protected the treasure of our science and language; even at that time he scored a series of the most illustrious intellectual victories which have filled even our arrogant enemies and proud conquerors with admiration.

After German soil was freed of the world conqueror, Gauss under the protection of five noble kings, under the care of a benevolent university board of curators completed a quiet, retired, absolutely unassuming life worthy of a great philosopher, which has been dedicated only to the highest intellectual purposes. He created under such external circumstances in the course of forty years a series of the grandest researches in the field of mathematics, astronomy, and physics, which without exception have opened up to us new paths in science. I do not want here to go into the details of these thoroughly reformatory investigations; I want only in a few words to emphasize the spirit which like a golden thread, like a noble string of pearls winds its way through all his works from beginning to end—it is the spirit of a pure, a completed science; by that I understand the science which is not created for earthly purposes, but that one which, like a brilliant star, like a glittering meteor falls from heaven, in order to transfigure with its rays the twilight of earthly life, the science which grants us the consolation, the firm confidence that the soul of man is not born in dust, in order to perish again in dust. This spirit of true science is opened up to us through Gauss' researches; he has bequeathed it as his great legacy, as his holy testament to our hard-pressed, severely tested Georgia Augusta, that he may accompany us through the most distant times, that he may precede us like a guardian angel on our pilgrimage, and with the torch of truth may illumine our paths, that he may grant us joy in the days of good fortune and consolation and calmness in times of need, so that he may protect and strengthen us, elevate our souls and ennoble our hearts. As long as this spirit is alive among us, that science, to which we have also dedicated our life, will be vigorously preserved among us, and our university will gloriously continue to exist even in the midst of winter storms; she will always bear new young buds, fresh fragrant blossoms, and ripe fruits. By the great man just deceased, from whom we part with tears, the path which we must tread in the future is mapped out and opened in the most splendid manner—it is our obligation to pursue it further. God! Almighty God! In this serious moment hear our sincere prayer and give us strength and manly courage, striving for such a goal, also to finish our course until that day when we die. Amen! Amen!

2. Valhalla: Posthumous Recognition and Honors

In a park of his native city of Brunswick, at the foot of a knoll now called the Gaussberg, is a splendid monument of Gauss by one of Germany's greatest sculptors, Fritz Schaper (1841–1919). It was erected in 1877 for the centenary of the mathematician, and the granite base bears this inscription:

Dedicated by a grateful posterity on the centenary of his birthday in his native city Brunswick to the sublime thinker who unveiled the most recondite secrets of the science of numbers and of space, who fathomed the laws of terrestrial and celestial natural phenomena and made them serviceable to the welfare of humanity.

The house in which Gauss was born in Brunswick served as a museum until it was destroyed in an air raid on October 15, 1944. Fortunately the contents had been removed and are now kept in the municipal library, along with the Gauss manuscripts. A street, a bridge,[1] and a school in Brunswick bear the name of Gauss, and a Gauss medal is conferred annually by the Scientific Society of Brunswick on some outstanding scientist. The Technical Institute of Brunswick has an oil painting and a bust of its most famous alumnus. The city of his birth has never been lax about honoring her greatest son. On several occasions publications about Gauss were subsidized by the city. In 1927 a special sesquicentennial celebration was held in Brunswick.[2] A Gauss Memorial Room in the old City Hall of Brunswick, serving as a museum, was officially opened on February 23, 1955, the hundredth anniversary of his death.

In the Deutsches Museum at Munich one finds in the Hall of Honor a full-length oil portrait of Gauss in his academic robe, by R. Wimmer. Under it is the following inscription by the astronomer Professor Martin Brendel:

His mind penetrated into the deepest secrets of number, space, and nature;
He measured the course of the stars, the form and forces of the earth;
He carried within himself the evolution of mathematical sciences of a coming century.

The city and the University of Göttingen have not lagged behind Brunswick in honoring Gauss. On the campus of the univer-

---

[1] On the bridge a bronze celestial globe exhibits the planetoid Ceres.
[2] The German government had an extensive Gauss-Weber exhibit at the Chicago World's Fair of 1893.

sity is a masterful monument of Gauss and Weber by Ferdinand Hartzer (1838–1906), erected in 1899. It represents the two scientists discussing their telegraph. Elaborate ceremonies were held for the unveiling. The only error in the monument lies in the fact that the two men appear to be almost of the same age, whereas actually Gauss was twenty-seven years older than Weber.

Perhaps the finest tribute ever paid Gauss was the publication of his *Collected Works,* which began in 1863 and extended to 1935. It was one of the most thorough jobs of editing ever done and was carried out by a considerable number of leading German scientists, each a specialist in his field. The publication was sponsored by the Royal Society of Sciences in Göttingen.

Banquets, lectures, and numerous other ceremonies celebrating the centenary and sesquicentennial of Gauss were held in Göttingen. A street in the city was named for him. The university set up the Gauss Archive to house his letters, manuscripts, and library. There are many mementos of him at the observatory and the department of geophysics. A bust is to be found in the main academic building. The heroic-sized white marble bust by Hesemann, generally considered the best, is housed in the university library. Special ceremonies were held in 1933 on the centenary of the telegraph, and a medal was issued. When the King ordered the Gauss medal soon after his death, he put on it this inscription: "George V, King of Hanover, to the Prince of Mathematicians," and added these words: "Academiae suae Georgiae Augustae decori aeterno."

Gauss was also honored in Berlin. On the Potsdam Bridge beside Helmholtz, Siemens, and Röntgen was a splendid monument of Gauss by Gerhard Janensch (1860–1933) showing him busy with the telegraph. On the façade of the Technical Institute at Charlottenburg beside Schinkel, Eytelwein, Redtenbacher, and Liebig is a fine sandstone bust of Gauss. The agricultural ministry has a bust of him by Janensch. Portraits of Gauss and Weber are to be found in the Reichspostmuseum.

The most imposing memorial is the Gauss Tower on Mt. Hohenhagen near the village of Dransfeld, a few miles from Göttingen. It is one vertex of the classical triangle, Brocken, Hohenhagen, Inselsberg, so important in the triangulation of the kingdom of Hanover. The tower is of basalt, topped by a red tile roof; it rises to a height of 120 feet, and from the observation platform the visitor has a splendid view of the rolling lands and forests of that region. It is a favorite spot for tourists. Nearby is a tavern, and also to be seen is a "Gauss stone" which he used in the survey. It was dedicated on July 29, 1911, before an audience of leading men of science and government. In the tower is a room containing relics, instruments, and a large marble bust of Gauss by Gustav Eberlein (1847–1926).

On the centenary of Gauss' death in 1955 the government of Nether-Saxony established at the University of Göttingen and at the Institute of Technology in Brunswick a Gauss Fund of 15,000 marks each per year for visiting scholars or professors in his fields. At the Technology Institute in Hanover, a new building of the regional geodetic survey called the Gauss House is to be opened and dedicated September, 1955.

When the German expedition under Professor Erich von Drygalski of Munich was sent to the South Pole in 1903, its three-mast schooner was named the *Gauss*. This ship was in later years purchased by the United States government and put into service under another name, along the North Pacific coast from California to Alaska. A volcanic mountain discovered by the expedition was called the "Gaussberg."

The name of Gauss occurs at so many places in mathematics and science that only a few instances where he has been honored will be mentioned.

The name "Gauss" was given to a unit used to measure the intensity of magnetic field.[3] It is the intensity produced by a magnetic pole of unit strength (a "Weber") at a distance of one centimeter. In 1932 the oersted was substituted for the gauss to represent the centimeter-gram-second unit of field intensity. The gauss is now the c.g.s. electromagnetic unit of induction. The

---

[3] The gaussage (whatever the unit may be) is the line integral of magnetic force around any closed curve.

term "gauss" may still be used if magnetic induction and magnetic field strength have the same dimensions.

In optics, the Gauss eyepiece is used in spectrometers and refractometers to set the axis of a telescope accurately at right angles to a plane polished surface. The Gauss eyepiece tube has an aperture in the side through which light is admitted to a piece of plane unsilvered glass at an angle of forty-five degrees to the axis of the telescope. The light is thus reflected past the cross wires and down the telescope tube to the plane polished surface. If the latter is exactly at right angles to the telescope axis, the light will be reflected back down the telescope and an image of the cross wires will be formed exactly coincident with the cross wires themselves. The observer must adjust the position of the telescope until this coincidence is obtained.

Gaussian logarithms are so arranged as to give the logarithm of the sum and difference of numbers whose logarithms are given. Gaussian logarithms are intended to facilitate the finding of the logarithms of the sum and difference of two numbers, the numbers themselves being unknown, but their logarithms being known, wherefore they are frequently called addition and subtraction logarithms.

The Gaussian method of approximate integration is one in which the values of the variable for which those of the function are given are supposed to be chosen at the most advantageous intervals.

A Gaussian period is a period of congruent roots in the division of the circle.

The hypergeometric series is also called the Gaussian series. A Gaussian function is the hypergeometric function of the second order.

The Gaussian sum is a sum of terms the logarithm of which is the square of the ordinal number of the term multiplied by $2\pi\sqrt{-1}$ times a rational constant, the same for all the terms.

In spherical trigonometry there are four important, frequently used formulas commonly called Gauss' analogies or Gaussian equations.

The name of Gauss is also frequently attached to a formula for approximate quadrature.

In geometry, Gauss' theorem relates to the curvature of surfaces; the measure of curvature of surface depends only on the expression of the square of a linear element in terms of two parameters and their differential coefficients. There is a certain expression involving the equation of the surface and the coordinate of a point, called the "Gauss curvature," which will be constant on the plane, the sphere, and certain other surfaces.

Gauss' name is always attached to the rule for finding the date of Easter, a formula which he established.

Two magnetic measurements are called "Gauss' A position" and "Gauss' B position."

During World War II the demagnetization of ships was called "degaussing." Counteracting coils were placed about the ship's compass so that it would respond normally to the earth's field against magnetic mines. Officers in charge were called "degaussing officers," and there were many so-called "degaussing stations."

The law of quadratic reciprocity of Legendre is also known as the fundamental theorem of Gauss.

As these lines are being written, word comes of several special ceremonies in 1955 in Göttingen and elsewhere, marking the hundredth anniversary of the death of Gauss. The occasion is to be observed by the issuance of a special commemorative postage stamp bearing a profile picture of Gauss.

Appendixes, Bibliography and Index

APPENDIX A

Estimates of His Services

H. J. S. Smith (1826–1883), one of England's leading mathematicians of the nineteenth century, gave the best appraisal of Gauss to be found:

If we except the great name of Newton (and the exception is one which Gauss himself would have been delighted to make) it is probable that no mathematician of any age or country has ever surpassed Gauss in the combination of an abundant fertility of invention with an absolute rigorousness in demonstration, which the ancient Greeks themselves might have envied. It may be admitted, without any disparagement to the eminence of such great mathematicians as Euler and Cauchy that they were so overwhelmed with the exuberant wealth of their own creations, and so fascinated by the interest attaching to the results at which they arrived, that they did not greatly care to expend their time in arranging their ideas in a strictly logical order, or even in establishing by irrefragable proof propositions which they instinctively felt, and could almost see to be true. With Gauss the case was otherwise. It may seem paradoxical, but it is probably nevertheless true that it is precisely the effort after a logical perfection of form which has rendered the writings of Gauss open to the charge of obscurity and unnecessary difficulty. The fact is that there is neither obscurity nor difficulty in his writings, as long as we read them in the submissive spirit in which an intelligent schoolboy is made to read his Euclid. Every assertion that is made is fully proved, and the assertions succeed one another in a perfectly just analogical order; there is nothing so far of which we can complain. But when we have finished the perusal, we soon begin to feel that our work is but begun, that we are still standing on the threshold of the temple, and that there is a secret which lies behind the veil and is as yet concealed from us . . . no vestige appears of the process by which the result itself was obtained, perhaps not even a trace of the considerations which suggested the successive steps of the demonstration. Gauss says more than once that, for brevity, he gives only the synthesis, and suppresses the analyis of his propositions. *Pauca sed matura* were the words with which he delighted to describe the character which he endeavored to impress upon his mathematical writings. ... If, on the other hand, we turn to a memoir of Euler's, there is a sort of free and luxuriant gracefulness about the whole performance, which tells of the quiet pleasure which Euler must have taken in each step of his work; but we are conscious nevertheless that we are at an immense distance from the severe grandeur of design which is characteristic of all Gauss' greater efforts. The preceding criticism, if just, ought not to appear wholly trivial; for though it is quite true that in any mathematical work the substance is immeasurably more important than the form, yet it cannot be doubted that many mathematical memoirs of our own time suffer greatly (if we may dare to say so) from a certain slovenliness in the mode of presentation; and that (whatever may be the value of their contents) they are stamped with a character of slightness and perishableness, which contrasts strongly with the adamantine solidity and clear hard modeling, which (we may be sure) will keep the writings of Gauss from being forgotten long after the chief results and methods contained in them have been incorporated in treatises more easily read, and have come to form a part of the common patrimony of all working mathematicians. And we must never forget that it is the business of mathematical science not only to discover new truths and new methods, but also to establish them, at whatever cost of time and labor, upon a basis of irrefragable reasoning.

The μαθηματικὸς πιθανολογῶν has no more right to be listened to now than he had in the days of Aristotle; but it must be owned that since the invention of the "royal roads" of analysis, defective modes of reasoning and of proof have had a chance of obtaining currency which they never had before. It is not the greatest, but it is perhaps not the least, of Gauss' claims to the admiration of mathematicians, that, while fully penetrated with a sense of the vastness of the science, he exacted the utmost rigorousness in every part of it, never passed over a difficulty, as if it did not exist, and never accepted a theorem as true beyond the limits within which it could actually be demonstrated.[1]

In the *Biographie universelle* (Michaud, Paris, 1856) in a sketch of Gauss, Wagener wrote: "... each work of his is an event in the history of science, a revolution, which, overturning the old theories and methods, replaces them by new ones, and advances science to a height which no one had ever before dreamed of."

---

[1] Presidential Address, *Proceedings* of the London Math. Soc., VIII, 18.

M. Marie gave one of the clearest estimates of Gauss' work which have come down to us:

The genius of Gauss is essentially original. If he treats a subject which has already claimed the attention of other scholars, it seems as if their works were wholly unknown to him. He has his own manner of approaching the problems, his own method, and his solutions are absolutely new. These solutions have the merit of being general, complete, and applicable to all the cases that can be included under the question. Unfortunately, the very originality of the methods, a particular mode of notation, and the exaggerated, perhaps affected, laconicism of his demonstrations, make the reading of Gauss' works extremely laborious.[2]

An obituary notice of Gauss published by the Royal Astronomical Society contains these statements:

The *Theoria motus* will always be classed among those great works, the appearance of which forms an epoch in the history of the science to which they refer. The processes detailed in it are no less remarkable for originality and completeness, than for the concise and elegant form in which the author has exhibited them. Indeed it may be considered as the textbook from which have been chiefly derived those powerful and refined methods of investigation by which the German astronomy of the present century is especially characterized.[3]

Cayley, a leading British mathematician of the nineteenth century, wrote: "All that Gauss has written is first rate; the interesting thing would be to show the influence of his different memoirs in bringing to their present condition the subjects to which they relate, but this is to write a History of Mathematics from the year 1800."

Isaac Todhunter, another important British mathematician of the period, had this to say: "Gauss' writings are distinguished for the combination of mathematical ability with power of expression; in his hands Latin and German rival French itself for clearness and precision."

---

[2] M. Marie, *Histoire des sciences mathématiques et physiques,* t. 11 (Paris, 1887), 110.
[3] *Monthly Notices* of the Royal Astronomical Society, Report of Council read February 8, 1856, Vol. XVI, No. 4, pp. 80-83.

Felix Klein (1849–1925), professor of mathematics at Göttingen in the generation after Gauss, gave an excellent appreciation of his predecessor:

> In him [Gauss] the historical epochs separate; he is the highest manifestation of the past, which he closes, at the same time the foundation of the new, which he penetrates deeply with his last rays more thoroughly and more actively than is perhaps vivid in the consciousness of the time. If I may be permitted to use a figure of speech: Gauss appears to me like the Zugspitze in the over-all picture of our Bavarian mountains, as it appears to the observer from the North. The peaks rising gradually from the east culminate in a gigantic Colossus which steeply descends into the depths of a new formation, into which its spurs reach out for many miles and in which the waters gushing from it generate new life.[4]

Theodor Mommsen once expressed great admiration for the part played by aged men in furthering science, and added these words:

> Our own experience tells us scientists above all, that great scientific accomplishment can be successful only in many years of restlessly continued work. It is perhaps correct that in the case of men like Gauss the great aperçus by which they advanced the knowledge of the world, all appeared in their youthful years; but the seed is only half of scientific activity, and the harvest time is no less indispensable if an important scientist is to fulfill his destiny.

Felix Klein called the years 1798–1807 Gauss' *Heldenzeit* (heroic period), that is, the period of crowning productivity. His precision and clearness in thinking as well as the simplicity of presentation remained with him in old age. He devoted his later years more to optics, electricity, magnetism, and geodesy, although he never completely stopped work in astronomy and pure mathematics.

In 1870 during the Franco-Prussian War the Royal Irish Academy at Dublin protested against the siege of Paris on account of the danger to treasures of art and science there and also called on the German universities to participate in this manifesto.

---

[4] F. Klein, *Vorlesungen über die Entwicklung der Mathematik*, Teil I (Berlin: Julius Springer, 1926), 62.

Richard Wilhelm Dove (1833–1907), professor of ecclesiastical law and proctor of the University of Göttingen, rejected this proposal very decidedly. Since Dove had said in his letter that the German people in their intellectual struggle were still trying to make true the proud word of Paracelsus: "Englishmen, Frenchmen, Italians, *ihr mir nach, nicht ich euch*," Karl Hillebrand, the well-known historian and cosmopolitan publicist, accused him of "German arrogance." In 1887 at the sesquicentennial jubilee of the University of Göttingen when Dove printed the above-mentioned letter along with other documents of the university's history, he attempted to justify himself by means of the following statement:

It is sufficient to call the one name *Gauss* in order to demonstrate that the victor's prize by the unanimous decision of educated nations has really been ungrudgingly given to great German scholars.

Frobenius, in his eulogy of Kronecker before the Berlin Academy of Sciences on June 29, 1893, expressed himself thus on the work of Gauss:

The theory of numbers, which in Diophantus and Fermat bore the character of an entertaining exercise in thinking, an intellectual game, after the preliminary work of Euler, Lagrange, and Legendre had been elevated by Gauss to the rank of a science. The prince of mathematicians called it the queen of mathematics, and it deserved this title not only by its high rank, but also by the proud abstractness in which it sat on the throne, far from all other fields of knowledge, far from the other mathematical disciplines. In his theory of circle division the genius of Gauss made algebra tributary to it.[5]

In writing of the relationship between music and mathematics, J. J. Sylvester, the great British mathematician, predicted the day "... when the human intelligence, elevated to its perfect type, shall shine forth glorified in some future Mozart-Dirichlet or Beethoven-Gauss—a union already not indistinctly foreshadowed in the genius and labours of a Helmholtz!"[6]

---
[5] *Berliner Abhandlungen*, 1893, p. 4.
[6] British Association Report, 1869, *Notices and Abstracts*, p. 7.

Moritz A. Stern, pupil of Gauss and later professor of mathematics at Göttingen, had this to say in an address delivered there on the occasion of his master's centenary: "As in the case of only few known to history, in Gauss mathematical depth of thought was united with the talent of the observer, with the genius' practical insight into the invention of new means of observation, and with the most complete skill and endurance in calculation. Like the lion in the fable, Germany at that time could say, I have borne only one, but this one is a lion."[7]

When Alexander von Humboldt returned from his American tour in 1804, he wrote the following year to Emperor Friedrich Wilhelm III: "The only man who can give new splendor to the Berlin Academy is named Carl Friedrich Gauss."

In his vivid manner of expressing himself the Duke of Cambridge once said: "One often criticizes Göttingen, but as long as we have the library and Gauss, we can let them scold."

Alexander von Humboldt, to whom he made this remark, replied thus: "I agree, but it is my duty to request your royal highness to change the order of rank of the treasures and to name first the foremost mathematician of our age, the great astronomer, the brilliant physicist."

In 1808 Schumacher wrote Gauss that if he could have the title "pupil of Gauss" he would never demand another title. In 1810 he wrote his master: "I sign myself like Dissen[8] always called me: Ἀεὶ ὁ περὶ τον Γαυσσιον."

Schumacher's letters are unusually filled with the most inordinate flattery and admiration of Gauss, much of which was justified and all of which was sincere. Young Eisenstein and Encke adopted a similar attitude.

Kummer, one of the leading mathematicians of the late nineteenth century, has given a clear and exact estimate of Gauss' achievements:

Among all the major and minor works of Gauss there is none which through new methods and new results did not start essential progress in

---

[7] *Denkrede auf C. F. Gauss*, Festrede Göttingen Univ. April 30, 1877, pp. 3–4.

[8] Ludolf Dissen (1784–1837), professor of classical philology in Göttingen.

the field concerned; they are masterpieces which bear in themselves that character of classicity which guarantee that they will be used and studied with diligence, for all times, preserved not merely as monuments of the historical development of science, but also by future generations of mathematicians of all nations, as the basis of every more penetrating study and as a rich source of fertile ideas. With these extraordinary advantages of the Gaussian writings, indeed partly even on account of them, their effect on mathematical studies, namely in Germany, was for a long time extremely slight. In presentation all Gaussian writings have that perfect clearness and precision which in an absolutely penetrating study excludes even the possibility of misunderstandings, but this presentation, just as the methods themselves, is not calculated to facilitate the study of Gaussian writings. This character of his writings was also in full harmony with Gauss' own character. At his sublime position in science, on which he stood isolated for a rather long time, he possessed so great an autarchy that he scarcely perceived the need of attracting others to him and educating them.[9]

Thirty years later G. Hauck paid tribute to some of Gauss' practical work: "Gauss did not disdain to work practically with the theodolite and then on the basis of this work created practical methods of precision which have become the common property of all the civilized world and with which he has presented to humanity one of its most valuable gifts. Thus geodesy has become a model for the rational, applied activity of a theoretical science. Every surveyor is anointed with a drop of Gaussian oil."[10]

In his article on Gauss in the *Allgemeine Deutsche Biographie*, Moritz Cantor stated that the *Disquisitiones arithmeticae* should be called the Magna Charta of the theory of numbers.

Highest praise of Gauss came from a French source: "For loftiness of character, as for the power of genius, he was the greatest of all."[11]

In closing this chapter it is pertinent to quote a remark made in 1950 by Albert Einstein, than whom no one is better qualified to evaluate the work of Gauss:

The importance of C. F. Gauss for the development of modern physical theory and especially for the mathematical fundament of the theory

---

[9] *Rektoratsrede*, Berlin Univ., August 3, 1869, pp. 8–9.
[10] *Jahresbericht, Deutsche Mathematiker-Vereinigung,* VIII (1899), 108.
[11] J. Bertrand, *Éloges académiques* (Nouvelle série; Paris, 1902), p. 314.

of relativity is overwhelming indeed; also his achievement of the system of absolute measurement in the field of electromagnetism. In my opinion it is impossible to achieve a coherent objective picture of the world on the basis of concepts which are taken more or less from inner psychological experience.[12]

On another occasion Einstein had this to say of Gauss:

The best that Gauss has given us was likewise an exclusive production. If he had not created his geometry of surfaces, which served Riemann as a basis, it is scarcely conceivable that anyone else would have discovered it. I do not hesitate to confess that to a certain extent a similar pleasure may be found by absorbing ourselves in questions of pure geometry.

---

[12] The last sentence gives Einstein's reaction to Gauss' views on spiritual matters and a spiritual order, as set forth in Chapter XXII.

APPENDIX B

Honors, Diplomas, and Appointments of Gauss

1. Matriculation papers, University of Göttingen, October 15, 1795.
2. Doctor's diploma, University of Helmstedt, July 16, 1799.
3. Full member, club in the Hotel d'Angleterre, Brunswick, August 5, 1802.
4. Corresponding member, Imperial Academy of Sciences in St. Petersburg, January 31, 1802.
5. Royal Society of Sciences in Göttingen, "amicus et familiaris litterarum commerciis conjunctus," November 13, 1802.
6. Institut National, class of mathematical and physical sciences, corresponding member of the section for geometry, Paris, January 24–30, 1804. Signed by Carnot, President, and Delambre, Secretary.
7. Fellow, Royal Society of London, April 12, 1804.
8. Call to University of Göttingen, July 25, 1807. Signed at Hanover by Brandes.
9. Corresponding member, Royal Academy of Sciences in Munich, April 6, 1808.
10. The mathematical class of the Royal Academy of Sciences in Berlin unanimously named Gauss a full member and called him there; the entire academy approved this election on April 18, 1810, and the King of Prussia added his approval.
11. Foreign member, Italian Society of Sciences, Verona, May 4, 1810. Signed by Octavius Cagnoli, Secretary.

12. Knight of the Order of the Westphalian Crown (Jérôme Napoléon), Göttingen, August 19, 1810. Signed by Minister of State Count von Fürstenstein.
13. Lalande Prize; French Academy, 1810.
14. Knight Commander's Cross, first class, of the Guelph Order. Notification signed by Count Münster at Hanover, December 28, 1815.
15. Letters patent conferring the title of Royal *Hofrath,* signed by King George III at Carlton House, November 29, 1816.
16. Full member, Society for the Advancement of All Natural Sciences in Marburg, March 26, 1817.
17. Knight's Cross, Dannebrog Order, April, 1817; Commander's Cross, June 28, 1840.
18. Foreign Member, Royal Bourbon Society, Academy of Sciences, Naples, September 4, 1818. Notification signed by Teodoro Monticelli. (Note by Gauss: Monticelli died at Naples, October 7, 1845.)
19. Honorary member, Society for the Propagation of Mathematical Sciences, Hamburg, October 17, 1818.
20. Honorary Member, Courland Society for Literature and Art, Mitan, June 1, 1819.
21. Foreign Member, Royal Society of Edinburgh, June 5, 1820.
22. Full foreign member, Royal Bavarian Academy of Sciences, October 23, 1820.
23. Foreign Associate, Institut de France, Royal Academy of Sciences, Paris, September 4, 1820.
24. Director of Geodetic Survey, Kingdom of Hanover, 1821.
25. Member, Royal Danish Society of Sciences, May 1, 1821. Notification signed at Copenhagen by Oersted, Secretary.
26. Foreign Member, Royal Swedish Academy of Sciences at Stockholm, June 5, 1821. Notification signed by Berzelius, Secretary. Diploma dated June 26, 1821.
27. Fellow of the American Academy of Arts and Sciences, Boston, Massachusetts, January 30, 1822. Notification signed by John Quincy Adams, President; John Thornton Kirk-

land, Vice-President; John Farrar and Josiah Quincy, Secretaries.
28. Prize, Copenhagen Society of Sciences, for memoir on conformal projection, 1822.
29. Raise in salary to 2,500 thalers, Board of Curators, University of Göttingen. Signed by Curator, Baron Karl Friedrich Alexander von Arnswaldt, December 29, 1824.
30. Appointed full professor, 1828.
31. Member, Royal Academy of Palermo, 1828.
32. Associate, Royal Astronomical Society of London, April 13, 1832. Notification signed by Francis Baily, President, and Augustus DeMorgan, Secretary.
33. Corresponding member, Academy of Sciences at Bologna, January 3, 1833.
34. Foreign member, Royal Academy of Sciences of Turin, January 20, 1833. Notification signed by Count Prosper Balbo, Minister of State and President.
35. Foreign Member, Royal Bohemian Society of Sciences, Prague, December 1, 1833.
36. Director of Hanoverian Board of Weights and Measures, 1836.
37. Honorary Member, Physical Union, Frankfort, August 17, 1836.
38. Chevalier of the Royal Order of the Legion of Honor, August 26, 1837.
39. Copley Medal, Royal Society of London, 1838.
40. Foreign correspondent, Royal Academy of Sciences and Belles-Lettres of Brussels, December 14, 1841. Signed by Quetelet.
41. Knight of the Peace Class of the order *pour le mérite* for sciences and arts, Berlin, May 31, 1842.[1]
42. Member, Royal Society of Sciences at Uppsala, June 3, 1843.
43. Honorary member, Royal Irish Academy, March 16, 1843. Diploma dated at Dublin, October 5, 1851.

---

[1] This was the highest honor conferred by the Kingdom of Prussia and later by the German Empire.

44. Knight of the Swedish Order of the North Star, Stockholm, October 14, 1844.
45. Member, first class, Royal Institute of Sciences, Belles-Lettres and Beaux Arts in Holland, Amsterdam, April 19, 1845. Signed by W. H. de Vriese, Secretary *ad interim*.
46. Letters patent conferring the honorary title of *Geheimer Hofrath,* dated at Göttingen, July 1, 1845. Signed by King Ernst August of Hanover.
47. Diploma of honorary member of the philosophical faculty, University of Prague, 1848.
48. Honorary doctorate and member of the Society of Sciences, University of Kasan, January 12, 1848.
49. Foreign honorary member, Imperial Academy of Sciences at Vienna, January 26, 1848.
50. Congratulatory diploma of the Royal Bavarian Academy of Sciences on the golden jubilee of his doctorate, Munich, July 10, 1849.
51. Commander's Cross, second class, Guelph Order, July 11, 1849. Notification signed at Hanover by King Ernst August.
52. Congratulations of the Royal Hanoverian University Board of Curators on the golden jubilee of his doctorate, Hanover, July 14, 1849.
53. Congratulatory letter on the golden jubilee from C. W. Wippermann, electoral Hessian Minister of Finance, Kassel, July 15, 1849.
54. Jubilee congratulations of the Collegium Carolinum, Brunswick, July 15, 1849.
55. Commander's Cross of the Order of Henry the Lion, conferred by Duke Wilhelm of Brunswick, July 16, 1849.
56. Jubilee congratulations of the University of Kasan, July, 1849 (on silk).
57. Renewal of doctor's diploma by the University of Göttingen, July, 1849.[2]

---

[2] Since the University of Helmstedt no longer existed.

HONORS, DIPLOMAS, AND APPOINTMENTS 355

58. Honorary citizenship of the city of Göttingen, July, 1849.
59. Jubilee congratulations of the Bavarian Academy of Sciences, July, 1849.
60. Honorary citizenship of the city of Brunswick, July, 1849.
61. Member, Spanish Academy of Sciences, Madrid, 1850.
62. Foreign honorary member, Imperial Russian Geographical Society, September 29, 1851.
63. Honorary member, Royal Church Deputation of the University, Göttingen, December 28, 1852 (in recognition of Gauss' work on the insurance fund for widows of professors).
64. Honorary member, Imperial Academy of Sciences at Vienna, diploma dated December 31, 1852 (elected January 26, 1848).
65. Member, American Philosophical Society in Philadelphia. Diploma dated January 21, 1853 (received February 26, 1853).
66. Knight Commander's Cross, first class, Guelph Order, Hanover, May 27, 1853. Diploma signed by King George.
67. Bavarian Order of Maximilian for Art and Science, November 28, 1853. Notification signed by Maria Josef Portner.

No Date

68. Notification of the Hanoverian Cabinet Ministry concerning a special gift of 1,000 thalers in gold to Gauss.
69. Cambridge Philosophical Society.
70. Dutch Society of Sciences, Harlem.
71. Imperial Royal Lombard Institute at Milan.
72. Royal Saxon Society of Sciences, Leipzig.

APPENDIX C

The Will of Gauss

The following document is kept in the Gauss collection of the Municipal Archive and Library of the City of Brunswick, Germany (Inventory C. F. Gauss No. 151, MS No. 8). It is in Gauss' handwriting, but without signature and date (believed to be about December, 1854):

In consideration of my advanced age and the continuing increase in strength and stubbornness of the complaints connected with it, it has seemed advisable to me to write down those explanations and terms which shall serve as a guide in settling my estate after my death.

§ 1.

Of my two marriages four children are still living, namely,

of the first marriage a son, Joseph, now government architect in Hanover,

of the second marriage two sons and a daughter, namely Eugene, merchant in St. Charles in North America in the state of Missouri; Wilhelm, farmer in the same state, his farm in Chariton County near New Brunswick.

Therese as my loyal nurse continually with me,

these my four children, eventually their legitimate descendants, I make my heirs, but under the following modifications:

§ 2.

My oldest son Joseph took possession of his maternal inheritance long ago. Likewise I paid out completely at the proper time to my sons of the second marriage their share in the estate of their mother according to the testamentary disposition, so that they have no further special claims —on the other hand the amount of the share of my daughter in her maternal estate has not yet been paid out to her and therefore before the division of my estate that amount (4,754 thalers gold plus interest) of course is to be deducted from it as an encumbering debt to my daughter.

§ 3.

However, since my daughter has obviously suffered an encroachment through delayed entering into the enjoyment of this sum, I therefore direct, in order to give her some indemnity for it, that the following parts of my estate shall finally also go to my daughter:

A. all furniture, in which shall be included

   (a) all linens
   (b) all silverware. As regards the latter, that part of the silverware which came from Therese's mother was bequeathed to my daughter as her full property in the will of the former, and therefore by the above disposition a division of the silverware according to origin is unnecessary [in pencil:] the silver medals do not belong in this rubric.

B. All books which are kept upstairs, since a large part consists of gifts which I made to Therese and which are consequently already her property.

§ 4.

If he so desires my oldest son may choose as a special souvenir up to 30 volumes of my books. [In pencil:] for whose education many very significant costs did not occur as in the case of his brothers.

§ 5.

Everything else that belongs to my estate shall be divided in equal shares among my four children. Hereby forbidding all judicial sealing and inventory, I convey the execution of this division to my oldest son Joseph, who is already provided by his brothers in America with full power of attorney for looking after all their interests in Europe, which he has also partially used in several cases.

The above document was held to be valid as Gauss' will and was followed in settling his estate. Weber, Listing, and Ludwig Schweiger (1803–1872), librarian and professor of literature, were appointed to appraise Gauss' library for the Hanoverian government. Klinkerfues prepared the catalogue.

Joseph Gauss thought that the appraisal of 2,500 thalers for the entire library and manuscripts was too low, and also too low as to single items. Yet he was willing to let it go for that low price, because it would not be scattered and would thus be permanently preserved in the University of Göttingen library. Joseph demanded the atlases of Württemberg, Hessen, and Hanover (the latter a gift from him to his father). He picked duplicates of his father's works and other books which Schweiger assessed very low. Joseph decided that if the amount of 2,500 thalers did not stand after his choice, he would not limit himself to 50 or 60 volumes. Actually the family finally removed 226 volumes.

The manuscripts and correspondence going to the library were to be only those of "scientific content." Joseph undertook the sorting of the letters. The instruments and telescopes were not listed in the inventory and did not go for the 2,500 thalers. Manuscripts of nonscientific content (references to finances or family matters) were given back to Joseph. In this way a number of important items, including Gauss' scientific diary, lay hidden among family papers until 1898. Before his death in 1927 Gauss' grandson Carl August finally placed in the permanent care of the archives in Brunswick and Göttingen practically all manuscript material touching on his ancestor.

Gauss' heirs refused a number of other offers from the outside in order that his literary papers might go to the university and be available to the editors of his *Collected Works*. Joseph agreed not to remove books with marginalia in Gauss' handwriting. Later the editors made a thorough study of all such marginalia and made many important discoveries in this way. Joseph claimed he would be willing to return any books he had taken if they were found on examination to have value for the scientific collection. Eight bookcases and three astronomical instruments were placed at the disposal of the observatory by Joseph.

On May 29, 1856, the Hanoverian government turned over the manuscripts and library to the observatory. All the titles not already in the university library were placed there, and the remainder were set up in the west wing of the observatory. Therese donated her share from the sale of the library, 6,000 marks, to cover the cost of bookcases and binding.

The task of accurately cataloguing Gauss' library and combining it with the observatory library did not begin until 1897. In this way the accession list grew from 3,769 to 5,865—a gain of 2,096 volumes. If we count brochures and other miscellaneous items, the total mounts to 11,424. In 1899 the library was in an orderly condition for the first time in forty-four years.

On September 22, 1855, Joseph made a division of the various medals and related articles which Gauss possessed. Those sent to Eugene and Wilhelm are designated "America." Except for three major orders which had to be returned to the princes who conferred them, Joseph's inventory reflects the following items:

1. The old Westphalian Order of the Crown, not returned because the kingdom no longer existed (America)
2. The small Dannebrog Order (private property; it was a gift from Schumacher and was only a miniature copy of the real order) (America)
3. A small buckle of the Guelph Order (Joseph)
4. An exact duplicate of #3 (Joseph)
5. Gold pocket watch with chain and signets (Therese)
6. A very old silver pocket watch
7. The gold Copley Medal of the Royal Society (Joseph)
8. The gold Dalberg medal (Frankfort) (America)
9. Silver medal of Olbers (America)
10. Silver medal of the Berlin scientists' convention (America)
11. Silver medal of Jean Paul (America)
12. Silver medal of Blumenbach (America)
13. Silver medal of the Pulkova observatory near St. Petersburg (Joseph)
14. Copper duplicate of #10 (Joseph)
15. Copper medal of Bonn University, 1848 (America)
16. Copper medal of the Danish Society of Sciences, 1842 (America)
17. Copper medal of Napier, Edinburgh (Joseph)
18. Copper medal of Olbers (Joseph)

19. Duplicate of #18 (America)
20. Copper medal of Leibniz (Joseph)

Therese received No. 5; Joseph considered No. 6 not worth sending, since Gauss never wore it. He also sent Wilhelm and Eugene two pairs of spectacles and a lorgnette which Gauss wore.

APPENDIX D

Children of Gauss

The eldest child of Gauss, (Carl)[1] Joseph, was born August 21, 1806, in Brunswick; his mother was Johanna Osthoff, the first wife of the mathematician. Joseph was named for Piazzi, the discoverer of Ceres. All the other children of Gauss were born in Göttingen. Joseph attended the high school in Göttingen and for a short time was a student at the university. Early in November, 1824, he enlisted in the Hanoverian Foot Artillery. Eventually he became a first lieutenant in a battalion of the foot artillery stationed in Stade. He soon became dissatisfied with the rather idle life of a soldier and also realized that chances of promotion were slight—his father's fame did not seem to help him in this respect.[2] Joseph had already served as his father's assistant in the triangulation of the Kingdom of Hanover, and Captain August Papen used him as a principal aide in preparing the topographical atlas of the Kingdom of Hanover and the duchy of Brunswick.

Joseph believed that by participating in the construction of the railroad in Hanover he could make good use of the knowledge he had thus gained. He considered it necessary to study the new railroad system in America, and that was the purpose of the trip which he took in 1836 at his own expense. A leave of absence was granted him, and he was away most of a year. In the United States he met leading statesmen, engineers, and military officers.

Returning home in December, 1836, Joseph resumed his mili-

---

[1] In the spring of 1827 he prefixed the name Carl.
[2] He ended twenty years' service as a first lieutenant.

tary service as adjutant of a battalion in Stade. After being engaged for several years, on March 18, 1840, he married Sophie Erythropel (1818–1883), daughter of a Stade physician.

Joseph played a prominent role in fighting the great Hamburg fire of May 5–8, 1842. A telegram was sent to the artillery battalion garrisoned in Stade, requesting cannon, large quantities of powder, and the necessary man power to blast the burning structures, in order to stop the fire from spreading. Major Pfannkuche was in command of the expedition, and Joseph Gauss was his adjutant. He reported fully to his father about these experiences in a letter dated May 22, 1842. The Senate of Hamburg conferred on him a medal with a white and red ribbon in recognition of these services.

In April, 1845, Joseph resigned from military service in order to participate in the construction of the Hanoverian railroad. At first he was in charge of the line from Burgstemmen to Hildesheim. In October, 1846, he received his final discharge from the Hanoverian army and simultaneously became the fourth member of the Hanoverian railroad directorate,[3] to which he had actually belonged since December, 1845. This was a goal toward which he had been working for eleven years. His new position was full of responsibility and he was soon overworked to such an extent that he aged prematurely.[4]

Joseph retired when he was about to be transferred to Münster in Westphalia after the annexation of Hanover by Prussia. Politically he was a loyal adherent of the royal house of Hanover, especially because the King had granted his father so many honors. His only son fought in the Franco-Prussian war, and Joseph wrote to his brother Wilhelm on August 12, 1871: "Since France's superiority has now been broken for a long time and in Germany the unity so long striven for is attained, we can be well satisfied with the result of the war. Really a better time for Germany than you ever knew seems to have dawned."

---

[3] He received the title *Oberbaurath* in 1856; in later years he was superintendent of the special department of telegraphs.

[4] Joseph served as administrator of his father's estate.

Joseph resembled his father in appearance and had many of Gauss' traits of character. He died in Hanover at the home on Wilhelmstrasse on July 4, 1873. His only child was Carl August Adolph Gauss (1849–1927), who lived in Hamlin, Germany.

* * *

Wilhelmine Gauss, second child of Carl Friedrich and his first wife, was born in Göttingen on February 29, 1808. She was named for his friend Olbers and always used the name Minna. Gauss stated that she was the image of her mother, and in 1837 Alexander von Humboldt wrote Gauss that she was beautiful. In mind and spirit she was also like her mother—clever, kindhearted, loving, open, and happy. While Gauss was on a trip, his second wife wrote thus to him about her stepdaughter on September 29, 1814: "The older she gets, the more lovable the little girl becomes, and certainly she is the crown of our children." When her stepmother's health failed, she loyally aided her in every way possible. She nursed her, carried on the correspondence with her father during his absence on the survey, kept house, and cared for her grandmother Gauss as well as two younger brothers and a sister, to whom she became a second mother.

The following incident is said to have occurred in 1813 when Minna was only five years old. Gauss had explained to his little daughter that the small, rosy clouds in the evening sky are called cirri or fleecy clouds (*Schäfchenwolken*). One evening when these clouds again appeared and Gauss prepared to leave the family circle and make some observations, Minna attempted to hold him back with the remark: "Papa, stay with us, the sky is so *belämmert* today."[5]

Minna very early had many suitors for her hand. Eventually her choice fell on Heinrich Ewald, young professor of theology and Oriental languages. They became engaged in February, 1830, and were married on September 15, 1830. It was a marriage which brought genuine happiness to both.

---

[5] *Belämmert* would be translated "belambed," but it sounds approximately like *"belemmert,"* meaning encumbered or befouled.

Ewald was a very absent-minded scholar, and his friends had been advising him to marry in order to have domestic order. Ewald agreed but did not know whom he should marry. Since he was well acquainted with both Gauss' daughters (Minna and Therese), it was settled that he should decide at the next evening tea in the Gauss home, for the two daughters would alternately serve tea. Ewald was to accept tea from the one who pleased him better, and a colleague would then discuss the whole matter with father Gauss. Ewald promised to do everything as agreed upon. On the way home from the evening tea when the matchmaker congratulated him and explained to him that Gauss had agreed to everything, Ewald did not know what it was all about. He remained just as ignorant of the matter when it was explained to him that his wooing was intended for Minna, from whom he had accepted tea. Furthermore it has been related that he could not be found on the day before the wedding and had to be summoned from the ladder in front of his book shelves. Again he had forgotten a nonacademic matter.

Relations between Gauss and Ewald were always extremely cordial. After Minna's death he married again, and Gauss fully approved. In 1850 a daughter was born to Ewald and his second wife. Gauss regarded her almost as his own grandchild; especially because she bore the name Minna. One day Ewald's young (second) wife had to go downtown to attend to some urgent errands. She found her husband busy with his Arabic grammar. Since the maidservant was not available, she asked her husband to baby sit, which he did against his will. When she came back, the baby had disappeared. Ewald seemed to know nothing about it. An anxious search was instituted. Owing to a gentle whimpering, the baby was finally found in a closed dresser drawer, where Ewald had placed it, thinking that was the safest place.

Unfortunately Minna's health was bad almost from the beginning of her marriage. It was lung trouble and one must assume that she contracted tuberculosis in nursing her stepmother. Treatments at Bad Ems and Franzensbad helped her, but of course were unable to cure her. The exile of Ewald and Minna to Tübingen in 1838 affected her mental condition unfavorably and thus caused her illness to become aggravated.[6] In Tübingen she was confined to the house most of the time. Her ardent desire was that political conditions would allow her and her husband to return to Göttingen, but she died on August 12, 1840, before this wish could be fulfilled. She had visited Göttingen in the autumn of 1838 and also of 1839, and carried on an extensive correspondence with her father and her sister Therese. The bond of love between Minna and her father was a strong one.

* * *

Louis, third child of Gauss, was born September 10, 1809, and died March 1, 1810. He was named for the astronomer Harding. Aftereffects of Louis' birth brought death to Gauss' first wife.

* * *

By far the most interesting child of Gauss and the one who deserves more attention here than the others was Peter Samuel Marius Eugenius Gauss—always known as Eugene. He was born in Göttingen on July 29, 1811, and reached the age of eighty-five, the last surviving child of the mathematician. The name Peter was taken from his maternal grandfather, since he was the first child of Gauss' second wife.

There is good reason to believe that Eugene was the most talented of Gauss' children, both in mathematics and languages. Certainly he inherited much from his father. In his earlier years his life assumed the proportions of romantic adventure.

Gauss at first built high hopes on what Eugene would do. As the infant reached boyhood he displayed far more than ordinary ability, especially in languages. His father once took a French book, examined him in the knowledge of French, and then said that he knew that language well enough and need not study it any further.[7] Another time Gauss took the boy from Göttingen to the town of Celle, to place him in a school. While stopping at an

---

[6] See Chapter XVI.
[7] In later years in America he spoke French fluently and was sometimes mistaken for a Frenchman.

inn Eugene stated to his father his delight in having solved some little problem in grammar. His father, with eyes brightened with pleasure, replied: "Yes, my son, the pleasure one gets from the solution of such problems is very great, but it is not to be compared with the similar pleasure one derives from the solution of mathematical problems."

When Eugene reached adolescence, it seems that Gauss did not want him or his brothers to attempt mathematics, for the father felt that none of them would surpass him, and he did not wish the name lowered. Apparently he felt the same way about any other line of scientific work, for, while Eugene, after completing the secondary school in Göttingen, desired to make the study of philology his lifework, the father wanted him to take up law. Hence he enrolled in the University of Göttingen law school.

Eugene indulged in the gay life of a Göttingen student of that day. His principal weakness was gambling. A scar on his face, though not conspicuous, bore witness of his participation in a student duel. He was tall and slender, had blue eyes, and in his youth intensely black hair, which became white in later years. In appearance Eugene strongly resembled his mother.

Eventually Eugene gave an elaborate supper to some of his fellow students and sent the bill to his father. A scene ensued when Gauss rebuked him for this. The hotheaded Eugene suddenly decided that he would leave Germany and go to the United States. He was naturally of a restless disposition and was guided by the spirit of adventure. He wished to be independent and to make his own way in a new country. Eugene started off without bidding the family good-by or making any preparation for the journey. His mother was seriously ill at the time, and Gauss felt that this behavior of his son hastened her death. In her will she placed a clause stating that he should not inherit his share unless he gave evidence of good conduct. At the proper time he was able to produce such evidence. Yet Gauss never fully forgave his son for this rash conduct, even though in later years there was reconciliation of a sort. They never saw each other again. The incident deeply affected Gauss for many years.

When Eugene left, Gauss wrote to his friend Schumacher, requesting him to have the Hamburg Chief of Police[8] institute a search for Eugene. This was in early September, 1830. But Eugene was not found in Hamburg. Gauss was so disturbed at the time that he stated he never wanted to see his son again. Later, of course, this strong emotion was modified.

Gauss followed Eugene, who had gone to Bremen, and urged him to return, at the same time telling him that he had brought his trunk and if he was determined to go to America he would furnish him ample funds. Father and son were guests in the home of Olbers. Eugene refused to return, and the two parted. Gauss felt sick when he returned home, so upset was he emotionally.

Eugene remained in Bremen from September 3 to October 13, 1830, spending most of his time in learning English. Olbers reported to Gauss that his behavior during those weeks was quite good, and also that a friend named Bredenkamp as well as a German merchant then established in Philadelphia would aid Eugene at the time of landing.

The crossing lasted from October 13 to late December, 1830, when Eugene landed in New York. Owing to a poor command of English as well as a lack of seriousness and desire to work at the time, Eugene did not find a job in New York or Philadelphia. By February, 1831, he had used up all the funds his father had placed at his disposal. In this serious situation Eugene enlisted as a private in the United States Army on April 19, 1831, for a five-year term. He was first attached to Company F of the First Regiment, U. S. Infantry, stationed at Fort Crawford, Wisconsin, under the command of Captain S. Loomis.[9]

On July 20, 1832, Loomis sent to Gauss a good-conduct statement for Eugene, needed in connection with his mother's estate. At Fort Crawford Eugene was placed in charge of the post library. Gauss was surprised to find attached to Loomis' certificate a letter of greeting from a young German named Heinrich Schliephacke,

---

[8] Senator Dammert.
[9] He was under Lieutenant Kingsbury.

a stranger to him, who had left home under similar circumstances in 1822 and had joined the American army. His old home was on the highway between Brunswick and Wolfenbüttel and thus only several miles from Gauss' birthplace. He had aided Eugene in his new surroundings and praised him in the letter as a good soldier, predicting rapid promotion.

Eugene was later transferred to Fort Snelling, Minnesota. The post was under the command of General Zachary Taylor, and Jefferson Davis was a young officer there. By accident the officers found out that Eugene was an educated man, and he was put in charge of the post library. About the close of his term of enlistment[10] his brother Joseph came to America to study railway construction. Joseph brought letters to General Winfield Scott and others and thought he could obtain for Eugene a commission in the regular army, if he desired it. But Eugene had other plans. Joseph did not succeed in seeing Eugene during his American sojourn.

At first Eugene had hope of getting a job in St. Louis, Missouri. If that plan failed, he desired to get a job on a southern plantation. He feared to leave the army because the future looked uncertain to him. In 1834 he had become first sergeant of his company, yet he was embittered when he reflected on what he might have been, and on the fact that men who were his equals in birth and education were now his superiors. He realized that he had been impetuous in leaving home.

On May 15, 1831, while he was still in a recruit depot near New York, Eugene wrote his father a letter full of remorse and contrition. At that time the life of a soldier seemed to be unbearable to him, yet in the end it was his salvation. In the letter just mentioned he stated that he would rather be a day laborer at one-fourth of the usual pay than to remain any longer as a soldier. A little later he wrote a second letter to his father asking for his aid in gaining a discharge, since his hope of dismissal on account of nearsightedness had vanished. Gauss refused in a letter which

---

[10] April 19, 1836.

is a remarkable combination of a sermon preaching repentance and an expression of hope for a better future. Eugene had begun to save a little of his pay. He wrote to his father: "Your name is well known even here in this wilderness."

During his army service Eugene became a pious Presbyterian and decided he wanted to be a missionary.[11] His brother Joseph was displeased by this, but Gauss was not. In later years Eugene made an intellectual study of Christianity that convinced him, aside from his spiritual conviction.[12] Many of the books in his library were theological. He conducted family worship in his home night and morning; it is said that his prayer was remarkable for clear statement and understanding and for unaffected earnestness and humility. He expressed satisfaction at having come to America, because otherwise he might never have professed the religion of Christ.

At the conclusion of his time in the army Eugene entered the employ of the American Fur Company, on the headwaters of the Mississippi and Missouri rivers. He was attached to the office at Prairie du Chien, Wisconsin, in charge of Hercules L. Dousman.[13] There he learned to speak the Sioux language with ease, and assisted a missionary named Pond in preparing a Sioux alphabet and in translating the Bible. At this time he met the French astronomer Nicollet.[14]

By the summer of 1838 Eugene was acting as clerk in the office of an insurance agent. He took a position as a private tutor the following spring. Following that, he returned to the Indians for a short time, in the service of the fur company.

Gauss was now convinced that Eugene had mended his ways and paid him his maternal inheritance. Eugene opened a store at St. Charles, Missouri, in 1840 and spent most of the balance of his life in that city. He finally settled down after having seen much of the country, with varied and sometimes dangerous experiences.

---

[11] See Chapter XXII.
[12] He read the New Testament in the original.
[13] Actually he spent most of this period at Fort Pierre, South Dakota.
[14] See Chapter XXI.

In St. Charles, Eugene Gauss had various business interests—flour milling, lumber, and so forth. He organized and was the first president of the First National Bank. On February 14, 1844, he married Henrietta Fawcett, who was born February 3, 1817, in Rockingham County, Virginia, near the city of Harrisonburg. Henrietta's father, Joseph Fawcett, was of Huguenot ancestry, and her mother, Lucretia Keyes, had moved to Missouri from Virginia some years before Henrietta's marriage to Eugene. Gauss was quite pleased when he heard of Eugene's marriage plans and wrote him a cordial, affectionate letter. Eugene and Henrietta had seven children. They built a substantial brick residence in St. Charles, with lawn, gardens, and orchard, where they lived in old-time southern comfort and plenty. This house still stands, though now somewhat changed and with less grounds, and for many years has been out of the family, having been sold in 1885. In St. Charles the friends and connections of Eugene were chiefly among the Virginians, Kentuckians, and Carolinians who went to the town and county after the early French settlement.

Eugene Gauss was a man of very strong will and naturally quick temper. He was a student all his life and after he became blind in his last years his wife read to him and discussed various subjects with him as his constant companion. In all his intellectual life he was genuine and thorough, yet his modesty was a prominent characteristic. He was not a man to push himself or make use of his father's name. A year before his death he spoke of his desire to study philology and thought that had he remained in Europe he would have been able to obtain a university professorship. He thought of the great difference this would have made in his life. Eugene was a man of shrewd judgment and often expressed wise views about his environment. He became quickly and thoroughly Americanized, and disapproved of the behavior of many German immigrants that he observed. The Revolution of 1848 interested him especially, and he made sharp observations on it in letters to his family in Germany.

When Eugene was over eighty years old and had become blind, he used to entertain himself by making long arithmetical calculations in his head. He computed the amount to which one dollar would grow, if compounded annually at the rate of 4 per cent interest from the time of Adam to that date (about 1894), assuming this to be 6,000 years. This, if in gold, would make a cubic mass so large that it would require light quadrillions of years to pass along one side of it. This mental computation was such as to be almost beyond belief. The only assistance Eugene had was from his son Theodore (1849–1895), who was asked to write down, at intervals during the several days he was so occupied, the results that marked the different stages of his work. Eugene arrived at his result by ordinary arithmetic. His son preserved the paper on which were written the long lines of figures which Eugene thought he might not be able to retain in his memory. On the sheet are several memoranda that are interesting. For instance, Eugene directed his son to write down the figures:

$$\begin{array}{r} 123456789057182178039 \\ 3680824926969613857 \\ \hline 123456789060863002965969613857 \\ \times \end{array}$$

The second line of figures was written down several days after the first and added to the upper one by Theodore. His father had directed him to begin the second line of figures by placing the figure 3 under the 7 of the upper line. In reading off the result of this addition Theodore read 7 in place of the 8 marked with an $\times$. Eugene detected the error and his son made the correction, showing that the blind and aged man was able to retain in his mind the long line of thirty figures. This wonderful computation shows an extraordinary memory, to say the least. Later he obtained the value of a cubic foot of gold as expressed in gold dollars. A professor of astronomy later checked the result and found it to be correct.

In 1885 Eugene moved from St. Charles to a four-hundred-acre farm on Highway 63 in Boone County five miles from Columbia, Missouri. There he died on Saturday, July 4, 1896, at the venerable age of eighty-five. The funeral took place from the First

Presbyterian Church at St. Charles. His widow lived on until November 24, 1909, thus attaining the great age of ninety-two. Eugene and Henrietta are buried in Oak Grove Cemetery, St. Charles, Missouri.

* * *

Wilhelm August Carl Matthias Gauss, fourth son of Gauss and second son of his second wife, was born on October 23, 1813, in Göttingen. He was always known as Wilhelm. His life was less romantic than that of his brother Eugene. At the age of fifteen he left the secondary school in Celle simultaneously with Eugene, in order to be a farmer. After the close of his apprentice period he had several administrative positions on farms. Since he was rather quarrelsome and vehement, he did not hold one of these jobs very long. Finally he held a mercantile position in Potsdam for over two years, but only because he wanted to wait for the repayment of a bond he had put up, which, however, led to several years of litigation. As early as 1832 he had plans of migrating, convinced that his funds were insufficient to purchase a farm in Germany. On August 21, 1837, Wilhelm married at Levern near Preussisch-Oldendorf Aletta Christiane Luise Fallenstein (1813–1883), daughter of the Reverend Heinrich Fallenstein, pastor in Levern, and Charlotte Friederike Amalie Bessel. His wife was a niece of the astronomer Bessel, who was an intimate friend of Gauss.

A brother of the bride joined the young couple, and the three went on a sailing vessel to New Orleans and from there traveled up the Mississippi to Missouri. At that time America made uncontrollable appeals to Wilhelm's imagination. As a matter of fact, he never became fully Americanized as did his brother Eugene. He was never quite happy in America and for many years played with the idea of returning to Germany. Wilhelm was very critical of life in America and of its people. He did not feel that a democracy is the best form of government, but favored a limited monarchy. Eventually he decided to remain in America for the sake of his eight children. His wife was never too happy in America; Eugene married an American, whereas Wilhelm had a German wife who did not adapt herself easily to the new environment. That and a difference in temperament will explain why one brother became a genuine American and the other did not. Wilhelm was a warm-hearted man of fine intellectual gifts, but in the home he was frequently very severe and unbending.

At first Wilhelm leased some land near St. Charles and soon afterward had his own farm nearby. But he was plagued by illness and crop failure to such an extent that he left his farm and in 1840 moved to Glasgow, in Howard County, Missouri, where he and his brother-in-law opened a store. Within a few years he had built up a sizable fortune. But farming appealed much more strongly to him than the life of a storekeeper. After a few years he bought and moved to a farm thirty miles from Glasgow near Brunswick, Missouri, where he lived until 1855. In that year Wilhelm moved to St. Louis and went into the wholesale boot and shoe business with his brother-in-law under the firm name Fallenstein and Gauss.

Wilhelm spent the remainder of his life in St. Louis. He acquired wealth and was recognized as one of the representative businessmen of St. Louis. In 1855 he took there with him a family of his slaves, as house servants. Before the Civil War he freed all of them, starting the father as a hack driver on his own account, by giving him a pair of horses and a carriage. To be an independent hack driver was the ambition of many a southern Negro of that time.

The Wilhelm Gauss residence was on California Avenue near Lafayette Avenue. It was a large house and had extensive grounds with orchard and gardens. After the death of Wilhelm's widow in 1883 it was sold to the Missouri Pacific Railroad, which used it as a hospital. Much later it was torn down and there is now a school playground where the house once stood.

Wilhelm died August 23, 1879; he and his family are buried in the Bellefontaine Cemetery, St. Louis, Missouri.

* * *

Henriette Wilhelmine Caroline Therese Gauss, youngest child of Carl Friedrich and his second wife, was born on June 9, 1816, in Göttingen. The name Wilhelmine was for Olbers, but she was always known as Therese. She resembled her mother in appearance, and her character was very similar to that of her mother. Therese was quite different from her half sister Minna. She was clever and kindhearted, but was also very sensitive and looked on the dark side of life. Therese was definitely an introvert. Her father and her sister Minna noticed this trait in her character; they could not understand her melancholy nature. She was probably maladjusted sexually. Therese and her sister Minna did not get along too well together. Therese was frequently ill, but much of her "illness" was probably of psychosomatic origin. It is believed that she had lung trouble, contracted from her mother. After Minna's death in 1840 her health seems to have improved.

After her sister moved to Tübingen in 1838 she took over the running of the Gauss household and cared for Gauss' aged mother. In his last years she nursed her father and was his constant companion, manifesting deep love for him and unusual loyalty to duty. She sacrificed everything for him and was tenderly and closely attached to him.[15] After the death of Minna she was his all.

In September, 1855, several months after her father's death Therese went to Switzerland. She was at first in Montreux, later in Vevey and on Lake Geneva. Her letters of that time are full of complaints about her health and about the fact that nobody in Göttingen had written to her. As a matter of fact, she had written to no one there.

On September 23, 1856, at Elsterwerda Therese married Constantin Wilhelm Staufenau (1809–1886), an actor and theatrical producer from Thuringia with whom she had corresponded uninterruptedly for fourteen years. The marriage seems to have been a happy one although Therese did not survive very long. The couple made their home in Dresden,[16] where she died February 11, 1864. There were no children.

---

[15] See Chapter XXIII.
[16] At Waisenhaus 271 and Karnlastrasse 4.

Therese's marriage led to strained relations between her and her brother Wilhelm. He had expressed himself rather freely in a letter to Joseph, which fell into her hands for some unexplainable reason. Wilhelm felt that Staufenau was marrying her for her money and said so in the letter. Eugene and Joseph perhaps had the same feelings, but she never knew it if they did. Hence her relations with them in her last years were more pleasant.

Staufenau married a second time on July 15, 1865, Miss Johanna Horack (1832–1891), daughter of his physician in Dresden. At her death she returned to the Gauss family money which had come from Therese.

Gauss was a tender father only to his daughters. It would be wrong to conclude that he did not have deep affection for his sons. True, he did not write frequently to them—except that in his last years he wrote numerous letters to Joseph. He was rather proud of the high official position which Joseph occupied, but also had real joy in the knowledge that his two sons in America were conquering life in a new world. He had many sorrows but at the expense of a great struggle armed himself with firm serenity before the world. He did not wish to appear weak before his fellow men, and he found outward serenity necessary for scientific work. His correspondence with his children, with other members of his family, and with intimate friends shows that he did not lack the deepest emotions.

APPENDIX E

Genealogy

This section contains information only on descendants of Gauss and his half brother. His ancestors have been covered fully by Rudolf Borch in *Ahnentafeln berühmter Deutscher* (Leipzig, 1929), Lieferung 1, p. 63. These charts are not a complete listing of the descendants. The decimal-letter system, explained in the *New England Historical and Genealogical Register* (LI, 305), has been adopted for arrangement of these data. This system of notation goes as follows:

Gebhard Dietrich Gauss, the earliest member of the family on this chart, is lettered *a;* his children are lettered *aa, ab,* and so forth. The children of his elder son are lettered *aaa, aab,* and so forth; and so on through each succeeding generation. For convenience in counting, a space, corresponding to a decimal point, is left after the fourth letter.

Descendants of Gebhard Dietrich Gauss

First Generation

a. Gebhard Dietrich Gauss, son of Jürgen Gauss and his wife, Cathrine Magdalene Eggeling, was b. about Feb. 13, 1744 at Brunswick, Germany, and d. there Apr. 14, 1808. He was a gardener, bricklayer, and official of the local burial association. m. (1) Apr. 28, 1768, Dorothea Emerenzia Warnecke (Sollerich), b. 1745 and d. Sept. 2, 1775. m. (2) at Velpke on Apr. 25, 1776, Dorothea Benze, dau. of Christoph Benze, stonemason, and his wife, Catharina Maria Crone; Dorothea

GENEALOGY 377

was b. at Velpke June 18, 1743, and d. at Göttingen Apr. 18, 1839.

SECOND GENERATION

aa. Johann Georg Heinrich Gauss, b. at Brunswick Jan. 24, 1769, and d. there Aug. 7, 1854. He followed the same occupation as his father. m. (1) at Brunswick Oct. 20, 1808, Marie Friederike Juliane Dannehl of that city. m. (2) at Brunswick Nov. 19, 1826, Christiane Sophie Regine Höber of Wolfenbüttel.

ab. Carl Friedrich Gauss, b. at Brunswick Apr. 30, 1777, and d. at Göttingen Feb. 23, 1855. Professor of mathematics and astronomy, director of the University of Göttingen Observatory. m. (1) at Brunswick Oct. 9, 1805, Johanna Elisabeth Rosina Osthoff, dau. of Christian Ernst Osthoff, master tanner, and his wife, Johanna Maria Christine Ahrenholz; the bride was b. Aug. 5, 1780, at Brunswick and d. Oct. 11, 1809. m. (2) at Göttingen Aug. 4, 1810, Friederica Wilhelmine (Minna) Waldeck, dau. of Johann Peter Waldeck, professor of law at the University of Göttingen, and his wife, Charlotte Auguste Wilhelmine Wyneken; the bride was b. Apr. 15, 1788, at Göttingen and d. there Sept. 12, 1831.

THIRD GENERATION

aaa. Caroline (Line) Magdalene Dorothee Gauss, b. at Brunswick Feb. 5, 1809, d. there March 29, 1870. m. at Brunswick May 20, 1832, Eduard Wilhelm Bauermeister, master nailsmith. (There were seven children of this marriage, and descendants are living today. It is not traced further here.)

aab. Georg Gebhard Albert Gauss, master plumber in Brunswick, b. May 20, 1811, and d. Aug. 24, 1879. m. at Brunswick Sept. 4, 1842, Johanne Dorothee Friederike Schäffer of Brunswick.

aba. (Carl) Joseph Gauss, Hanoverian artillery officer and later director of Hanoverian railroads. b. at Brunswick Aug. 21, 1806, and d. at Hanover July 4, 1873. m. at Stade March 18, 1840, Sophie Friederike Erythropel, dau. of August Christian

Erythropel and Amalie Dorothea Johannes; the bride was b. Jan. 20, 1818, at Stade and d. April 6, 1883, at Hanover.

abb. Wilhelmine (Minna) Gauss, b. Feb. 29, 1808, at Göttingen and d. Aug. 12, 1840, at Tübingen. m. at Grone Sept. 15, 1830, Georg Heinrich August Ewald, theologian and professor of Oriental languages at Göttingen and later at Tübingen, son of Heinrich Andreas Ewald, master linen weaver in Göttingen, and his wife, Catharina Maria Ilse; Ewald was b. Nov. 16, 1803, at Göttingen and died there May 4, 1875. On Dec. 27, 1845, at Darmstadt he married (2) Auguste Friederike Wilhelmine Schleiermacher, dau. of Andreas August Ernst Schleiermacher, *Oberfinanzrat* and librarian in Darmstadt, and his wife Caroline Luise Maurer; Ewald's second wife was b. March 23, 1822, at Darmstadt and d. Dec. 19, 1897, at Göttingen. They had one daughter, Caroline Therese Wilhelmine Ewald, who was b. May 9, 1850, at Göttingen and d. there May 7, 1917; she never married.

abc. Louis Gauss, b. Sept. 10, 1809, at Göttingen and d. there March 1, 1810.

abd. Peter Samuel Marius Eugenius (Eugene) Gauss, flour miller, lumber dealer, and bank president at St. Charles, Mo. He was b. July 29, 1811, at Göttingen and d. on his farm in Boone County, Mo., July 4, 1896. Came to U.S.A. in 1830 after studying law at the University of Göttingen. m. Feb. 14, 1844, at St. Charles, Mo., Henrietta Fawcett, dau. of Joseph Fawcett and Lucretia Keyes; Henrietta was b. Feb. 3, 1817, in Harrisonburg, Va., and d. Nov. 24, 1909, on the Gauss farm in Boone County, near Columbia, Mo.

abe. Wilhelm August Carl Matthias Gauss, farmer in Germany, and near St. Charles, Mo., merchant in Glasgow, Mo., farmer in Chariton Co., Mo., and finally wholesale shoe dealer in St. Louis, Mo., b. Oct. 23, 1813, at Göttingen and d. Aug. 23, 1879, at St. Louis. Came to U.S.A. in 1837. m. at Levern, Prussia, Aug. 21, 1837. Aletta Christiane Luise Fallenstein, dau. of Rev. Heinrich Fallenstein and his wife, Charlotte

GENEALOGY 379

Friederike Bessel, a niece of the astronomer Bessel; the bride was b. at Levern Apr. 20, 1813, and d. at St. Louis, Mo., Sept. 15, 1883.

abf. Henriette Wilhelmine Caroline Therese Gauss, b. at Göttingen June 9, 1816, and d. at Dresden Feb. 11, 1864. m. at Elsterwerda Sept. 23, 1856, Constantin Wilhelm Staufenau, actor and stage director in many localities, then retired in Zörbig and later Dresden, son of Carl Philipp Staufenau, teacher in Weissenfels, and his wife, Johanna Dorothee Künzel; he was b. Feb. 17, 1809, at Weissenfels and d. Nov. 14, 1886, at Dresden. On July 15, 1865, at Kötzschenbroda C. W. Staufenau m. Johanna Carolina Maria Horack, dau. of Johann Carl Horack, physician in Dresden, and Marie Caroline Rumpelt; she was b. Aug. 16, 1832, at Dresden and d. there July 17, 1891. Both marriages of Staufenau were childless.

FOURTH GENERATION

aaba. Georg Christian Albert Gauss, master plumber, b. in Brunswick June 1, 1843, and d. there Jan. 27, 1907. m. at Brunswick July 3, 1870, Caroline Wilhelmine (Minna) Riechelmann of Gross Schwülper.

abaa. Carl August Adolph Gauss, owner of landed estate in Lohne near Gross˙ Burgwedel, later retired in Hamlin, b. Apr. 10, 1849, in Hanover, d. Jan. 22, 1927, in Hamlin. m. at Stolzenau Sept. 4, 1874, Anna Sophie Johanne Ebmeier, dau. of Diedrich Ebmeier, farmer in Stolzenau, and his wife, Henriette Delius; the bride was b. at Stolzenau Feb. 8, 1850, and d. at Hamlin July 27, 1900.

abab. Unnamed dau. of Joseph Gauss, stillborn Jan. 24, 1853.

abac. Unnamed dau. of Joseph Gauss, stillborn Nov. 17, 1854.

abda. Charles Henry Gauss, b. Aug. 14, 1845, at St. Charles, Mo., and d. there Jan. 18, 1913. m. Charlotte Elizabeth Johns (1850–1938) of St. Charles.

abdb. Theresa Gauss, b. at St. Charles, Mo., May 21, 1847. d. in infancy.

abdc. Theodore Gauss, b. at St. Charles, Mo., Dec. 14, 1849. d. 1895, unmarried.

abbd. Robert Gauss, b. at St. Charles, Mo., Sept. 1, 1851. Editor of the Denver *Republican.* d. at Denver, Colo., Jan. 19, 1913. Unmarried.

abde. Virginia Gauss, b. Sept. 18, 1853, at St. Charles, Mo. d. Feb. 1930 at Columbia, Mo. Lived in Columbia, Mo.; unmarried.

abdf. Eugene Gauss, Jr., b. at St. Charles, Mo. Oct. 10, 1856. d. Dec. 10, 1951, at La Crescenta, Calif. Lived in Columbia, Mo.; unmarried.

abdg. Albert F. Gauss, b. Dec. 2, 1862, at St. Charles, Mo. m. Josephine Morison (of Scotch descent) on June 20, 1897, at Austin, Texas. They moved to Calif. in 1904, and Albert d. Oct. 30, 1953, at Alhambra, Calif. There were no children.

abea. Charles Frederick Gauss, b. at St. Charles, Mo., May 30, 1838, and d. Dec. 2, 1913, at St. Louis, Mo. m. (1) on Dec. 5, 1861, at St. Louis, Mo., Mary Josephine Lamereaux (1842–1875), dau. of Moses and Adele Guion Lamereaux. m. (2) at St. Louis on May 14, 1879, Ida Helene Smith (d. Feb. 24, 1932), dau. of Charles H. and Catherine B. Smith of Bourbon County, Ky.

abeb. Maria Sophia Theresa Gauss, b. Aug. 18, 1840, at Glasgow, Mo., and d. there Sept. 7, 1841.

abec. Oscar William Gauss, b. March 20, 1842, at Glasgow, Mo. and d. 1918 at Boonville, Mo. Was a Presbyterian clergyman. m. Aug. 5, 1869, at St. Louis, Mo., Esther Gill (b. May 10, 1841).

abed. Mary Louise Gauss, b. Nov. 30, 1844, on a farm in Chariton County, near Brunswick, Mo., d. at St. Louis, Mo., in 1925. Unmarried.

abee. John Bernard Gauss, b. Sept. 30, 1847, on a farm in Chariton County, near Brunswick, Mo., d. Oct. 5, 1886, at St. Louis, Mo. m. Nov. 10, 1875, at St. Louis, Mo., Anna Heermans of that city.

abef. William Theodore Gauss, b. July 1, 1851, on a farm in Chariton County, near Brunswick, Mo. Lived in Colorado

Springs and d. there Nov. 14, 1928. m. at Pittsfield, Ill., on June 24, 1876, Helen Worthington, b. Jan. 29, 1855, at Pittsfield and d. at Colorado Springs in 1933.

abeg. Louis Frederick Gauss (twin), b. Aug. 30, 1855, on a farm in Chariton County, near Brunswick, Mo. d. June 17, 1908, at St. Louis, Mo.

abeh. Joseph Henry Gauss (twin), b. Aug. 30, 1855, on a farm in Chariton County, near Brunswick, Mo. Retired Presbyterian clergyman, lives at Newburgh, Ind. m. (1) at St. Louis, Mo., on Oct. 23, 1882, Annie Gill, b. Feb. 15, 1850, at Mexico, Mo. and d. Aug. 5, 1908, at St. Louis, Mo. m. (2) a widow, Mrs. Frederick Townsley, née Olive Montgomery, on Aug. 30, 1910, at St. Louis, Mo.

FIFTH GENERATION

aaba a. Theodor Robert Gebhard Gauss, b. March 28, 1871, at Brunswick and d. there Aug. 15, 1871.

aaba b. Theodor Robert Gebhard Gauss, b. June 22, 1872, and d. there Apr. 17, 1873. (Named for deceased brother.)

aaba c. Minette Georgine Albertine Gauss, b. at Brunswick Aug. 4, 1875, and lives there as a widow. m. on Apr. 24, 1898, Wilhelm Georg August Böttger (1870–1917), a hardware merchant in Brunswick.

abaa a. Carl Joseph Gauss, b. Oct. 29, 1875, at Lohne, near Gross Burgwedel. m. Feb. 6, 1919, at Düsseldorf Emilie Auguste Magdalene Bingel, b. Jan. 24, 1886, at Raurel near Kastrop, dau. of Rudolf Bingel and Mathilde Hohendahl, and widow of Arthur Lindenberg (1873–1918). Dr. C. J. Gauss was professor of gynecology and director of the Women's Clinic at the University of Würzburg. Retired and lives now in Bad Kissingen.

abaa b. Sophie Friederike Elisabeth Gauss, b. at Lohne Oct. 30, 1876, and d. there Feb. 8, 1878.

abaa c. Helene Charlotte Sophie Alfriede Gauss, b. Dec. 15, 1878, at Lohne. m. at Hamlin Apr. 5, 1907, Johann (Hans) Otto Kuno Nöller, b. Jan. 11, 1873, at Ronnenberg, near Hanover, son of Georg Nöller and Wilhelmine Hartleben. He was district judge, later notary, in Gummersbach, then in Bonn, where they now live.

abaa d. Carl Louis Harry Gauss, b. at Lohne Dec. 16, 1880, d. Oct. 27, 1913, as captain of artillery at Wurzen in Saxony. m. at Wurzen Sept. 30, 1905, Elisabeth Marie Baessler, b. at Wurzen Dec. 19, 1882, dau. of Carl Baessler, factory owner in Wurzen, and Johanna Schroeder.

abaa e. William August Gauss, b. May 20, 1885, at Hamlin, committed suicide March 13, 1939, while lieutenant colonel and district commander at Hamlin. m. June 28, 1913, Clara Marianne Dürbig, b. Nov. 13, 1892, at Plaussig, near Leipzig, dau. of Theodor Dürbig, landed-estate owner in Plaussig, and Charlotte Kabitzsch.

abda a. Blanche Lindsay Gauss, b. Oct. 18, 1870, at St. Charles, Mo., and d. Feb. 26, 1950, at Miami, Fla. Unmarried.

abda b. Henrietta Gauss, b. Apr. 4, 1872, at St. Charles, Mo., and d. Sept. 12, 1872, at Sedalia, Mo.

abda c. Eugene Gauss III, b. Nov. 12, 1873, at St. Charles, Mo., and d. Dec. 11, 1891, at San Antonio, Tex. Unmarried.

abda d. Anne Durfee Gauss, b. June 27, 1876, at Sedalia, Mo., and d. Apr. 25, 1932, at St. Charles, Mo. Unmarried.

abda e. Charles Frederick Gauss, b. Apr. 9, 1878, at Sedalia, Mo., and d. there June 27, 1879.

abda f. Martha Gauss, b. Sept. 24, 1879, at Sedalia, Mo., and d. there Dec. 17, 1882.

abda g. Henry Gauss, b. June 18, 1881, at Sedalia, Mo., and d. there June 19, 1881.

abda h. John Montgomery Gauss, b. Nov. 29, 1882, at Sedalia, Mo., and d. May 12, 1932, at St. Charles, Mo. Unmarried.

abda i. Virginia Fawcett Gauss, b. Nov. 28, 1883, at St. Charles, Mo., d. there 1955. Unmarried.

abda j. Matthew Johns Gauss, b. Apr. 11, 1887, at San Antonio, Tex., and lived at St. Charles, Mo. m. Oct. 30, 1923, **Mary Gladden Grant** of Fulton, Mo. d. Sept. 18, 1954.

abda k. Lois E. Gauss, b. Oct. 3, 1888, at San Antonio, Tex., and lives in Miami, Fla. m. March 4, 1918, at St. Charles, Mo.,

Arnold Simmons, son of Josiah Phelps Simmons and his wife Elizabeth of Richmond, Ky.

abda l. Minna Waldeck Gauss, b. July 2, 1892, at San Antonio, Tex., and lives in New Orleans, La. She was a schoolteacher before her marriage on Oct. 15, 1930, at St. Charles, Mo., to Rev. Fred L. Reeves, a Presbyterian minister.

abea a. Adele Louisa Gauss, b. Sept., 1862, at St. Louis, Mo., and d. there Feb. 15, 1948. m. at St. Louis in 1885 Charles W. Bullen, b. Feb. 7, 1854, at Louisville, Ky., and d. at St. Louis Apr. 2, 1897.

abea b. Emma Josephine Gauss, b. Nov. 1864 at St. Louis, Mo., and d. there Aug. 24, 1953. m. May 12, 1887, Samuel Hart Young, b. Apr. 12, 1851, at St. Louis.

abea c. Louise Aletta Gauss, b. Feb., 1867, at St. Louis, Mo., and d. there Feb. 21, 1953. m. at St. Louis Dec., 1889, Louis G. Kies of Cleveland, Ohio. He d. in 1897 or 1898.

abea d. Mary (Mamie) Guion Gauss, b. at St. Louis, Mo., in Nov., 1869, and lives in Shreveport, La. m. in March, 1893, J. Paul Annan.

abea e. Sarah (Sadie) Lamereaux Gauss, b. at St. Louis, Mo., Sept., 1873. d. Nov. 16, 1954. m. (1) at St. Louis in 1896 George S. Tenney of New York, who d. in 1910. m. (2) Apr. 17, 1914, Frank Schiffmacher.

abec a. Aimee Esther Gauss, b. May 5, 1871, at St. Louis, Mo., and d. in 1908 at Greeley, Colo. Unmarried.

abec b. Marianne Gauss, b. Nov. 5, 1873, at Boonville Mo. Lives in Greeley, Colo. Unmarried.

abec c. Frances Louise Gauss, b. Oct. 19, 1875, at Cape Girardeau, Mo., and d. Oct. 15, 1887, at Jefferson City, Mo.

abec d. Theresa Gauss, b. Sept. 30, 1877, at Boonville, Mo. m. in 1907 George D. Robinson of Lamar, Colo.

abec e. Charlotte Gauss, b. Aug. 23, 1879, at Boonville, Mo. Lives in Greeley, Colo. Unmarried.

abec f. Louise Gauss, b. July 1, 1881. Lives in Greeley, Colo. Unmarried.

abec g. Oscar Gauss, b. Apr. 15, 1884, at Boonville, Mo. Lives in Greeley, Colo. Unmarried.

abee a. Philip William Gauss, b. Oct. 10, 1876, at St. Louis, Mo. m. Dec. 14, 1916, Ethel White. Lived in Port Arthur, Tex. and d. there Dec. 13, 1954.

abee b. Charles William Gauss, b. at St. Louis, Mo. in Jan., 1879. Unmarried.

abee c. Ralph Bernard Gauss, b. Jan., 1881, at St. Louis, Mo., and d. there Dec. 2, 1909. Unmarried.

abee d. Mabel Aletta Gauss (1883–1890), St. Louis, Mo.

abee e. John Bernard Gauss, Jr., b. 1885 at St. Louis, Mo. Lived in Tex.; d. 1914.

abef a. Carl Friedrich Gauss, b. Oct. 19, 1878, at St. Louis, Mo., and d. June 4, 1929, at Littleton, Colo. m. Dec. 12, 1914, Anne Palmer Griffith.

abef b. (Minna) Helen Worthington Gauss, b. Apr. 9, 1881, at St. Louis, Mo. Lives in Colorado Springs, Colo. Unmarried. The name Minna was dropped and Helen Worthington assumed at the death of *abefd*.

abef c. (William) Theodore Worthington Gauss, b. Sept. 4, 1884, at St. Louis, Mo. Lives in San Diego, Calif. m. Sept. 20, 1913, Gladys Olivia Robinson.

abef d. Helen Worthington Gauss, b. July 7, 1887, at St. Louis, Mo., and d. there Feb. 8, 1889.

abeh a. Esther Mary Gauss, b. Oct. 20, 1883, at St. Louis, Mo., and lives in Duarte, Calif.

abeh b. Henry Fallenstein Gauss, b. Apr. 21, 1885, at Pernambuco, Brazil. m. June 8, 1912, Myrtle Elizabeth Leisk, b. Oct. 29, 1891, at Milwaukee. Is research professor and head of the Department of Mechanical Engineering at the University of Idaho.

abeh c. Frank Evans Gauss, b. Oct. 25, 1887, at St. Louis, Mo., m. Apr. 5, 1924, Martha Dorothy Meyer.

abeh d. Paul William Gauss, b. July 6, 1889, at St. Louis, Mo. m. May 27, 1913, Ruby Fay Tomlinson. Presbyterian clergyman in Youngstown, Ohio.

abeh e. Annie Gauss, b. Jan. 20, 1891, at St. Louis, Mo. m. at St. Louis Apr. 4, 1921, Andrew Hays Kean, b. May 19, 1887, at St. Louis and d. Nov. 16, 1953.

abeh f. Janet Lee Gauss, b. July 18, 1912, at St. Louis, Mo., m. at University City, Mo., June 14, 1941, Milton Schrader, b. June 27, 1912, at St. Louis, son of Albert Ferdinand Schrader and his wife Estella Walker.

abeh g. Josephine Henrietta Gauss, b. July 10, 1918, at St. Louis, Mo. m. at St. Louis, Nov. 27, 1941, George Garven, b. July 2, 1916, at St. Louis, son of George E. and Mae M. Garven.

SIXTH GENERATION

aaba aa. Albert Böttger, b. 1899 in Brunswick, Germany. Had a hardware store in Leipzig. Kidnapped by the Russians in 1945 and fate unknown after that date.

aaba ab. Wilhelm Böttger, b. 1901 in Brunswick. d. 1946 of tuberculosis. Had a country store in Lengede. m. and had one daughter.

abaa ca. Klaus Wilhelm Nöller, b. at Gummersbach, Germany, May 20, 1908, and d. there Aug. 8, 1908.

abaa cb. Hans Heinrich Nöller, b. at Gummersbach, Germany, July 15, 1909. Was prisoner of war (World War II) in Canada and lost sight of one eye.

abaa cc. Klaus Kuno Nöller, b. Sept. 30, 1910, at Gummersbach, Germany. Missing as prisoner of war in Rumania since 1945.

abaa cd. Eberhard Georg Nöller, b. at Gummersbach, Germany, Nov. 25, 1911. Returned to Bonn after being prisoner of war in World War II.

abaa da. Carl Harry Gauss, b. July 20, 1906, at Wurzen, Germany. d. Feb. 17, 1932, as a result of being accidentally shot while serving as an artillery officer in Halberstadt.

abaa db. Unnamed son, stillborn, at Hanover, June 23, 1908.

abaa dc. Hanspeter Carl Joseph Gauss, always known as Peter, b. June 7, 1909, at Hanover. He married and had three children. Was a veterinary in Wendelheim (Kreis Alzey, Rheinhessen). Since 1945 missing in action in Rumania.

abaa dd. Johanne Christa Gauss, b. at Wurzen March 30, 1912. The only German descendant of Gauss mathematically gifted. m. mining engineer Köhler and lives in Peine, Hanover.

abaa ea. Harry Joseph Gauss, b. at Leipzig, Germany, Sept. 5, 1918. d. in Lapland (Fischer Islands) in 1940 as lieutenant of infantry.

abaa eb. Stillborn son, Hamlin, Dec. 24, 1924.

abaa ec. Senta, b. Sept. 13, 1925. Lives with her mother in Hamlin.

abda ja. Mary Elizabeth (Bettie) Gauss, b. Aug. 26, 1924, at St. Charles, Mo. m. June 28, 1948, Harold W. Wilson, Jr., a banker. Lives in Horton, Kan.

abda jb. Matthew Johns (Jack) Gauss, Jr., b. Feb. 21, 1927, at St. Charles, Mo. Unmarried. Lieutenant (jg), U.S.N., in charge of PCE 902, ship training naval reservists operating out of Milwaukee, Wis.

abda jc. David Warren Gauss, b. Sept. 3, 1936, at St. Charles. Student at Westminster College, Fulton, Mo.

abda ka. Elizabeth Johns Simmons, b. May 18, 1922, in Miami, Fla. m. Oct. 2, 1944, Clifford Arnold Burchsted, son of Charles F. and Emma Burchsted of Quincy, Mass.

abda kb. Lois Winston Simmons, b. Sept. 25, 1924, in Miami, Fla. m. Apr. 16, 1949, Robert Devore Chambless, son of Guy and Betty Chambless of Miami, Fla.

abea aa. Adele Gauss Bullen, b. July 23, 1886, at St. Louis Mo. m. Apr. 12, 1910, Clifford R. Croninger of St. Louis.

abea ab. Charles Gauss Bullen, 1888–1893, St. Louis.

abea ac. Mary Helene Bullen, b. 1895 and d. 1935. m. Clifford Raymond Garrison.

abea ba. Marie Hart Young, b. Feb., 1888 at St. Louis. m. Nov. 12, 1916, Robert J. Gartside.

abea bb. Ida Helene Young, b. Nov., 1889. m. (1) on Oct. 25, 1913, William Jefferson Hill who d. Aug. 25, 1914. m. (2) Apr. 18, 1925, Clifford C. Jones.

abea ca. Gertrude Louise Kies, b. May 7, 1895. m. Oct. 21, 1914, George P. Whitelaw II, who was b. March 7, 1890. She d. 19??.

abea da. Ruth Annan, b. 1895. m. 1921, Randolph Mayer, b. 1891. Lives in Shreveport, La.

abea db. Margaret Annan, b. 1895. m. 1921, Jessie C. Layfield, b. 1890. Lives in Shreveport, La.

abea dc. Paul Gauss (Jack) Annan, b. 1911. m. 1935, Virginia Knighton, b. 1909. Lives in Shreveport, La.

abea ea. Janet Tenney, b. 1897 at St. Louis. m. 1921, Clarence E. Smith, b. 1893.

abea eb. George Gauss Tenney, b. 1900 at St. Louis. m. 1925, Maude Crowley.

abec da. Aimee Esther Robinson, b. Oct. 7, 1908, at Lamar, Colo. m. March, 1927, Oran Palmer.

abee aa. Philip William Gauss, Jr., b. July 13, 1918. m. at Port Arthur, Texas, Dec. 21, 1946, Frances Lillian Barlow, b. Dec. 17, 1923, dau. of Elton Merritt and Eunice (Stewart) Barlow. Lives in Austin, Texas.

abee ab. Robert White Gauss, b. Jan. 8, 1921. m. Oct. 3, 1953, at Harlingen, Texas, Bonita Little, b. July 25, 1927, dau. of Homer D. and May (Matthews) Little. Live in Port Arthur, Texas.

abef ca. Theodore Worthington Gauss, Jr., b. July 12, 1914, at Colorado Springs.

abef cb. Robert Parker Gauss, b. June 16, 1918, at Colorado Springs.

abef cc. William Wharton Gauss, b. Feb. 4, 1926, at Colorado Springs. m. Joan Hollabaugh Apr. 26, 1952.

abeh ba. William Henry Gauss, b. Oct. 29, 1913, at St. Louis. m. (1) Jan. 9, 1933, Rachel Julia Chenoweth. b. Apr. 9, 1917, at Boston, Mass. dau. of Professor C. W. Chenoweth, head of Philosophy Department at University of Idaho. m. (2) Lorraine Boone, b. Oct. 9, 1919, at Hazard, Ky.

abeh bb. Joseph Henry Gauss, b. Aug. 24, 1915, at St. Louis. m. Dorothy Lenfest, b. Aug. 25, 1914, at Missoula, Mont.

abeh bc. Mary Louise Gauss, b. Sept. 18, 1920, at Akron, Ohio. m. Robert Driscoll, b. May 19, 1920, at Spokane, Wash.

abeh bd. Phyllis Caroline Gauss, b. Aug. 26, 1924, at Tomahawk, Wis. m. Loyd Skiles, b. March 16, 1924, at Denton, Tex.
abeh ca. Charles Frederick Gauss, b. Jan. 26, 1925, at St. Louis. Married; no children.
abeh cb. Marianne Gauss, b. July 8, 1928. Unmarried.
abeh da. Dorothy Marie Gauss, b. Apr. 20, 1914, at Odessa, Mo. m. Harry Carlson.
abeh db. Marjorie Gill Gauss, b. Aug. 6, 1920, at Parma, Idaho. m. Reed.
abeh ea. Andrew Hays Kean, Jr., b. May 20, 1922, at Bridgeport, Conn. m. Apr. 30, 1949, at Ardmore, Pa., Mary Kelso, dau. of Samuel Kennedy Kelso. Lives in Berkeley Heights, N.J.
abeh fa. Ellen Montgomery Schrader, b. Jan. 26, 1944, at St. Louis.
abeh fb. Martha Walker Schrader, b. Apr. 7, 1951, at Oak Park, Ill.
abeh ga. Kathleen Garven, b. Apr. 10, 1948, at St. Louis.
abeh gb. George MacCloy Garven, b. Jan. 13, 1953, at St. Louis.
abeh gc. Robert Lee Garven, b. Apr. 21, 1954, at St. Louis.

SEVENTH GENERATION

aaba aba. Erika Böttger, b. 1930 at Brunswick, Germany.
abaa dca. Carl Ulrich Gauss, b. May 5, 1936, at Wendelheim (Kreis Alzey) Reinhessen, Germany. Last bearer of the Gauss name in Germany.
abaa dcb. Gudrun Gauss, b. Aug. 15, 1937, in Wendelheim, Germany.
abaa dcc. Rolf Gauss, 1939–1941. Was mentally deficient.
abda jaa. Nancy Elizabeth Wilson, b. July 25, 1950.
abda jab. Douglas Grant Wilson, b. May 27, 1953.
abea aaa. Adele Bullen Croninger, b. March 23, 1920, at St. Louis. Cancer research specialist at Washington University in St. Louis.
abea baa. Marie Louise Gartside, b. May 1, 1918. m. 1942, John Harris Bates.

abea bab. Dorothy Young Gartside, b. Feb. 24, 1921. m. 1942, Carl William Riesmeyer.
abea caa. Charles Wilgers Whitelaw II, b. Aug. 17, 1917.
abea cab. George P. Whitelaw III, b. 1919. Married.
abea daa. Randolph Annan Mayer, b. 1919. m. 1943, Isobel Janet Kronzer, b. 1918.
abea dab. Paul Robards Mayer, b. 1920. m. Mary Virginia Adsit, b. 1920.
abea dac. William Gauss Mayer, b. 1927. m. Tommye Eloise Watson, b. 1927.
abea dad. James Reade Mayer, b. 1923. m. Olive Maxine Bradley, b. 1923.
abea dae. Robert Lewis Mayer, b. 1925. m. Aino.
abea dca. Martha Annan, b. 1937.
abea eaa. Georgia Tenney Smith, b. June 14, 1922.
abea eab. Janet Merriam Smith, b. Feb. 25, 1924. m. 1947, Robert K. Weaver, b. 1921.
abea eac. Carolyn Smith, b. Oct. 3, 1929. m. Burke.
abea ead. Austin Tenney Smith, b. Feb. 13, 1933.
abea eae. Joan Tenney, b. Oct. 5, 1926.
abef cca. Robert Parker Gauss, b. Nov. 14, 1954.
abeh baa. Bradford William Gauss, b. Apr. 19, 1934, at Moscow, Idaho.
abeh bab. Jo Ann Gauss, b. Oct. 13, 1936, at Moscow, Idaho.
abeh bac. Curtis Henry Gauss, b. Sept. 2, 1938, at Moscow, Idaho.
abeh bad. Sandra Jane Gauss, b. March 27, 1940, at Moscow, Idaho.
abeh bae. Sharon Marsha Conley Gauss, b. May 26, 1943. at Huntington, W. Va.
abeh bba. Joseph Charles Gauss, b. July 3, 1941, at Ft. Benning, Ga.
abeh bbb. Michael Gauss, b. Feb. 23, 1948, at Erie, Pa.
abeh bca. Daniel Robert Driscoll, b. Jan. 9, 1945, at Moscow, Idaho.
abeh bcb. Robert Wallace Driscoll, b. May 30, 1947, at Moscow, Idaho.

abeh bcd. Harry Michael Driscoll, b. Feb. 19, 1951, at Moscow, Idaho.
abeh bda. Lynn Elizabeth Skiles, b. Oct. 21, 1947, at Spokane, Wash.
abeh bdb. Susan Leigh Skiles, b. Aug. 22, 1952, at Moscow, Idaho.
abeh daa. Paul William Carlson, b. Sept. 4, 1941, at Youngstown, Ohio.
abeh dba. Judy Ann Reed, b. Dec. 6, 1947, at Cleveland, Ohio.
abeh dbb. Barbara Reed, b. March 13, 1950, at Cleveland, Ohio.
abeh dbc. Katherine Reed, b. Apr. 30, 1953, at Cleveland, Ohio.
abeh eaa. Richard Andrew Kean, b. Sept. 23, 1950, at North Plainfield, N.J.
abeh eab. Elizabeth Kean, b. Nov. 30, 1952, at Berkeley Heights, N.J.

EIGHTH GENERATION

abea baaa. Carolyn Lamereaux Bates, b. 1944.
abea baab. Stewart Elkin-Harris Bates, b. 1947.
abea baac. Cynthia Louise Bates, b. 1948.
abea baba. William Duncan Riesmeyer, b. 1946.
abea daaa. Randolph Annan Mayer, Jr.
abea daab. Janet Mayer, b. 1946.
abea daba. Paul Robards Mayer, Jr., b. 1948.
abea daca. Carolyn Eloise Mayer, b. 1947.
abea dacb. Jeannie Mayer.
abea dacc. Carl Guion Mayer, b. July 16, 1954.
abea dada. James Mayer, Jr.
abea daea. Robert Mayer, Jr.
abea eaba. Kenneth Andrew Weaver, b. Oct. 17, 1951.
abea eaca. Richard Lawrence Burke, b. Oct. 3, 1953.

APPENDIX F

Chronology of the Life of Carl F. Gauss

1777	April 30, born in Brunswick.
1784	Enters St. Katharine's School in Brunswick.
1786	Enters Büttner's arithmetic class. Büttner ordered him a textbook from Hamburg.
1787	Friendship with Bartels. They study the binomial theorem and infinite series together.
1788	Bartels leaves the Büttner school. Gauss enters second class of the "gymnasium." Exhibits great talent in languages.
1790	Enters the senior class of the "gymnasium."
1791	Presented at court to the Duke of Brunswick. Minister of state Geheimrat Feronçe von Rotenkreuz presents him a table of logarithms.
1792	February 18, enters the Collegium Carolinum, supported by the Duke of Brunswick. Perfects himself in ancient and modern languages. Studies the works of Newton, Euler, and Lagrange.
1795	March, discovered by induction the fundamental theorem for quadratic residues (already published by Legendre in 1785). October 11, leaves Brunswick. October 15, registers as student in the University of Göttingen. Application of his method of least squares.

1796	March 30, discovers inscription of the regular polygon of seventeen sides in a circle.
	April 8, proof that -1 is the quadratic residue of all primes of the form $4n + 1$ and nonresidue of those of the form $4n + 3$. First proof of the fundamental theorem for quadratic residues.
	April 29, generalization of this theorem for composite numbers.
	June 22, beginning of investigations on binary quadratic forms.
	July 27, began the second proof of the fundamental theorem of quadratic residues.
1797	January 8, beginning of research on the lemniscate function.
	February 4, second proof for the number 2 as quadratic residue or nonresidue.
	July 22, theorem that the product of two integral rational functions of one variable with fractional coefficients and unity as coefficient of the highest degree is a function in which not all coefficients can be whole numbers.
	October 1, discovery of the principles on which the proof of the fundamental theorem for rational algebraic functions is based.
1798	April, proof that the only possible character of classes of forms are all possible.
	September 29, leaves the University of Göttingen. Returns to Brunswick and prepares his major work in the theory of numbers. Uses the University of Helmstedt library and works with Pfaff, in whose home he is a guest.
	Autumn, beginning of research on the composition of binary quadratic forms.

CHRONOLOGY

1799	February 14, beginning of research on ternary quadratic forms.
	July 16, receives doctor of philosophy degree, University of Helmstedt, *in absentia*. Dissertation contains first proof of the fundamental theorem of algebra. Later proofs in 1815, 1816, 1849.
1800	January, receives Legendre's essay on the theory of numbers.
	February 13, discovers that the number of simplest ternary quadratic forms is finite.
	Spring, research on elliptic functions.
	May, publishes formula for determining the date of Easter.
1801	January 1, Piazzi discovers Ceres.
	September 29, publishes his *Disquisitiones arithmeticae*.
	December, calculates first elliptic elements of Ceres.
1802	Summer, observations of Pallas.
	September 5, offered the directorship of the St. Petersburg observatory.
1803	January 20, decides to remain in Brunswick.
	Summer, visits Olbers in Bremen.
1804	Further work in astronomy.
1805	October 9, marries Johanna Osthoff,
1806	August 21, birth of his son Joseph.
1807	July 25, official call to the University of Göttingen. Accepts.
	November 21, arrives in Göttingen with his family.
1808	February 29, birth of his daughter Minna.
	April 14, death of his father.

	Autumn, Schumacher goes to Göttingen to study under Gauss.
1809	Publication of *Theoria motus*, his major work in astronomy.
	September 10, birth of his son Louis.
	October 11, death of his wife.
1810	March 1, death of his son Louis.
	August 4, marries Minna Waldeck.
	Autumn, Gerling, Nicolai, Möbius, and Encke go to Göttingen in order to study under Gauss. Efforts to get him to accept a professorship in Berlin. Interest in optics.
1811	Summer, research on comets.
	July 29, birth of his son Eugene.
1812	Memoir on hypergeometric series published.
1813	October 23, birth of his son Wilhelm.
1814	Publication of his memoir on a new method of approximate integration.
1815	New proof of the fundamental theorem of algebra.
1816	June 9, birth of his daughter Therese.
	September 17, moves into the new Göttingen observatory.
1817	His mother makes her home with him.
1818	Commissioned to survey the Kingdom of Hanover.
1819	Publication of memoir on method of least squares.
1820	Memoir on the new meridian circle at Göttingen.
1821	Invention of the heliotrope.
1822 to 1826	Geodetic survey of the Kingdom of Hanover.

1827	Publication of memoir on the theory of curved surfaces.
1828	Attends scientific convention in Berlin. Guest of Alexander von Humboldt. Made full professor.
1829	Research on mechanics and fluids in a state of equilibrium.
1830	Son Eugene goes to America.
	Work on theory of capillarity.
1831	Weber appointed professor of physics in Göttingen.
	Studies crystallography.
	September 12, death of his wife.
1832	Research on magnetism and electricity.
1833	Easter, operation of the electromagnetic telegraph in collaboration with Weber. Publication of basic memoir on magnetism.
	July, dean of the philosophical faculty for one year.
1834	August 31, death of Harding at the Göttingen observatory.
	December 19, Goldschmidt appointed to position at the Göttingen observatory.
1835	Publication of memoir on magnetic observations.
1836	Invents the bifilar magnetometer.
	Founding of the magnetic union.
1837	September, centenary jubilee of the University of Göttingen. Humboldt his house guest.
	October 29, son Wilhelm goes to America.
	December, son-in-law Ewald exiled.
1838	Receives the Copley Medal from Royal Society of London.

Daughter Minna moves to Tübingen.

May 30, first grandchild born, near St. Charles, Missouri.

Studies Russian.

1839 April 18, death of his mother.

December 6, becomes secretary of the Royal Society of Göttingen.

1840 Studies Sanskrit.

Publishes *Atlas of Terrestrial Magnetism*.

August 12, daughter Minna dies.

Work on theory of the potential.

1841 July, dean of philosophical faculty for one year.

Publishes proof of Legendre's theorem in spherical trigonometry.

1842 May, son Joseph Gauss participates in fighting the great fire in Hamburg.

1843 Publication of memoir in geodesy (also 1846).

1844 Publishes elliptic elements of the orbit of Faye's comet.

1845 January, Goldschmidt appointed associate professor at the observatory.

Research on comets.

July, dean of the philosophical faculty for one year.

Lightning destroys Gauss-Weber telegraph line.

1846 Publication of second memoir on geodesy.
Riemann studies under him.

1848 Observations of Neptune and Iris.

Revolution in Germany. Gauss favors the conservatives.

1849	July 16, celebration of the golden jubilee of his attaining the doctorate. Last proof of the fundamental theorem of algebra.
	July 26, Lindenau visits Gauss.
1850	Dedekind and Moritz Cantor study under Gauss.
1851	February 15, Goldschmidt dies.
	Last regular astronomical observations.
	Klinkerfues studies under him.
1852	Ernst Schering and Alfred Enneper study under him.
1853	Observations of Psyche.
	Investigates table rapping.
1854	June 16, visits railway construction between Göttingen and Kassel.
	July 31, attends opening of railroad at Göttingen.
	August 7, death of his brother Georg Heinrich.
1855	February 23, his death.
	February 26, his burial.

APPENDIX G

Books Borrowed by Gauss From the University of Göttingen Library During His Student Years

In the winter semester of 1795–1796 Professor Arnold Heeren signed for Gauss since he originally planned to study philology. His local address is indicated thus: "at the home of Blum, Weenderstrasse." Records of the library are missing for the summer semester of 1796. For the winter semester of 1796–1797 and the following semesters Professor Lichtenberg signed for him. His local address is indicated thus: "at the home of the widow Vollbaum." Some of these entries are illegible.

WINTER SEMESTER, 1795–1796

1795	Oct. 18	Richardson, *Clarissa*, Vols. I & II
	24	Lambert, *Beiträge zum Gebrauch der Mathematik* (3 vols., Berlin, 1765–1772)
	25	Lucian, *Opera*, Vol. I
	30	Sahlstädt, *Svensk grammatika*
	Nov. 5	Lucian, *Opera*, Vols. II, III, IV
	12	*Memorie della Società italiana*, Vol. I
	14	Lalande, *Astronomie*, Vol. I
	20	" ", Vols. II & III
	Dec. 1	*Leipziger Magazin zur Naturkunde* (3 vols.; Funck, Leske, Hindenburg)
	16	Cicero, *De Officiis*, Vols. III & IV
	20	*Miscellanea Taurinensia*, Vols. I-III
	21	Robertson, *History of Charles V*, Vol. I
	24	Lesage, *Gil Blas*, Vols. I, III, IV
	31	Lucian, *Opera*, Vols. VIII-IX

1796	Jan. 7	Hasselquist, *Resa till heliga land*
	22	Nouvelles Mémoires de l'Académie de Berlin, I-IV (1770–1774, Lagrange)
		Sparrmann, *Resa* ...
	25	Richardson, *Clarissa*, Vol. IV
	27	Lange, *Deutsche Sprachlehre*, Vols. I, II
	28	Klemnos (?), *Underjorske resa*
	Feb. 9	Mémoires de l'Académie de Berlin, V-VII (1775–1777)
	12	Sahm (?), *Forsög om folkens* ... 2 vols.
	15	Kalm, *Resa till Norra America*, Vols. I, II
	24	Mémoires de l'Académie de Berlin, 1779–1781, 1783
	Mar. 2	Kalm, *Resa till Norra America*, Vol. III

WINTER SEMESTER, 1796–1797

1796	Nov. 1	Hell, *Ephemeriden 1796*
	10	Maier, *Opera inedita*, Vol. I
	16	Sejour, *Traité analytique des mouvements célestes*, Vols. I & II
	17	Mémoires de l'Académie de Berlin, 1753–1757
		Mémoires de l'Académie de Berlin, 1758–1760
	19	Mémoires de l'Académie de Berlin, 1761–1765
		Sammlung astronomischer Tafeln, 3 vols.
		Vince, *On Practical Astronomy*
	24	Cousin, *Introduction à l'astronomie physique*, 1787
	25	Cousin, *Leçons du calcul différentiel et intégral*, 1777
		Mémoires de l'Académie de Berlin, 1766–1769
	26	Philosophical Transactions, 1780–1782 (4 vols.)
	30	Lalande, *Astronomie*, Vol. I
	Dec. 7	Ariosto, *Orlando furioso*, Vols. I & II
	12	Mémoires de l'Académie de Berlin, 1760–1763
	16	Euler, *Opuscula Analytica*, Vols. I & II
	26	Cousin, *Introduction à l'astronomie physique*

1797 Jan.	2	Lambert, *Photometria*
	4	*Miscellanea Taurinensia*, Vols. I-IV
		Waring, *Meditationes algebraicae*
	6	*Mémoires de l'Académie de Berlin*, 1766–1768
	10	Löwenson, *Berettning af en Sjöreas* (?)
	14	Euler, *Institutiones calculi integralis*, Vol. II
	18	Landen, *Mathematical Memoirs*, Vols. I & II
	20	*Mémoires de l'Académie de Paris* (Berlin?) 1769–1771
	23	*Miscellanea physico-mathematica*, Vol. V
	26	Belidor, *Architecture*
1797 Feb.	9	*Mémoires de l'Académie de Paris*, No. 1, 2 (1772); 1773
	13	*Mémoires de l'Académie de Paris*, 1784–1785
	18	" " " " ", 1776–1778
	13	Bode, *astronomisches Jahrbuch*, 1788–1789; 1791–1793
	20	Bode, *astronomisches Jahrbuch*, 1794, 1799 and supplement
	22	De Luc, *Lettres, phys. et morales*, Vols. II-IV
	25	Cotes, *Harmonia mensurarum*
Mar.	2	*Mémoires de l'Académie de Paris*, 1779–1783
	4	*Recueil des pièces qui ont remporté le prix de l'Académie des Sciences* (Paris)
	6	De Luc, *Lettres phys. et morales*, Vol. V
	12	Eschenburg, *Beispielsammlung zur Theorie und Litteratur der schönen Wissenschaften*
	15	*Mémoires de l'Académie de Paris*, 1784–1785, 1786
	17	*Mémoires de l'Académie de Paris*, 1780–1781
	24	Cassini, *Voyage en Allemagne*

SUMMER SEMESTER, 1797

1797 May	1	*Nouvelles Mémoires de l'Académie de Berlin*, 1780–1782
		Lasius, *Beschreibung des Harzes*, 2 vols.

BOOKS BORROWED FROM LIBRARY

	4	Halber, *Physiologie*, Vols. I-III
	5	Zimmermann, *Beobachtungen . . . einer Reise*
	9	Zeulert (?), *Naturgeschichte des Unterharzes*
		Memorie della Società italiana, Vol. IV
	10	Euler, *Institutiones calculi integralis*
	17	Haller, *Magic*, Vol. I
	20	Eschenburg, *Beispielsammlung zur Theorie und Literatur der schönen Wissenschaften*, Vol. VIII
		Engel, *Philosophie für die Welt*, Vols. I & II
	22	Joh. G. Zimmermann, *Von der Einsamkeit* (1773)
	24	*Mémoires de l'Académie de Berlin*, 1768–1769
	27	Haller, *Gedichte*
June	12	Reimarus, *Vom Blitze*
	13	*Nouveaux Mémoires de l'Académie de Berlin*, 1781–1782
	18	Kraft, *Foreläsning over Mekanik*, 2 vols.
	20	*Mélanges de litérature de la Société de Turin*
		Philosophical Transactions, No. 1, 2 (1789); No. 1, 2 (1792)
	28	Metastasio, *Opere*, Vols. III-V
July	4	Löhlein, *Klavierschule*
	7	Schiller, *Thalia*, Vols. I & II
	23	Euler, *Nova Theoria Musicae*
Aug.	1	Clairaut, *Théorie de la lune*
	2	Euler, *Opuscula* (1746), 2 vols.
	4	Lagrange, Mécanique (1787)
	12	Naturw. nation.⎰ Magazin (?), Magazin für Naturkunde (?), Vols. VII-IX
	14	Gellebrand, *Trigonometria britannica* (1658)
	17	Pitiscus, *Trigonometria* (1600)
	21	" *Tabulae trigonometricae*
	23	Cousin, *Leçons du calcul différentiel*
	31	*Mémoires de l'Académie de Paris*, 1748
Sept.	7	" " " " " , 1746–1747

Winter Semester, 1797–1798

1797	Oct.	26	Bach, *über die beste Art, Klavier zu spielen*
		30	Pitiscus, *Mecanic. trigonom.*
	Nov.	6	Löhlein, *Klavierschule*
		7	*Mémoires de l'Académie des Sciences* (Turin), 1666–1669
		9	Rousseau, *Oeuvres*, Vols. XIX-XXII
		12	*Mémoires de l'Académie de Paris*, Tomes 5, 6
		15	*Deutsches Museum*, Vols. I & II
		21	*Mémoires de l'Académie*, Vols. VII, VIII, IX
		24	*Deutsches Museum*, 1786 ?
			1780 ?
		28	*Mémoires de l'Académie des Sciences*, Tomes 10, 11
		30	*Acta eruditorum*, ann. 1769
			Commentarii nov. Academiae Petropolitanae, 7, 8, 18
	Dec.	5	*Mémoires de l'Académie . . . 1699*, 1801–1802
		6	*Deutsches Museum*, 1786–1787
		2–4	Hugenius, *Opera*, Vols. I-III
		7	Fermat, *Opera*
		10	*Deutsches Museum*, Vol. I (1779); Vols. I & II (1788)
		12	*Mémoires de l'Académie de Berlin* (Paris?), 1775
			Margary (?), *Anleitung zum Klavierspielen*
		17	*Deutsches Museum*, July-Aug., 1789
		23	*Göttinger Musenalmanach*, 1770–1773
		28	Euler, *Opuscula*, 2 vols.
		29	*Göttinger Musenalmanach*, 1774–1777
		31	*Mémoires de l'Académie de Paris*, 1700–1703
1798	Jan.	4	*Göttinger Musenalmanach*, 1778–1779
			Nouveaux Mémoires de l'Académie de Berlin, 1775
		6	*Ephemerides astronomicae*, 1798
			Göttinger Musenalmanach, 1789–1790

BOOKS BORROWED FROM LIBRARY 403

	16	Göttinger Musenalmanach, 1791–1795
		Histoire de l'Académie de Paris, 1704–1706
	23	Mémoires de l'Académie de Berlin ...
	30	" " " " ", 1708
		Deutsches Museum, 1783. Vols. I & II
	31	Mémoires de l'Académie des Inscriptions, 1785–1786
Feb.	5	Deutsches Museum, 1784. Vols. I & II
	9	Newton, Opera omnia, Vols. I & II
		Mémoires de l'Académie de Paris, 1709–1710
	13	" " " " Berlin, 1773–1775
	20	" " " " Paris, 1711–1713
Mar.	2	" " " " ", 1714–1715

SUMMER SEMESTER, 1798

1798	May	5	Mémoires de l'Académie de Paris, 1784–1785
		14	Bernoulli, Recueil pour les astronomes, Vols. I-III and Supplement
		15	Saussure, Voyage ... Tome I
		20	" " dans les Alpes, Tome 2
		26	Mémoires de l'Académie de Paris, Tomes 5, 6
		27	Mémoires de l'Académie de Paris, Tomes 3, 4
		28	Lagrange, Mecanique analytique
	June	14	Mémoires présentés a l'Académie, Vols. VII, VIII
		15	Brigard, du Massacre de Bartholomé 2b04 (?)
		20	Mémoires de l'Académie de Berlin, 1781–1782
		29	" " " " Paris, Tomes 9, 10
	July	9	Mémoires présentés a l'Académie, Tome XI
		24	Cousin, Introduction à L'étude de l'Astronomie
		29	Nova acta Academia Petropolitanae, Part I (1783); 1789; b.5?
		30	Hupel, Nordische Nachrichten (Leipzig), Vol. I
			Rognal, Histoire des établissements ..., Tome 1

Aug. 6 Rognal, *Histoire des établissements* . . . , Tome 3
16 Rognal, *Histoire des établissements* . . . , Tome 6
8 *Mélanges de mathématiques et de physique,* Vols. I & II (D'Alembert)
12 *Mémoires de l'Académie de Berlin,* 1767–1768

APPENDIX H

Courses Taught by Gauss

Semester	Title of Course
Summer 1808	Astronomy
Winter 1808	Astronomy, with exercises in observing and astronomical calculation
	Theory of the Movement of Comets
Summer 1809	Advanced Study of the Movements of Celestial Bodies According to the *Theoria motus*
Winter 1809	Individual Topics in the Theory of Numbers
	Theoretical Astronomy
Summer 1810	Rudiments of Astronomy
	Practical Astronomy (*privatissime*)
Winter 1810	Rudiments of Astronomy
	Theory of the Movement of Comets
	Practical Astronomy (*privatissime*)
Summer 1811	Rudiments of Astronomy
	Theory of Eclipses, Occultations, and Transits
	Practical Astronomy (*privatissime*)
Winter 1811	Rudiments of Theoretical Astronomy
	Advanced Study of the Movements of the Planets
	Practical Astronomy (*privatissime*)
Summer 1812	Rudiments of Astronomy
	Theory of the Movement of Comets
Winter 1812	Rudiments of Theoretical Astronomy
	Calculation of Planet Perturbations
	Practical Astronomy (*privatissime*)

Semester	Title of Course
Summer 1813	Rudiments of Theoretical Astronomy
	Calculation of Eclipses, Occultations, and Transits
	Practical Astronomy (*privatissime*)
Winter 1813	Rudiments of Theoretical Astronomy
	Theory of Planet Perturbations
	Practical Astronomy (*privatissime*)
Summer 1814	Theoretical Astronomy
	Theory of the Movement of Comets
	Practical Astronomy (*privatissime*)
Winter 1814	Rudiments of Theoretical Astronomy
	Theory of Occultations, Eclipses, and Transits
	Practical Astronomy (*privatissime*)
Summer 1815	Theoretical Astronomy
	Calculation of the Movements of Comets
	Practical Astronomy (*privatissime*)
Winter 1815	Rudiments of Theoretical Astronomy
	Theory of Occultations, Eclipses, and Transits
	Practical Astronomy (*privatissime*)
Summer 1816	Theoretical Astronomy
	Some Principal Topics in Astronomical Calculation
Winter 1816	Rudiments of Theoretical Astronomy
	Theory of the Movement of Comets
	Practical Astronomy (*privatissime*)
Summer 1817	Theoretical Astronomy
	Practical Astronomy (*privatissime*)
Winter 1817	Theory of the Movement of Comets and the Determination of their Orbits by Observation
	The Calculation of Eclipses, Occultations and Transits
	Practical Astronomy (*privatissime*)
Summer 1818	Theoretical Astronomy
	Doctrine of the Perturbations of the Movement of Planets
	Practical Astronomy (*privatissime*)

COURSES TAUGHT BY GAUSS

Semester	Title of Course
Winter 1818.	Theory of the Movement of Comets
	Calculation of Eclipses, Occultations, and Transits
	Practical Astronomy (*privatissime*)
Summer 1819	The Use of the Calculus of Probabilities in Applied Mathematics
	Theoretical Astronomy
Winter 1819	Theory of Astronomy
	Theory of the Movement of Comets
	Practical Astronomy (*privatissime*)
Summer 1820	Theoretical Astronomy
	Principal Theories of Astronomical Calculation
Winter 1820	Theory of the Movement of Planets and Comets
	Practical Astronomy (*privatissime*)
Summer 1821	(on leave of absence for geodetic survey)
Winter 1821	Theory of the Movement of Comets
	Practical Astronomy (*privatissime*)
Summer 1822	(on leave of absence for geodetic survey)
Winter 1822	Theory of the Movement of Comets
	Practical Astronomy (*privatissime*)
Summer 1823	(on leave of absence for geodetic survey)
Winter 1823	Use of the Calculus of Probabilities in Applied Mathematics
	Practical Astronomy (*privatissime*)
Summer 1824	(on leave of absence for geodetic survey)
Winter 1824	(not announced in catalogue)
Summer 1825	(on leave of absence for geodetic survey)
Winter 1825	Theory of the Motion of Comets
	Practical Astronomy (*privatissime*)
Summer 1826	Theory of the Motion of Celestial Bodies
	Practical Astronomy (*privatissime*)
Winter 1826	Use of the Calculus of Probabilities in Applied Mathematics
	Practical Astronomy (*privatissime*)

Semester	Title of Course
Summer 1827	General Theory of Curved Surfaces
	Practical Astronomy (*privatissime*)
Winter 1827	Use of Calculus of Probabilities in Applied Mathematics
Summer 1828	Theory of the Movement of Comets
Winter 1828	Instruments, Observations, and Calculations Used in Higher Geodesy
	Practical Astronomy (*privatissime*)
Summer 1829	Theory of the Movements of Comets
	Instruments, Observations, and Calculations Used in Higher Geodesy
Winter 1829	Use of Calculus Probabilities in Applied Mathematics
	Practical Astronomy (*privatissime*)
Summer 1830	Theory of the Movements of Comets
	Instruments, Observations, and Calculations Used in Higher Geodesy
Winter 1830	Calculation of Perturbations of Planets and Comets
	Practical Astronomy (*privatissime*)
Summer 1831	Theory of the Movements of Planets and Comets
	Instruments of Geodesy
Winter 1831	Use of Calculus of Probabilities in Applied Mathematics, Especially Astronomy, Geodesy, and Crystallography
	Practical Astronomy (*privatissime*)
Summer 1832	Theory of Motion of Planets and Comets
	Instruments of Geodesy
Winter 1832	Theory and Practice Observing Magnetic Phenomena
	Practical Astronomy (*privatissime*)
Summer 1833	Theory of Numerical Equations
	Instruments of Geodesy

COURSES TAUGHT BY GAUSS

Semester	Title of Course
Winter 1833	Use of Calculus of Probabilities
	Practical Astronomy
Summer 1834	Practical Astronomy
Winter 1834	Same as Winter, 1833–1834
Summer 1835	Practical Astronomy
Winter 1835	Method of Least Squares and Its Application to Astronomy, Higher Geodesy, and Natural Science
	Practical Astronomy
Summer 1836	Theory of Observation of Magnetic Phenomena
	Practical Astronomy
Winter 1836	Method of Least Squares
	Practical Astronomy
Summer 1837	Theory of Observation of Magnetic Phenomena
	Practical Astronomy
Winter 1837	Method of Least Squares
	Practical Astronomy
Summer 1838	Theory of Observation of Terrestrial Magnetic Phenomena
	Practical Astronomy
Winter 1838	Method of Least Squares
	Practical Astronomy
Summer 1839	Same as 1838
Winter 1839	Same as Winter, 1836–1837
Summer 1840	Special Topics in Dynamics
	Practical Astronomy
Winter 1840	Practical Astronomy
Summer 1841	Practical Astronomy with Preliminary Explanation of Principles of Dioptrics (*privatissime*)
Winter 1841	Same as winter, 1835–1836
Summer 1842	Practical Astronomy
Winter 1842	Same as winter, 1835–1836
Summer 1843	Practical Astronomy

Semester	Title of Course
Winter 1843	Same as winter, 1835–1836
Summer 1844	Instruments, Measurements, and Calculations of Higher Geodesy
	Practical Astronomy
Winter 1844	Same as winter, 1835–1836
Summer 1845	Same as summer, 1844
Winter 1845	Same as winter, 1835–1836
Summer 1846	Same as summer, 1844
Winter 1846	Same as winter, 1835–1836
Summer 1847	Same as summer, 1844
Winter 1847	Same as winter, 1835–1836
Summer 1848	Same as summer, 1844
Winter 1848	Same as winter, 1835–1836
Summer 1849	Same as summer, 1844
Winter 1849	Same as winter 1835–1836
Summer 1850 to Winter 1854	Same as winter, 1835–1836

APPENDIX I

Doctrines, Opinions, Theories, and Views

In his memoir *Erdmagnetismus und Magnetometer* Gauss gave a definition on which he based his work; at least he showed the principle on which he operated: "By explanation the scientist understands nothing except the reduction to the least and simplest basic laws possible, beyond which he cannot go, but must plainly demand them; from them however he deduces the phenomena absolutely completely as necessary."[1]

How clearly he adhered to this austere ideal is shown in the following passage of a letter to Schumacher dated November 7, 1847:

In general I would be cautious against . . . plays of fancy and would not make way for their reception into scientific astronomy, which must have a quite different character. Laplace's cosmogenic hypotheses belong in that class. Indeed, I do not deny that I sometimes amuse myself in a similar manner, only I would never publish such stuff. My thoughts about the inhabitants of celestial bodies, for example, belong in that category. For my part I am (contrary to the usual opinion) convinced (which in such things one calls conviction) that the larger the cosmic body, the smaller are the inhabitants and other products. For example, on the sun trees, which in the same ratio would be larger than ours, as the sun exceeds the earth in magnitude, would not be able to exist, for on account of the much greater weight on the surface of the sun, all branches would of themselves break off, in so far as the materials are not of a sort entirely heterogenous with those of the earth.

In a memoir on mechanics Gauss gave another example of how closely he followed the principle of exact definition:

---

[1] *Collected Works*, V (1877), 315–316.

As is well known the principle of virtual velocities transforms all statics into a mathematical assignment, and by Dalembert's principle for dynamics the latter is again reduced to statics. Although it is very much in order that in the gradual training of science and in the instruction of the individual the easier precedes the more difficult, the simpler precedes the more complicated, the special precedes the general, yet the mind, once it has arrived at the higher standpoint, demands the reverse process whereby all statics appears only as a very special case of mechanics.[2]

Gauss gave his views on certain extensions of mathematics in a letter to Schumacher dated May 15, 1843:

All . . . new systems of notation are such that one can accomplish nothing by means of them which would also not be accomplished without them; but the advantage is that when such a system of notation corresponds to the innermost essence of frequently occurring needs, each one who has entirely made it his own, even without the equally unconscious inspirations of the genius, which nobody can conquer, can solve the problems belonging in that category, indeed can mechanically solve them, just as mechanically in cases so complicated that without such an aid even the genius becomes powerless. Thus it is with the invention of calculating by letters in general; thus it was with the differential calculus; thus it is also (even though in partial spheres) with Lagrange's calculus of variations, with my calculus of congruences, and with Möbius' (barycentric) calculus. Through such conceptions innumerable problems which are otherwise isolated and every time demand new (minor or major) efforts of the spirit of invention, become equally an organic realm.

A letter to Schumacher on September 1, 1850, contains Gauss' ideas on convergent and divergent series:

It is the character of mathematics of modern times (in contrast to antiquity) that through our language of signs and nomenclature we possess a lever by which the most complicated arguments can be reduced to a certain mechanism. Thereby science has won infinitely in richness, in beauty, and in solidity, but, as the business is usually carried on, has lost just as much. How often that lever is applied only mechanically, although the authorization for it in most cases implies certain tacit hypotheses. I demand that in all use of the system of notation, in all uses of a concept one shall remain conscious of the original conditions, and never

---

[2] *Ibid.*, 25–26.

regard as (one's) property any products of the mechanism beyond the clear authorization. But the usual course is that one claims for the analysis a character of generality and expects of the *other* of the results thus produced, not yet recognized as proved, that it shall prove the opposite. But one may make this demand only of the one who for his part *maintains* that a result is wrong, but not of the one who recognizes as unproved a result which rests on a mechanism whose original essential conditions do not tally in the present case. Thus it is very often with divergent series. Series have a clear meaning when they converge; this clarity of meaning disappears with this condition, and nothing is essentially changed whether one uses the word sum or value. . . . Instead of the above comparison of a machine take that of paper money. This can be used advantageously for great works, but the usage is sound, if I am not mistaken, in being able to convert it at any moment into hard cash.

As a footnote to the above remarks we should add here a sentence found in a letter to Bessel dated May 5, 1812: "As soon as a series ceases being convergent, its sum as *sum* has no meaning."

The mathematician Georg Cantor called the following passage in a letter from Gauss to Schumacher (July 12, 1831) a *horror infiniti* and a sort of nearsightedness: "The use of an infinite magnitude (quantity) as a *completed one* is never permitted in mathematics. The infinite is only a façon de parler, while one really speaks of limits which certain ratios approach as closely as one desires, while others are permitted to increase without limitation."

Writing to his friend Bolyai on September 2, 1808, Gauss touched on his favorite study: "It is always noteworthy that all those who seriously study this science [the theory of numbers] conceive a sort of passion for it."

Certainly the reason for his preference just mentioned can be found in a letter of Gauss to Schumacher on September 17, 1808: "I have the vagary of taking a lively interest in mathematical subjects only where I may anticipate ingenious association of ideas and results recommending themselves by elegance or generality."

In a letter to Olbers on March 21, 1816, Gauss gave vent to some views about the famous theorem of Fermat:

I confess that the Fermat theorem as an isolated one has little interest for me, for a multitude of such theorems can easily be set up, which one can neither prove nor disprove. But I have been stimulated by it to bring out again several old ideas for a great *extension* of the theory of numbers. Of course this theory belongs to the things where one cannot predict to what extent one will succeed in reaching obscurely hovering distant goals. A happy star must also rule, and my situation and so manifold distracting affairs of course do not permit me to pursue such meditations as in the happy years 1796-1798 when I created the principal topics of my *Disquisitiones arithmeticae*. But I am convinced that if good fortune should do more than I may expect, and make me successful in some advances in that theory, even the Fermat theorem will appear in it only as one of the least interesting corollaries.[3]

The following passage in a letter to Schumacher dated January 1–5, 1845, is Gauss' answer to a charge frequently made against mathematicians:

It may be true that people who are *merely* mathematicians have certain specific shortcomings; however that is not the fault of mathematics, but is true of every exclusive occupation. Likewise a *mere* linguist, a *mere* jurist, a *mere* soldier, a *mere* merchant, and so forth. One could add to such idle chatter that when a certain exclusive occupation is often *connected* with certain specific shortcomings, it is on the other hand almost always free of certain *other* specific shortcomings.

It has often been alleged that Gauss actually did not like to teach, but this statement needs some qualification and clarification. The truth is that Gauss did not enjoy giving elementary instruction, and unfortunately he found most of his students poorly prepared for advanced work. He felt that the most desirable situation would be to lecture on research in which one was engaged at the time. The following passage in a letter to Olbers dated October 26, 1802, gives a rather full explanation of his views on this matter:

I have a true aversion to teaching. The perennial business of a professor of mathematics is only to teach the ABC of his science; most of the few pupils who go a step further, and usually, to keep the metaphor, remain in the process of gathering information, become only *Halbwisser*,[4] for

---

[3] This hope of Gauss was never realized.
[4] One who has superficial knowledge of a subject.

the rarer talents do not want to have themselves educated by lecture courses, but train themselves. And with this thankless work the professor loses his precious time. At the home of my excellent friend Pfaff, with whom I lived several months,[5] I saw how few fragmentary hours for his own work he has left from the *public* and *private* lectures, the preparations for them and from other occupations connected with the office of a professor. Experience also seems to corroborate this. I know of no professor who really would have done *much* for science, other than the great *Tobias Mayer,* and in his time he rated as a bad professor. Likewise, as our friend Zach has often noted, in our days those who do the best for astronomy are not the salaried university teachers, but so-called dilettanti, physicians, jurists, and so forth.

And in that attitude, if the colors should perhaps be somewhat too dark, I would infinitely prefer the latter, rather than the former, if I had the choice of only two. With a thousand joys I would accept a nonacademic job for which industriousness, accuracy, loyalty, and such are sufficient without specialized knowledge, and which would give a comfortable living and sufficient leisure, in order to be able to sacrifice to my gods. For example, I hope to get the editing of the census, the birth and death lists in local districts, not as a job, but for my pleasure and satisfaction, to make myself somewhat useful for the advantages which I enjoy here.

Writing to Bessel on December 4, 1808, Gauss told what his difficulties were:

To the distracting occupations belong especially my lecture courses which I am holding this winter for the first time, and which now cost much more of my time than I like. Meanwhile I hope that the second time this expenditure of time will be much less, otherwise I would never be able to reconcile myself to it, even practical (astronomical) work must give far more satisfaction than if one brings up to B a couple more mediocre heads which otherwise would have stopped at A!

Again to Bessel he wrote on January 7, 1810: "This winter I am teaching two courses for three listeners, of whom one is only moderately prepared, one scarcely moderately prepared, and the third lacks preparation as well as ability. Those are the *onera* of a mathematical profession."

In the winter of 1811 Gauss was teaching one course for one

---

[5] In Helmstedt, 1799–1800.

student whom he called "most simple." The previous winter he had reported to Schumacher that he had several very capable students in two courses. The story is told that Gauss once saw a student staggering through the streets of Göttingen. When the student saw the professor approaching, he attempted to straighten himself out, but did not succeed and cut quite a figure. Gauss scrutinized him and threatened by pointing his finger at him, smiling as he said: "My young friend, I wish that science would intoxicate you as much as our good Göttingen beer!"

Concerning the gifted students Gauss expressed himself very clearly in a letter to Schumacher on October 2, 1808:

In my opinion instruction is very purposeless for such individuals who do not want merely to collect a mass of knowledge, but are mainly interested in exercising (training) their own powers. One doesn't need to grasp such a one by the hand and lead him to the goal, but only from time to time give him suggestions, in order that he may reach it himself in the shortest way.

On May 21, 1843, Gauss wrote to Weber about some of the pitfalls which can trap even one of the greatest mathematicians, and frequently do:

In the last two months I have been very busy with my own mathematical speculations, which have cost me much time, without my having reached my original goal. Again and again I was enticed by the frequently intersecting prospects from one direction to the other, sometimes even by will-o'-the-wisps, as is not rare in mathematical speculations.

Gauss held at least one basic view in common with Goethe; the following passage in a letter to Bolyai on September 2, 1808, might well have been uttered by Faust:

It is not knowledge, but the act of learning, not possession but the act of getting there, which grants the greatest enjoyment. When I have clarified and exhausted a subject, then I turn away from it, in order to go into darkness again; the never satisfied man is so strange—if he has completed a structure, then it is not in order to dwell in it peacefully, but in order to begin another. I imagine the world conqueror must feel thus, who, after one kingdom is scarcely conquered, stretches out his arms again for others.

Gauss was not by any means indifferent to the practical applications of his research and to the material value of such applications. But he insisted that the evaluation of scientific work should not be based on such standards. He once said that science should be the friend of practicality but not its slave. In the introduction to his work on magnetism he wrote: "Science, even though advancing material interests, cannot be limited by these, but demands equal effort for all elements of its research."[6]

Gauss used to maintain that mathematics is far more a science for the eye rather than the ear. The eye surveys something at a glance and comprehends it, owing to the splendid language of signs which have been perfected in mathematics, as in no other science, especially with reference to the operations to be carried out. On the long road of translation of signs into words, this will never be presented to the ear so precisely that it can be made sufficiently accessible to the listener, in order to be pictured by those less familiar with the subject. Thus Gauss was uncommonly painstaking in the choice of designations, definitions, and even in the choice of individual letters. This attitude is shown in the following passage of a letter to Bessel on November 21, 1811:

> The $\sin^2 \phi$ is annoying to me every time, although even Laplaces uses it; if one fears that $\sin \phi^2$ could be ambiguous (which occurs perhaps never or very rarely if one were speaking of $\sin [\phi^2]$). Well now, let one write $(\sin \phi)^2$, but not $\sin {}^2\phi$, which by analogy ought to mean $\sin (\sin \phi)$.

Gauss felt that the expression in the case of equal roots that "an equation of the nth degree has n roots" is a "façon de parler" which one may tolerate. He illustrated this point in a letter to Schumacher on April 30, 1830 (his fifty-third birthday):

> *In verbis simus faciles;* on the other hand I always demand that the mathematician always remain conscious of the *things,* whereby of course nothing unsuitable can ever be built on the phrase. Our modern mathematics has, with respect to language, a completely different character from that of ancient mathematics; every moment one permits such modes of expression as must be understood *cum grano salis.*

---

[6] *Collected Works,* V (1877), 121.

A letter to Schumacher on September 17, 1808, expressed Gauss' high hopes in the field of functions. He was a young man then, and unfortunately all of these hopes were never realized. The passage runs thus:

> We now know how to operate with circular and logarithmic functions, just as with one times one, but the splendid gold mine which the interior of higher functions contains is still almost terra incognita. Formerly I did much work on the subject and some day shall publish my own major work on it, of which I have already given a hint in my *Disquisitiones arithmeticae*, p. 593. One is astonished at the superabundant richness in new, highly interesting truths and relations which such functions offer (where among other things belong also those with which the rectification of the ellipse and hyperbola is connected).

In 1811 Gauss stated that there are no true controversies in mathematics. In a letter to Schumacher on October 4, 1849, he expressed more fully his views on this subject: "I have a great antipathy against being drawn into any sort of polemic, an antipathy which is increased with every year for reasons similar to those which Goethe mentions in a letter to Frau von Wolzogen."

In the reference mentioned, Goethe said that "in advanced years when one must proceed so economically with one's time, one becomes most annoyed on account of wasted days."[7]

In one of his valuable flashes of insight Gauss declared: "I have the result, only I do not yet know how to get to it." In this utterance we see above all that he emphasizes a lightninglike intuition. He has possession of a thing, which is, however, not yet his own, and which can only become his own when he has found the way to it. From the point of view of elementary logic, this is certainly contradictory; but methodologically, by no means. Here it is a question of *Erwirb es um es zu besitzen!* This makes necessary a series of further intuitions along the road of invention and of construction.

Gauss repudiated certain proofs of earlier algebraists as being "not sufficiently rigorous," and replaced them by more rigorous proofs. This means that even in mathematics, what appears to

---

[7] Frau von Wolzogen, *Literarischer Nachlass*, I (Leipzig, 1848), 445.

one investigator as flawless, strict, and evident, is found by another to have gaps and weaknesses. Absolute correctness belongs only to identities, tautologies, that are absolutely true in themselves, but cannot bear fruit. Thus at the foundation of every theorem and of every proof there is an incommensurable element of dogma, and in all of them taken together there is the dogma of infallibility that can never be proved nor disproved. That is Gauss' view. Today there is common agreement on this point.

Coming to more personal matters, we find that Gauss in 1803, while he was still a bachelor, uttered the dictum that only he who is a father has the full right of citizenship on earth. After the birth of his six children he must certainly have felt that he was a full-fledged citizen. A year earlier he had said that marriage is like a lottery in which there are many blanks and few winning tickets, and continued by expressing the hope that he would not draw a blank. In 1812 Bessel reminded him that he united ideally interest in science and interest in a woman, adding that he hoped to imitate Gauss in this respect. Shortly afterward Gauss wrote these words to Bessel: "Certainly you too will find out that among all the goods of life happiness springs from a well-chosen marriage which is the greatest and purest that crowns all others."

In 1807 when Gauss calculated the orbit of a planetoid and gave it the name of Vesta, Bessel congratulated him and stated that it was particularly pleasant because it showed to which goddess he offered sacrifices.

Bibliography

1. Publications of Gauss

(This list does not include book reviews and similar notices.)

1. Doctoral dissertation: *Demonstratio nova theorematis omnem functionem algebraicam rationalem integram unius variabilis in factores primi vel secundi gradus resolvi posse.* Helmstedt: C. G. Fleckeisen, 1799.
2. "Berechnung des Osterfestes," *Monat. Coresp.* (ed. Zach), II (1800), 121–130.
3. "Neigung der Bahn der Ceres," *Monat. Coresp.* (ed. Zach), IV (1801), 649.
4. *Disquisitiones arithmeticae.* Leipzig: Gerhard Fleischer, Jr., 1801. French translation: *Recherches arithmétiques,* by A. C. M. Poulet-Delisle. Paris, 1807. German translation: *Untersuchungen über höhere Arithmetik,* by H. Maser. Berlin: Julius Springer, 1889.
5. "Sur la division de la circonférence du cercle en partes égales," *Soc. philom. bull.* (Paris), III (1802), 102–103.
6. "Berechnung des jüdischen Osterfestes," *Monat. Coresp.* (ed. Zach), V (1802), 435–437.
7. "Vorschriften, um aus der geocentrischen Länge und Breite eines Himmelskörpers, dem Orte seines Knotens, der Neigung der Bahn, der Länge der Sonne und ihrem Abstande von der Erde abzuleiten: des Himmelskörpers heliocentrische Länge in der Bahn, wahren Abstand von der Sonne und wahren Abstand von der Erde," *Monat. Coresp.* (ed. Zach), V (1802), 540–546.
8. "Erste Elemente der Pallas," *Monat. Coresp.* (ed. Zach), V (1802).
9. "Störungsgleichungen für die Ceres," *Monat. Coresp.* (ed. Zach), VI (1802), 387–389.
10. "Tafeln für die Störungen der Ceres," *Monat. Coresp.* (ed. Zach), VII (1803), 259–275.
11. "Einige Bemerkungen zur Vereinfachung der Rechnung für die geocentrischen Oerter der Planeten," *Monat. Coresp.* (ed. Zach), IX (1804), 385–400.

12. "Ueber die Grenzen der geocentrischen Oerter der Planeten," *Monat. Corresp.* (ed. Zach), X (1804), 173–191.
13. "Erste Elemente der Juno," *Monat. Corresp.* (ed. Zach), X (1804), 464–552.
14. "Theorematis arithmetici demonstratio nova," *Comment.* (Göttingen), XVI (1804–1808), 69–74.
15. "Der Zodiacus der Juno," *Monat. Corresp.* (ed. Zach), XI (1805), 225–228.
16. "Ueber den zweiten Cometen von 1805," *Monat. Corresp.* (ed. Zach), XIV (1806).
17. "Ephemeride für den Lauf der Ceres," *Monat. Corresp.* (ed. Zach), XV (1807), 154–157.
18. "Beobachtungen der Pallas," *Monat. Corresp.* (ed. Zach), XV (1807), 377–378.
19. "Erste und zweite Elemente der Vesta," *Monat. Corresp.* (ed. Zach), XV (1807), 596–598.
20. "Allgemeine Tafeln für Aberration und Nutation," *Monat. Corresp.* (ed. Zach), XVII (1808), 312–317.
21. "Beobachtungen der Juno, Vesta und Pallas," *Monat. Corresp.* (ed. Zach), XVIII (1808), 83–86, 173–175, 182–188, 269–273.
22. "Aus zwei beobachteten Höhen zweier Sterne, deren Rectascensionen und Declinationen als gegeben angesehen werden, und den entsprechenden Zeiten der Uhr, die entweder nach Sternzeit geht oder deren Gang während der Beobachtungen als bekannt angenommen wird, den Stand der Uhr und die Polhöhe zu bestimmen," *Monat. Corresp.* (ed. Zach), XVIII (1808), 277–293; XIX (1809).
23. "Summatio quarundam serierum singularium," *Comment.* (Göttingen), I (1808–1811).
24. "Disquisitio de elementis ellipticis Palladis," *Comment.* (Göttingen), I (1808–1811).
25. "Beobachtungen der neuen Planeten," *Monat. Corresp.* (ed. Zach), XX (1809), 78–79.
26. *Theoria motus corporum coelestium in sectionibus conicis solem ambientium.* Hamburg: Perthes and Besser, 1809. English translation by Charles Henry Davis. Boston: Little, Brown and Company, 1857. French translation by Edm. Dubois, 1864. German translation by Carl Haase, Hanover, 1865. There was also a Russian translation.
27. "Summarische Uebersicht der zur Bestimmung der Bahnen der beiden neuen Hauptplaneten angewandten Methoden," *Monat. Corresp.* (ed. Zach), XX (1809), 197–224.
28. "Fortgesetzte Nachrichten von dem neuen Hauptplaneten Vesta," *Monat. Corresp.* (ed. Zach), XIX (1809), 407–410, 504–515.

29. "Pallas- und Vesta-Beobachtungen," *Monat. Corresp.* (ed. Zach), XXI (1810), 276–280.
30. "Bestimmung der grössten Ellipse, welche die vier Seiten eines gegebenen Vierecks berührt," *Monat. Corresp.* (ed. Zach), XXII (1810), 112–121.
31. "Elemente der Pallas," *Monat. Corresp.* (ed. Zach), XXII (1810), 400–403; XXIII (1811), 97–98; XXIV (1811), 449–465.
32. "Tafeln für die Mittagsverbesserung," *Monat. Corresp.* (ed. Zach), XXIII (1811), 401–409.
33. "Beobachtungen des Cometen," *Monat. Corresp.* (ed. Zach), XXIV (1811), 180–182.
34. "Elemente des zweiten Cometen von 1811," *Monat. Corresp.* (ed. Zach), XXIV (1811).
35. "Observationes cometae secundi ann. 1813 in observatorio Gottingensi factae, adjectis nonnullis annotationibus circa calculum orbitarum parabolicarum," *Comment.* (Göttingen), II (1811–1813); *Nouv. Ann. math.*, XV (1856), 5–17.
36. "Disquisitiones generales circa seriem infinitam

$$1 + \frac{\alpha \cdot \beta}{1 \cdot \gamma} x + \frac{\alpha(\alpha+1)\beta(\beta+1)}{1 \cdot 2 \cdot \gamma(\gamma+1)} x^2 + \frac{\alpha(\alpha+1)\beta(\beta+1)(\beta+2)}{1 \cdot 2 \cdot 3 \cdot \gamma(\gamma+1)(\gamma+2)} x^3,"$$

Comment. (Göttingen), II (1811–1813). German translation by Heinrich Simon. Berlin: Julius Springer, 1888.
37. "Theoria attractionis corporum sphæroidicorum ellipticorum homogeneorum methodus nova tractata," *Comment.* (Göttingen), II (1811–1813). German translation by von Lindenau in *Monat. Corresp.* (ed. Zach), XXVIII (1813), pp. 37–57, 125–234. Also edited by A. Wangerin as No. 19 in Ostwald's *Klassiker der exakten Wissenschaften*.
38. "Neue Methode aus der Höhe zweier Sterne die Zeit und die Polhöhe zu bestimmen," (1809), *Astron. Jahrb.* (ed. Bode), 1812, pp. 129–143. German translation by Ludwig Harding of *"Methodus peculiaris elevationem poli determinandi."* Göttingen, 1808.
39. "Ueber die Tafel für die Sonnen-Coordinaten in Beziehung auf den Aequator," *Monat. Corresp.* (ed. Zach), XXV (1812), 23–36.
40. "Parabolische Elemente des zweiten Cometen von 1811," *Monat. Corresp.* (ed. Zach), XXV (1812), 95–97.
41. "Sternbedeckungen," *Monat. Corresp.* (ed. Zach), XXV (1812), 206–207.
42. "Pallas-Beobachtungen," *Monat. Corresp.* (ed. Zach), XXVI (1812), 199–203; XXVIII (1813), 197–198.
43. "Juno-Beobachtungen," *Monat. Corresp.* (ed. Zach), XXVI (1812), 297–299.

44. "Tafel zur bequemern Berechnung des Logarithmen der Summe oder Differenz zweier Grössen, welche selbst nur durch ihre Logarithmen gegeben sind," *Monat. Corresp.* (ed. Zach), XXVI (1812), 498–528.
45. "Ueber Attraction der Sphäroiden," *Monat. Corresp.* (ed. Zach), XXVII (1813), 421–431.
46. "Beobachtungen mit einem 12-zolligen Reichenbach'schen Kreise zur Bestimmung der Polhöhe der Göttinger Sternwarte," *Monat. Corresp.* (ed. Zach), XXVII (1813), 481–484.
47. "Verzeichniss von Stern-Declinationen," *Monat. Corresp.* (ed. Zach), XXVIII (1813), 97–99.
48. "Beobachtungen des zweiten Cometen vom Jahre 1813, angestellt auf der Sternwarte zu Göttingen, nebst einigen Bemerkungen über die Berechnung parabolischer Bahnen," *Monat. Corresp.* (ed. Zach), XXVIII (1813), 501–513. "Observationes cometae secundi anni 1813." German translation by Nicolai in *Gött. gel. Anz.*, 1814, 25.
49. "Achte Opposition der Juno," *Monat. Corresp.* (ed. Zach), XXVIII (1813), 574–578.
50. "Nachricht von dem Reichenbach'schen Repetitionskreise und dem Theodolithen," *Gött. gel. Anz.*, 1813.
51. "Beobachtungen usw. zur Bestimmung der Polhöhe der Göttinger Sternwarte," *Monat. Corresp.* (ed. Zach), XXVII (1813).
52. "Methodus nova integralium valores per approximationem inveniendi," (1814), *Comment.* (Göttingen), III (1814–1815), 39–76; *Nouv. Ann. math.*, XV (1856), 109–129, 207–211, 315–321.
53. "Demonstratio nova altera theorematis omnem functionem algebraicam rationalem integram unius variabilis in factores reales primi vel secundi gradus resolvi posse," *Comment.* (Göttingen), III (1814–1815), 107–134, 135–142; *Nouv. Ann. math.*, XV (1856), 134–139.
54. "Ejusdem theorematis demonstratio, tertia," *ibid.*
55. "Eigenthümliche Darstellung der Pfaff'schen Integrationsmethode," *Gött. gel. Anz.*, 1813.
56. "Bestimmung der Genauigkeit der Beobachtungen," *Zeitschr. für Astr.* (ed. Lindenau and Bohnenberger), (1816), 185–197.
57. "Theorematis fundamentalis in doctrina de residuis quadraticis demonstrationes et ampliationes novae," *Comment.* (Göttingen), IV (1816–1818), 3–20.
58. "Determinatio attractionis, quam in punctum quodvis positionis datae exerceret planeta, si ejus massa per totam orbitam, ratione temporis, que singulae partes describuntur, uniformiter esset dispertita," *Comment.* (Göttingen), IV (1816–1818), 21–48.

59. "Ueber die Differenz der Polhöhe, wenn sie aus Sonnenbeobachtungen oder Nordsternbeobachtungen mit dem Multiplications-Kreise abgeleitet wird," *Zeitschr.* (ed. Lindenau), IV (1817), 119–131.
60. "Beobachtungen des Polarsterns in der untern Culmination auf der Göttinger neuen Sternwarte," *Zeitschr.* (ed. Lindenau), IV (1817), 126–131.
61. "Ueber die achromatischen Doppelobjective besonders in Rücksicht der vollkommnern Aufhebung der Farbenzerstreuung," *Zeitschr.* (ed. Lindenau), IV (1817), 345–351; *Annal.* (ed. Gilbert), LIX (1818), 188–195.
62. "Ueber einige Berichtigungen an Bordaischen Wiederholungskreisen," *Zeitschr.* (ed. Lindenau), V (1818), 198–211.
63. "Nachricht von d. Repsold'schen Meridiankreise," *Gött. gel. Anz.* (1818).
64. "Cometenbeobachtungen," *Zeitschr.* (ed. Lindenau), V (1818), 276–277.
65. "Distances au zénith de quelques étoiles, observées à Göttingen, . . ." *Monat. Corresp.* (ed. Zach), II (1819), 53–61.
66. "Vom Reichenbach'schen Meridiankreise," *Gött. gel. Anz.* (1820), p. 905. English translation by Herschel in the *Memoirs* of the Astronomical Society, Vol. I.
67. "Von d. Reichenbach'schen Mittagsfernrohr," *Gött. gel. Anz.*, 1819.
68. "Theoria combinationis observationum erroribus minimis obnoxiae," *Comment.* (Göttingen) V. (1819–1822), 33–62, 63–90. French translation: *Méthode des moindres carrés*, by J. L. F. Bertrand. Paris: Mallet-Bachelier, 1855.
69. "Vom Reichenbach'schen Meridiankreise," *Gött. gel. Anz.*, 1820.
70. "Ueber den Repsold'schen Meridiankreis, Beobachtung des Uranus, Saturns und der Pallas, . . ." (1818), *Astron. Jahrb.* (ed. Bode), 1821, pp. 212–216.
71. "Vom Heliotropen, und den ersten damit angestellt. Versuchen," *Gött. gel. Anz.*, 1821.
72. "On the new Meridian Circle at Göttingen," (1820), *Astron. Soc. Mem.*, I (1822), 129–134.
73. "Lettre sur le héliotrope réflecteur," *Monat. Corresp.* (ed. Zach), VI (1822), 65–69; *Quart. Journ. Sci.*, XIII (1822), 421–422.
74. *Kartenprojektion* (1822). Ed. by A. Wangerin as No. 55 in Ostwald's *Klassiker*.
75. "Über das Heliotrop," *Annal.* (ed. Poggendorff), IX and XVII.
76. "Anwendung der Wahrscheinlichkeitsrechnung auf eine Aufgabe der praktischen Geometrie: Die Lage eines Punktes aus den an

demselben gemessenen horizontalen Winkeln zwischen anderen Punkten von genau bekannter Lage zu finden." *Astr. Nachr.*, Vol. I (1823), col. 81–86.
77. "Aus drei der Lage nach bekannten Puncten die des vierten zu finden," *ibid.*
78. "Ueber die Hannoversche Gradmessung (Heliotrop)," *Astr. Nachr.*, Vol. I (1823), col. 105–106, 441–444.
79. "Theoria residuorum biquadraticorum," *Comment.* (Göttingen), VI (1823–1827), 27–56; *Comment., secunda* VII (1828–1831), 89–148; *Gött. gel. Anz.*, 1831, p. 625.
80. "Supplementum theoriae combinationis observationum erroribus minime obnoxiae," *Comment.* (Göttingen), VI (1823–1827), 57–98; VII, 89–148.
81. "Disquisitiones generales circa superficies curvas," *Comment.* (Göttingen), VI (1823–1827), 99–146; *Nouv. Ann. math.*, XI (1852), 195–252. French translation by Liouville (Paris, 1850), in the fifth edition of Monge's *Application de l'analyse à la géométrie.* Another French translation by Tiburce Abadie, *Nouv. Ann. math.*, 11 (1852). A third French translation by E. Roger. Paris, 1855; 2d ed., Grenoble, 1870 and Paris, 1871. German translation by O. Böklen. Stuttgart, 1884. Another German translation by A. Wangerin. Leipzig, 1889; 2d ed., Leipzig, 1900. Hungarian translation by Szíjártó Miklós. Budapest, 1897. English translation by J. C. Morehead and A. M. Hiltebeitel. Princeton, N.J.: Princeton University Press, 1902.
82. "Neue Methode die gegenseitigen Abstände der Fäden in Meridian-Fernröhren zu bestimmen," *Astr. Nachr.*, Vol. II (1824), col. 371–376.
83. "Allgemeine Auflösung der Aufgabe: Die Theile einer gegebenen Fläche auf einer andern gegebenen Fläche so abzubilden, dass die Abbildung dem Abgebildeten in den Kleinsten Theilen, ähnlich wird," *Astron. Abhandl.* (ed. Schumacher), III (1825), 1–30; *Phil. Mag.*, IV (1828), 104–113, 206–215.
84. "Vertheidigung Pasquich's Ehrenrettung," *Astr. Nachr.*, Vol. III (1825), col. 78–89.
85. "Beobachtungen des Cometen von 1824 in Göttingen," *Astr. Nachr.*, Vol. III (1825), col. 179–182.
86. "Methode, mittlere Lufttemperatur zu bestimmen," *Annal.* (ed. Poggendorff), IV (1825).
87. "Über die Umlaufszeit des Biela'schen Cometen, . . ." *Astr. Nachr.*, IV (1826).
88. "Chronometrische Längenbestimmungen," *Astr. Nachr.*, Vol. V. (1827), col. 227–234, 234–240, 245–248.

89. "Ueber die vortheilhafte Anwendung der Methode der Kleinsten Quadrate," *Astr. Nachr.*, Vol. V (1827), col. 230.
90. "Die Berichtigung des Heliotrops," *Astr. Nachr.*, Vol. V (1827), col. 329–334.
91. "Beobachtungen des von Pons im Luchs entdeckten Cometen," *Astr. Nachr.*, Vol. VI (1828), col. 43–44.
92. "Pallasbeobachtungen," *Astr. Nachr.*, Vol. VI (1828), col. 67-68.
93. "Lehrsatz über den Zusammenhang der Anzahl der positiven und negativen Wurzeln einer algebraischen Gleichung mit der Anzahl der Abwechselungen und Folgen in den Zeichen der Coefficienten," *Crelle's Journ.*, III (1828), 1–4.
94. *Bestimmung des Breitunterschiedes zwischen den Sternwarten von Göttingen und Altona durch Beobachtungen am Ramsdenschen Zenithsector.* Göttingen: Vandenhoek and Ruprecht, 1828.
95. "Principia generalia theoriae figurae fluidorum in statu aequilibrii," *Comment.* (Göttingen), VII (1828–1831) 39–88. German translation by Rudolf H. Weber, ed. by H. Weber as No. 135 in Ostwald's *Klassiker.*
96. "Ueber ein neues allgemeines Grundgesetz in der Mechanik: 'Die Bewegung eines Systems materieller, auf was immer für eine Art unter sich verknüpfter Punkte, deren Bewegung zugleich an was immer für äussere Beschränkungen gebunden sind, geschieht in jedem Augenblick in möglich grösster Uebereinstimmung mit der freien Bewegung, oder unter möglich kleinstem Zwange, indem man als Maas des Zwanges, den das ganze System in jedem Zeittheilchen erleidet, die Summe der Produkte aus dem Quadrate der Ablenkung jedes Punkts von seiner freien Bewegung in seine Masse betrachtet,'" *Crelle's Journ.*, IV (1829), 232–235; *Nouv. Ann. math.*, IV (1845), 477–479.
97. "Zusätze zu Seeber's Werke über d. ternären quadrat. Formen," *Gött. gel. Anz.*, 1831.
98. "Beobachtung der Sonne, der Ceres, und der Pallas in Göttingen," *Astr. Nachr.*, Vol. VII (1829), col. 15–16; Vol. VIII (1831), col. 321-322.
99. "Intensitas vis magneticae terrestris ad mensuram absolutam revocata," *Comment.* (Göttingen), VIII (1832–1837), 3–44; *Annal. de Chimie*, LVII (1834), 5–69; *Astr. Nachr.*, Vol. X (1833), col. 349–360. Italian translation by Paolo Frisiani in *Effemeride astronomiche di Milano*, 1839, pp. 3–132 (first suppl.). German translation in Poggendorff's *Annalen der Physik*, XXVII and XXVIII (1833), 241–273, 591–614. Second

BIBLIOGRAPHY 427

German translation by E. Dorn, No. 53 in Ostwald's *Klassiker der exakten Wissenschaften*. French translation by Arago in *Annales de physique*. English translation in the *Proceedings of the Royal Society*, No. 11. There was also a Russian translation in the scientific series of the University of Moscow. "Multiplicator," *Annal.* (ed. Poggendorff), XXVII and XXVIII; also in *Braunschw. Magazin* and *Gött. Taschenbuch*.

100. "Beobachtungen der magnetischen Variation in Göttingen und Leipzig am 1. and 2. Oct. 1834," *Annal.* (ed. Poggendorff), XXXIII (1834), 426–432.
101. "Nachrichten über d. magnet. Observatorium in Göttingen," *Gött. gel. Anz.*, 1834.
102. "Nouvelles observations magnétiques faites à Goettingue," *L'Institut*, III (1835), 300–302.
103. "Bericht von neuerlich in Göttingen angestellten magnetischen Beobachtungen," *Annal.* (ed. Poggendorff), XXXIV (1835), 546–556.
104. "Beobachtungen der magnetischen Variation am 1. April 1835 von fünf Oertern," *Annal.* (ed. Poggendorff), XXXV (1835), 480–481; *Bibl. Univ.*, I (1836), 345–346.
105. "Erdmagnetismus und Magnetometer," *Jahrb.* (ed. Schumacher), 1836, pp. 1–47.
106. "Ueber die Berichtigung der Schneiden einer Waage," *Astr. Nachr.*, XIV (1837), col. 241–244.
107. "Neue Methode z. Berichtigung d. Waagen," *Gött. gel. Anz.*, 1837.
108. "Das in den Beobachtungsterminen anzuwendende Verfahren," in Gauss, *Resultate* (1837), pp. 34–50.
109. "Auszug aus dreijährigen täglichen Beobachtungen der magnetischen Declination zu Göttingen," *Resultate* (ed. Gauss), 1837, pp. 50–62.
110. "Ueber ein neues zunächst zur unmittelbaren Beobachtung der Veränderungen in der Intensität des horizontalen Theils des Erdmagnetismus bestimmtes Instrument (Bifilarmagnetometer)," *Resultate*, (ed. Gauss), 1838, p. 1–19; *Scientif. Mem.* (ed. Taylor), II (1841), 252–267.
111. "Anleitung zur Bestimmung der Schwingungsdauer einer Magnetnadel," *Resultate* (ed. Gauss), 1838, pp. 58–80.
112. "Dioptrische Untersuchungen," *Abhandl. Ges. Wiss.* (Göttingen), I (1838–1841), 1–20; *Annal. de Chimie*, XXXIII (1851), 259–294; *Scientif. Mem.* (ed. Taylor), III (1843), 490–498. English translation by Oscar Faber, in *Ferrari's Dioptric Instruments*. London: H. M. Stationery Office, 1919. See also William Hal-

lows Miller, *Abstract of Some of the Principal Propositions of Gauss' Dioptric Researches.* London, 1843.
113. "Allgemeine Theorie des Erdmagnetismus," *Resultate* (ed. Gauss), 1839, pp. 1–57, 146–148; *Scientif. Mem.* (ed. Taylor), II (1841), 184–251, 313–316.
114. "Allgemeine Lehrsätze in Beziehung auf die im verkehrten Verhältnisse des Quadrats der Entfernung wirkenden Anziehungs- und Abstossungskräfte," *Resultate,* (ed. Gauss), 1840, pp. 1–51; *Journ. Math.* (Liouville), VII (1842), 273–324; *Scientif. Mem.* (ed. Taylor), III (1843), 153–196. Edited by A. Wangerin as No. 2 in Ostwald's *Klassiker.*
115. *Atlas des Erdmagnetismus.* (With W. E. Weber and C. W. B. Goldschmidt.) Leipzig: Weidmannsche Buchhandlung, 1840. Reprinted in *Werke,* Vol. XII.
116. "Ueber ein Mittel, die Beobachtung von Ablenkungen zu erleichtern," *Resultate* (ed. Gauss), 1840, pp. 52–62; *Scientif. Mem.* (ed. Taylor), III (1843), 145–152.
117. "Mondfinsterniss," *Astr. Nachr.,* Vol. XVIII (1841), col. 143–144.
118. "Elementare Ableitung eines zuerst im Legendre aufgestellten Lehrsatzes der sphärischen Trigonometrie," *Crelle's Journ.,* XXII (1841), 96; *Journ. Math.* (Liouville), VI (1841), 273–274.
119. "Zur Bestimmung des Constanten des Bifilarmagnetometers," *Resultate* (ed. Gauss), 1841, pp. 1–25.
120. "Vorschriften zur Berechnung der magnetischen Wirkung, welche ein Magnetstab in der Ferne ausübt," *Resultate* (ed. Gauss), 1841, pp. 26–34.
121. "Untersuchungen über Gegenstände der höheren Geodäsie." *Abhandl.* (Göttingen), II (1842-1844, 1845), 3–45; III (1847), 3–43; *Nachrichten* (Göttingen), 1846, pp. 210–217; *Nouv. Ann. math.,* 1851, pp. 363–364. Edited by J. Frischauf as No. 177 in Ostwald's *Klassiker.*
122. "Theorie des Erdmagnetismus verglichen mit A. Erman's Beobachtungen," *Astr. Nachr.,* XIX (1842).
123. "Ueber die Berechnung der Anomalien aus Elementen vermittelst Burckhardt's Tafeln," *Astr. Nachr.,* Vol. XX (1843), col. 299–300.
124. "Ueber die Anwendung des Magnetometers zur Bestimmung der absoluten Declination," *Resultate* (ed. Gauss), 1843, pp. 1–9. (Vol. VI., 1841.)
125. "Beobachtungen der magnetischen Inclination in Göttingen in Jahre 1841," *Resultate* (ed. Gauss), 1843, pp. 10–61; *Scientif. Mem.* (ed. Taylor), III (1843), 623–665.

126. "Beobachtungen des Faye'schen Cometen." *Astr. Nachr.,* Vol. XXI (1844), col. 221–222.
127. "Elliptic Elements of the Orbit of Faye's comet," *Astr. Nachr.,* Vol. XXI (1844), col. 235-238.
128. "Beobachtung der Mondfinsterniss vom 31-ten Mai 1844," *Astr. Nachr.,* Vol. XXII (1845), col. 31–32.
129. "Cometenbeobachtungen," *Astr. Nachr.,* Vol. XXII (1845), col. 113–116, 165–168, 189–192, 277–280.
130. "Beobachtungen von Le Verrier's Planeten," *Astr. Nachr.,* Vol. XXV (1847), col. 82–83.
131. "Ueber die Limiten des Zodiakus, eines die Sonne nach Kepler's Gesetzen umkreisenden Himmelskörpers," *Astr. Nachr.,* Vol. XXVI (1848), col. 1–4.
132. "Beobachtungen des Neptun und der Iris," *Astr. Nachr.,* Vol. XXVI (1848), col. 153–154, 241–242.
133. "Beobachtungen des neuen Planeten (Iris)," *Astr. Nachr., Vol.* XXVI (1848), col. 173–176.
134. "Beobachtungen der Iris," *Astr. Nachr.,* Vol. XXVI (1848), col. 197–198.
135. "Beobachtungen und Elemente der Flora," *Astr. Nachr.,* Vol. XXVI (1848), col. 367–368.
136. "Beobachtungen des Graham's Planet," *Astr. Nachr.,* Vol. XXVII (1848), col. 235–236, 265–266.
137. "Auszug über die Berechnung trigonometrischer Messungen," (1847), *St. Petersb. Acad. Sci. Bull.,* Vol. VI (1848), col. 257–266.
138. "Beiträge zur Theorie der algebraischen Gleichungen," *Abhandl.* (Göttingen), IV (1848–1850), 3–34; *Nouv. Ann. math.,* X (1851), 165–174.
139. "Ueber den von Hind entdeckten veränderlichen Stern im Ophiuchus," *Astr. Nachr.,* Vol. XXVIII (1849), col. 23–24.
140. "Meridianbeobachtungen der Victoria: Berechnung der Bahn des Hind'schen Cometen," *Astr. Nachr.,* Vol. XXXI (1851), col. 305–306.
141. "Einige Bemerkungen zu Vega's 'Thesaurus Logarithmorum,' " *Astr. Nachr.,* Vol. XXXII (1851), col. 181–188.
142. "Exercice numérique sur les équations du premier degré," *Nouv. Ann. math.,* X (1851), 359–361.
143. "Elemente der Psyche, nebst Bemerkungen über die Bestimmung der Bahn dieses Planeten," *Astr. Nachr.,* Vol. XXXV (1853), col. 17–20; XXXVI (1853), col. 301–302.
144. "Beobachtungen auf der Göttinger Sternwarte," *Astr. Nachr.,* Vol. XXXVII (1854), col. 197–198; 409–410.

145. "Entdeckung des Cometen I, 1854," *Astr. Nachr.*, Vol. XXXVII (1854), col. 363–364.
146. "Beobachtungen und Elemente des neuesten Cometen III, 1854," *Astr. Nachr.*, Vol. XXXVIII (1854), col. 93–94, 353–354.
147. "Beobachtungen auf der Göttinger Sternwarte," *Astr. Nachr.*, Vol. XXXIX (1855), col. 161–162.
148. "Recherches dioptriques," *Journ. Math.* (Liouville), I (1856), 9–43.
149. "Die orthogonale Transversale and die Brennlinie der zurückgeworfenen Strahlen für die gemeine Cycloide, wenn die einfallenden Strahlen der Axe derselben parallel sind, und für die logarithmische Spirale, wenn die einfallenden Strahlen vom Pol derselben ausgehen," *Archiv* (ed. Grunert), XXX (1858), 121–134.
150. "Sehr einfache Bestimmung eines bekannten Integrals," *Archiv* (ed. Grunert), XXX (1858), 229–230.
151. "Results of the Observations made by the Magnetic Association in the Year 1836," (with W. E. Weber), *Scientif. Mem.* (ed. Taylor) II (1841), 20–25 (Transl.).
152. "Anziehung eines elliptischen Ringes. Nachlass zur Theorie des arithmetisch-geometrischen Mittels und der Modulfunktion," translated and edited by H. Geppert as No. 225 in Ostwald's *Klassiker*.
153. "Die vier Beweise der Zerlegung ganzer algebraischer Funktionen usw." (1799–1849), edited by Eugen Netto as No. 14 in Ostwald's *Klassiker*.
154. "Sechs Beweise der Fundamentaltheorems über quadratische Reste," edited by Eugen Netto as No. 122 in Ostwald's *Klassiker*.
155. "Abhandlungen über die Prinzipien der Mechanik," edited by Philip E. B. Jourdain as No. 167 in Ostwald's *Klassiker*.

COLLECTED WORKS

Carl Friedrich Gauss' Werke. Sponsored by the Royal Society of Sciences at Göttingen. General editors: E. Schering, F. Klein, M. Brendel, and L. Schlesinger; assisted by R. Fricke, P. Stäckel, E. Wiechert, C. Schaefer, A. Galle, and H. Geppert. Vols. I–XII. Published successively by Friedrich Andreas Perthes (Gotha), B. G. Teubner (Leipzig), and Julius Springer (Berlin), 1863–1933.

2. ABOUT GAUSS

This is not a complete Gauss bibliography, but it is believed that no important item has escaped. A few minor errors are inevitable in an extensive listing of this sort, the first of its kind on Gauss. It has not been deemed necessary to cite here a few of the works referred to in footnotes. This bibliography is intended as a working guide for others who are especially interested in Gauss. Although some publications dealing with Gauss' scientific work are included, the primary emphasis is on those of a biographical nature. In a few instances it was difficult to decide whether an item should be classified as a book or a pamphlet. Standard abbreviations have been used, and it is hoped that they will be intelligible to all. An effort has been made to be consistent in the use of such abbreviations. Where there are several items under the name of one author, they are arranged chronologically. This is not a complete roster of sources consulted, but all those which yielded valuable information are to be found here.

MANUSCRIPTS AND RELATED MATERIAL

Gauss Archive, University of Göttingen.
Gauss Collection, Municipal Library and Archive, Brunswick, Germany.
Dunnington Collection, Natchitoches, La.
William T. Gauss Collection, Colorado Springs, Colo. (Parts of this collection have been donated to Harvard University and Princeton University.)

BOOKS

Ahrens, Wilhelm. *Mathematiker-Anekdoten.* Leipzig: B. G. Teubner, 1920. (Gauss, p. 8.)
Arnim, Max. *Corpus Academicum Gottingense.* Göttingen: Vandenhoeck and Ruprecht, 1930.
Auwers, Georg Friedrich Julius Arthur von (ed.). *Briefwechsel zwischen Gauss und Bessel.* Leipzig: Wilhelm Engelmann, 1880.
Bell, E. T. *Men of Mathematics.* New York: Simon and Schuster, 1937. (Gauss, pp. 218–269.)
Bieberbach, Ludwig. *Carl Friedrich Gauss, ein deutsches Gelehrtenleben.* Berlin: Keil Verlag, 1938.
Böttiger, C. A. *Franz Volkmar Reinhard.* Dresden: Arnoldische Buch- und Kunsthandlung, 1813.
Boncompagni, Baldassare (Prince), ed. *Cinq lettres de Sophie Germain à C. F. Gauss.* Berlin, 1880.

Borch, Rudolf. "Ahnentafel des Mathematikers C. F. Gauss." In *Ahnentafeln berühmter Deutscher*. Leipzig, 1929, Lieferung 1, p. 63.

———— *Alexander von Humboldt*. Berlin: Druckhaus Tempelhof, 1948.

Bruhns, C. *Johann Franz Encke, sein Leben und Wirken*. Leipzig: Ernst Julius Günther, 1869.

Bruhns, K. (ed.). *Briefe zwischen A. v. Humboldt und Gauss*. Leipzig: Wilhelm Engelmann, 1877.

Cajori, Florian. *The Chequered Career of Ferdinand Rudolph Hassler*. Boston: Christopher Publishing House, 1929.

Davies, T. Witton. *Heinrich Ewald, Orientalist and Theologian*. London: T. Fisher Unwin, 1903.

Davis, Charles Henry. *The Computation of an Orbit from Three Complete Observations. From the Theoria Motus of C. F. Gauss*. Cambridge, Mass.: Metcalf and Company, 1852.

Dunnington, G. Waldo. *Carl Friedrich Gauss. Inaugural Lecture on Astronomy and Papers on the Foundations of Mathematics*. Baton Rouge: Louisiana State University Press, 1937.

Dyck, Walter von. *Georg von Reichenbach*. Munich, 1912.

Engel, Friedrich, and Stäckel, Paul. *Die Theorie der Parallellinien von Euklid bis auf Gauss*. Leipzig: B. G. Teubner, 1895.

Feyerabend, Ernst. *Der Telegraph von Gauss und Weber im Werden der elektrischen Telegraphie*. Berlin: Reichspostministerium, 1933.

Fick, R., and von Selle, Götz. *Briefe an Ewald*. Göttingen: Vandenhoeck and Ruprecht, 1932.

Hänselmann, Ludwig. *Carl Friedrich Gauss, zwölf Kapitel aus seinem Leben*. Leipzig: Duncker and Humblot, 1878.

Kistner, A. *Deutsche Meister der Naturwissenschaft und Technik*. Vol. I. Munich: Kösel and Pustet, 1925. (Gauss, pp. 74–88.)

Klein, Felix (ed.). *Gauss' wissenschaftliches Tagebuch 1796–1814*. Berlin: Weidmannsche Buchhandlung, 1901.

———. *Vorlesungen über die Entwicklung der Mathematik im 19. Jahrhundert*. Teil I. Berlin: Julius Springer, 1926. (Gauss, pp. 6–62.)

Klein, F., Brendel, M., and Schlesinger, L. (eds.). *Materialien für eine wissenschaftliche Biographie von Gauss*. Vols. I–VIII. Leipzig: B. G. Teubner, 1911–1920.

Klingenberg, Wilhelm, and others. *Gaussgedenkband*. Leipzig: Julius Springer, 1955.

Koenigsberger, Leo. *Carl Gustav Jacob Jacobi*. Leipzig: B. G. Teubner, 1904.

Kowalewski, Gerhard. *Grosse Mathematiker*. Munich and Berlin: J. F. Lehmann, 1938. (Gauss, pp. 247–273.)

Kück, Hans. *Die Göttinger Sieben*. Berlin: Emil Ebering Verlag, 1934.

Lenard, Philipp. *Great Men of Science*. London: G. Bell and Sons, 1933. (Gauss, pp. 240–247.)
Leutner, K. *Deutsche, auf die wir stolz sind*. Berlin: Verlag der Nation, 1954.
Mack, Heinrich. *C. F. Gauss und die Seinen*. Brunswick: E. Appelhans and Company, 1927.
Mahrenholz, Johannes. *Anekdoten aus dem Leben deutscher Mathematiker*. Leipzig: B. G. Teubner, 1936. (Gauss, pp. 22–29.)
Müller, Franz Johann. *Johann Georg von Soldner, der Geodät*. Munich: Kastner and Callwey, 1914.
Munro, J. *Heroes of the Telegraph*. New York and Chicago: Fleming H. Revell Company, 1892.
Peters, C. A. F. (ed.). *Briefwechsel zwischen C. F. Gauss und H. C. Schumacher*. Vols. I–VI. Altona: Gustav Esch, 1860–1865.
Prasad, Ganesh. *Some Great Mathematicians of the Nineteenth Century: Their Lives and Their Works*. Vol. I. Benares, India: Mahamandal Press, 1933. (Gauss, pp. 1–67.)
Riecke, Eduard. *Wilhelm Weber*. Göttingen: Dieterich, 1892.
Sartorius von Waltershausen, Wolfgang. *Gauss zum Gedächtniss*. Leipzig: S. Hirzel, 1856.
———. "C. F. Gauss." In *Göttinger Professoren*, pp. 205–229. Gotha: F. A. Perthes, 1872.
Schaefer, Clemens (ed.). *Briefwechsel zwischen Carl Friedrich Gauss und Christian Ludwig Gerling*. Berlin: Otto Elsner, 1927.
Schering, Ernst. *C. F. Gauss und die Erforschung des Erdmagnetismus*. Göttingen: Dieterich, 1887.
Schilling, C. (ed.). *Briefwechsel zwischen Olbers und Gauss*. 2 vols. Berlin: Julius Springer, 1900–1909.
Schmidt, Franz, and Stäckel, Paul (eds.). *Briefwechsel zwischen C. F. Gauss und Wolfgang Bolyai*. Leipzig: B. G. Teubner, 1899.
Schur, Wilhelm. *Beiträge zur Geschichte der Astronomie in Hannover*. Berlin: Weidmannsche Buchhandlung, 1901.
Selle, Götz von. *Die Georg-August-Universität zu Göttingen, 1737–1937*. Göttingen: Vandenhoeck and Ruprecht, 1937.
———. *Universität Göttingen, Wesen und Geschichte*. Göttingen: Musterschmidt Verlag, 1953.
Süss, Wilhelm. *Bestimmung einer geschlossenen konvexen Fläche durch die Gauss'sche Krümmung*. 1932.
Thomas, R. Hinton. *Liberalism, Nationalism, and the German Intellectuals, 1822–1847: An Analysis of the Academic and Scientific Conferences of the Period*. Cambridge, England: W. Heffer and Sons, 1951.
Trier, Betty. *Sayings of Gauss and Bessel*. Northfield, Minn.: Carleton College, 1905.

Turnbull, H. W. *The Great Mathematicians.* London: Methuen and Company, 1929. (Gauss, pp. 108–118.)
Valentiner, Wilhelm (ed.). *Briefe von C. F. Gauss an B. Nicolai.* Karlsruhe, 1877.
Voit, Max. *Bildnisse Göttinger Professoren 1737–1937.* Göttingen: Vandenhœck and Ruprecht, 1937.
Weber, Heinrich. *Wilhelm Weber,* Breslau: Eduard Trewendt, 1893.
Wellhausen, Julius. *Heinrich Ewald.* Berlin: Weidmannsche Buchhandlung, 1901.
Worbs, Erich. *Carl Friedrich Gauss: Ein Lebensbild.* Leipzig: Koehler and Amelang, 1955.

PAMPHLETS

Ambronn, Leopold. *Der Hohe Hagen und C. F. Gauss' Beziehungen zu demselben.* Göttingen: Tageblatt Press, 1911.
Aufruf zur Errichtung eines Standbildes für C. F. Gauss. Brunswick: Das Comite für Herstellung eines Gauss-Standbildes, Dec., 1876.
Dedekind, Richard. *Gauss in seiner Vorlesung über die Methode der kleinsten Quadrate.* Berlin: Weidmannsche Buchhandlung, 1901.
Faber, Georg. *Bemerkungen zu Sätzen der Gauss'schen theoria combinationis observationum.* Munich, 1922. See also *Sitzungsberichte d. bayr. Akad. d. Wiss. z. München.* Math.-Phys. Kl., 1922, pp. 7–23.
Festlieder zum Gauss-Commers. Göttingen: Hofer, Apr. 30, 1877.
Feyerabend, Ernst. *An der Wiege des elektrischen Telegraphen.* Deutsches Museum, Abhandlungen und Berichte. Berlin: VDI-Verlag, 1933.
Frischauf, Johannes. *Die Gauss-Gibbssche Methode der Bahnbestimmung eines Himmelskörpers aus drei Beobachtungen.* Leipzig: Wilhelm Engelmann, 1905.
Gieseke, G. *Der Gaussturm auf dem Hohen Hagen bei Dransfeld.* Göttingen: Tageblatt Press, 1911.
Goldscheider, Franz. *Über die Gauss'sche Osterformel.* Berlin: R. Gaertner, 1896, 1899. Parts I and II.
Kowalewski, A. *Newton, Cotes, Gauss, Jacobi. 4 grundlegende Abh. über Interpolation und genäherte Quadratur.* Leipzig, 1916.
Mack, Heinrich. *Das Gauss Museum in Braunschweig.* Brunswick: Johann Heinrich Meyer, 1929.
Mangoldt, Hans Carl Friedrich. *Bilder aus der Entwicklung der reinen und angewandten Mathematik während des 19ten Jahrhunderts mit besonderer Berücksichtigung des Einflusses von C. F. Gauss.* Aachen: LaRuelle (Jos. Deterre), 1900.
Mathé, Franz. *Carl Friedrich Gauss.* Leipzig: Weicher, 1906. (2d ed.; Feuer-Verlag, 1923.)

Naumann (ed.). *Carl Friedrich Gauss: Festschrift der Gausschule zum 100sten Todestag.* Brunswick: Richard Norek KG, 1955.
Oppermann, Albert. *Carl Friedrich Gauss.* Gausschule, fünfter Jahresbericht. Brunswick, 1914. 27 pages.
Riebesell, P. (ed.). "Briefwechsel zwischen Gauss und Repsold," *Mitteilungen der math. Ges. in Hamburg,* VI, No. 8 (Sept., 1928), 398–431.
Schering, Ernst. *C. F. Gauss' Geburtstag nach hundertjähriger Wiederkehr.* Göttingen: Dieterich, 1877.
Schlesinger, Ludwig. *Der junge Gauss.* Giessen: Alfred Löpelmann, 1927.
Smend, Rudolf. *Die Göttinger Sieben.* Göttingen: Musterschmidt Verlag, 1951.
Sommer, Hans Zincke. *Festrede zur 100-jährigen Jubelfeier des Mathematikers Gauss.* Brunswick, Apr. 30, 1877.
Stäckel, Paul, *Eine von Gauss gestellte Aufgabe des Minimums.* Halle, 1917.
Stern, Moritz A. *Denkrede auf C. F. Gauss zur Feier seines hundertjährigen Geburtstages.* Göttingen: W. Fr. Kaestner, 1877.
Voss, Aurel. *Carl Friedrich Gauss.* Darmstadt: Arnold Bergstraesser, 1877.
Wienberg, Margarete. *Gauss-Erinnerungen für die lieben Kinder.* Brunswick: Friedrich Bosse, 1928.
Wietzke, A. "C. F. Gauss' Beziehungen zu Bremen," *Abh. Nat. Ver. Bremen,* XXVII, Heft 1, pp. 125–142.
Winnecke, F. A. T. *Gauss, ein Umriss seines Lebens und Wirkens.* Brunswick: Friedrich Vieweg, 1877.
Wittstein, Theodor. *Gedächtnissrede auf Carl Friedrich Gauss.* Hanover: Hahn'sche Buchhandlung, 1877.

ARTICLES

"Abends am Gaussberge," *Braunschweigische Anzeigen,* No. 151 (June 30, 1880) p. 1,199.
Ahrens, Wilhelm. "C. F. Gauss im Spiegel der Zeitgenossen und der Nachwelt," *Grimme Natalis Monatsschrift,* II (June, 1914), 428–444.
———. "Die kleinen Planeten und Gauss' Kinder," *Das Weltall,* XXIV, Heft 11 (Aug., 1925), 205–213.
———. "Gauss' amerikanische Nachkommen," *Das Weltall,* XXIV, Heft 12 (Sept., 1925), 231.
———. "Sophie Germain und Gauss," *Das Weltall,* XXVI (1926), 8–11. *Also in Kölnische Zeitung,* Apr. 1, 1926, and *Neue Zürcher Zeitung,* No. 28 (1926), 3.
Archibald, R. C. "Gauss and the Regular Polygon of 17 Sides," *Am. Math. Monthly,* XXVII (July-Sept., 1920), 323–326.

"Aus dem Herzogthume" (Gauss-Feier), *Braunschweigische Anzeigen*, No. 100, (May 1, 1877), 1,266–1,267.
"Aus der Jugend eines mathematischen Genies," *Hamburger Nachrichten*, Apr. 28, 1927. (Morgenausgabe.)
B. ––––– W. J. "Grundsteinlegung zum Gaussdenkmal in Braunschweig," *Weser-Zeitung*, May 1, 1877.
Baltzer, R. "Zu Gauss' hundertjährigem Geburtstag," *Im neuen Reich*, I, No. 18 (1877), 681–684.
Bauer, M. "Algebraische Behauptung von Gauss," *Jahresbericht d. dt. Math.-Vereinigung*, XXVII, 348.
Bechstein, O. "Gauss," *Prometheus*, XXVI, 331.
Berger, Karl R. "Zwei für die Optik Unsterbliche," *Der Augenoptiker*, No. 4, (Apr. 25, 1952), 5–6.
Berger, R. "Zur nichteuklidischen Geometrie," *Archiv für Metallkunde*, I, Heft 5 (May, 1947), 200.
"Berliner Gauss-Festcommers." Reported in all Berlin newspapers, May 2, 1877.
Bernhardt, Alois. "Gedanken um ein Kulturdenkmal" (The Grave of Gauss), *Hannoversche Allgemeine Zeitung*, Aug. 27, 1949.
Beutel, E. "Gauss' 150jähriger Geburtstag," *Schwäbische Merkur*, Apr. 29, 1927.
Birck, O. "Die unveröffentlichte Apexberechnung von C. F. Gauss," *Vierteljahrsschrift der astron. Ges.*, LXI (1926), 46–74.
Bloch, O. "Geometrie d. Gauss'schen Zahlenebene," *Verhandlungen d. schweizer. Naturforsch. Ges.*, Vol. 98, II, Sect. 1, p. 105.
Bock, W. "Zusatz z. d. Artikel 129 d. Disq. arith. von Gauss," *Mitt. d. Math. Ges. in Hamburg*, VI, 307–309.
Bölsche, Carl. "C. F. Gauss," *Illustrirte Deutsche Monatshefte*, Feb., 1857, pp. 255–260.
Boncompagni, Baldassare (Prince), ed. *Lettera inedita di Carlo Federico Gauss a Sofia Germain*. Firenze, 1879.
–––––––. "Sur les lettres de Sophie Germain à Gauss," *Nouvelle Corresp. Math.* (Brussels), Tome VI (Sept., 1880).
Boring, E. G. "Use of the Gaussian Law," *Science*, N. S. LII (Aug. 6, 1920), 129–130.
Born, M. "Gauss," *Hannoversch. Kurier*, Apr. 29, 1927.
Bothe, W. "Gültigkeitsgrenzen d. Gauss'schen Fehlergesetzes für unabhängige Elementarfehlerquellen," *Zeitschrift für Physik*, IV (1921), **161-177.**
Brendel, Martin. "Aufforderung," *Astr. Nachr.* Vol. 146, No. 3,504 (24), (June 18, 1898), 439. (Concerns Gauss' *Collected Works*.)
–––––––. "Das Gauss-Archiv," *Jahresbericht der Deutschen Mathematiker-Vereinigung*, XII, Heft 1, pp. 61–63. Leipzig: B. G. Teubner, 1903.

Cajori, Florian. "Carl Friedrich Gauss and His Children," *Science*, N. S. IX, No. 229 (May 19, 1899), 697–704.

———. "Notes on Gauss and His American Descendants," *Pop. Sci. Monthly*, LXXXI (Aug. 1912), 105–114.

Cantor, Moritz B. "Carl Friedrich Gauss," *Allgemeine Deutsche Biographie*, VIII (1878), 430–445.

———. "Carl Friedrich Gauss," *Neue Heidelberger Jahrbücher*, 1899, pp. 234–255.

Cayley, Arthur. "The Gaussian Theory of Surfaces," *Proc. London Math. Soc.*, XII, 187.

"C. F. Gauss," *Kölnische Zeitung*, Apr. 30, 1877.

———, *Leipziger Illustr. Zeitung*, Apr. 21, 1877.

———, *Hannover. Courier*, May 2, 1877. (Abendausgabe.)

"C. F. Gauss zu seinem 80. Todestage am 23. Februar," *Braunschw. Landeszeitung*, Feb. 23, 1935.

"C. F. Gauss, der grösste Mathematiker aller Zeiten, zu seinem 80. Todestag am 23. Februar," *Göttinger Zeitung*, Feb. 23, 1935.

"C. F. Gauss beim Schulfest," *Braunschweiger Neueste Nachrichten*, March 24, 1935.

"C. F. Gauss," *Nachrichten-Ztg. d. Ver. dtsch. Ingenieure*, VII (1927), No. 17.

"C. F. Gauss und Oldenburg," *Nachrichten für Stadt und Land Oldenburg*, Apr. 29, 1927.

"Charles Frederic Gauss," *Proc. Royal Soc.* VII, No. 17 (Nov. 30, 1855), pp. 589–598. London: Taylor and Francis, 1855. Obituary notice.

"Correspondenz aus Frankfurt a. M. über die Buonarroti- und Gaussfeier," *Augsburger Allgemeine Zeitung*, May 7, 1877, p. 1,946.

Danzer, O. *Über die Sätze von Gauss und Pohlke*. Weimar, 1918. See also *Sitzungsberichte d. Akad. d. Wiss. Vienna. Math.-naturw. Kl.*, II Abt., Vol. 127, IIa, p. 1,701.

Darmstaedter, Ludwig. "Carl Friedrich Gauss," *Die Umschau*, XXXI, Heft 18 (Apr. 30, 1927), 355–358.

"Der 'Pythagoras' als Signal zum Mars, eine phantastische Idee des berühmten Göttinger Mathematikers Gauss," *Göttinger Tageblatt*, May 17, 1935.

"Die Einweihung des Gausszimmers, Wilhelmstr. 30," *Braunschw. Neueste Nachrichten*, XV, No. 102 (May 2, 1911).

"Die Enthüllung des Gauss-Denkmals" *Braunschweigische Anzeigen*, No. 150 (June 29, 1880), 1,191–1,192.

"Die Feier der Universität Göttingen zum hundertjährigen Jubiläum der elektrischen Telegraphie," *Elektrische Nachrichten-Technik*, XI (1934), Heft 1, pp. 33–35.

"Die Gaussfeier in Braunschweig und die Göttinger Gaussfeier," *Augsburger Allgemeine Zeitung*, May 9, 1877. (Beilage.)

"Die Gaussgedenkfeier im Schloss," *Braunschweiger Neueste Nachrichten*, No. 101 (May 1, 1927).

Döbritzsch, H. "Zur Ableitung des Fehlergesetzes von Gauss auf Grund der Hayenschen Hypothese d. Elementarfehler," *Astron. Nachr.*, Vol. 236, p. 276.

Dohse, Fritz E. "Die Nachkommen von Gauss in den USA," *Braunschweigische Zeitung*, Jan. 20, 1951. Also appeared in *Göttinger Tageblatt*.

Domke, O. "Gauss'sches Auflösungsverfahren," *Bautechnik*, 1927. (Beilage 227.)

Dove, Alfred. "Bruhns, Humboldt und Gauss," *Im neuen Reich*. I, No. 20 (1877), 770–779.

Dunnington, G. Waldo. "The Sesquicentennial of the Birth of Gauss," *Scientific Monthly*, Vol. 24 (May, 1927), 402–414.

———. "The Gauss Archive and the Complete Edition of His Collected Works," *Math. News Letter*, Vol. 8 (Feb., 1934), 103–107.

———. "The Historical Significance of C. F. Gauss in Mathematics and Some Aspects of His Work," *Math. News Letter*, Vol. 8 (May, 1934), 175–179.

———. "Gauss' Disquisitiones arithmeticae and His Contemporaries in the Institut de France," *Nat. Math. Mag.*, Vol. 9 (Apr., 1935), 187–192.

———. "Gauss' Disquisitiones arithmeticae and the French Academy of Sciences," *Scripta math.*, Vol. 3, No. 2 (Apr., 1935), 193–196.

———. "Gauss' Disquisitiones arithmeticae and the Russian Academy of Sciences," *Scripta math.*, Vol. 3 (Oct., 1935), 356–358.

———. "Jean Paul und Carl Friedrich Gauss," *Monatshefte für Deutschen Unterricht*, Nov., 1935, pp. 268–272.

———. "Carl Friedrich Gauss, zu seinem 81-ten Todestag," *Göttinger Tageblatt*, Feb. 21, 1936.

———. "A Gauss Statuette," *Scripta math.*, Apr., 1936, pp. 199, 195.

———. Über Gauss und Jean Paul," *Jean Paul Blätter*, Vol. 13, No. 1 (Apr., 1938), 2–3.

———. "Note on Gauss' Triangulations," *Scripta mathematica*, XX, No. 1 (March, 1954), 108–109.

———. "Biographische Streiflichter zum 100–jährigen Todestag von Gauss," *Göttinger Tageblatt*, Feb. 19, 1955.

———. "Der Geist einer vollendeten Wissenschaft" (Sartorius' funeral oration for Gauss), *Göttinger Tageblatt*, Feb. 28, 1955.

———. "Das Problem der Unsterblichkeit bei Gauss und Jean Paul," *Hesperus*, March, 1955.

Ebeling, (?), "C. F. Gauss," *Praktische Schulphysik*, VII (1927), 121–126.
"Ein Füllhorn von Ehrungen," *Braunschw. Nachrichten*, Feb. 21, 1955.
"Ein Gauss-Weber Jubiläum," *Göttinger Zeitung*, Apr. 15, 1933.
"Ein König im Reiche der Zahlen," *Gartenlaube*, No. 17 (1877), 278.
Elsner, H. "Gaussens Reise nach München und Benediktbeuern 1816," *Centralztg. f. Optik u. Mechanik*, XLVIII (1928), 18–20.
E. M. "Ehrungen im Geiste von Gauss," *Braunschw. Zeitung*, Feb. 19, 1955.
———. "Wissenschaft trägt das Erbe von Gauss weiter," *Braunschw. Zeitung*, Feb. 21, 1955.
Endtricht, Hugo. "Die Erfindung des Telegraphen," *Göttinger Tageblatt*, Apr. 30, 1927, pp. 3–4.
———. "100 Jahre Telegraphie, ein Rückblick," *Göttinger Tageblatt*, May 1, 1933.
"Excellenz Stephan beglückwünscht Wilhelm Weber," *Göttinger Zeitung*, Dec. 9, 1933.
Faust, Heinrich. "Fürst der Mathematiker," *Frankfurter Allgemeine Zeitung*, Feb. 22, 1955.
Fladt, K. "C. F. Gauss," *Aus Unterricht und Forschung*, III, 11–16, 60–65.
Foerster, G. "Das Gauss'sche Fehlergesetz," *Astron. Nachr.*, Vol. 208, p. 379.
Fricke, R. "C. F. Gauss," *Braunschweigische Staatszeitung*, Vol. 183, No. 100 (Apr. 30, 1927).
"50jähriges Jubiläum des ersten Telegraphen im Jahre 1883," *Göttinger Zeitung*, Dec. 2, 1933.
Gaede, W. "Beiträge zur Kenntnis von Gauss' praktisch-geodätischen Arbeiten," *Zeitschrift für Vermessungswesen*, XIV (1885).
Galle, Andreas. "Gauss und Kant," *Das Weltall*, XXIV, Heft 10 (July, 1925), 194–200.
———. "Eine Äusserung von Gauss über Kant," *Das Weltall*, XXIV, 230.
"Gauss als Naturforscher," *Augsburger Allgemeine Zeitung*, LV (1877).
"Gauss Descendants Cherish Award of George V," *The Missourian Magazine*, XXII, No. 59 (Nov. 9, 1929), 1, 3.
"Gauss-Gedenkraum wird morgen eingeweiht," *Braunschw. Nachrichten*, Feb. 19, 1955.
"Gauss und sein Werk," *Vermessungstechnische Rundschau*, VI (1929), 177–182.
"Geburt des Telegraphen," *Volksblatt der Vossischen Zeitung*, No. 113 (Apr. 24, 1933). See also *Zeitschrift für phys. chem. Unterricht*, XI (1927), 265.

Genocchi, Angelo. "Il carteggio di Sofia Germain e Carlo Federico Gauss," *Comptes rendus* (Academy of Turin), June 20, 1880.

Gieseke, G. "Die Grundsteinlegung zum Gaussturm auf dem Hohenhagen," *Monatsblatt für die Gemeinde Dransfeld*, Sept., 1909.

———. "Die Einweihung des Gaussturmes," *Monatsblatt für die Gemeinde Dransfeld*, Sept., 1911.

Gieseke, (?) Geheimrat. "Carl Friedrich Gauss," *Niedersachsen*, XV, No. 21 (Aug. 1, 1910), 374–377.

Ginzel, F. K. "Zur Erinnerung an C. F. Gauss," *Illustrirte Zeitung* (Leipzig), Apr. 21, 1877. Also in *Göttinger Zeitung*, Apr. 30, 1877.

Govi, Gilberto. "Gauss and Sophie Germain," *Comptes rendus* (Royal Academy of Naples), June, 1880.

Günther, P. "Die Untersuchungen von Gauss in der Theorie der elliptischen Functionen," *Nachrichten von der kgl. Ges. d. Wiss.* (Göttingen), No. 2 (1894), 92–105.

Hahne, F. "Die Gauss-Sammlung im Vaterländischen Museum zu Braunschweig," *Braunschweiger Genealogische Blätter*, Nos. 3–5 (Jan., 1927), 50–54.

Halstead, G. B. "Gauss and Lobachevsky," *Science* IX, No. 232, pp. 813–817.

Hammer, E. "Gesch. d. Abbildungslehre: ein Gauss'scher Satz," *Kartogr. Zeitschrift* (Vienna), VII, 86.

———. "Zwei Sätze von Gauss," *Zs. f. Vermessungswesen*, LIII, 126.

Hänselmann, Ludwig. "Eine Erinnerung an C. F. Gauss," *Braunschweigische Anzeigen*, No. 278 (Nov. 28, 1876), 3,367–3,368.

Hecke, E. "Reciprocitätsgesetz u. Gauss'sche Summe in quadrat. Zahlkörpern," *Nachr. d. Ges. d. Wiss. z. Gött.*, Math.-phys. Kl., 1918, pp. 265–278.

Heckmann, O. "Unveröffentl. Apex-Berechnungen von Gauss," *Die Himmelswelt*, XXXVI, 124.

Heffter, Lothar. "C. F. Gauss zum 100sten Todestag," *Neue deutsche Hefte*, Feb., 1955.

Herglotz, Gustav. "Letzte Eintragung im Gauss'schen Tagebuch," *Bericht über d. Verhandl. d. sächs. Ges. d. Wiss. zu Leipzig*, Math.-Phys. Kl., LXXIII, 271–277.

Hesemann, Friedrich. "Ein Urteil Alex. v. Humboldt's über Gauss," *Göttinger Leben*, II, No. 19, (June 15, 1927).

———. "Wie sah Gauss in Wirklichkeit aus?" *Göttinger Zeitung*, Die Spinnstube, 1927, p. 146.

Hofmann, H. W. "Carl Friedrich Gauss," *Archiv für Metallkunde*, I, Heft 5 (May, 1947), 198–199.

Holden, Edward S. "Phenomenal Memories," *Harper's Monthly Magazine*, Nov., 1901, p. 907.

Hundertmark, Heinz. "Akademie und Universität ehrten das Gedächtnis von C. F. Gauss," *Göttinger Tageblatt,* Feb. 21, 1955.

Irmisch, Linus. "An C. F. Gauss," *Braunschweigische Anzeigen,* No. 149 (June 27, 1880), 1,181–1,182. A poem.

Jaeger, Wilhelm. "Gauss und das absolute Massystem," *Forschungen und Fortschritte,* IX, No. 10 (Apr. 1, 1933), 143–144.

Jaernevelt, Gustaf. "Zur synthetischen Axiomatik der Gaussischen Geometrie auf einer regulären Fläche," *Suomalaisen Tiedeakatemian Toimituksia* (Helsinki), Sarja A, 34, 2 (1931).

Jelitai, József. "Briefe von Gauss und Encke im Ungarischen Landesarchiv," *Magyar Tudományos Akadémia Matematikai és Természettudományi Ertesitöje* (Budapest), LVII (1938), 136–144.

Jordan, Robert. "Die verlorene Ceres," *Neueste Nachrichten* (Brunswick), May 1, 1927. A *Gauss-Novelle* on his engagement to Johanna Osthoff.

Jungk, Robert. "Ein totes Gehirn—das weiter wirkt," *Süddeutsche Zeitung,* No. 45, Feb. 23, 1955.

Kasten, Willi. "Zum 100sten Todestag von C. F. Gauss," *14 Tage Göttingen,* No. 4, 1955; pp. 6–11.

Kirchberger, Paul. "C. F. Gauss," *Hamburger Nachrichten,* Apr. 14, 1927. (Abendausgabe.)

Klein, Felix. "Gauss' wissenschaftliches Tagebuch 1796-1814," *Math. Annalen,* LVII, No. 1.

Klein, F., and Born, M. "Bericht über den Stand der Herausgabe von Gauss' Werken," 1–25 *Nachrichten von der kgl. Ges. d. Wiss.* Göttingen, 1898–1934.

Klein, F., and Schwarzschild, K., "Ueber das in der Festschrift des hundertfünfzigjährigen Bestehens der kgl. Gesellschaft der Wissenschaften mit dem Gaussischen Tagebuch reproducierte Porträt des '26-jährigen Gauss,'" *Nachrichten von der kgl. Ges. d. Wiss.* Göttingen, 1903. Geschäftliche Mittheilungen, Heft 2.

Kobold, Hermann. "Ein Brief von Gauss an Argelander," *Astr. Nachr.,* Vol. 183, No. 4,380 (Dec. 20, 1909).

Köhler, W., and Strauch, H., "Im Geiste von Gauss," *Archiv für Metallkunde,* I, Heft 5 (May, 1947), 197.

"König der Mathematiker," *Uranus-Kalender,* 1931, p. 85.

Krause, A. "Gauss'sche Formel z. Berechnung d. Ostersonntags," *Prometheus,* XXXI, 236.

Krylov, A. N. "On Gauss' Memoir 'Intensitas vis magneticae,'" *Archives of the History of Science and Technology* (Leningrad), III (1934), 183–192. In Russian. One of a series of papers devoted to Gauss, read at the Dec. 28, 1932, meeting of the Inst. of the Hist. of Sci. and Technology.

Kueser, A. "Gauss'sche Theorie d. übertragen a. d. Mayer'schen Problem d. Variationsrechnung," *Journal f. reine u. angew. Mathematik,* Vol. 146, pp. 116–127.

Küstermann, (?) "Fürsten der Mathematik: Gauss, Volta, Laplace," *Gartenlaube,* 1927, p. 176.

Lammer, A. "C. F. Gauss in Bremen," *Nordwest.,* No. 33 (1878), 270–278.

Landau, E. "Vorzeichen d. Gauss'schen Summe," *Nachr. d. Ges. d. Wiss. zu Göttingen,* Math.-Phys. Kl., 1928, p. 19.

Langner, Hans Joachim. "Wegweiser zum verlorenen Stern," *Braunschw. Zeitung,* Feb. 19, 1955.

Larmor, Joseph, "C. F. Gauss and His Family Relatives," *Science,* Vol. 93, No. 2,422 (May 30, 1941), 523–524.

"La tour de Gauss," *Le Camp de Göttingen (Göttinger Zeitung),* No. 5 (March 15, 1915).

Liebmann, H. "Hilberts Beweise d. Sätze über Flächen festen Gauss'schen Krümmungsmasses," *Math. Zs.,* XXII, 26–33.

Loewy, Alfred. "Ein Ansatz von Gauss zur jüdischen Chronologie aus seinem Nachlass," *Jahresbericht der Deutschen Math. Vereinigung,* XXVI, Heft 9–12 (1917), 304–322.

———. "Eine algebraische Behauptung von Gauss," *Jahresbericht d. dt. Math.-Vereinigung,* XXVI, 101–109. See also XXX, 147–153.

Lorey, Wilhelm. "Friedrich Ludwig Wachter," *Sachsen-Altenburgischer vaterländischer Geschichts- und Hauskalender,"* 1934.

———. "Über eine von Gauss erwähnte Mischungsaufgabe und deren Lösung aus einer Laplaceschen Differenzgleichung mit ausgiebiger Verwendung einer Rechenmaschine," *Festschrift zu Ehren von Georg Höckner,* pp. 166–174. Berlin: E. S. Mittler und Sohn, 1935.

Ludendorff, H. "Ein Gauss betreffender Brief von Olbers," *Vierteljahrsschrift der astron. Ges.,* LXII, Heft 1 (1927).

Lüring, Bruno. "Der Fürst der Mathematik," *Die Nation,* No. 8, Feb. 26, 1955.

Lüroth, Jacob. "Zur Erinnerung an C. F. Gauss," *Zeitschrift für Vermessungswesen,* VI, Heft 4 (1877), 201–210. First published in the supplement to the *Allgemeine Zeitung,* No. 55 (1877).

MacDonald, T. L. "The Anagram of Gauss," *Astron. Nachr.,* Vol. 241, p. 31.

Mack, Heinrich. "Die Verleihung des Ehrenbürgerrechts der Stadt Braunschweig an C. F. Gauss," *Braunschw. Neueste Nachrichten,* XXIX, No. 101 (May 1, 1925), 2–3.

———. "C. F. Gauss und Braunschweig," *Braunschweigisches Addressbuch für 1927.*

———. "Vorfahren und Nachkommen von C. F. Gauss," *Braunschweigische Landeszeitung,* No. 118 (Apr. 30, 1927), 9.

———. "C. F. Gauss," *Braunschw. Neueste Nachrichten*, May 1, 1927. (Blatt 2.)

———. "Ein Sohn von C. F. Gauss als Helfer bei der Bekämpfung des Hamburger Brandes im Jahre 1842," *Braunschw. Neueste Nachr.*, May 14, 1927.

———. "Das wieder zu Tage gekommene Jugendbild von C. F. Gauss," *Braunschw. Neueste Nachr.*, July 23, 1929, p. 16.

———. "Das Gehirn von C. F. Gauss," *Braunschw. Neueste Nachr.*, Oct. 5, 1930, p. 10.

———. "Die Nachkommen von Carl Friedrich Gauss," *Göttinger Blätter für Geschichte und Heimatkunde Südhannovers*, I, N. F. Heft 3 (1935), 19–28.

Maennchen, Ph. "Interpolationsversuch d. jugendl. Gauss," *Jahresbericht d. dt. Math.-Vereinigung*, XXVIII, 80–84.

———. "Lösung d. rätselhaften Gauss'schen Anagramms," *Unterrichtsblätter f. Mathematik und Naturwissenschaften*, XL (1934), 104.

Mansion, Paul. "Gauss contre Kant sur la géometrie non Euclidienne," *Verhandlungen des III. Internationalen Kongresses für Philosophie*. Heidelberg: Carl Winter, 1908.

Meder, Alfred. "Direkte und indirekte Beziehungen zwischen Gauss und der Dorpater Universität." *Archiv für Gesch. d. Math., d. Naturwiss. und d. Technik*, XI, Heft 1-2, pp. 62–67.

Meitner-Heckert, Karl. "Vom Wunderkind zum Zahlenkönig," *Grimme Natalis Monatsschrift*, Heft 3-4 (March-Apr., 1927), 92–95.

Metzner, Karl. "Zum 150. Geburtstage von Gauss," *Zeitschrift für den physikalischen und chemischen Unterricht*, Vol. 40, Heft 4 (July-Aug., 1927), 180–188.

Michel, H. "C. F. Gauss' Beurteilung der Erfindung des Telegraphen," *Grimme Natalis Monatsschrift*, Heft 5-6 (May-June, 1926), 191.

Michelmann, Emil. "Gauss und Weber," *Göttinger Zeitung*, Apr. 29, 1933.

"Michelmann kommt!" *Göttinger Tageblatt*, May 1, 1933. The Gauss-Weber telegraph.

Moderhack, Richard. "Das ehemalige Gaussmuseum in Braunschweig," *Braunschweiger Kalender*. Joh. Heinr. Meyer, 1948.

Müller, Otto. "Gauss, princeps mathematicorum," *Württembergische Schulwarte*, III, 343–347.

Nagy, H. v. Sz. "Topologisches Problem bei Gauss," *Math. Zs.*, XXVI, 579-592.

Nickel, K. "Herleitung d. Abbildungsgleichung d. Gauss'schen konformen Abbildungen d. Erdellipsoids in d. Ebene," *Zs. f. Vermessungswesen*, LV, 493–496.

"Obituary Notice of C. F. Gauss," *Monthly Notices of the Royal Astron. Soc.*, XVI, No. 4 (1856), 80–83.

Oppermann, E. "C. F. Gauss," *Preussische Lehrer-Zeitung*, No. 52 (1927).

"Ostern 1833," *Göttinger Zeitung*, Apr. 15, 1933. The Gauss-Weber telegraph.

Ostrowski, A. "Zum letzten u. 4. Gauss'schen Beweise d. Fundamentalsatzes d. Algebra," *Nachr. d. Ges. Wiss. z. Gött.*, Math.-Phys. Kl., 1920. Beiheft, pp. 50–58.

"Pilgerfahrt über den Atlantik," *Göttinger Tageblatt*, No. 189, Aug. 17, 1950. H. W. Gauss in Göttingen.

Pl. "Sein Geist drang in die tiefsten Erkenntnisse," *Braunschw. Zeitung*, Feb. 21, 1955.

Pohl, Robert W. "Die Gauss-Weber Feier der Universität," *Göttinger Nachrichten*, I, No. 213 (Nov. 23, 1933).

———. "Jahrhundertfeier des elektromagnetischen Telegraphen von Gauss und Weber," *Kgl. Ges. d. Wiss. zu Göttingen*, Jahresbericht über das Geschäftsjahr 1933-1934, pp. 48–56.

Polya, G. "Anschaul.-exper. Herleitung d. Gauss'schen Fehlerkurve," *Zeitschrift f. d. Math. r. naturw. Unterricht*, LII, 57–65.

———. "Das Gauss'sche Fehlergesetz," *Astron. Nachr.*, Vol. 208, pp. 185–191, and Vol. 209, p. 111.

———. "Herleitung d. Gauss'schen Fehlergesetzes aus einer Funktionalgleichung," *Math. Zeitschrift*, XVIII, 96–108.

Rautmann, H. "Bedeutung d. Gauss'schen Verteilungsfunktion f. d. klin. Variationsforschung," *Zs. f. Konstitutionslehre*, XIII, 450–476.

Ricci, G. "C. F. Gauss," *Jahresbericht der Deutschen Math.-Vereinigung*, XI (1902), 397.

"Richtungsreduktionen bei d. Gauss'schen konform. Abbildung," *Zs. f. Vermessungswesen*, 1933, p. 624.

Riese, (?) von. "C. F. Gauss." *Kölnische Zeitung*, No. 80 (March 21, 1855). (Beilage.)

Sack, C. W. "Carl Friedrich Gauss, einige Züge aus seinem Leben," *Braunschweigisches Magazin*, 1857. Seven installments.

Sack, Friedrich. "C. F. Gauss und die Lüneburger Heide," *Niedersachsen*, XXXVIII (July, 1933), 346–351.

Salomon, Ludwig. "C. F. Gauss," *Hannover. Courier*, Apr. 29, 1877.

Schaefer, Clemens. "Gauss' Investigations on Electrodynamics," *Nature*, Aug. 29, 1931.

———. "Ein Briefwechsel Zwischen Gauss, Fraunhofer und Pastorff," *Nachrichten von der Ges. d. Wiss. zu Göttingen*, 1934, pp. 57–75.

Schering, Ernst. "Gauss und Sophie Germain," *Nachrichten der kgl. Ges. der Wiss. zu Göttingen*, June 23, 1880, pp. 367–369.

Schimank, Hans. "Carl Friedrich Gauss," *Abhandl. d. Braunschw. Wiss. Ges.*, II (1950), 123–140.

Schlesinger, Ludwig. "Über Gauss Jugendarbeiten zum arithmetisch-geometrischen Mittel," *Jahresbericht d. Deutschen Math.-Vereinigung*, XX, Heft 11-12 (Nov.-Dec., 1911), 396–403.

———. "Eine wissenschaftliche Biographie von C. F. Gauss," *Literaturblatt der Frankfurter Zeitung*, No. 109 (June 16, 1912). (Erstes Morgenblatt.)

Schmidt, Adolf. "Gauss als Physiker, insbesondere als Erdmagnetiker," *Bericht über die Tätigkeit des preussischen meteorologischen Instituts*, 1927, pp. 41–46.

Schneider, Heinrich. "Neue Gaussiana aus der Landesbibliothek zu Wolfenbüttel," *Grimme Natalis Monatsschrift*, Sept., 1923, pp. 489–494.

———. "Weiteres zu C. F. Gauss' wissenschaftlichem Briefwechsel," *Grimme Natalis Monatsschrift*, Feb., 1924, pp. 57–62.

———. "Carl Friedrich Gauss und die Göttinger Sieben," *Grimme Natalis Monatsschrift*, Heft 1-2 (Jan.-Feb., 1926), 7–15.

———. "Ein unveröffentlichter Brief von C. F. Gauss," *Braunschweigisches Magazin*, No. 6 (Nov.-Dec., 1926), 92–94.

———. "C. F. Gauss zum Gedächtniss." *Braunschweigisches Magazin*, No. 3 (1927), 43.

Schneidewin, Max. "Eine persönliche Erinnerung an C. F. Gauss," *Deister- und Weserzeitung* (Hameln), May 17, 1927.

Stäckel, Paul. "Ein Brief von Gauss an Gerling," *Nachrichten von der kgl. Ges. d. Wiss.*, Heft 1. Göttingen, 1896.

———. "Vier neue Briefe von Gauss," *Nachrichten d. kgl. Ges. d. Wiss.* Göttingen, 1907.

Stäckel, Paul, and Engel, Friedrich. "Gauss, die beiden Bolyai und die nichteuklidische Geometrie," *Math. Annalen*, Vol. 49 (1898), 149–206.

Stender, Emil. "Eine Orientierungstafel auf dem Wilseder Berge," *Hamburger Nachrichten*, July 18, 1925. (Erste Beilage.)

Sterne, Carus and Cantor, Moritz. "C. F. Gauss," *Augsburger Allgemeine Zeitung*, 1877.

Sticker, Bernhard. "Vor 100 Jahren starb C. F. Gauss," *Bayreuther Tageblatt*, Feb. 25, 1955.

Stieda, Wilhelm. "Die Berufung von Gauss an die Kaiserliche Akademie in St. Petersburg," *Jahrbücher für Kultur und Geschichte der Slaven*, N. F. Vol. III, Heft 1 (1927), 79–103.

———. "Carl Friedrich Gauss und die Seinen," *Grimme Natalis Monatsschrift*, Heft 11-12 (Nov.-Dec., 1928), 487–492.

"Stiftungsurkunde über das Gaussdenkmal," *Braunschweigische Anzeigen*, No. 151 (June 30, 1880), 1,198.

Stökl, H. "C. F. Gauss," *Dtsche. Internierten-Ztg.*, Heft 11-12, pp. 2–7.

Svyatski, D. O. "Letters of C. F. Gauss to the Academy of Sciences in St. Petersburg, 1801-1807 (N. I. Fuss)," *Archives of the History of Science and Technology*, III, 209–238. Leningrad, 1934. In German, with Russian translations and notes by D. O. Svyatski.

Tait, P. G. "Gauss and Quaternions," *Proc. Royal Soc. of Edinburgh*, Dec. 18, 1899.

Teege, H. "Algebraisch. Beweis f. d. Vorzeichen d. Gauss'schen Summen," *Mitt. d. Math. Ges. in Hamburg*, V, Heft 8, pp. 283–289; and VI, 281–288.

"Theoria motus," *Nürnberger Literaturblatt*, No. 17–18 (1821). Review.

Tietze, H. "Gauss-Green-Stokes'sche Integralsätze," *Journal für reine u. angew. Math.*, Vol. 153, pp. 141–158.

Timerding, H. E. "Kant und Gauss," *Kant-Studien*, XXVIII, Heft 1-2 (1923), 16–40.

Tucker, Robert, "Carl Friedrich Gauss," *Nature*, XV, No. 390 (Apr. 19, 1877), 533–537.

———. "C. F. Gauss," *Nature*, XXI (1880), 467.

Veen, S. C. van "C. F. Gauss," *Mathematica, Tijdschrift voor Studeerenden*, III (1934), 1–14.

"Verleihung der Gauss-Medaille," *Deister- und Weserzeitung*, Jan. 27, 1950.

Video. "C. F. Gauss—menschlich gesehen," *Braunschw. Nachrichten*, Feb. 20, 1955.

"Vor 115 Jahren: 'Fernsehen' vom Brocken zum Inselsberg; dem Fernsehpionier Gauss zum Gedächtnis," *Göttinger Tageblatt*, July 18, 1935.

Wagner, K. W. "Biografia de C. F. Gauss," *Boletin Mat. Buenos Aires*, III, 25. Correction to it, pp. 47–48.

Wagner, Rudolf. "Zur Erinnerung an C. F. Gauss," *Hannoversche Zeitung*, No. 122 (March 13, 1855), IV, 44. Also *Zugaben*, same newspaper, Feb. 28, and March 2 and 4, 1855.

———. "Vor Gauss' Sarge am 26. Feb. 1855," *Braunschweiger Neueste Nachrichten*, XXXV, No. 45 (Feb. 22, 1931). A poem.

Wattenberg, Hermann. "Der Gaussturm auf dem Hohenhagen bei Dransfeld," *Niedersachsen*, XVI, No. 21 (Aug. 1, 1911), 439–442.

"Wie Gauss entdeckt wurde," *Göttinger Tageblatt*, May 22, 1935.

Wietzke, A. "Zur Lösung eines rätselhaften Gauss'schen Anagramms," *Astron. Nachr.*, Vol. 240, p. 403.

———. "C. F. Gauss in Bremen und Lilienthal." *Weser-Zeitung*, Vol. 84, No. 231 A (Apr. 29, 1927), 1.

———. "C. F. Gauss und Bremen," *Bremer Nachrichten*, Vol. 185, No. 119 (May 1, 1927), 1.

———. "C. F. Gauss, zu seinem 75. Todestag," *Göttinger Tageblatt*, No. 44 (Feb. 21, 1930).

———. "Das Jugendbildnis von Gauss wieder aufgefunden!" *Die Umschau*, XXXIV, Heft 8 (Feb. 22, 1930), 154.

———. "Das wiederaufgefundene Jugendbild von C. F. Gauss," *Jahresbericht d. Mathem.-Vereinigung*, XLI, 2. Abt. Heft 1–4 (1931).

Williams, Henry Smith, "Astronomical Progress of the Century," *Harper's Monthly Magazine*, March, 1897, pp. 540–548.

Wincierz, Kurt. "Gauss' unsterbliches Werk," *Braunschw. Neueste Nachrichten*, May 1, 1927.

Winkler, Bruno. "Der Schritt des Mädchens," *Neue Badische Landeszeitung* (Mannheim), Apr. 28, 1927. Romantic version of the heliotrope.

Witt, G. "Nomogramme der Gauss'schen Gleichungen," *Astron. Nachr.*, Vol. 199, p. 257.

Wittke, Paul. "C. F. Gauss," *Göttinger Tageblatt*, Apr. 30, 1927, p. 3.

Wülfing, E. A. "Gauss'sche Hauptebenen," *Neues Jb. f. Mineral., Geol. r. Palaeontologie*, N. F. I. XLVIII, Beilage B, pp. 310–324.

Youmans, W. J. "Sketch of Carl Friedrich Gauss," *Pop. Sci. Monthly*, Sept., 1898, pp. 694–698.

Zacharias, Max. "C. F. Gauss," *Technik und Kultur*, XIII (1922), 159.

Zeitschrift für Vermessungswesen, Gauss-Heft, Feb., 1955.

Zimmermann, Paul. "Gauss' Zulassungsgesuch zur Promotion," *Braunschweigisches Magazin*, No. 15 (July 16, 1899), 113–117.

———. "Zum Gedächtniss an C. F. Gauss," *Braunschweigisches Magazin*, No. 16 (July 30, 1899). Supplement to *Br. Anz.*, No. 209, pp. 124–127.

———. "C. F. Gauss' Briefe an seine Tochter Minna und deren Gatten H. A. Ewald," *Braunschweigisches Magazin*, Dec., 1915 (Wolfenbüttel), pp. 133–141.

———. "Neue kleine Beiträge zu C. F. Gauss' Leben und Wirken," *Grimme Natalis Monatsschrift*, Heft XI (1921). 12 pages.

"Zur Erinnerung an C. F. Gauss," *Göttinger Zeitung*, No. 16 (Apr. 15, 1877), 121–124.

"Zwei Briefe von Gauss," *Im neuen Reich*, Jan., 1877.

An Introduction to Gauss's Mathematical Diary

On March 1796, when still 18, Gauss opened a mathematical diary that he maintained as a private record of his discoveries for the next several years. It begins with his celebrated discovery that the regular 17-gon can be constructed by ruler and compass alone. Forty-nine entries were made in that year, most of them to do with number theory and algebra, as were the bulk of the thirty-three entries of 1797, although the lemniscatic integral (to be defined below) also makes its appearance. Gauss's interests then shift away from number theory to probability only to return in 1800 to the elaboration of a general theory of elliptic integrals. By the end of 1801 the demands of astronomy take over, and the final entries lack the enthusiasm and profusion of the first, but by 1805, when the entries resume, Gauss was 28, an established mathematician and astronomer, and, by October, a married man. He no longer needed the diary.

In any case, his use for the diary seems to have been psychological reassurance. Discoveries may have been recorded to establish independence, even priority, or simply to capture the flood of ideas sweeping over the young man's mind. The bulk of the entries reveal a marvellous eye for the significant in mathematics, but (as Klein noted in his introductory essay) some are hopelessly vague. Few record how a problem was solved, and the lack of even a clue to the solution is sometimes so dramatic that on one occasion Gauss himself was unable to reconstruct the discovery later (see [141]). Gauss was plainly excited to see that he could compare himself favourably with mathematicians of earlier generations, notably Euler and Legendre (Schlesinger's essay on him makes this very clear, see Gauss, *Werke* X.2). But above all the diary is testimony to the rapidity with which ideas came to Gauss at the start of his career, so that they seem almost to crowd one another out (which was Klein's explanation of why several of Gauss's best ideas were never published at all).

Number theory

Gauss's discovery of the constructibility of the regular 17-gon derives from an already considerable grasp of the algebraic nature of the problem. He thought of the 17 vertices as the 17 roots of $z^{17} - 1 = 0$, of which one is trivially $z = 1$, and the other 16 satisfy $z^{16} + z^{15} + \cdots + z + 1 = 0$. Gauss's crucial observation, which he described in detail in the *Disquisitiones* §354, is that since 17 is a prime and $16 = 2^4$ the calculation of the roots reduces to the successive solution of 4 quadratic equations. Solving quadratics can be done geometrically by ruler and compass, so the 17-gon is constructible geometrically, and Gauss more or less showed how it can be done by writing down the 4 quadratic equations explicitly. He also gave the complex coordinates of each point to ten decimal places, showing that he did regard the plane of complex numbers geometrically in advance of Argand and Wessel. But what is more remarkable is that Gauss was in sight of a far-reaching theory of equations of the form $z^n - 1 = 0$. We can see some of the steps he took in the direction of the theory of cyclotomy, as it is called, in later entries in the diary, and it is not surprising that on making this discovery Gauss decided to study mathematics and not philology.

The second entry in the diary is equally dramatic for it concerns quadratic reciprocity, a topic somewhat slighted by Dunnington. One says that an integer a is a (quadratic) residue mod p if the congruence $x^2 \equiv a \pmod{p}$ has a solution (where p is prime). The quadratic reciprocity theorem asserts that, given two odd primes p and q then p is a residue mod q if and only if q is a residue mod p unless both are of the form $4n + 3$, in which case p is a residue mod q if and only if q is not a residue mod p. There are auxiliary statements concerning the prime 2 and the number -1: 2 is a residue mod p if and only if p is of the form $8k \pm 1$, and -1 is a residue mod p if and only if p is of the form $4k - 1$. Because a product ab is a residue mod p if and only if the factors are either both residues or both not residues, it is easy to determine whether any given number is a residue modulo a prime.

Dunnington noted that whatever the situation concerning prior discovery of the quadratic reciprocity theorem, the difficult task was to come up with a proof, and that Gauss was the first to do. His first proof, however, is generally agreed to be unpleasant, and Gauss speedily found a second (see entry [16]). This invoked the theory of binary quadratic forms (which are expressions of the form $Ax^2 + 2Bxy + Cy^2$).

The central problems in this theory are to find what integers can be written in this way for specified integers A, B, and C, and when two given quadratic forms represent the same set of integers. Gauss called two binary quadratic forms $Ax^2 + 2Bxy + Cy^2$ and $A_1 x_1^2 + 2B_1 x_1 y_1 + C_1 y_1^2$ properly equivalent if one is obtained from the other by writing

$$\begin{pmatrix} x_1 \\ y_1 \end{pmatrix} = \begin{pmatrix} a & b \\ c & d \end{pmatrix} \begin{pmatrix} x \\ y \end{pmatrix},$$

where a, b, c, d are integers with $ad - bc = 1$. He then showed that proper equivalence classes of forms with the same square-free discriminant $(B^2 - AC)$ obey a composition law (in modern terminology, they form a finite commutative group). The idea of using composition of forms to analyse families of quadratic forms is profound; Edwards calls it Gauss's great contribution to the theory. Put in modern language, the crucial observation is that there is a subgroup consisting of the squares in this group. Gauss, following Legendre, called the cosets of this subgroup the *genera*. He distinguished the genera by their 'total character' (D.A. §231) which is indeed their character in the modern sense of group representation theory, whence the name.

Gauss gave several illustrative examples in the *Disquisitiones Arithmeticae* (§230, 231). For example, the form $10x^2 + 6xy + 17y^2$ has discriminant $3^2 - 10 \cdot 17 = -161 = -1 \cdot 7 \cdot 23$. The numbers which this form can represent must be non-residues mod 7 and non-residues mod

23. The odd numbers this form represents must also be congruent to 1 mod 4. These three pieces of information make up the total character of the class defined by the form $10x^2 + 6xy + 17y^2$, and one sees that a system of plus and minus 1's has been associated to it according as the numbers it represents are or are not residues modulo the primes that divide the discriminant (that is how it 'discriminates'). Around 1880 Weber and Dedekind began to interpret the Gaussian total character as a homomorphism from the group of forms under consideration to the subgroup $\{\pm 1\}$ of the nonzero complex numbers.

The fact that half the assignable characters cannot be ascribed to a quadratic form in which A, $2B$, and C have no common divisor (D.A. §261) is the basis of the second proof of quadratic reciprocity (D.A. §262). The connection can be made because the theorem in §261 was established by Gauss independently of the law of quadratic reciprocity, so an argument by contradiction based upon the theorem allows Gauss to deduce the law: if the law was false forms can easily be written down for all assignable characters.

How much of this theory was apparent to Gauss on June 27, 1796 is not clear. Some of an extant preliminary version of the *Disquisitiones Arithmeticae*, called the *Analysis Residuorum*, appears in Gauss, *Werke*, II, 199–240, and more was found in Berlin by U. Merzbach (see [Merzbach, 1981]). Unfortunately, the relevant chapter V is still missing, but we do know Gauss wrote it at least four times in 1797 and 1798, because he said so in a letter to W. Bolyai (quoted in [Merzbach, 1981] p. 175). One supposes that in 1796 Gauss saw how the proof would go, if not all the details.

In mid-May 1801 Gauss found a fifth way to prove the quadratic reciprocity theorem, when he saw how it was connected with what are now called Gauss sums

$$\sum_n \exp 2\pi i \frac{n^2}{p}.$$

The proof itself continued to elude him until August 30th 1805 (when he wrote to Olbers that 'As lightning strikes the puzzle was solved'). It depends on showing that if

then
$$r = \cos\frac{2\pi}{n} + i\sin\frac{2\pi}{n},$$

$$1 + r + r^4 + r^9 + \cdots + r^{(n-1)^2} = \begin{cases} +\sqrt{n}, & n \equiv 1 \pmod{4} \\ +i\sqrt{n}, & n \equiv 3 \pmod{4} \end{cases}.$$

It is fairly easy to show that the sum is $\pm\sqrt{n}$ or $\pm i\sqrt{n}$ in each case; the difficulty lies in determining the sign, and it is this problem that occupied Gauss until 1805. For a good recent survey of Gauss sums which explains this and other matters, see Berndt and Evans [1981].

Among the final entries in the diary are some, [130–138], that show Gauss was preoccupied in 1807 with extending the theory to cubic and biquadratic residues, and incidentally thereby discovered a sixth proof of the quadratic reciprocity theorem. When Gauss published his papers on biquadratic residues, in 1825 and 1831 (*Werke,* II, 65–92 and 93–148), he observed that the theorems were best stated in terms of imaginary numbers, or, as we would call them, Gaussian integers. By then Jacobi had discovered the cubic reciprocity law, and in the 1840s Eisenstein also found it, together with a proof and connections with the theory of elliptic functions. A brief account of these developments and the unfortunate priority dispute they provoked is given in Collison [1977] and Weil [1974, 1976].

Gauss's legacy to the next generation of German mathematicians was not just the remarkable theorems about numbers that he discovered. It was also the deep and often surprising inter-connections he found, and the profundity of the proofs involved. Even when expositors like Dirichlet made Gauss's theory more comprehensible, by casting out binary forms in favour of quadratic integers, they accepted and endorsed its importance, and as a result the *Disquisitiones Arithmeticae* started a tradition, which continues to the present day, which finds the theory of numbers to be one of the most profound and important branches of mathematics. Early nineteenth century neo-humanists such as Alexander von Humboldt, who valued learning for its own sake and trusted applications to follow naturally, had no trouble sharing Gauss's taste in this matter. Neither, perhaps, does the large modern public who thrilled to the final resolution of Fermat's Last Theorem by Andrew Wiles in 1996.

Elliptic function theory

As early as September 9, 1796, another major theme emerged in the diary, that of elliptic functions, which may be thought of as a far-reaching generalisation of the familiar trigonometric functions. Gauss worked on and off on the theory of functions of a complex variable and in particular elliptic functions for the next 35 years, but he published very little of what he found. Instead he left a profusion of notes and drafts, which were collected in the Gauss *Werke*, where they have been thoroughly analysed by Schlesinger and others.

In September 1796 Gauss considered the so-called lemniscatic integral

$$z = \int_0^x \frac{dt}{\sqrt{1-t^4}}$$

which gives the arc length of the lemniscate. In January 1797 he read Euler's posthumous paper [1786] on elliptic integrals, and learned that if

$$A = \int_0^1 \frac{dx}{\sqrt{1-x^4}} \quad \text{and} \quad B = \int_0^1 \frac{x^2 dx}{\sqrt{1-x^4}}$$

then $AB = \pi/4$. Gauss now began 'to examine thoroughly the lemniscate'. By analogy with the integral for *arcsin*, he regarded the integral above as defining x as a function of z, and he eventually called this function *sl* for *sinus lemniscaticus*. He pursued the analogy with the trigonometric functions, defining the analogue of the cosine function, which he denoted $cl(x)$, and noting early on that the equation for $sl(3x)$ in terms of $sl(x)$ is of degree 9, whereas the equation for $\sin(3x)$ as a function of $\sin(x)$ is only a cubic equation ($\sin(3x) = 3\sin x - 4\sin^3 x$). On March 19 he noted in his diary 'Why dividing the lemniscate into n parts leads to an equation of degree n^2. The reason is that the roots are complex. He thereupon regarded *sl* and *cl* as functions of a complex variable.

He had already seen that, as a real function, sl was a periodic function with period $2\omega = 4A$. It followed from the addition law for sl and cl that the complex function sl had two distinct periods, 2ω and $2i\omega$, and so, when m and n are integers, $sl(x + (m + in)2\omega) = sl(x)$, the first occurrence of the Gaussian integers, $m + in$. This observation enabled Gauss to write down all the points where sl or cl were zero or infinite, and so to write them as quotients of two infinite series. Gauss wrote $sl(x) = M(x)/N(x)$ and did one of his provocative little sums. He calculated $N(\omega)$ to 5 decimal places and $\log N(\omega)$ to 4, and noted that this seemed to be $\pi/2$. He wrote 'Log hyp this number = 1.5708 = $1/2\pi$ of the circle?' [*Werke*, X.1, 158]. In his diary (March 29, 1797, nr. 63) he checked this coincidence to 6 decimal places, and commented that this 'is most remarkable and a proof of this property promises the most serious increase in analysis'.

In July 1798 an improved representation of M and N led Gauss back to the calculation of ω, and he commented in his diary (July, nr. 92): 'we have found out the most elegant things exceeding all expectations and that by methods which open up to us a whole new field ahead.' Entry was delayed, however, by a year, and the path unlocked by the arithmetico-geometric mean. This is defined for two real numbers a and b as follows: set $a_0 = a$ and $b_0 = b$, and recursively

$$a_{n+1} = \frac{1}{2}(a_n + b_n) \quad \text{and} \quad b_{n+1} = \sqrt{a_n b_n}.$$

Then it is easily seen that the two sequences (a_n) and (b_n) converge to the same value, called the arithmetico-geometric mean of a and b, which may be written $M(a, b)$ here. On May 30, 1799 Gauss wrote in his diary (nr. 98): 'We have found that the arithmetico-geometric mean between 1 and 2 is π/ω to 11 places, which thing being proved a new field in analysis will certainly be opened up.' The proof was still eluding Gauss even in November 1799, as a letter from Pfaff to Gauss makes clear (*Werke*, X.1, 232).

In early 1800 Gauss could show that

$$M(1, 1+x) = 1 + \frac{x}{2} - \frac{x^2}{16} + \frac{x^3}{32} - \frac{21x^4}{1024} + \cdots.$$

This reminded Gauss of similar formulae for the arc-length of an ellipse as a function of its eccentricity, and so he could connect his ideas with existing algorithms for computing such things, known to Lagrange (for elliptical arc-length) and Legendre (for elliptic integrals in their own right). The connection with the arithmetico-geometric mean function was, however, original with Gauss, and was to prove most instructive. Gauss soon found this striking power series:

$$\frac{1}{M(1+x,1-x)} = 1 + \frac{x^2}{4} + \frac{9x^4}{64} + \frac{25x^6}{256} + \cdots.$$

From it, he deduced that the function $1/M(1+x, 1-x)$ satisfies the differential equation

$$Y''x(x^2-1) + Y'(3x^2-1) + xY = 0,$$

and noted that another independent integral of this differential equation is $1/M(1,x)$. An ingenious series of calculations (described in Cox, [1984], a masterly account of Gauss's work, well worth considering on several counts) led Gauss to the result he had long sought: $M(1,\sqrt{2}) = \pi/\varpi$.

More results followed until on May 6 he believed he had led the theory of elliptic integrals 'to the summit of universality' (Diary entry nr. 105) only to find by May 22 that the theory was 'greatly increased and unified', and was becoming 'most beautifully bound together and increased infinitely' (nr. 106). Gauss had begun to study the general elliptic function with a real modulus k:

$$z = \int_0^x \frac{dt}{\sqrt{(1-t^2)(1-k^2t^2)}}.$$

As before, Gauss regarded this integral as defining x as a complex function of z, and, as other authors were to later, he allied this approach to one that developed elliptic functions (complex functions having one real and one imaginary period) directly as quotients of entire functions, without reference to an integral. The natural conjecture is that these two approaches describe the same objects. To show that a function defined via an elliptic integral is indeed a quotient was, for Gauss, a straightforward generalisation of the lemniscatic case. It was much harder to show that every doubly-periodic function defined as a quotient arises as the inverse of a suitable elliptic integral, but using his theory of the arithmetico-geometric mean Gauss was able to establish this.

The whole question becomes much harder when doubly-periodic functions are admitted having two complex periods (to avoid trivialities, the quotient of the periods must not be real). Now the arithmetico-geometric mean becomes complex, and complex numbers have two square roots, so there is a choice to be made at every stage of the procedure. However, on 3 June 1800 Gauss wrote that the connection between the infinitely many means has been completely cleared up (Diary entry nr. 109). Two days later he remarked "We have now immediately applied our theory to elliptic transcendents" (nr. 110), which confirms that Gauss generalised his theory of elliptic functions by making the arithmetico-geometric mean into a complex function. Unfortunately, nothing survives from 1800 to indicate what Gauss's complete solution was. Later work by Gauss, in the 1820s, shows that Gauss did not start from elliptic integrals with a complex modulus k but worked intensively with power series in complex variables.

Gauss was clear from an early stage in his work on function theory about the importance of the complex domain. As he wrote in a famous letter to Bessel (December 18, 1811, in *Werke*, X.1, 366, but not quoted in Dunnington) he asked of anyone who wished to introduce a new function into analysis to explain:

if he would apply it only to real quantities, and imaginary values of the argument appear so to speak only as an offshoot, or if he agrees with my principle that the imaginaries must enjoy equal rights in the domain of quantities with the reals. Practical utility is not at issue here, but for me analysis is an independent science that through the neglect of any fictitious quantity loses exceptionally in beauty and roundness and in a moment all truths, that otherwise would be true in general, have necessarily to be encumbered with the most burdensome restrictions.

Gauss was confident that a viable theory of a complex variable required only the representation of complex numbers as points in the plane. In his *Disquisitiones Arithmeticae* of 1801 he had illustrated his discoveries in number theory with that representation, for example in his theory of the ruler and compass construction constructibilty of the regular 17-gon. So he had in mind a geometrical, rather than a formal or purely algebraic theory of functions of a complex variable.

Dunnington did not comment critically on Gauss's lofty remark that Abel's work in the 1820s brought him about one-third of the way. Schlesinger suggested (Gauss *Werke*, X.2, 184) that by 1828 Gauss's mostly unpublished theory came conveniently in three parts: 'The first third was the general theory of functions arising from the [hypergeometric series], the second the theory of the arithmetico-geometric mean and the modular function, and finally the third, which Abel published before Gauss, was the theory of elliptic functions in the strict sense.' Gauss would presumably have appreciated Abel's theory of when the division equations are solvable algebraically, and the investigation of the divisibility of the lemniscate by ruler and compass would have struck him most forcefully, inspired as it was by the hint he had dropped in his *Disquisitiones Arithmeticae*. Gauss's judgement was based on his observation that what Abel presented was only an account of elliptic functions with a real modulus. But had Gauss chosen to comment a year later on Jacobi's *Fundamenta Nova*, he would have seen a general theory of the transformations of elliptic functions surpassing anything he had written down. He would also have found something more like his theory of theta functions, but again only in the context of elliptic functions with real modulus. He might also have become aware of Abel's account of the same power series in his paper [1828]. In that paper Abel also allowed the modulus to become purely imaginary, but in all respects his theory of a complex variable was, like Jacobi's, entirely formal.

A fairer comparison of the work of Gauss and Abel would be that they were proceeding in the same direction, although Abel's theory (and indeed Jacobi's) lacked a good way of writing the new functions (such as Jacobi's theta series were to provide), a rigorous theory of convergence, and an explanation of double periodicity. Gauss, but not Abel, had a theory of the modular function (which expresses the modulus as a function of the periods, $k = k(K/K')$) and knew the connection with differential equations (such as Legendre's). The theory of transformations is novel, and was better developed by Abel than Gauss (not in his *Recherches* but in his later *Précis*).

Dunnington briefly mentioned Gauss's published paper of 1812 on the hypergeometric series[1]

$$F(\alpha,\beta,\gamma,x) = 1 + \left(\frac{\alpha \cdot \beta}{1 \cdot \gamma}\right)x + \left(\frac{\alpha(\alpha+1)\beta(\beta+1)}{(1 \cdot 2)\gamma(\gamma+1)}\right)x^2 + \cdots$$

which Gauss treated as a complex function of a complex variable x, but dependent on real parameters α, β, γ. Using his newly-developed theory of this function, Gauss gave an immediate proof of the formula that had intrigued him for so long, $A \cdot B = \pi/4$, but he revealed none of his theory of the arithmetico-geometric mean and elliptic functions. He also wrote, but did not publish, a study of the differential equation that the hypergeometric series satisfies, which is called the hypergeometric equation:

$$(x-x^2)\frac{d^2F}{dx^2} + (\gamma - (\alpha + \beta + 1)x)\frac{dF}{dx} - \alpha\beta F = 0.$$

A special case of it, as Gauss knew, is Legendre's differential equation ($\alpha = \beta = 1/2$ and $\gamma = 1$).

Gauss's analysis of this equation shows very clearly that he appreciated the distinction between F as a function and F as an infinite series. The former is defined for all finite values of the variable except 0 and 1, while the latter is only defined when the variable is less than 1 in absolute value. But the series, when defined, takes a unique value for each value of the variable, whereas the function does not. Another important point Gauss did not make was that he had a powerful reason for studying the hypergeometric series, for although the differential

---

[1] For its prehistory see Schlesinger's essay and for the later history of this important equation Gray [2000].

equation satisfied by the functions $1/M(1+x, 1-x)$ and $1/M(1,x)$ is not the hypergeometric equation, on making the transformation $x^2 = z$ the differential equation becomes the equation

$$z(1-z)\frac{d^2y}{dx^2} + (1-2z)\frac{dy}{dz} - \frac{y}{4} = 0$$

which is the hypergeometric equation with $\alpha = \beta = 1/2$, $\gamma = 1$ (in fact, Legendre's equation).[2] This special case is mentioned in Gauss's unpublished notes of 1809, (*see Werke* X.1, 343).

Gauss's ideas about the meaning of a complex integral were connected by Schlesinger with his discussion of the hypergeometric series, because Gauss, although inspired by Euler's work, did not follow Euler's representation of the series as an integral. Schlesinger connected this with Gauss's earlier refusal to invert the general elliptic integral with complex modulus, and ascribed it to an awareness that a complex integral may well define a many-valued function of its upper end-point. Indeed, in Gauss's letter to Bessel already quoted, Gauss observed that the value of a complex integral depends on the path between its end points, and then he wrote:

> The integral $\int \phi x \cdot dx$ along two different paths will always have the same value if it is never the case that $\varphi x = \infty$ in the space between the curves representing the paths. This is a beautiful theorem whose not-too-difficult proof I will give at a suitable opportunity.... In any case this makes it immediately clear why a function arising from an integral $\int \phi x \cdot dx$ can have many values for a single value of x, for one can go round a point where $\varphi x = \infty$ either not at all, or once, or several times. For example, if one defines $\log x$ by $\int dx/x$, starting from $x = 1$, one comes to $\log x$ either without enclosing the point $x = 0$ or by going around it once or several times; each time the constant $+2\pi i$ or $-2\pi i$ enters; so the multiplicity of logarithms of any number are quite clear.

---

[2] See Gauss's unpublished notes of 1809, *Werke* X.1, 343 for a mention of this special case.

This is generally regarded as the first crucial insight into the integration of functions of a complex variable.

Gauss and non-Euclidean geometry

The claim, made on Gauss's behalf, that he was a, or even the, discoverer of non-Euclidean geometry is very hard to decide because the evidence is so slight. It is nonetheless implicit in the excellent commentaries of Stäckel and Dombrowski [1979], as it is in Reichardt's book [1976] and the broader but slighter survey by Coxeter [1977]. Proponents of this view, with Dunnington, are happy to tie documents written in the late 1820s and 1830s to cryptic claims made by Gauss for early achievements, and to equally elusive passages from the 1810s. In fact, the evidence points in another direction. It suggests that Gauss was aware that much needed to be done to Euclid's *Elements* to make them rigorous, and that the geometrical nature of physical space was regarded by Gauss as more and more likely to be an empirical matter, but in this his instincts and insights were those of a scientist, not a mathematician.

Gauss was 22 when he confided to Wolfgang Bolyai that he was doubtful of the truth of geometry. He had already found too many mistakes in other people's arguments in defence of the parallel postulate to be so confident any longer in their conclusion. He had begun to consider the fundamental assumptions of geometry at least two years earlier, in July 27, 1797, when he wrote in his mathematical diary only too cryptically that he had 'demonstrated the possibility of a plane'. It is tempting to connect this with fragments of arguments dating from 1828 to 1832 in which Gauss investigated whether the locus of a line perpendicular to a fixed line and rotating about that fixed line has all the properties of a plane, because, in a famous letter to Bessel of January 1829, where Gauss claims to have harboured these thoughts for almost 40 years, he wrote that "apart from the well-known gap in Euclid's geometry, there is another that, to my knowledge no-one has noticed and which is in no way easy to alleviate (although possible). This is the definition of a plane as a surface that contains the line joining any two of its points. This definition contains more than is necessary for the determination of the surface, and tacitly involves a theorem which must first be proved..." One knows from the later history of geometry, most clearly from the remarks of Pasch (Pasch [1882]) that trying to spell out what exactly elementary Euclidean geometry is about is extremely difficult.

By 1808 Gauss was aware that in the hypothetical non-Euclidean geometry similar triangles are congruent, and therefore there is an absolute measure of length. But at this stage, according to Schumacher, he found this conclusion absurd, and therefore held that the matter was still unclear. As he put it: "In the theory of parallels we are no further than Euclid was. This is the shameful part of mathematics, that sooner or later must be put in quite another form". Evidently he did not then feel confident in a non-Euclidean geometry. By 1816 he had shifted his opinion to accommodate an absolute measure of length as paradoxical but not self-contradictory (Gauss to Gerling, April 1816, in *Werke*, **8**, pp. 168–169) and now he thought it would be remarkable if Euclid's geometry was not true, because then we would have an a priori measure of length, such as the length of the side of an equilateral triangle with angle $59°59'59.99999''$. As Dunnington correctly observed, being remarkable is consistent with being attractive. But still there is no evidence that Gauss deduced anything specific about the new geometry.

In 1816 we do get a glimpse of what Gauss knew as reported by his former student Wachter. On a certain (unspecified) hypothesis, Wachter wrote to Gauss, the opposite of Euclidean geometry would apparently be true, which would involve us with an undetermined constant, a sphere of infinite radius which nonetheless lacks some properties of the plane, and the use of a transcendent trigonometry that probably generalises or underpins spherical trigonometry. Gauss now, as he wrote to Olbers, was coming "ever more to the opinion that the necessity of our geometry cannot be proved, at least not with human understanding. Perhaps in another life.... but for now geometry must stand, not with arithmetic which is pure a priori, but with mechanics." (Gauss to Gerling, April 1816, in Gauss's *Werke*, **8**, pp. 177).

This passage has been quoted more often than it has been understood. How would a proof beyond human understanding differ from a proof that does not surpass human understanding? Would it be some argument, compelling even to God, that made Euclidean geometry the right geometry for Space? Arithmetic, it seems, has an apodictic status, a truth of a higher kind than the truth of geometry, which is down there with mechanics. But the passage does not say that there *are* two geometries at some logical level and some experiment must choose between them. It says that knowledge is lacking. Gauss did not claim to possess knowledge of a new geometry, which surely means that even the ideas he was discussing with Wachter he considered to be hypothetical, and capable of turning out to be false. The 'transcendent trigonometry' is usually taken to be the hyperbolic trigonometry appropriate to non-Euclidean geometry, but there is very little evidence to support any interpretation. Accordingly, when Gauss replied to Schweikart in March 1819 that he could "do all of astral geometry once the constant is given" we cannot be sure what, precisely, Gauss had formulae for. The only one dated to this period is the one in his reply to Schweikart for the maximum area of a triangle in terms of Schweikart's Constant (the maximum altitude of an isosceles right-angled triangle). And all that his correspondence with Taurinus reveals is that, by 1824, Gauss was more comfortable than ever with the idea of a new geometry.

Far from being an exhaustive exploration of the problem of parallels, most of Gauss's work before 1831 on the fundamentals of geometry is firmly in a style one can call classical. Point, line, plane, distance, and angle are taken as undefinable, primitive terms with, perhaps, a set of obscure relationships between them, which is in need of elucidation. Gauss's study of the problem of parallels is mostly in this vein, and Bessel's letter to Gauss of 1829 was further encouragement to Gauss to state that geometry has a reality outside our minds whose laws we cannot completely prescribe a priori. This is entirely consistent with the classical formulation. The concepts of point, line, plane and so forth are formed as all scientific concepts are, and the task of the mathematician is to get them truly clear in the mind. This might involve teasing out tacit assumptions and fitting them up with proofs, or it might call for the elaboration of new ideas about a hitherto unsuspected species of geometry, which might nonetheless turn out to be (for some value of an unknown constant) the true geometry of space. This was also the approach of Gerling, Crelle and Deahna (see Zormbala [1996]).

Adherence to the classical formulation denies trigonometric methods a fundamental role. And in fact there is very little evidence of Gaussian contributions to trigonometry in non-Euclidean geometry before the letter to Schumacher of 12 July 1831, where he says that the circumference of a semi-circle is $1/2 \pi k(e^{r/k} - e^{-r/k})$ where k is a very large constant that is infinite in Euclidean geometry.

In particular, there is no evidence that Gauss derived the relevant trigonometric formulae from the profound study of differential geometry that occupied him in the 1820s. What he did say in the *Disquisitiones generales circa superficies curvas* is summed up in what he regarded as one of the most elegant theorems in the theory of curved surfaces: "The excess over 180° of the sum of the angles of a triangle formed by shortest lines on a concavo-concave surface, or the deficit from 180° of the sum of the angles of a triangle formed by shortest lines on a concavo-convex surface, is measured by the area of the part of the sphere which corresponds, through the direction of the normals, to that triangle, if the whole surface of the sphere is set equal to 720 degrees."

It is tempting to suppose that Gauss connected this elegant theorem with the study of non-Euclidean geometry, by considering a concavo-convex surface of constant negative Gaussian curvature. Even if we do, we must still note that Gauss did not develop the trigonometry of triangles on surfaces of constant (positive or negative) curvature until after 1840, when he had read Lobachevskii's *Geometrische Untersuchungen*. Moreover, he did not have an example of a surface of constant negative curvature to hand; there is every reason to suppose that Minding was the first to discover one. Minding's example, moreover has a number of topological properties that rule it out as a model of space, notably self-intersecting geodesics and pairs of geodesics that meet in more than one point.

But even if (in contradiction to Hilbert's later theorem about surfaces of constant negative curvature) Gauss had found a surface of constant negative curvature in space, it would only establish that there is a surface in space whose intrinsic geometry is non-Euclidean. It would not establish that space could be non-Euclidean, because space is three dimensional. There is no sign that Gauss had any of the concepts needed to formulate a theory of differential geometry in three dimensions. Nor did he have any of the essential mathematical machinery for doing differential geometry in three or more dimensions, without which almost nothing useful can be said.

What weight can the two formulae be made to carry? They are not difficult to obtain and manipulate if, as for example Taurinus did, one assumes that non-Euclidean geometry is described by the formulae of hyperbolic trigonometry—a natural enough assumption. To introduce hyperbolic trigonometry into the study of non-Euclidean geometry properly is, as Bolyai and Lobachevskii found, a considerable labour of which no trace remains in Gauss's work. It is more plausible to imagine that he made the assumption, but did not derive it from basic principles. So, perhaps by 1816, Gauss was convinced of ideas like these:

there could be a non-Euclidean geometry, in which the angle sum of triangles is less than π,

the area of triangles is proportional to their angular defect and is bounded by a finite amount,

the trigonometric formulae for this geometry are those of hyperbolic trigonometry, and the analogy with spherical geometry and trigonometry extends to formulae for the circumference and area of circles.

Such a position is unsatisfactory, to Gauss and to us, because it is purely and simply an analogy with spherical geometry. But whereas spherical geometry starts with a sphere in the non-Euclidean case there is no surface. And even if one found non-Euclidean geometry on a surface in three-dimensional Euclidean space, it would not follow that three-dimensional space was non-Euclidean, any more than the existence of spheres in three-dimensional space forces the conclusion that space is a three-dimensional sphere.

It becomes clear that a mathematician persuaded of the truth of non-Euclidean geometry and seeking to convince others is almost driven to start by looking for, or creating, non-Euclidean three-dimensional space, and to derive a rich theory of non-Euclidean two-dimensional space from it—as Bolyai and Lobachevskii did, but not Gauss. The only hint we have that he explored the non-Euclidean three-dimensional case is the remark by Wachter, but what Wachter said was not encouraging: "Now the inconvenience arises that the parts of this surface are merely symmetrical, not, as in the plane, congruent; or, that the radius on one side is infinite and on the other imaginary" and more of the same. This is a long way from saying, what enthusiasts for Gauss's grasp of non-Euclidean geometry suggest, that this is the Lobachevskian horosphere, a surface in non-Euclidean three-dimensional space on which the induced geometry is Euclidean.

Gauss, by contrast, possessed a scientist's conviction in the possibility of a non-Euclidean geometry which was no less, and no greater, than that of Schweikart or Bessel. The grounds for his conviction are greater, but still insubstantial, because he lacks almost entirely the substantial body of argument that gives Bolyai and Lobachevskii their genuine claim to be the discoverers of non-Euclidean geometry.

Did Gauss then, as a scientist, make an empirical test of the matter? This is one of the most discussed questions in the whole subject of Gauss and non-Euclidean geometry. Those who believe that he did quote Sartorius von Waltershausen's reminiscence, where on p. 81, he states that Gauss did check the truth of Euclidean geometry on measurements of the triangle formed by the mountains Brocken, Hohenhagen, and Inselsberg (BHI), and found it to be approximately true. This claim was most recently advanced by Scholz [1992], on the basis of a number, quoted by von Waltershausen elsewhere in his reminiscence, relating to the very close agreement between the measurements of this triangle and the predictions of Euclidean geometry (once the mountain tops are treated as three points on a sphere). Scholz concludes that "there is no longer any reason to doubt that Gauss *himself* conducted such a test of the angle sum theorem." (Scholz [1992], p. 644).

Those who dispute that Gauss made such a test argue that the problem that occupied Gauss, and figures so prominently at the end of the *Disquisitiones generales circa superficies curvas*, is the question of the spheroidal or spherical shape of the Earth, and that von Waltershausen was simply confused about the hypothesis that Gauss found to be approximately confirmed. This is the opinion of Miller [1972].

The most thorough analysis of the question was made by Breitenberger [1984]. He confronts the question: 'if von Waltershausen was not simply confused in some way, what was he saying?' and he gives it an elegant answer. Surveying Hanover threw up many triangles and many numbers (a figure of a million is sometimes mentioned). Conclusions were drawn (and maps made) on numbers which are the result of many calculations, and at every stage discrepancies between real and expected results lay within expected error bounds (Gauss analysed the errors quite carefully). Not only was Euclidean geometry never called into question, because the errors were only what was to be expected, each calculation amounted to a tacit defence of Euclidean geometry. But the measurements of the BHI triangle were not fed into such a mill. They show that, within experimental error, space is described by Euclidean geometry. To be sure "as a single instance it proves very little, but it has been designed so as to be transparent, and hence it will drive a point home" (Breitenberger [1984], p. 288). Newton dropped an apple in conversation to similar purpose and effect. The myth, Breitenberger concludes, is that the BHI triangle was surveyed as part of a *deliberate* test of Euclidean geometry. But it did incidentally show that Euclidean geometry is true to within the limits of the best observational error of the time. Put that way, the gap between Scholz and Breitenberger may be quite small.

Gauss's Mathematical Diary

Between 1796 and 1814 Gauss kept an informal record of his mathematical discoveries. After his death it remained in the possession of the Gauss family until, in 1898, a grandson of Gauss gave it to Stäckel for use in the preparation of the later volumes of Gauss's *Werke*. It was first published in the *Mathematische Annalen* (LVII), 1903, with a brief introduction and some notes by Klein, the editor-in-chief of the *Werke*. It was then re-published in the *Werke* (X.1, 1917, 485-574), with extensive notes supplied by the team of editors, the diary itself being reprinted in facsimile form as a supplement.

1796

[1] The principles upon which the division of the circle depend, and geometrical divisibility of the same into seventeen parts, etc.

[1796] March 30 Brunswick.

[2] Furnished with a proof that in case of prime numbers not all numbers below them can be quadratic residues. April 8 ibid.

[3] The formulae for the cosines of submultiples of angles of a circumference will admit no more general expression except into two periods. [Or, see Johnsen [1986], pp. 168–169: The formulae for the cosines of submultiples of angles of a circumference will admit a more general expression only in terms of both of the two periods.] April 12 ibid.

[4] An extension of the rules for residues to residues and magnitudes which are not prime. April 29 Göttingen.

[5] Numbers which can be divided variously into two primes.

May 14 Göttingen.

[6] The coefficients of equations are given easily as sums of powers of the roots. May 23 Göttingen.

[7] The transformation of the series $1 - 2 + 8 - 64 + \cdots$ into the continued fraction

$$\cfrac{1}{1+\cfrac{2}{1+\cfrac{2}{1+\cfrac{8}{1+\cfrac{12}{1+\cfrac{32}{1+\cfrac{56}{1+128\,\cdots}}}}}}} = \cfrac{1-1+1\cdot 3-1\cdot 3\cdot 7+1\cdot 3\cdot 7\cdot 15\cdot+\cdots}{1+\cfrac{1}{1+\cfrac{2}{1+\cfrac{6}{1+\cfrac{12}{1+28\,\cdots}}}}}$$

and others. May 24 Göttingen.

[8] The simple scale in series which are recurrent in various ways is a similar function of the second order of the composite of the scales.

May 26.

[9] A comparison of the infinities contained in prime and compound numbers. May 31 Göttingen.

[10] A scale where the terms of the series are products or even arbitrary functions of the terms of arbitrarily many series. June 3 Göttingen.

[11] A formula for the sum of factors of an arbitrary compound number: general term

$$\frac{a^{n+1}-1}{a-1}.$$

June 5 Göttingen.

[12] The sum of the periods when all numbers less than a [certain] modulus are taken as elements: general term $[(n+1)a - na]a^{n-1}$.

June 5 Göttingen.

[13] Laws of distributions. June 19 Göttingen.

[14] The sum to infinity of factors $= \pi^2/6 \cdot$ sum of the numbers.

June 20 Göttingen.

[15] I have begun to think of the multiplicative combination (of the forms of divisors of quadratic forms). June 22 Göttingen.

[16] A new proof of the golden theorem all at once, from scratch, different, and not a little elegant. June 27.

[17] Any partition of a number a into three $\square$ gives a form separable into three $\square$. July 3.

[17a] The sum of three squares in continued proportion can never be a prime: a clear new example which seems to agree with this. Be bold!
July 9.

[18] EUREKA. number = $\Delta + \Delta + \Delta$. July 10 Göttingen.

[19] Euler's determination of the forms in which composite numbers are contained more than once. [July Göttingen.]

[20] The principles for compounding scales of series recurrent in various ways. July 16 Göttingen.

[21] Euler's method for demonstrating the relation between rectangles under line segments which cut each other in conic sections applied to all curves. July 31 Göttingen.

[22] $a^{2^n \mp 1(p)} \equiv 1$ can always be solved. August 3 Göttingen.

[23] I have seen exactly how the rationale for the golden theorem ought to be examined more thoroughly and preparing for this I am ready to extend my endeavours beyond the quadratic equations. The discovery of formulae which are always divisible by primes: $\sqrt[n]{1}$ (numerical).
August 13 ibid.

[24] On the way developed $(a+b\sqrt{-1})^{m+n\sqrt{-1}}$ August 14.

[25] Right now at the intellectual summit of the matter. It remains to furnish the details. August 16 Göttingen.

[26] $(a^p) = (a)$ mod p, a the root of an equation which is irrational in any way whatever. [August] 18.

[27] If P, Q are algebraic functions of an indeterminate quantity which are incommensurable. One is given $tP + uQ = 1$ in algebra as in number theory. [August] 19 Göttingen.

[28] The sums of powers of the roots of a given equation are expressed by a very simple law in terms of the coefficients of the equation (with other geometric matters in the *Exercitiones*). [August] 21 Göttingen.

[29] The summation of the infinite series $1 + \dfrac{x^n}{1\cdots n} + \dfrac{x^{2n}}{1\cdots 2n}$, etc.

same [day August 21.]

[30] Certain small points aside, I have happily attained the goal namely if $p^n \equiv 1 \pmod{\pi}$ then $x^\pi - 1$ is composed of factors not exceeding degree n and therefore a sum of conditionally solvable equations; from this I have deduced two proofs of the golden theorem.

September 2 Göttingen.

[31] The number of different fractions whose denominators do not exceed a certain bound compared to the number of all fractions whose numerators or denominators are different and less then the same bound when taken to infinity is $6:\pi^2$

September 6

[32] If $\int^{[x]} dt/\sqrt{(1-t^3)}$ is denoted $\prod:x = z$ and $x = \Phi:z$ then

$$\Phi:z = z - \frac{1}{8}z^4 + \frac{1}{112}z^7 - \frac{1}{1792}z^{10} + \frac{3z^{13}}{1792 \cdot 52} - \frac{3 \cdot 185 z^{16}}{1792 \cdot 52 \cdot 14 \cdot 15 \cdot 16} \cdots$$

September 9.

[33] If $\Phi : \int dt/\sqrt{(1-t^n)} = x$ then

$$\Phi:z = z - \frac{1 \cdot z^n}{2 \cdot n+1}A + \frac{n-1 \cdot z^n}{4 \cdot 2n+1}B - \frac{nn-n-1[z^n]}{2 \cdot n+1 \cdot 3n+1}C \cdots$$

[34] An easy method for obtaining an equation in y from an equation in x, if given $x^n + ax^{n-1} + bx^{n-2} \cdots = y$.

September 14.

[35] To convert fractions whose denominator contains irrational quantities (of any kind?) into others freed of this inconvenience.

September 16.

[36] The coefficients of the auxiliary equation for the elimination are determined from the roots of the given equation.

Same day.

[37] A new method by means of which it will be possible to investigate, and perhaps try to invent, the universal solution of equations. Namely by transforming into another whose roots are $\alpha\rho' + \beta\rho'' + \gamma\rho''' + \cdots$ where $\sqrt[n]{1} = \alpha, \beta, \gamma$, etc. and the number n denotes the degree of the equation.

September 17

[38] It seems to me the roots of an equation $x^n - 1$ [$= 0$] can be obtained from equations having common roots, so that principally one ought to solve such equations as have rational coefficients.

September 29 Brunswick.

[39] The equation of the third degree is this:
$$x^3 + xx - nx + \frac{n^2 - 3n - 1 - mp}{3} = 0$$
where $3n + 1 = p$ and m is the number of cubic residues omitting similarities. From this it follows that if $n = 3k$ then $m + 1 = 3l$, if $n = 3k \pm 1$ then $m = 3l$. Or $z^3 - 3pz + pp - 8p - 9pm = 0$. By these means m is completely determined, $m + 1$ is always $\square + 3\square$. October 1 Brunswick.

[40] It is not possible to produce zero as a sum of integer multiples of the roots of the equation $x^p - 1 = 0$.   ☉ October 9 Brunswick.

[41] Obtained certain things concerning the multipliers of equations for the elimination of certain terms, which promise brightly.

☉ October 16 Brunswick.

[42] Detected a law: and when it is proved a system will have been led to perfection. October 18 Brunswick.

[43] Conquered GEGAN. October 21 Brunswick.

[44] An elegant interpolation formula. November 25 Göttingen

[45] I have begun to convert the expression $1 - \frac{1}{2^\omega} + \frac{1}{3^\omega}$ into a power series in which ω increases. November 26 Göttingen

[46] Trigonometric formulae expressed in series. By December.

[47] Most general differentiations. December 23.

[48] A parabolic curve is capable of quadrature, given arbitrarily many points on it. December 26.

[49] I have discovered a true proof of a theorem of Lagrange.

December 27.

1797

[50]

$$\int \sqrt{\sin x}\, dx = 2 \int \frac{yy\, dy}{\sqrt{1 - y^4}} \qquad \int \sqrt{\tan x}\, dx = 2 \int \frac{dy}{\sqrt[4]{1 - y^4}} \quad yy = \frac{\sin}{\cos} x$$

$$\int \sqrt{\frac{1}{\sin x}}\,dx = 2\int \frac{dy}{\sqrt{1-y^4}}$$

1797 January 7

[51] I have begun to examine thoroughly the [elastic] lemniscatic curve which depends on $\int dx/\sqrt{(1-x^4)}$. January 8.

[52] I have spontaneously discovered the ground for Euler's criterion. January 10.

[53] I have invented a way to reduce the complete integral $\int dx/\sqrt[n]{1-x^n}$ to quadratures of the circle. January 12.

[54] An easy method for determining $\int \frac{x^n\,dx}{1+x^m}$.

[55] I have found a distinguished supplement to the description of polygons. Namely, if $a, b, c, d, \ldots$ are the prime factors of the prime number p diminished by one, then to describe a polygon of p sides nothing other is required than that:
  1) the indefinite arc is divided into parts $a, b, c, d, \ldots$
  2) and that polygons of $a, b, c, d, \ldots$ sides be described.

January 19 Göttingen

[56] Theorems on the residues $-1, \mp 2$ proved by similar methods to the rest. February 4 Göttingen.

[57] Forms $aa + bb + cc - bc - ac - ab$, pertaining to the divisors, coincide with this: $aa + 3bb$. February 6.

[58] An amplification of the penultimate proposition on page 1, namely
$$1 - a + a^3 - a^6 + a^{10} \cdots = \cfrac{1}{1+\cfrac{a}{1+\cfrac{a^2-a}{1+\cfrac{a^3}{1+\cfrac{a^4-a^2}{1+\cfrac{a^5}{1+\text{etc.}}}}}}}$$

From this, all series where the exponents form a series of the second order are easily transformed. February 16.

[59] Established a comparison between integrals of the form $\int e^{-t^\alpha} dt$ and $\int du / \sqrt[\beta]{(1+u^\gamma)}$.

[60] Why dividing the lemniscate into n parts leads to an equation of degree n^2. March 19.

[61] On powers of the integral $\int_0^1 \frac{dx}{\sqrt{1-x^4}}$ depends
$$\sum \left(\frac{mm + 6mn + nn}{(mm+nn)^4} \right)^k.$$

[1797 March.]

[62] The lemniscate is divisible geometrically into five parts.

March 21.

[63] Amongst many other properties of the lemniscatic curve I have observed:

The numerator of the decomposed sine of the doubled arc is = 2 · Numerator denominator sine × numerator denominator cosine of the simple arc.

In fact the denominator = (numerator sine)^4 + (denominator sine)^4.

Now if the denominator for the arc π^l is θ, then the denominator of the sine of the arc $k\pi^l = \theta^{kk}$.

Now $\theta = 4.810480$, the hyperbolic logarithm of which number is = 1.570796, i.e., $\pi/2$ which is most remarkable and a proof of which property promises the most serious increase in analysis. March 29.

[64] I have discovered even more elegant proofs for the connection of divisors of the form $\square - \alpha, +1$ with $-1, \mp 2$. June 17 Göttingen.

[65] I have perfected a second deduction of the theorem on polygons.

July 17 Göttingen.

[66] It can be shown by both methods that only pure equations need be solved.

[67] We have given a proof of what on October 1st was discovered by induction. July 20.

[68] We have overcome the singular case of a solution of the congruence $x^n - 1 = 0$ (evidently when the auxiliary congruence has equal

roots), which troubled us for so long, with most happy success, from solutions to congruences when the modulus is a power of a prime number. July 21

[69] If $x^{\mu+\nu} + ax^{\mu+\nu-1} + bx^{\mu+\nu-2} + \cdots + n$ (A)
is divided by
$x^\mu + \alpha x^{\mu-1} + \beta x^{\mu-2} + \cdots + m$ (B)
and all the coefficients a, b, c, etc. in (A) are integer numbers and indeed all the coefficients in (B) are rational then these will also be integers and the ultimate m a divisor of the ultimate n. July 23.

[70] Perhaps all products of $(a + b\rho + c\rho^2 + d\rho^3 + \cdots)$, where ρ denotes all primitive roots of the equation $x^n = 1$ can be reduced to the form $(x - \rho y)(x - \rho^2 y)\cdots$. For example

$$\left(a+b\rho+c\rho^2\right)\times\left(a+b\rho^2+c\rho\right)=(a-b)^2+(a-b)(c-a)+(c-a)^2$$

$$\left(a+b\rho+c\rho^2+d\rho^3\right)\times\left(a+b\rho^3+c\rho^2+d\rho\right)=(a-c)^2+(b-d)^2$$

$$\left(a+b+c\rho^2+d\rho^3+e\rho^4+f\rho^5\right)\times=(a+b-d-e)^2$$

$$-(a+b-d-e)(a-c-d-f)+(a-c-d-f)^2$$

$$=(a+b-d-e)^2+(a+b-d-e)(b+c-e-f)+(b+c-e-f)^2$$

Seen February 4.

This is false. For it would follow from this that two numbers of the form $(x - \rho y)$ have a product of the same form, which is easily refuted.

[July.]

[71] Demonstrated that the several periods of the roots of the equation $x^n = 1$ cannot have the same sum. July 27 Göttingen.

[72] I have demonstrated the possibility of the plane.

July 28 Göttingen.

[73] What we wrote on July 27 involved an error: but happily we have now worked out the thing more successfully since we can prove that none of the periods can be a rational number. August 1.

[74] How one should attach the signs in doubling the number of periods. [August.]

[75] I have found the number of prime functions by a most simple analysis. August 26.

[76] Theorem: If $1 + ax + bxx +$ etc. $+ mx^\mu$ is a prime function with respect to the modulus p, then $d + x + x^p + x^{p^2} +$ etc. $+ x^{p^{\mu-1}}$ is divisible by this function with respect to this modulus, etc. etc. August 30.

[77] Showed, and the way to much more is laid open by the introduction of multiple moduli. August 31.

[78] August 1 more generally adapted to any moduli. September 4

[79] I have uncovered principles by which the resolution of congruences according to multiple moduli is reduced to congruences with respect to a linear modulus. September 9.

[80] Proved by a valid method that equations have imaginary roots.

Brunswick October.

Published in my own dissertation in the month of August 1799.

[81] New proof of Pythagoras' theorem. Brunswick October 16.

[82] Considered the sum of the series
$$x - \frac{1}{2}x^2 + \frac{1}{12}x^3 - \frac{1}{144}x^4 + \cdots$$
and showed that $= 0$, if
$$2\sqrt{x} + \frac{3}{16}\frac{1}{\sqrt{x}} - \frac{1}{1024}\frac{1}{\sqrt{.3x}} \cdots = \left(k + \frac{1}{4}\right)\pi.$$

1798

[83] Setting $l(1+x) = \phi'(x)$; $l(1 + \phi'(x)) = \phi''(x)$; $l(1 + \phi''(x)) = \phi'''(x)$ etc., then $\phi'x = \sqrt[3]{\dfrac{1}{\frac{3}{2}i}}.$ Brunswick April.

[84] Classes are given in any order, and from this the representability of numbers as three squares is reduced to solid theory.

Brunswick April 1.

[85] We have found a genuine proof of the composition of forces.

Göttingen May.

[86] The theorem of Lagrange on the transformation of functions extended to functions of any variables. Göttingen May.

[87] The series $1+\frac{1}{4}+\left(\frac{1}{2}\cdot\frac{1}{4}\right)^2+\left(\frac{1}{2}\cdot\frac{1}{4}\cdot\frac{1}{6}\right)^2+$ etc. $=\frac{4}{\pi}$ connected to the general theory of series of sines and cosines of angles increasing arithmetically.   June.

[88] The calculus of probabilities defended against Laplace.

                                                  Göttingen June 17.

[89] So solved the problem of elimination that nothing extra could be desired.   Göttingen June.

[90] Various rather elegant results concerning the attraction of a sphere.

                                                  June or July.

[91a] $1+\frac{1}{9}\frac{1\cdot3}{4\cdot4}+\frac{1}{81}\frac{1\cdot3\cdot5\cdot7}{4\cdot4\cdot8\cdot8}+\frac{1}{729}\frac{1\cdot3\cdot5\cdot7\cdot9\cdot11}{4\cdot4\cdot8\cdot8\cdot12\cdot12}$

$=1.02220\ldots=\frac{1.3110\ldots}{3.1415\ldots}\sqrt{6}\left[=\frac{\varpi}{2}\frac{1}{\pi}\sqrt{6}\right]$.   July.

[91b] $\arcsin\text{lemn}\sin\phi-\arcsin\text{lemn}\cos\phi=\varpi-\frac{2\phi\varpi}{\pi}$

$\sin\text{lemnisc}[\phi]=0.95500598\sin[\phi]$
$\qquad\qquad\qquad-0.0430495\sin 3[\phi]$
$\qquad\qquad\qquad+0.0018605\sin 5[\phi]$
$\qquad\qquad\qquad-0.0000803\sin 7[\phi]$

$\sin^2\text{lemn}[\phi]=0.4569472=\frac{\pi}{\varpi\varpi}-\cdots\cos 2[\phi]$

$\arcsin\text{lemn}\sin\phi=\frac{\varpi}{\pi}\phi+\left(\frac{\varpi}{2}-\frac{2}{\tilde{\omega}}\right)\sin 2\phi+\left(\frac{11}{2}\frac{\varpi}{\pi}-\frac{12}{\varpi}\right)\sin 4\phi$

$\sin^5[\phi]=0.4775031 \qquad \sin[\phi]+0.03\ldots[\sin 3\phi]\ldots$

[92] On the lemniscate, we have found out the most elegant things exceeding all expectations and that by methods which open up to us a whole new field ahead.   Göttingen July

[93] Solution of a problem in ballistics.   Göttingen July.

[94] Edited the completed theory of comets.   Göttingen July.

[95] New things in the field of analysis opened up to us, namely investigation of a function etc. October.

1799

[96] We have begun to consider higher forms.   Brunswick February 14.

[97] We have discovered new exact formulas for parallax.
                                                                      Brunswick April 8.

[98] We have proved that the arithmetico-geometric mean between 1 and $\sqrt{2}$ is $\pi/\tilde{\omega}$ to 11 places, which thing being proved a new field in analysis will certainly be opened up.   Brunswick May 30

[99] We have made exceptional progress in the principles of Geometry.
                                                                      Brunswick September.

[100] We have discovered many new things about the value of the arithmetico-geometric mean.   Brunswick November.

[101] We had already discovered long ago that the arithmetico-geometric mean was representable just as a quotient of two transcendental functions: now we have discovered the second of these functions is reducible to integral quantities.   Helmstadt December 14.

[102] The arithmetico-geometric mean itself is an integral quantity. Proved.   December 23.

1800

[103] Succeeded in determining the reduced forms in the theory of ternary forms.   February 13.

[104] The series $a\cos A + a'\cos(A + \phi) + a''\cos(A + 2\phi)$ + etc. leads to a limit if a, a', a'', etc. form a progression without change of sign which converges continually to 0. Proved.   Brunswick April 27.

[105] We have led the theory of transcendental quantities:

$$\int \frac{dx}{\sqrt{(1-\alpha xx)(1-\beta xx)}}$$

to the summit of universality.   Brunswick May 6.

[106] Succeeded in making a great increase in this theory, Brunswick May 22, by which at once all the preceding, not just the theory of the

arithmetico-geometric means, is most beautifully bound together and increased infinitely. [May 22.]

[107] On about these days (May 16) we most elegantly resolved the problem of the chronology of the Easter Feast. Published in Zach's *Comm Liter* August 1800 p. 121, 223. [May 16.]

[108] Succeeded in reducing the numerator and denominator of the lemniscatic sine (interpreted most universally) to integral quantities, at once derived from true principles the development in infinite series of all functions of lemniscates which can be thought of; a most beautiful invention which is not inferior to any of the above. Moreover in these days we have discovered the principles according to which the arithmetico-geometric series must he interpolated, so that it is now possible to exhibit terms in a given progression corresponding to any rational index by algebraic equations. Late May or June 2, 3.

[109] Between two given numbers there are always infinitely many means both arithmetico-geornetric and harmonico-geometric the mutual connection of which we have fortunately completely cleared up.

Brunswick June 3.

[110] We have now immediately applied our theory to elliptic transcendents. June 5.

[111] Finished the rectification of the ellipse in three different ways.

June 10.

[112] We have invented a totally new numerico-exponential calculus.

June 12.

[113] We have solved the problems in the calculus of probabilities concerning continued fractions which at one time were attacked in vain.

October 25.

[114] November 30. This has been a good day, in which it has been given to us to determine the number of classes of binary forms by three methods, thus

 1) by infinite products
 2) by infinite sums
 3) by finite sums of cotangents or logarithms of sines.

Brunswick.

[115] Dec 3. We have discovered a fourth method and the most simple of all for negative determinants which is derived only from the number of numbers ρ, ρ', etc. if $Ax + \rho$, $Ax + \rho'$, etc. are linear forms of divisors of forms $\square + D$. The same.

1801

[116] Proved that it is impossible to reduce the division of the circle to an equation of lower degree than our theory suggests.

Brunswick April 6 [1801]

[117] These days we have learned to determine the Jewish Easter by a new method. April.

[118] A method for proving the fifth fundamental theorem has been found by means of a most elegant theorem in the division of the circle, thus

$$\sum {\sin \brace \cos} \frac{nn}{a} P = + \begin{vmatrix} \sqrt{a} & 0 \\ \sqrt{a} & +\sqrt{a} \end{vmatrix} \begin{vmatrix} 0 & +\sqrt{a} \\ 0 & 0 \end{vmatrix} \begin{vmatrix} +\sqrt{a} \\ 0 \end{vmatrix}$$

according as $a \equiv 0\ 1 \quad 2\ 3 \pmod 4$ substituting for n all numbers from 0 to $(a-1)$. Brunswick mid-May.

[119] A new, most simple and expeditious method for investigating the elements of the orbit of celestial bodies. Brunswick mid-September.

[120] We are attacking the theory of the motion of the moon. August.

[121] We have discovered many new formulae most useful in theoretical astronomy. Month of October.

1805

[122] In the following years 1802, 1803, 1804 astronomical work took up the greatest part of my free time, first of all carrying out the calculations concerning the theory of new planets. From this it happened that in those years this catalogue was neglected. And those days in which it was possible to make some increase to mathematics are forgotten.

[123] The proof of the most charming theorem recorded above, May 1801, which we have sought to prove for 4 years and more with every

effort, at last perfected. *Commentationes recentiores*, 1.

1805 August 30.

[124] We have further worked out the theory of interpolation.

1805 November

1806

[125] We have discovered a new and most perfect method for determining the elements of the orbit of a body moving about the sun from two heliocentric positions. 1806 January.

[126] We have carried forward to the highest degree of perfection a method for determining the orbit of a planet from three geocentric positions. 1806 May.

[127] A new method for reducing the ellipse and hyperbola to a parabola. 1806 April.

[128] At about the same time we finished off the resolution of the function $\frac{x^p - 1}{x - 1}$ into four factors.

1807

[129] A new method for determining the orbit of a planet from four geocentric positions of which the last two are incomplete.

1807 January 21.

[130] Began the theory of cubic and biquadratic residues.

1807 February 15.

[131] Further worked out and completed. Proofs thereto are still wanted.

[132] The proof of this theory discovered by most elegant methods so that it is totally perfect and nothing further is desired. By this residues and quadratic non-residues are illustrated exceptionally well at the same time. February 22.

[133] Theorems, which attach to the preceding theory, most valuable additions, provided with an elegant proof (namely for which primitive

roots b itself must be taken positive and which negative, $aa + 27bb = 4p$; $aa + 4bb = p$). February 24.

[134] We have discovered a totally new proof of the fundamental theorem based on totally elementary principles. May 6.

1808

[135] The theory of division into three periods (Art 358) reduced to far simpler principles. 1808 May 10.

[136] The equation $X - 1 = 0$ which contains all primitive roots of the equation $x^n - 1 = 0$ cannot be decomposed into factors with rational coefficients. Proved for composite values of n. June 12.

[137] I have attacked the theory of cubic forms, solutions of the equation $x^3 + ny^3 + n^2z^3 - 3nxyz = 1$. December 23.

1809

[138] The theorems for the cubic residue 3 proved with elegant special methods by considering the values of $\frac{x+1}{x}$ where the three always have the values $\alpha, \alpha\varepsilon, \alpha\varepsilon\varepsilon,$ with the exception of two which give $\varepsilon, \varepsilon\varepsilon$ but these are $\frac{1}{\varepsilon-1} = \frac{\varepsilon\varepsilon-1}{3}, \frac{1}{\varepsilon\varepsilon-1} = \frac{\varepsilon-1}{3}$ with product $\equiv \frac{1}{3}$.

1809 January 6.

[139] Series pertaining to the arithmetico-geometric mean further evolved. 1809 June 20.

[140] We have finished the division into five by the arithmetico-geometric mean. 1809 June 29

1812

[141] The preceding catalogue interrupted by unjust fate a second time resumed the beginning of the year 1812. In the month November 1811 we succeeded in giving a purely analytical proof of the fundamental theorem in the study of equations but because nothing of this was put to paper an essential part of it was completely forgotten. This, sought for

unsuccessfully for a rather long time, we however have happily rediscovered. 1812 February 29.

[142] We have discovered an absolutely new theory of the attraction of an elliptical spheroid on points outside the body.

Seeberg 1812 September 26.

[143] We also solved the remaining part of the same theory by new and remarkably simple methods. October 15 Göttingen.

1813

[144] The foundation of the general theory of biquadratic residues which we have sought for with utmost effort for almost seven years but always unsuccessfully at last happily discovered the same day on which our son is born. 1813 October 23 Göttingen.

[145] This is the most subtle of all that we have ever accomplished at any time. It is scarcely worthwhile to intermingle it with mention of certain simplifications pertaining to the calculation of parabolic orbits.

1814

[146] I have made by induction the most important observation that connects the theory of biquadratic residues most elegantly with the lemniscatic functions. Suppose $a + bi$ is a prime number, $a - 1 + bi$ divisible by $2 + 2i$, the number of all solutions to the congruence

$$1 \equiv xx + yy + xxyy \pmod{a + bi}$$

including $x = \infty, y = \pm i, x = \pm 1, y = \infty$ is $= (a-1)^2 + bb$. 1814 July 9.

Commentary on
Gauss's Mathematical Diary

1. See the introduction.

2. This assertion is trivial, and is regarded as a slip of the pen. Klein and Bachmann conjectured plausibly that Gauss meant to assert that for every prime $p \geq 5$, there is a prime q, $q < p$, such that p is not a quadratic residue mod q. This theorem (for $p = 4n + 1$) is described by Gauss in a hand-written note to his own copy of the *Disquisitiones Arithmeticae* (§130) as having been discovered on April 8, 1796, and is an essential step on the road to Gauss's first proof of the theorem of quadratic reciprocity.

3. Johnsen [1968] observed that if one defines $a_j := \cos 2j\pi/p$, $1 \leq j \leq (p-1)/2$, as the cosines of the submultiples of the angles, and sets $r := \cos 2\pi/p + i\sin 2\pi/p$, then Gauss called $d_1 := \sum_{(k/p)=1} r^k$ and $d_2 := \sum_{(k/p)=-1} r^k$ the periods (D.A. § 343), and the entry refers to the way the a_j are linearly independent over the rational numbers, $\mathbb{Q}$, but may become linearly dependent over $\mathbb{Q}(d_1)$. This has implications for the irreducibility of the cyclotomic polynomials.

4. Generalised quadratic reciprocity, see *Disquisitiones Arithmeticae* [D.A.] § 133.

5. Related to Goldbach's conjecture.

7. Both the series and the continued fractions diverge. Euler had begun the study of these transformations in his [1754/55]. See also no. 58.

8. If $G(x)$ is a polynomial of degree at most $n-1$, and $G(x)(1 + a_1 x + a_2 x^2 + \cdots + a_n x^n)$ is developed as a power series in x, say $s_0 + s_1 x + s_2 x^2 + \cdots$, then $s_{n+k} = a_1 s_{n+k-1} + a_2 s_{n+k-2} + \cdots + a_n s_k$, $k = 0, 1, 2,\ldots$.

De Moivre called $a_1, \ldots, a_k$ the scale (Index, or Scala) of the series [De Moivre. 1730, 22].

9. Presumably Gauss was approaching the prime number theorem which he later said, in a letter to Encke (*Werke*, II, 444), that he had suspected when reading Lambert's tables in 1792 or 1793.

11. If an integer N is written as a product of primes, $N = \prod a_i^{n_i}$, then

$$\sum_{d|N} d = \prod_{d|N} \frac{a^{n_i+1} - 1}{a_i - 1}.$$

12. Period is defined in D.A. §46 as follows. If p is a prime, not dividing N, and d is the exponent of N (the least integer such that $N^d = 1 \pmod{p}$), then the period is the set of powers $1, N, \ldots, N^{d-1} \pmod{p}$. Bachmann argued that Gauss meant to write $((n+1)n - n)$ for

$$\sum_{d\delta = p-1} \phi(d)\delta = ((n+1)n - n)a^{n-1}$$

and this expression counts the number (= *Summa*) of periods.

14. Gauss claimed, correctly, that $\sum_{n \leq x} d(n)$ is asymptotically equal to $\pi^2 x/6$ where $d(n)$ denotes the number of divisors of n.

15. In Gauss's theory of binary quadratic forms, D.A. §287, a divisor of $x^2 + Ay^2$ is a prime p such that a multiple, pm, is equal to $x^2 + Ay^2$ for some relatively prime integers x, y, and also a quadratic form $px^2 + 2qxy + ry^2$ with $q^2 - pr = -A$.

16. The second proof of the 'golden' theorem. See his marginal note to the D.A. §262, where this date is also confirmed, printed in *Werke*, 1, 476.

17. The symbol □ denotes a square number, the forms considered are undoubtedly binary quadratic forms. See D.A. §§279, 280.

17a. Struck out in the diary and scarcely legible; omitted from the *Werke* edition. If

$$x : y : z = 1 : \frac{m}{n} : \frac{m^2}{n^2}$$

so x^2, y^2, and z^2 are in continued proportion, then
$$m^4 + m^2n^2 + n^4 = (m^2 + mn + n^2)(m^2 - mn + n^2)$$
which is never prime unless $m = n = 1$, when $m^2 + mn + n^2 = 3$ and $m^2 - mn + n^2 = 1$.

18. Every number is a sum of three triangular numbers (the triangular numbers are $\sum_1^N n$). Fermat conjectured that every number is the sum of 3 triangular numbers, 4 squares, 5 pentagonal numbers and so on, a result first proved by Cauchy in 1815.

19. Euler investigated [1752/53] when a number of the form $4n + 1$ is prime as part of a general investigation in which he showed that all such primes are a sum of two squares, unlike those of the form $4n+3$.

20. See no. 8.

21. Euler [1748. II, §§92, 93] gave a simple proof of a theorem of Apollonius [*Conics*, III, §§17, 19, 22]: if AB and $A'B'$ are two chords of a conic meeting at O, then the ratio $\dfrac{OA \cdot OB}{OA' \cdot OB'}$ is a constant independent of the position of O.

22. Obscure.

23. The details are taken up in the *Analysis Residuorum*, *Werke*, II, 230–234. Gauss had seen a connection between quadratic and cyclotomic questions, which he then investigated until he achieved success on September 2nd, when it yielded two more proofs of the golden theorem.

25. There is no evidence as to what the matter was, except that, like nos. 22, 23, 26, 27 it is underlined in red and so may refer to the same family of ideas.

26. See *Werke* II, 224; (*a*) is a polynomial in x with a root $x = a$, (a^p) is a polynomial whose roots are the pth powers of the roots of (*a*), p is a prime.

28. The *Exercitiones* were published in *Werke* X.1 138–143, with notes by Schlesinger.

29. The series satisfies the differential equation $d^n y/dx^n = y$ with the initial conditions that, at $x = 0$, $y = 1$, $\dfrac{dy}{dx} = 0 = \cdots = \dfrac{d^{n-1}y}{dx^{n-1}}$.

30. Further proofs of the golden theorem. The 'small points' are taken up in no. 68. See also Gauss's Analysis Residuorum, *Werke* II, 230–234.

31. If $A(n)$ denotes the number of fractions a/b in lowest terms such that $b \leq n$, then $A(n) = \sum_{i=1}^{n} \phi(i)$, where ϕ is Euler's ϕ function. If $B(n)$ denotes the number of fractions a/b with $1 \leq a \leq b \leq n$, then $B(n) = \tfrac{1}{2}n(n+1)$. Gauss asserted that $\lim_{n \to \infty} \dfrac{A(n)}{B(n)} = \dfrac{6}{\pi^2}$, a result first published by Dirichlet in 1849.

32. The first appearance in the diary of the inversion of elliptic integrals. Gauss, as was customary in his day, used x for both the variable of integration and tacitly, its upper limit.

33. Here A stands for the first term of the series (i.e., z), B the second (i.e., $\dfrac{1 \cdot z^n}{2(n+1)}$), C the third (i.e., $\dfrac{(n-1)z}{4(2n+1)} \cdot \dfrac{1 \cdot z^n}{2(n+1)} z$), etc. This formalism was introduced by Newton in his study of the binomial series [1676 = *Correspondence* II, 130–132].

34. The Tschirnhausen transformation replaces x in an equation $x^n + ax^{n-1} + bx^{n-2} \ldots = 0$ by y, with a view to simplifying it until it becomes solvable.

37. The Lagrange resolvent of a polynomial equation. Gauss hoped to use it to solve arbitrary polynomial equations. Loewy noted that by 1797 Gauss had become convinced that this was impossible, a view he implied in his dissertation and in the D.A. §359.

38. Gauss wants to reduce the study of $x^n - 1 = 0$ to the study of the equations $x^{a_i^{n_i}} - 1 = 0$, where $n = \prod a_i^{n_i}$.

39. Here m denotes the number of solutions of $x^3 - y^3 \equiv 1 \pmod{p}$.

40. The irreducibility of the cyclotomic polynomial with prime exponent.

43. Schlesinger [Gauss, *Werke,* X.1, part 2, 291], Biermann [1963] and Schuhmann [1976] have all offered conjectures. In the absence of other evidence none of these can be conclusive.

44. Loewy conjectured the Lagrange interpolation formula is meant.

45. Also considered by Euler, and subsequently, by Riemann. The expression is equal to $(1 - 2^{1-\omega})\zeta(\omega)$ where ζ is the Riemann Zeta function.

47. Differentiation with arbitrary indices, presumably, but no other trace remains.

48. Parabolic curves are those of the form $y = a_0 x^n + a_1 x^{n-1} + \cdots + a_n$.

50, 51. Elliptic integrals; 'elastic' struck out in original.

52. Euler's criterion (strictly, Newton's) concerns integrals of the form $\int x^m \left(a + bx^n\right)^{\mu/\nu} dx$. See Newton *Mathematical Papers* 1670–1673, III, p. 375.

55. If $p - 1 = abc\ldots$, then the construction of a regular p-gon depends on the ability to divide a given angle into $a, b, c, \ldots$ equal parts, and hence on the cyclotomic equations $x^a - 1 = 0$, $x^b - 1 = 0,\ldots$.

56. See D.A. § 145.

57. If $a^2 + b^2 + c^2 - bc - ac - ab = \alpha$, then $(2a - b - c)^2 + 3(b - c)^2 = 4\alpha$. So every odd divisor of the former divides $x^2 + 3y^2$. Conversely, if p divides the latter, then there is an odd integer A such that $A^2 \equiv -3 \pmod{p}$ and, if $B = 1$, $A^2 + 3B^2 \equiv 0 \pmod{p}$, so if $a = 0$, $b = (A + B)/2$, $c = (A - B)/2$, then p divides $a^2 + b^2 + c^2 - bc - ac - ab$.

59. Schlesinger conjectured that these are definite integrals taken between 0 and ∞. Indeed [53], [54], and [59] are all readily expressible in terms of beta and gamma functions, as he pointed out.

60. The division of the lemniscatic arc into n equal pieces depends on an equation of degree n^2 (unlike the division of the circular arc); $n^2 - n$ of the roots are complex. Gauss inferred that the lemniscatic functions are doubly periodic, which is the real breakthrough.

61. A slip of the pen: for $mm + 6mn + nn$ read $m^4 - 6m^2n^2 + n^4$.

$$S_k = {\sum_{m,n}}' \left\{ \frac{m^4 - 6m^2n^2 + n^4}{(m^2+n^2)^4} \right\}^k = \frac{1}{2} {\sum_{m,n}}' \left\{ \frac{1}{(m+ni)^4} + \frac{1}{(m-ni)^4} \right\}^k$$

where the sums are taken over all m, n not both zero. Each S_k can be written as a sum of terms $s_k = {\sum_{m,n}}' \frac{1}{(m+ni)^{4k}}$ and s_k depends on

$$\Pi = \int_0^1 \frac{dx}{\sqrt{1-x^4}}.$$

The series s_k are nowadays called Eisenstein s_k series, and were introduced by Eisenstein [1847]. They play a crucial role in the Weierstrassian theory of elliptic functions. The connection between the s_k and the periods is explained in Weil [1976].

62. By 'geometrically' Gauss meant by ruler and compass. The hint Gauss dropped about this in D.A. §335 greatly inspired Abel in his work on elliptic functions (see Ore [1971. p. 16]). It also makes clear that Gauss was considering the 'arithmetic' and 'analytic' aspects of elliptic functions together almost from the start.

63. So, if $\sin \operatorname{lemn} \phi = \frac{\Pi(\phi)}{N(\phi)}$ and $\cos \operatorname{lemn} \phi = \frac{\mu(\phi)}{\nu(\phi)}$, then

$$M(2\phi) = 2M(\phi)N(\phi)\mu(\phi)\nu(\phi), \quad N(2\phi) = M(\phi)^4 + N(\phi)^4.$$

Gauss was working towards a representation of elliptic functions as quotients (of theta functions).

64. This relates to the considerations in the D.A. §147-150 on how the divisors of $x^2 - \alpha$ depend on α, when $\alpha = 4n \pm 1, 4n \pm 2$.

65, 66. The method depended on the theory of Lagrange resolvents, see also no. 37.

67. The proof appears in the D.A. §358.

69. D. A. § 42.

70. Klein and Schlesinger pointed out that the last lines contain slips of the pen and should read, in part,

$$-(a+b-d-e)(a-c-d+f)+(a-c-d+f)^2$$
$$=-(a+b-d-e)(b+c-e-f)+(b+c-e-f)^2$$

71. See no. 73. 71, 73, 75–78 are again on cyclotomy.

72. This refers to Gauss's interest in the foundations of Euclidean geometry. In a letter to W. Bolyai, 6 March 1832 [Gauss, *Werke* VIII, 224] Gauss indicated that the usual definition of a plane presumed too much.

73. Bachmann pointed out that for Gauss to have made an error in no. 71 it is most likely that n should be taken as composite and periods interpreted in the sense of Kummer [1856] and Fuchs [1863].

74. The e periods with f terms may he transformed into $2e$ periods with $f/2$ terms and the members of the new periods are determined from the old ones by a quadratic equation. This concerns the relationship between Gauss sums of order n and $2n$, nowadays treated by the Hasse-Davenport theorem, see [Berndt, Evans, 1981, 122].

78. For Aug. 1 read Aug. 31 (no. 77).

80. Gauss received his doctorate with the first of his four proofs of the fundamental theorem of algebra, [*Werke* III, 1 -30].

81. The proof is by means of similar triangles inscribed in a semicircle with hypotenuse as radius, given in full in *Werke* X.1, 524-525.

82. Schlesinger pointed out that for $\sqrt{.3x}$, read $\sqrt[3]{x}$. The series is the Bessel function

$$\sum_{j=1}^{\infty} \frac{x^j (-1)^{j+1}}{j!(j-1)!}$$

c.f. Watson [1962].

83. Schlesinger's lengthy analysis fails to suggest what function l might be, but it is neither $\log(1+x)$ nor the function

$$l(1+x) = x - \frac{x^2}{4} + \frac{x^3}{8} - \frac{x^4}{112} + \cdots$$

which Gauss elsewhere denoted l.

84. Klein and Bachmann pointed out that for classes read genera, in conformity with D.A. §287.

85. Nothing in the *Nachlass* suggests what the proof might have been.

86. Schlesinger conjectured, on the basis of a letter from Gauss to Hindenburg (8 October 1799, *Werke* X.1, p. 429) that the generalisation was inspired by Laplace's proof of Lagrange's theorem.

87. The series $(a^2 + b^2 - 2ab\cos\theta)^n = \sum a_n \cos n\theta$ was studied by Ivory [1798]. If x denotes the eccentricity of an ellipse with major axis 1, then $1 - x^2\cos^2\phi = a^2 + b^2 - 2ab\cos 2\phi$ where

$$a = \frac{1+\sqrt{1-x^2}}{2}, \quad b = \frac{1-\sqrt{1-x^2}}{2}, \quad x^2 = \frac{4ab}{(a+b)^2},$$

so x is the ratio of the geometric and arithmetic means of a and b. If $n = 1/2$, then

$$\int_0^\pi (1 - x^2 \cos^2 \phi)^{1/2}$$
$$= a\pi \left(1 + \left(\frac{1}{2}\right)^2 \frac{b^2}{a^2} + \left(\frac{1 \cdot 1}{2 \cdot 4}\right)^2 \frac{b^4}{a^4} + \left(\frac{1 \cdot 1 \cdot 3}{2 \cdot 4 \cdot 6}\right)^2 \frac{b^6}{a^6} + \cdots\right),$$

which yields Gauss's series when $x = 1$, $a = 1/2 = b$.

88. Klein and Schlesinger noted that Gauss, in a letter to Olbers (24 January 1812 = *Werke* VIII, p. 140), subsequently dated his first investigations into the method of least squares by means of this entry. Laplace's error method was described in his [1793].

91. a, b These entries are imperfectly presented in the diary but they show Gauss was accumulating numerical evidence leading to 92.

92. Gauss refers to his discovery of the Fourier series expansions of his functions $P(\phi)$ and $Q(\phi)$ [*Werke* III, 465], which play the role of theta-functions in his theory of elliptic functions.

$$P(x) := 1 + 2x + 2x^4 + \cdots + 2x^{n^2} + \cdots$$
$$Q(x) := 1 - 2x + 2x^4 + \cdots + (-1)^n 2x^{n^2} + \cdots$$

$$R(x) := 2x^{1/4} + 2x^{9/4} + 2x^{25/4} + \cdots + 2x^{n^2/4} + \cdots$$

and Gauss observed, amongst "One hundred theorems on the new transcendents", that the arithmetico-geometric mean of $P(x)^2$ and $Q(x)^2$ is always 1. These last observations date from 1818, see Geppert [1927]. Klein's and Schlesinger's lengthy comments are particularly worthy of note here, and on [95].

94. Gauss applied his theory of elliptic functions to show that the attraction exerted by a planet on an arbitrary point is equal to the attraction exerted by a distribution of mass along the orbit whose density is proportional to the time the planet spends traversing each part of the orbit, Gauss [1818].

95. Gauss sought trigonometric expansions of $\log P$ and $\log Q$, and on the basis of numerical calculations conjectured a relation between the periods of lemniscatic integrals and the arithmetico-geometric mean (see no. 98), as Klein and Schlesinger pointed out.

97. Gauss was much interested in the parallax of the moon.

98. As Klein and Schlesinger pointed out this entry may represent a conclusion or a conjecture. Euler had shown [1768, § 334] that

$$\int_0^1 \frac{dx}{\sqrt{1-x^4}} \cdot \int_0^1 \frac{x^2 \, dx}{\sqrt{1-x^4}} = \frac{\pi}{4} \quad \text{and} \quad \varpi = 2\int_0^1 \frac{dx}{\sqrt{1-x^4}}.$$

Stirling [1730–57] had calculated the value of ϖ to 17 decimal places, and had also taken 1 and $\sqrt{2}$ as specific values in computations for the rectification of the ellipse. A letter from Gauss to Pfaff (24 November 1799 = *Werke* X.1, 232) makes it clear that Gauss did not yet have a proof that $M(\sqrt{2},1) = \pi/\varpi$.

99. Stäckel connected this with Gauss's work on the area of a triangle, a topic of central importance for the study of Euclid's parallel postulate. Gauss wrote to Bessel on the matter on 16 December 1799, see *Werke*, VIII, 159.

100–102. Klein and Schlesinger conjectured that these entries refer to Gauss's discovery that the reciprocal of the arithmetico-geometric mean is a solution of a linear differential equation. In fact the arithmetico-geometric mean satisfies Legendre's equation and its inverse function is Gauss's modular function. See Geppert pp. 40–42.

104. A variant of Dirichlet's test, valid provided $\phi \neq 0$.

105, 106. These signify Gauss's realisation of the importance of the arithmetico-geometric mean for the general elliptic integral, and not just for the lemniscatic case.

107. The first to give arithmetical rules for determining Easter was Lambert [1776]. Gauss was familiar with much of Lambert's work but one cannot be certain of an influence in this respect.

108. Only Gauss's later treatments [*Werke* III, 401, 473] have survived; they date from 1825 and 1827.

109. The multiplicity follows from allowing the arithmetico-geometric mean to become a complex function.

110. Elliptic transcendents means elliptic integrals of the first kind.

112. Klein and Schlesinger connected this with the calculation of powers of e.

113. The problem is, given M between 0 and 1, written as a continued fraction

$$M = \cfrac{1}{a' + \cfrac{1}{a'' + \cdots}}.$$

what is the probability $P(n, x)$ that the fraction

$$\cfrac{1}{a^{(n+1)} + \cfrac{1}{a^{(n+2)} + \cdots}}$$

lies between 0 and x. Gauss discussed this problem with Laplace in a letter dated 1812. He found that $P(1, x) = \Psi(x) - \Psi(0)$, where $\Psi(x) = \frac{d}{dx} \log \Pi(x)$ and $\Pi(x)$ is Gauss's factorial function, $\Pi(x) = \Gamma(x + 1)$, and

$$\lim_{n \to \infty} P(n, x) = \frac{\log(1 + x)}{\log 2}. \qquad \text{See [Khinchin §15].}$$

114, 115. See *Werke* II, 285, 286.

116. See D.A. §§365, 366, where this assertion is repeated, but again without a proof. Loewy in his commentary supplied a proof using only the techniques available to Gauss, i.e., avoiding Galois theory.

118. See no. 123. For a good discussion of Gauss sums see [Berndt and Evans, 1981].

119. Gauss tackled the problem of locating Ceres, observed for the first time (by Piazzi) on January 1801 and lost 42 days later when it went too near to the sun. His successful solution was published in Zach's *Comm. Liter.* for December 1801 = *Werke,* VI, 1874, 199–204.

120–122. The theory of planetary orbits, their observation, and the treatment of observational errors was congenial to Gauss, being useful, as it seemed to him, and offering much scope for his calculating prowess.

125. See *Theoria Motus* §§ 88–97.

126. See *Theoria Motus*, II, §§ 115–163.

127. See *Theoria Motus* §§ 33ff.

129. See *Theoria Motus,* II, §§ 164–171.

130–133. Gauss's Theoria Residuorum Biquadraticorum, Commentatio prima of 1825 = *Werke* II, 65–92.

134. The sixth proof of the golden theorem.

135. Art. 358 of the D.A. as rederived in Disquistionum circa aequationes puras ulterior evolutio, *Werke* II, 243.

137. This was taken up in a long and interesting paper by Eisenstein [1844].

138. The quantity ε is not a cube root of unity. but a rational root of the congruence $\varepsilon^2 + \varepsilon + 1 = 0 \pmod{p}$ where p is a prime of the form $3n + 1$.

146. Gauss wrote '=' for '≡'. This, the final entry, has become the most famous. Bachmann commented that the connection between lemniscatic functions with the theory of biquadratic residues remained to be cleared up. A. Weil pointed out [1974, 106] that the substitution $z = y(1 - x^2)$ reduces $1 = x^2 + y^2 + x^2 y^2$ to $z^2 = 1 - x^4$, which makes the connection with lemniscatic functions clear. When taken mod p it is also a question about biquadratic residues, and the investigation of $ax^4 - by^4 \equiv 1 \pmod{p}$ is connected with higher Gauss sums. Weil remarks that it was while studying Gauss's two papers on biquadratic residues that he was led eventually to formulate the Weil conjectures. It is interesting to see that Gauss, by counting infinity, is already thinking of the curve projectively, when the curve is $z^2y^2 = y^4 - x^4$. Had one written $\tilde{z}^2 = z$, the projective curve would have been $\tilde{z}^4 = y^4 - x^4$. Fermat knew, and indeed published a proof [1659], that there are no solutions in integers to $y^2 = z^4 + x^4$ and thus to $\tilde{z}^4 = y^4 - x^4$, so this curve is truly momentous in its significance for mathematics.

Annotated Bibliography

Cyclotomy, number theory, algebra

Gauss's claim to have provided the first rigorous proof of the fundamental theorem of algebra, in 1799, has been found excessive by recent historians (see Gilain [1991] and Baltus [1998]). On the one hand, Gauss's own argument makes claims about algebraic curves that could not be proved until Ostrowski rescued them in 1927. On the other, it is easy enough, using modern techniques, to rescue one of d'Alembert's proofs of the fundamental theorem of algebra. Gauss claimed that d'Alembert's proof showed only that, *if* the roots of a polynomial equation exist, *then* they must be of the form $a + b\sqrt{-1}$, but not that such roots exist. An argument, familiar in embryo to Cauchy and made explicit by Kronecker, shows that one may always adjoin numbers of this form to a field containing the coefficients of a given polynomial so as to factorise the polynomial. Gauss's estimate of the originality of his early work is therefore doubly in question.

On the other hand, none would dispute the power and originality of Gauss's work on the theory of numbers. Dunnington chose to skirt this achievement, and it would take many pages to navigate through it. Better, then, to cite a few works that illuminate the topic from an accessible modern standpoint that Gauss would also recognise: Berndt and Evans' paper [1981], the book by Berndt, Evans, and Williams [1998], and the book by Ireland and Rosen [1982]. Karsten Johnsen has essayed several difficult passages in Gauss's diary, and three of his papers are listed in the bibliography. Readers will also enjoy Waterhouse's papers, one of which, [1994], is a reconstruction of the interaction between

Germain and Gauss that led him to produce a surprisingly large counterexample to one of her remarks.

Geometry and topology

Particular mention should be made of Epple's charming explanation [1999] of a passage in Gauss's *Nachlass*, where Gauss introduced the linking number of two curves in the context of electro-magnetic theory. As Epple points out, this is but one illustration of the roots knot theory has in nineteenth century physics (see also Nash [1999]).

Statistics

Dunnington's treatment is unashamedly pro-Gauss, and in this he has the support of Sprott among the historians of statistics, and the distinguished statistician R.A. Fisher (as quoted in the opening pages of Sprott [1978]). Gauss's priority in the discovery of the method of least squares is generally agreed, and if his claim that he discovered it in 1794, at the age of 17, is hard to find good evidence for it does not mean that it is wrong. Dutka [1996] showed that Gauss was certainly using the method only a few years later, in 1799.

However, as Stigler argued (see especially his book [1986], pp. 141–143) there was also a vigorous French tradition, exemplified by Legendre and Laplace, and Legendre never conceded priority to Gauss. Stigler went on to argue that Gauss's first published defence of it, in 1809, was deeply flawed, because he combined both a non sequitur and vicious circle in the same argument, and was subsequently rejected by Gauss himself. Therefore, the result was better derived from the Central Limit Theorem by Laplace, and indeed by Gauss himself in 1821, and Gauss's appreciation of its generality may be due to his reading of Legendre [1805]. However, Waterhouse [1990] looked again at Gauss's argument, and found that it was neither circular nor vitiated by a gap in its reasoning, although it was true that Gauss did shift his position in response to Laplace [1811], when he discussed what estimate of the 'losses' due to error was the most suitable (expected square error or expected absolute error). Gauss, like his contemporaries, took the arithmetic mean of several equally good observations to be the right estimate of the true value, and deduced that the corresponding distribution of errors would then be that described by the normal distribution. He gave this distribution the same 'axiomatic' status as the ubiquitous assumption that the arithmetic mean was the right estimate to take, because the normal distribution followed, as a mathematical fact, from the assumption about the arithmetic mean.

Stigler also claims that the fame of Gauss's method is due to its absorption by Laplace, but this downplays its use within the German astronomical community. As Darrigol ([2000], p. 44) notes, the Germans were much more committed to error analysis than the French in this period, and astronomy seems to have been the science that set the highest standards of accuracy in the period, as much of Gauss's own work illustrates.

Telegraph and magnetism

The accounts given in Garland [1979] and Darrigol [2000] (pp. 50–54) are much more thorough mathematically than Dunnington's. Two pictures, reproduced from Garland's account (Figures 6a and b, pp. 16–17), give a very clear impression of just how good Gauss's survey of terrestrial magnetism actually was. Darrigol also argues that it was Gauss's insistence on the value of absolute units that drove Weber's originality, which "implied new kinds of instruments with simple geometry and high sensitivity, and required the verification of the laws according to which these instruments were analysed." (Darrigol [2000], p. 74).

Baltus, C. 1998 Lagrange and the fundamental theorem of algebra, *Proceedings of the Canadian Society for the History and Philosophy of Mathematics* **11**, 85–96.

Berndt, B.C. and Evans, R.J. 1981 The determination of Gauss sums, *Bulletin of the American Mathematical Society* **5.2**, 107–130.

Berndt, B.C., Evans, R.J. and Williams, K.S. 1998 *Gauss and Jacobi Sums*, Wiley-Interscience, New York.

Biermann, K.R. 1963 Zwei ungeklärte Schlüsselworte von C.F. Gauss, *Monatsberichte der Deutschen Akademie der Wissenschaften zu Berlin* **5**, 241–244.

Biermann, K.-R. 1970 Carl Friedrich Gauss in Autographenkatalogen. *Schriften zur Geschichte der Naturwissenschaften, Technik und Medizin* **7.1**, 60–65.

Biermann, K-R 1971 Zu Dirichlets geplantem Nachruf auf Gauss, *NTM Schriften zur Geschichte der Naturwissenschaften Technik und Medizin* **8** (1), 9–12.

Biermann, K.-R. 1983 C. F. Gauss als Mathematik- und Astronomiehistoriker, *Historia Mathematica* **10.4**, 422–434. A survey, written by Gauss, of the previous 100 years.

Biermann, K-R. 1990 (ed.) Carl Friedrich Gauss, Der "Furst der Mathematiker" in *Briefen und Gesprächen*, Leipzig, Urania-Verlag.

Breitenberger, E. 1984 Gauss's Geodesy and the Axiom of Parallels, *Archive for History of Exact Sciences* **29**, 273–289.

Breitenberger, E. 1993 Gauss und Listing: Topologie und Freundschaft, *Mitteilungen der Gauss-Gesellschaft, Göttingen* **30**, 3–56.

Bühler, W.K. 1981 *Gauss: A Biographical Study*. New York, Springer-Verlag, (bibliography lists volumes of correspondence between Gauss and others, up to 1981).

Collison, M.J. 1977 The origins of the cubic and biquadratic reciprocity laws, *Archive for History of Exact Sciences* **17.1**, 63–69.

Cox, D.A. 1984 The arithmetic-geometric mean of Gauss, *Enseignement Mathématique* (2) **30**, 275–330.

Cox, D A. 1985 Gauss and the arithmetic-geometric mean, *Notices of the American Mathematical Society* **32** (2), 147–151.

Coxeter, H.S.M. 1977 Gauss as a geometer, *Historia Mathematica* **4.4**, 379–396.

Darrigol, O. 2000 *Electrodynamics from Ampère to Einstein*, Oxford University Press, Oxford.

De Moivre A. 1730 *Miscellanea analytica de seriebus et quadraturis*, London.

Dick, W. R. 1992 Otto Struve über Carl Friedrich Gauss. *Mitteilungen der Gauss-Gesellschaft, Göttingen*, **29**, 43–51.

Dick, W. R. 1993 Martin Bartels als Lehrer von Carl Friedrich Gauss, *Mitteilungen der Gauss-Gesellschaft, Göttingen*. **30**, 59–62.

Dieudonné, J. 1962 L'oeuvre mathématique de C. F. Gauss, Paris, Conférence du Palais de la Découverte D. 79.

Dieudonné, J. 1978 Carl Friedrich Gauss: a bicentenary, *Southeast Asian Bulletin of Mathematics* **2.2**, 61–70.

Dombrowski, P. 1979 150 Years after Gauss' *Disquisitiones generales circa superficies curvas, astérisque*, **62** (with the original text of Gauss and an English translation by A. Hiltebeitel and J. Morehead).

Dutka, J. 1996 On Gauss' priority in the discovery of the method of least squares, *Archive for History of Exact Sciences* **49.4**, 355–370.

Eisenstein, G. 1844 Geometrischer Beweis des Fundamentaltheorems für die quadratischen Reste, *Journal für die reine und angewandte Mathematik* **28**, 246–248, in *Mathematische Werke*, 1975, **1**, 164–166.

Eisenstein, G. 1847 Genaue Untersuchungen der unendlichen Doppelproducte, *Journal für die reine und angewandte Mathematik* **35**, 153–247, in *Mathematische Werke*, 1975, **1**, 357–478.

Epple, M. 1999 Geometric aspects in the development of knot theory, I.M. James (ed.), *History of topology*, North-Holland, 301–357.

Euler, L. 1748 *Introductio in analysin infinitorum*, in *Opera Omnia*, series 1, vols. 8–10.

Euler, L. 1752/53 De numeris, qui sunt aggregata duorum quadratorum, *Novi Comm. Acad. Petrop.* **4**, 3–40 in *Opera Omnia*, series 1, **2**, 295–327.

Euler, L. 1768 Institutiones Calculi integralis, I, in *Opera Omnia*, series 1, **11**.

Fermat, P. 1659 Letter to Carcavi, *Oeuvres*, II, 431–436.

Festschrift zum 200. Geburtstag von Carl Friedrich Gauss, *Mitteilungen der Gauss-Gesellschaft, Göttingen* **14**. Göttingen, Verlag Erich Goltze, 1977.

Forbes, E.G. 1978 The astronomical work of Carl Friedrich Gauss (1777–1855), *Historia Mathematica* **5.2**, 167–181.

Fuchs, L.I. 1863 Über den Perioden, welche aus den Wurzeln der Gleichung $\omega^n = 1$ gebildet sind, wenn n eine zusammengesetze Zahl ist, *Journal für die reine und angewandte Mathematik* **61**, 374–386, in *Gesammelte Mathematische Werke*, **1**, 1904, 53–67.

Fuchs, W. 1972 Das arithmetisch-geometrische Mittel in den Untersuchungen von Carl Friedrich Gauss. *Mitteilungen der Gauss-Gesellschaft, Göttingen* **9**, 14–38.

Fuchs, W. 1978 Die Leiste-Notizen des jungen Gauss. *Mitteilungen der Gauss-Gesellschaft, Göttingen* **15**, 19–38.

Fuchs, W. 1980, 1981, 1982 Zur Lehre von den Kongruenzen bei C. F. Gauss I, II, III, *Mitteilungen der Gauss-Gesellschaft, Göttingen* **17**, 14–29; **18**, 49–62; **19**, 63–78.

Garland, G. D. 1979 The contributions of Carl Friedrich Gauss to geomagnetism, *Historia Mathematica* **6.1**, 5–29.

Geppert, H. 1927 *Bestimmung der Anziehung eines elliptischen Ringes: Nachlass zur Theorie des Arithmetischen-Geometrischen Mittels und der Modulfunktion von Carl Friedrich Gauss*, Ostwald Klassiker 225, Leipzig.

Gilain, C. 1991 Sur l'histoire du théorème fondamental de l'algèbre: théorie des équations et calcul intégral, *Archive for History of Exact Sciences* **42.2**, 91–136.

Goe, G. and van der Waerden, B. L. 1972 Comments on Miller's 'The myth of Gauss's experiment on the Euclidean nature of physical space', *Isis* **63**, 345–348. With a reply by Arthur I. Miller, *Isis* **65** (1974), 83–87.

Gray, J.J. 1984 A commentary on Gauss's mathematical diary, 1796–1814, with an English translation, *Expositiones Mathematicae* **2**, 97–130 (reproduced with minor alterations in this volume).

Gray, J.J. 2000 *Linear differential equations and group theory from Riemann to Poincaré,* Birkhäuser, Boston and Basel, second edition with three new appendices and other additional material.

Hall, T. 1970 *Carl Friedrich Gauss A Biography*. Translated from the Swedish by Alfred Froderberg. Cambridge, MA, MIT Press.

Ireland, K. and Rosen, M. 1982 *A Classical Introduction to Modern Number Theory*, Springer Verlag, New York and Berlin.

Ivory, J. 1798 A new series for the rectification of the ellipsis; together with some observations on the evolution of the formula $(a^2 + b^2 - 2ab\cos\phi)^n$. *Trans Royal Soc. Edinburgh* **4**, part II, 177–190.

Johnsen, K. 1982 Bemerkungen zu einer Tagebuchnotiz von Carl Friedrich Gauss. *Historia Mathematica* **9.2**, 191–194.

Johnsen, K. 1984 Zum Beweis von C. F. Gauss für die Irreduzibilität des p-ten Kreisteilungspolynoms, *Historia Mathematica* **11.2**, 131–141.

Johnsen, K. 1986 Remarks to the third entry in Gauss' diary, *Historia Mathematica* **13.2**, 168–169.

Khinchin, A. Ya. 1964 *Continued Fractions*. tr. H. Eagle, University of Chicago Press, Chicago and London.

Kummer, E.E. 1856 Theorie der idealen Primfaktoren der complexen Zahlen, welche aus den Wurzeln der Gleichung $\omega^n = 1$ gebildet sind, wenn n eine zusammengesetze Zahl ist, *Abh. der Wissenschaften zu Berlin, Math. Abt.*, 1–47 in *Coll. Papers*, I, 583–629.

Lambert, J.H. 1776 *Einige Anmerkungen über die Kirchenrechnung*, Astron, Jahrbuch für das Jahr 1778, Berlin, 210.

Laplace, P.S. 1793 Sur quelques points du Système du monde, *Mémoire de l'Académie royale des Sciences de Paris*, in *Oeuvres Complètes*, XI, 477–558.

Laubenbacher, R. and Pengelley, D. 1994 Eisenstein's misunderstood geometric proof of the quadratic reciprocity theorem, *College Mathematics Journal* **25**, 29–34.

May, K.O. 1972 Biography of Gauss in *Dictionary of Scientific Biography* (New York 1970-1990) vol. 5, 298–315.

Merzbach, U. 1981 An early version of Gauss's *Disquisitiones Arithmeticae, Mathematical perspectives,* J. Dauben (ed.) 167–177, Academic Press, New York.

Merzbach, Uta 1984 C. *Carl Friedrich Gauss: a Bibliography.* Wilmington, DE: Scholarly Resources.

Miller, A. I. 1972 The myth of Gauss' experiment on the Euclidean nature of physical space, *Isis* **63**, 345–348.

Monna, A. F. (ed.) 1978 *Carl Friedrich Gauss 1777–1855. Four Lectures on His Life and Work.* Utrecht, Rijksuniversiteit Utrecht, Mathematical Institute.

Müürsepp, P. 1977 Gauss' letter to Fuss of 4 April 1803, *Historia Mathematica* **4.1**, 37-41.

Müürsepp, P. 1978 Gauss and Tartu University, *Historia Mathematica* **5.4**, 455–459.

Nash, C. 1999 Topology and physics: a historical essay, I.M. James (ed.), *History of topology*, North-Holland, 359–415.

Neumann, O. 1981 ed. *Mathematisches Tagebuch,* 1796–1814. With a historical introduction by Kurt-R. Biermann, translated from the Latin by Elisabeth Schuhmann, with notes by Hans Wussing. Third edition. Ostwalds Klassiker der Exakten Wissenschaften, **256**. Leipzig, Akademische Verlagsgesellschaft Geest & Portig K.-G.

Newton, I. 1676 Epistola Prior in *The Correspondence of Isaac Newton*, II 32–41, Cambridge University Press, 1959.

O'Hara, J.G. 1983 Gauss and the Royal Society: the reception of his ideas on magnetism in Britain (1832–1842), *Notes and Records of the Royal Society of London* **38** (1), 17–78.

Ore, Ø. 1971 Abel, *Dictionary of Scientific Biography,* I, 11–18, Scribner's, New York.

Pasch, M. 1882 *Vorlesungen über neuere Geometrie*, Teubner, Leipzig.

Reich, K. 1977 *Carl Friedrich Gauss 1777/1977*. Munich, Moos.

Reich, K. 1998 Gauss' Theoria motus: Entstehung, Quellen, Rezeption. *Mitteilungen der Gauss-Gesellschaft, Göttingen* No. 35, 3–15.

Reichardt, H. 1976 *Gauss und die nicht-euklidische Geometrie*, Teubner, Leipzig.

Reichardt, H. 1983 Gauss, in H. Wussing and W. Arnold (eds.), *Biographien bedeutender Mathematiker*, Berlin.

Reichardt, H. ed. 1957 C. F. Gauss Gedenkband anlässlich des 100. Todestages am 25. Februar 1955, Leipzig.

Rowe, D.E. 1988 Gauss, Dirichlet and the Law of Biquadratic Reciprocity, *The Mathematical Intelligencer* **10**, 13–26.

Rüdiger, T. 1994 Mathematics in Göttingen (1737–1866), *The Mathematical Intelligencer* **16.4**, 50–60.

Schaaf, W.L. 1964 *Carl Friedrich Gauss, Prince of Mathematicians.* Immortals of Science series. New York, Franklin Watts.

Schlesinger, L. 1917 Über Gauss's Arbeiten zur Funktionentheorie, Gauss's *Werke*, X.2.

Schneider, I. 1981a Carl Friedrich Gauß (1777–1855) Arbeiten im Rahmen der Wahrscheinlichkeitsrechnung: Methode der kleinsten Quadrate und Versicherungswesen, in Schneider (ed.) [1981], pp. 143–172.

Schneider, I. 1981b Herausragende Einzelleistungen: Kreisteilungsgleichung, Fundamentalsatz der Algebra und Konvergenzfragen, pp. 37–63 (*see* Schneider, [1981a]).

Schneider, I. 1990 Gauss' Contributions to Probability Theory, in: M. Behara (ed.), Symposia Gaussiana, Series A: Mathematics and Theoretical Physics, Vol. I, , Berlin-Toronto-Sao Paulo, pp. 72–84.

Schneider, Ivo (ed.) 1981 *Carl Friedrich Gauß (1777–1855) Sammelband von Beitragen zum 200. Geburtstag von C. F. Gauss,* Minerva Publikation, München.

Scholz, E. 1992 Gauss und die Begründung der "höheren" Geodäsie, S.S. Demidov, M. Folkerts, D.E. Rowe, C.J. Scriba, (eds.) *Amphora*, Birkhäuser Verlag, Basel, Boston and Berlin, 631–648.

Schuhmann, E. 1976 Vicimus GEGAN, Interpretationsvarianten zu einer Tagebuchnotiz von C.F. Gauss, *Naturwiss. Tech. Medezin.* **13.2**, 17–20.

Sheynin, O. 1994 C.F. Gauss and geodetic observations, *Archive for History of Exact Sciences* **46.3**, 253–283.

Sheynin, O. 1995 Helmert's work in the theory of errors, *Archive for History of Exact Sciences* **49.1**, 73–104.

Sheynin, O.B. 1979 C.F. Gauss and the theory of errors, *Archive for History of Exact Sciences,* **20.1**, 21–72.

Sprott, D.A. 1978 Gauss's contribution to statistics, *Historia Mathematica* **5.2**, 183–203.

Stäckel, P. 1917 Gauss als Geometer, in Gauss *Werke*, X.2, Abh. 4, separately paginated.

Stewart, G.W. 1995 Gauss, statistics, and Gaussian elimination, *J. Comput. Graph. Statist.* **4** (1), 1–11.

Stigler, S.M. 1977 An attack on Gauss, published by Legendre in 1820, *Historia Mathematica* **4.1**, 31–35.

Stigler, S.M. 1981 Gauss and the invention of least squares, *Annals of Statistics.* **9** (3), 465–474.

Stigler, S.M. 1986 *The History of Statistics. The Measurement of Uncertainty before 1900*, Cambridge, Mass.-London.

Stirling, J. 1730–57 *Methodus Differentialis,* etc. London.

Szénassy, B. 1980 Remarks on Gauss's work on non-Euclidean geometry (Hungarian), *Mat. Lapok* **28** 1–3, 133–140.

Waltershausen, W.S. von 1966 *Gauss, a Memorial* translated from the German by Helen W. Gauss, (Colorado Springs, Colorado).

Waltershausen, W.S. von 2000 Gesprochen auf der Terrasse der Sternwarte an Gauss offenem Sarge (Feb. 26, 1855). Transcribed by Stefan Kramer. *Mitteilungen der Gauss-Gesellschaft, Göttingen* No. 37, 101–103.

Waterhouse, W.C. 1979 Gauss on infinity, *Historia Mathematica* **6.4**, 430–436.

Waterhouse, W.C. 1986 A neglected note showing Gauss at work, *Historia Mathematica* **13.2**, 147–156.

Waterhouse, W.C. 1990 Gauss's first argument for least squares, *Archive for History of Exact Sciences* **41.1**, 41–52.

Waterhouse, W.C. 1994 A counterexample for Germain, *American Mathematical Monthly* **101**, 140–150.

Watson, G.N. 1962 *Theory of Bessel Functions,* 2nd ed. Cambridge University Press, London.

Weil, A. 1974 Two lectures on number theory, past and present, *Enseignement Mathématique*, **20**, 87–110, in *Oeuvres scientifiques*, III, 279–303, Springer Verlag, New York.

Weil, A. 1976 Review of *Mathematische Werke* by Gotthold Eisenstein, Chelsea, New York, in *Bulletin of the American Mathematical Society* **82.5**, 658–663, in *Oeuvres scientifiques*, III, 398–403, Springer Verlag, New York.

Worbs, E. 1955 *Carl Friedrich Gauss Ein Lebensbild*. Leipzig, Koehler & Amelang.

Wussing, H. 1976 *Carl Friedrich Gauss*. (2nd ed.) Biographien Hervorragender *Naturwiss. Tech. Medezin.* **15**, Leipzig, Teubner.

Wussing, H. 1999 Implicit group theory in the domain of number theory, especially Gauss and the group theory in his *Disquisitiones arithmeticae* (1801), in Circe Mary Silva da Silva (ed), *III Seminário national de história da matemática,* Brasil, 114–125.

Zormbala, K. 1996 Gauss and the definition of the plane concept in Euclidean elementary geometry, *Historia Mathematica* **23.4**, 418–436.

Index

Abel, Niels Henrik, 112, 210, 211, 217, 251n, 255, 273, 274
Academy, Royal Irish, 346
Academy of Sciences, Bavarian, 151; Russian (St. Petersburg), 151, 289
Academy of Sciences in Berlin. *See* Prussian Academy of Sciences
Academy of Sciences in Breslau, 275
Account of the Rev. John Flamsteed, the first Astronomer Royal, by Francis Bailey, 250
Achenwall, Gottfried, 23
Adams, J. C., 285
Adams, John Quincy, 281
Adrain, Robert, 20
Aegidienstrasse 5, Brunswick address of Gebhard Gauss, 88
Ahort, maker of telescope, 75
Ahrenholz, Johanna Maria Christine, mother of Johanna Osthoff Gauss, 62
Ahrens, J. H., 143
Airy, G. B., 20, 157, 280
Alaska, 388
Albers, H. C., geodesist, 119
Albrecht, Wilhelm Eduard, signer of Göttingen protest to King over revocation of liberal constitution, 197; dismissed from the university, 197; recipient of honorary doctorate, 198; to Leipzig, 198; Gauss' feeling in regard to, 201
Alembert, d'. *See* D'Alembert
Alexander, sailing vessel, 204
"Algebraic Equations," Gauss' "Contributions to the Theory of," 276
Aller, Gauss survey of, 137, 138
Allgemeine Literaturzeitung, notice from, pertaining to discovery of heptadecagon, 28
"Allgemeine Theorie des Erdmagnetismus," 158; translation of, 262

"Als ich ein Junggeselle war," a song, 228
Alsen, Island of, Gauss survey of, 118
Altenburg, Germany, 267, 268
Altona (home of H. C. Schumacher), 82, 88n, 125, 129, 131, 135, 283, 285
America, 204, 219, 236, 240, 247, 281, 283, 284, 285, 286, 372
American Academy of Arts and Sciences in Boston, 281
American Almanac, 242
American Fur Company, 369
American Journal of Science and Arts, ed. by B. Silliman, 283
American literature, 241
Ammensen, Gauss survey of, 122
Ampère, André Marie, 153, 163
Anacreon, 241
Andrä, astronomer, 20
Angelrodt, N. N., merchant in St. Louis, Mo., 328
Angos, d'. *See* D'Angos
Angstrom, A. J., 287
Annalen, Poggendorff's, 154
Annuaire, Quetelet's, 234
Ansbach, Germany, 269
Apel, mechanic, 286
Apensen, Gauss survey of, 131
Appendix scientiam spatii absolute veram exhibens, by Johann Bolyai, 183, 184
Arago, Dominique François Jean, 218, 295
Archimedes, 28n, 211, 218, 231, 273, 275, 303
Aristotle, 241, 314
Arithmetic. *See* Gauss, Carl Friedrich, Research
Arnswaldt, Karl Friedrich Alexander, Baron von, 118, 121, 191, 197

507

Arrest, d'. *See* D'Arrest
Astor Library in New York, 281
"Astral magnitudes, theory of," 181
Astronomical Journal, ed. by B. A. Gould, 283, 284
Astronomical Society, Royal, 345
Astronomische Nachrichten, ed. by H. C. Schumacher, 88n, 209, 211, 223, 273, 275
Astronomy. *See* Gauss, Carl Friedrich, Research *and* Later Years
Athens, Greece, 225
Atlas des Erdmagnetismus, by Gauss and Weber, 159
Auerstedt, Germany, 80, 81
Augsburg, Germany, 109
Aula, administrative building, University of Göttingen, 191, 192, 276
Aulus Gellius, 241
Austrian bank stock, 235
Austrian bonds, 247
Austrian metal league, 237
Autopsy on Gauss' body, 323, 324

Babbage, Charles, 280n
Bach, Johann Sebastian, 23, 229
Bache, A. D., American friend of Gauss, 282, 286, 287
Bacon, Francis, 313
Baden-Baden, Germany, 134
Baden bonds, 237
Baden, survey of, 134
Bad Rehburg, Germany, 75, 115
Baeyer, General, 138
Bailey, Francis, 250
Bamberg, Germany, 270
Bancroft, George, 281
Banks, Sir Joseph, President, Royal Society of London, 119
Bank stock, Austrian, 235; of Brussels, 235; of Vienna, 237
Barlhof, living quarters of Gauss, 126, 131
Bartels, Johann Christian Martin, 13, 14, 188, 189, 255, 290
Bauermeister, Eduard Wilhelm, husband of Gauss' niece, 9

Baum, Wilhelm, physician to Gauss, 75, 263, 319, 321, 323, 328
Baumann, instrument maker, 75, 106
Bavarian Academy of Sciences, 151
Bäzendorf, Gauss' lodgings, 128
Becker, August, 272
Beethoven-Gauss, 347
Behrend, Friedrich, godfather of C. F. Gauss, 61
Beigel, 114
Belgian bonds, 237
Bellmann, song writer, 229
Benediktbeuren, seat of optical works, 107, 108, 109
Benfey, Theodor, 225
Benze, Andreas, founder of Benze line, 9, 10
Benze, Christoph, maternal grandfather of Gauss, 8, 9
Benze, Christoph Andreas, 10
Benze, Dorothea. *See* Gauss, Dorothea (Benze)
Benzenberg, Johann Friedrich, 26, 253
Berchtesgaden, Germany, 108, 109
Bergen, Gauss survey of, 126
Bergmann, F. C., prorector, University of Göttingen, 191, 192, 197
Berlin, 136, 194, 195, 199, 245, 283
Berlin, Academy of Sciences in. *See* Prussian Academy of Sciences
Berlin, University of, 198, 303, 313
Bernoulli, Daniel, physicist, 160, 270
Bertram heliotrope, 124
Bertrand, J., editor, 114n, 349
Bessel, Charlotte Friederike Amalie (Fallenstein), mother of Wilhelm Gauss' wife, 372
Bessel, Friedrich Wilhelm, astronomer, 20, 208, 313, 318, 419; biographical data of, 76; beginning of friendship with Gauss, 76–77; disapproves of Gauss' time spent on survey, 120; visits Gauss at Zeven, 132; his error in calculation discovered by Gauss, 171; his niece marries Gauss' son, 194; urges Gauss to greater speed in

INDEX 509

Bessel, Friedrich Wilhelm (cont'd.) publication, 211; bows to authority of Gauss, 215; his death, 226. See also Gauss, Carl Friedrich, Correspondence and Friendships
Bible, 303, 304n, 305, 307, 308, 311, 314
Bienaymé, 20
Biermiller, painter of copy of Jensen portrait of Gauss, 207
Biester, Gauss' assistant, 131
Bilk, Observatory of, 288
Biot, Jean Baptiste, physicist and astronomer, 153
Bird's wall quadrant, 96
Black Forest, Germany, 134
Blankenhayn, Germany, 80
Blankenese, Gauss survey of, 128
Blätter der Wahrheit, magazine, 263
Blohm, inspector, 128
Blücher, Field Marshal von, 80
Blunt, Edward, 286
Bode, Georg Heinrich, 219
Bode, Johann Elert, 49, 50, 51, 55, 58, 219
Bodleian Library, Oxford, 199
Bogenhausen, Observatory of, 270
Boguslawski, P. H. L. von, 216
Böhmer, Georg Ludwig, German jurist at Göttingen, 23, 246n
Bohnenberger, J. G. F., astronomer, 134, 140, 141, 162
Boileau, French critic and poet, 241
Boineburg, Gauss survey of, 116
Bois-Reymond, Emil du, 296
Bolotoff, on Russian pronunciation, 238
Bölsche, Carl, 192
Bolyai, Johann, son of Wolfgang, 174; research on theory of parallels by, 178, 183; publication by, on theory of parallels, 183; Gauss' estimate of, 184, 215; letter from his father concerning theory of parallels, 184; priority of Lobachevsky's published work over that of, 187; his contribution to liberation from Euclidean tradition, 189
Bolyai von Bolya, Wolfgang, 26, 27, 28, 29, 30, 176, 177, 178, 183, 184, 187, 215, 226, 240, 312. See also Gauss, Carl Friedrich, Friendships and Correspondence
Bombay, magnetic station at, 157
Bond, Geo. P., son of W. C. Bond, 284, 285
Bond, W. C., 284
Bonds, Austrian, 247; Baden, 237; Belgian, 237; Bückeberg municipal, 237; Hanoverian, 237; Prussian, 237; Russian, 237; Swedish, 237; of Württemberg, 237
Bonn, University of, 198
Boone County, Mo., 371
Borch, Rudolf, 4
Borda circle, 118
Börsch, editor, 114n
Borstorf (apples), 216
Böse, name recalled by Gauss, 309
Boston Almanac, 242
Boston, Mass., 281
Bottel, Gauss survey of, 130, 131
Bougainville, 32
Bouguer, astronomer, 124
Bourke, von, Danish envoy in London, 119
Bowditch, Nathaniel, 281, 282, 285
Braak, Gauss survey of, 121
Brahms, Johannes, 229
Bramstedt, birthplace of H. C. Schumacher, 88n
Brandenburg, survey of, 121
Brandes, Ernst, 85, 289
Brandes, Heinrich Wilhelm, 26, 140
Braubach's *Seelenlehre*, 310
Braunschweiger Monatsschrift, publication of Gauss–Zimmermann letters in, 16, 17n
Braunschweigisches Magazin, publication of Gauss' articles relative to Easter date formula in, 70
Bredenkamp, friend of Eugene Gauss, 367
Brehmer, Friedrich, sculptor, 322

Breithorn, Gauss survey of, 126, 127
Breitkopf, printer, 252
Bremen, Germany, 50, 55, 56, 62, 74, 76, 85, 87, 93, 128, 129, 130, 132, 133, 138, 204, 236, 251, 262, 269, 309, 367
Bremerlehe, Gauss survey of, 133
Brendel, Martin, 336
Breslau, Germany, 26, 157
Breslau, Academy of Sciences in, 275; University of, 275
Brewster, Sir David, 151, 280
Brigard, 241
Brillit, Gauss survey of, 130, 133
British Museum, London, 199
Brocken, Germany, 115, 116, 124, 125, 128, 129, 337
Broitzen, Gauss survey of, 114
Brückmann, Urban Friedrich Benedikt, 34
Brühl, Count, 290
Brune's mortality tables, 234
Brünnow, F. F. E., 288
Bruns, librarian at Helmstedt, 35
Brunswick, 4, 5, 6, 13, 30, 33, 34, 37, 82, 93, 115, 145, 165, 176, 259, 276, 278, 306, 320, 335, 336
Brunswick, Duchy of, trigonometric survey of, 137, 361
Brunswick, Duke of. *See* Ferdinand, Carl Wilhelm
Brunswick dialect, 260
Brussels bank stock, 235
Brüttendorf, Gauss survey of, 130, 132
Buch, Leopold von, 293
Bückeberg municipal bonds, 237
Budget of Paradoxes, by Augustus De Morgan, 250
Bullerberg, Gauss survey of, 128, 130
Burckhardt, J. K., 50, 51, 52
Bürg, Viennese astronomer, 70, 71, 72, 115
Burr, Aaron, 280
Busche, General von der, 143
Büttner, J. G., first teacher of Gauss, 12, 13
Buzengeiger, a teacher, 269

Byron, Lord, Gauss' reaction to his mode of thought, 243; weight of his brain, 324

Cajori, Florian, comments of, on *Theoria motus,* 90
"Calculations and Notes concerning the Observations at the Meridian Circles," notebook of Gauss, 110
Calendar. *See* Easter date formula; Jewish calendar
California, 338
Cambridge, Adolf Friedrich, Duke of, 143, 148
Cambridge University, 167, 254, 276, 284
Campe, Joachim Heinrich, 37
Cannstadt, Germany, 243
Cantor, Georg, 413
Cantor, Moritz, 260, 275; comments on Gauss' Latin, 38; reminiscences of Gauss, 257, 258; reputation of, 259; his estimate of Gauss, 349
Cape of Good Hope, magnetic station at, 157
Carnot's *Géométrie de position,* translated by H. C. Schumacher, 221
Caroc, Captain, assistant to Gauss in Lüneburg survey, 118
Carolineum. *See* Collegium Carolinum
Carpenter's *Mental Physiology,* 230
Catholicism, Roman, 302
Cauchy, A. L., 40, 273, 274
Cayley, Arthur, 345
Celle, Germany, 50, 125, 365, 372
Ceres (Ferdinandea), 306, 310, 336n; discovery of, 49; studies concerning, 50–54; sentiment for, 77; manuscript material concerning, acquired by Astor Library, 282. *See also* Cogswell, J. G.
Ceylon, magnetic station on, 157
Charkov, Russia, 180
Chauvenet, 20
Chemnitz, song writer, 229
Chicago World's Fair, 336
Christianity, 369

INDEX 511

Christie, 157
Cicero, 241
Cincinnati College, 283
City College of New York, 282
Clairaut, 71
Clausen, Thomas, 131, 227
Clausius, 274, 311
Clausthal, 30, 31, 176
Cogswell, J. G., 281; acquisitions by, of Gauss publications and manuscripts for Astor Library, 281, 282
Colborn, Hanover, 270
Collected Works of C. F. Gauss, published in Gauss' honor, 337; Gauss papers preserved for, 358; *passim*
"Collection of Knots, A," MS note by Gauss, 222
Collegium Carolinum (Carolineum), 13, 14, 259, 306; curricula and importance of, 15–18, 21; influence of, upon Gauss, 17–21; congratulatory letter of, to Gauss on occasion of his jubilee, 278
Columbia, Mo., 371
Condorcet, Marquis de, 241
Connaissance des temps, journal, 256
"Contributions to the Theory of Algebraic Equations," Gauss' jubilee lecture, 276
Cooper, James Fenimore, 241, 285, 287
Cooper, Thomas, 286
Copenhagen, Denmark, 116
Copenhagen Society of Sciences, 163
Copernican system, 262, 263
Cordova, Argentina, site of national observatory, 284
Cornelius Nepos, 241
Correspondence. See Gauss, Carl Friedrich, Correspondence
Coulomb, Charles Augustin de, 162
Crabb, George, *English Grammar* by, 241
Cracow, Poland, magnetic station at, 157
Credit union of Württemberg, 236
Crelle's Journal, scientific articles published in, 45, 69n, 185

Crofton, Professor, 91n
Crystallography. See Gauss, Carl Friedrich, Research
Cumberland, Ernst August, Duke of, 143
Curtius Rufus, 241
Curved surfaces. See Gauss, Carl Friedrich, Research
Cuvier, G. C. L. D., Baron, weight of his brain, 324
Cyclometric equations, Gauss' theory of, 28–30

Daguerreotypes of Gauss, 324
Dahlmann, Friedrich Christoph, 195, 196; signer of Göttingen protest to King over revocation of liberal constitution, 197; dismissed from university, 197, 198; to Bonn, 198
Dalberg, Karl Theodor Anton Maria, Baron von, Grand Duke of Frankfort, Prince Primate, Chamberlain of Worms, Archbishop of Regensburg, gift to Gauss from, 87; medal to Gauss from, 92
D'Alembert, Jean le Rond, 36, 71
Dammert, Senator, Hamburg chief of police, 367
D'Angos, Chevalier, 254
Danish triangulation, 133
Dante, weight of his brain, 324
Danzig, 179
Darmstadt, Germany, 134, 265
D'Arrest, H. L., 285, 286
Davis, Charles Henry, Rear Admiral, USN, translator of *Theoria motus* into English, 90n, 282
Davis, Jefferson, 368
Davy, John, 309n
Davy, Sir Humphrey, 309
Dedekind, Richard, pupil of Gauss, 48, 258; his reminiscences of Gauss, 259–261; his close association with Gauss, 267, 324
De Foncenex, mathematician, 36
Deister, Gauss survey of, 125, 126, 127
Delambre, J. B. J., 70, 90n, 93

De Luc, Swiss geologist, 241
De Morgan, Augustus, 250, 254, 294
Denmark, survey of, 88n, 116, 118, 121
Denver, Colo., 324
Descartes, René, 296, 311, 313
"Descartes' Rule of Signs," paper by Gauss, 45
Detmold, Germany, 240
Deutsches Museum at Munich, 336
De Villers, Charles F. D., 240
Dictionaries: Flügel's English-German, 241; C. P. Reiff's Russian-French, 238; Schmidt's Russian-German, 238n; *Johnson's Dictionary of the English Language*, 241; French-Polish, 242
Didaskalia, journal, 247
Diogenes Laërtius, 241
Diophantus, 43
Dioptrische Untersuchungen, by C. F. Gauss, 171, 172
Dirichlet, P. G. Lejeune, 46, 187, 217, 218, 245, 261, 271, 275, 276, 302, 323, 324, 347
Disquisitiones arithmeticae, first major work of C. F. Gauss, 37–45, 208, 210, 211, 218, 251, 289, 295, 304, 349
Disquisitiones generales circa seriem infinitam, memoir by C. F. Gauss, 104, 164–166
Dissen, Ludolf, 194
Doktor Katzenbergers Badereise, by Jean Paul, 239, 240
Donati's comet, 285
Dopmeyer, C., sculptor, 322
Dorpat, University of, 289
Dousman, Hercules, 369
Dove, R. W., 347
Dransfeld, Germany, 122, 337
Dräseke, 18
Dresden, Germany, 266, 291, 374
Drobisch, M. W., 314
Drude, F. L. H., Gauss' childhood teacher, 279
Drygalski, Erich von, 338
Dublin, Ireland, 254, 262
Dublin, magnetic station at, 157
Dublin, University of, 289
Dudley Observatory, Albany, N. Y., 283
Dühring, Eugen, 251, 273, 274, 311
Earth magnetism. See Gauss, Carl Friedrich, Research
Easter date formula, Gauss' discovery of, 69, 70
Eberlein, Gustav, sculptor, 338
Ebert, teacher, 17, 18
Eckermann, J. P., 302
Eckhardt, 134
Edinburgh, Scotland, 250
Education of Gauss. See Gauss, Carl Friedrich, Education
Eggelings, Katharine Magdalene (wife of Jürgen Gauss), 5
Eichhorn, Johann Albrecht Friedrich, of Wertheim, 26
Einstein, Albert, 316; on importance of Gauss, 349, 350
Eisenstein, Gotthold, pupil of Gauss, 45, 212, 267, 274, 275, 283, 307, 308, 348
Elba, Napoleon's exile at, 258
Elbhöhe, site of private Observatory of Repsold, 97
Electricity and magnetism. See Gauss, Carl Friedrich, Research
Elliptic modular functions, Gauss' mastery of, 47
Elmhorst, Gauss survey of, 130
Elsterwerda, Germany, 374
Emerson, William, 281
Emperius, Johann Ferdinand Friedrich, 18
Encke, Johann Franz, pupil of Gauss, 20, 91, 110, 111, 117, 124, 132, 156, 169, 210, 226, 253, 262, 267, 272, 275, 283, 285, 303, 348
Ende, von, astronomer, 50, 114
English Grammar, by George Crabb, 241
English literature, 239
Enneper, Alfred, pupil of Gauss, 272

INDEX 513

Epailly, military director of surveys in occupied zone of Germany, 115, 117, 121
Ephemerides, Nautical Almanac, 70
Erfurt, Germany, 80
Erlangen, University of, 23, 268, 269, 270
Erman, G. A., 245
Ernesti, 17
Ernst August, Duke of Cumberland, successor to William IV, 195; revokes liberal constitution of 1832, 195; reasons for this act, 196; character of new king, 196; reaction to his decree, 196–198
Ertel, Traugott Lebrecht, owner of instrument firm, 108
Erythropel, Sophie. See Gauss, Sophie
Eschenburg, Arnold Wilhelm, 25; his friendship with Gauss, 26, 240, 278
Eschenburg, August, son of A. W. Eschenburg, 279
Eschenburg, J. J., father of A. W. Eschenburg, 18, 279. See also Eschenburg, Arnold Wilhelm
Esenbeck, Nees von, 314
Espy, James P., 283
Essai philosophique sur les probabilités, by Laplace, 258
Essai sur la théorie des nombres, treatise by Legendre, 48
Essenrode, Germany, 4
Eucken, Rudolf, 315
Euclid, system of, 41, 175, 177, 231, 274. See also Non-Euclidean
Euler, Leonhard, 18, 27, 36–55 *passim*, 71, 72, 104, 165, 168, 221, 293, 347
Euler's *Nova theoria musicae*, 229
Euripides, 241
Europe, 204, 216, 247, 284, 370
European languages, 237
Evangelical Lutheran Church, 300
Everett, Edward, 281
Ewald, Auguste (Schleiermacher), second wife of Heinrich Ewald, 265

Ewald, Caroline Therese Wilhelmine (daughter of Heinrich and Auguste Ewald), birth and christening of, 265; her family history and death, 266
Ewald, Georg Heinrich August von (son-in-law of Gauss), scholarship of, 145; his marriage to Minna Gauss, 146; his academic interests, 146; publications of, 146; as signer of Göttingen protest to King over revocation of liberal constitution, 197; dismissed from university, 197, 198; effect upon his way of life, 198; to Tübingen, 198, 200; to England, 199; his life in Tübingen, 205, 206; recalled to Göttingen, 244, 265; death of wife Minna, 206; his second marriage, 265, 364; daughter born to, 265, 364; his family history, 266; delivered Gauss' funeral sermon, 326, 331–333; his absentmindedness, 364. See also Gauss, Carl Friedrich, Correspondence *and* Friendships; Göttingen, University of; Göttingen Seven
Experiments on the Laws of Gravity, the Resistance of the Air, and the Rotation of the Earth, treatise by Johann Friedrich Benzenberg, assisted by Gauss, 26
Eytelwein, J. A., 337

Falkenberg, Gauss survey of, 125, 126, 129
Fallenstein, Aletta Christiane Luise, wife of Wilhelm Gauss, 372
Fallenstein, Rev. Heinrich, father of Wilhelm Gauss' wife, 372
Family of C. F. Gauss. See Gauss, Carl Friedrich, Family Background *and* And His Family
Faraday, Michael, English physicist, 153, 162
Farbenlehre, by Goethe, 242
Farrar, John, 281

Fawcett, Henrietta, wife of Eugene Gauss, 370
Fawcett, Joseph, father of Henrietta (Eugene Gauss' wife), 370
Feldberg, Gauss survey of, 138
Ferdinand, Carl Wilhelm, Duke of Brunswick, 198, 231, 274; friend of Zimmermann, 14; as patron of Gauss, 14–16; as patron of Ide, 25; as patron of Illiger, 59; continues patronage to Gauss, 32, 35, 41, 43, 60, 78, 79; during gathering war storm, 78, 79; as leader of Prussian forces, 79, 80; his last days, 80–83
Ferdinandea. *See* Ceres
Flamsteed, John, astronomer, 51, 250
Fleischer, Gerhard, 42
Flora, planetoid, 223
Flügel's *Practical Dictionary of the English and German Languages*, 241
Föhr, Danish island, 267
Foncenex, de. *See* De Foncenex
Forbes, David, 250n
Förster, August, pathologist, 323
Fort Crawford, Minn., 311
Foucault, Jean Bernard Léon, 227
Frankfurt am Main, 87, 261, 270
Franklin, Benjamin, 280
Frauenberg, Gauss survey of, 138
Fraunhofer, Joseph von, 106, 107, 108, 169, 307
Fraunhofer heliometer, 96, 106
Frederick the Great, 272
Frederick VI, King of Denmark, 120
Freiburg, Germany, 140, 157
French literature, 241
French-Polish. *See* Dictionaries
French Revolution, 243
Freytagswerder, Gauss survey of, 116
Friedrich Wilhelm, son of Carl Wilhelm Ferdinand, Duke of Brunswick, 82
Friedrich Wilhelm III, King of Prussia, 79

Friends of Gauss. *See* Gauss, Carl Friedrich, Friendships
Fries, J. F., 313, 314, 315
Frisia, Germany, 138
Frobenius, on work of Gauss, 347
Frothingham's *Siege of Boston*, 285
Fuchs, K. H., 323
Fundamental Theorem of Gauss, 39
Fuss, Nikolaus von, 58, 287, 289

Galileo, 4
Galois, Evariste, 48
Garlste, Gauss survey of, 130, 133
Garssen, Gauss survey of, 125, 126
Gärtner, teacher at Carolineum, 17
Gatterer, German historian, 23
Gauss, Andreas, 4, 5
Gauss, Carl August Adolph, son of Joseph Gauss, grandson of C. F. Gauss, 205, 358
Gauss, Carl Friedrich:

FAMILY BACKGROUND AND EARLY LIFE

Ancestors, 4–10; birth of Carl Friedrich, 8; anecdotes of early childhood, 11; first school and early precocity, 12, 13; friendship of teachers, 14; progress in studies, 15

EDUCATION

Private study under patronage of the duke, 12–16. *See also* Ferdinand, Carl Wilhelm, Duke of Brunswick
At Carolineum, 16–21. *See also* Zimmermann, E. A. W.
At Göttingen, 21, 24, 175; student friends, 25, 26, 27; influence of teachers, 26, 27; mathematical discoveries by Gauss while a student there, 28–30, 37; departure, 32
Waiting period after Göttingen, 32–35
At Helmstedt, 35; receives doctor's degree, 35; value of doctoral the-

INDEX 515

Gauss, Carl Friedrich (cont'd.) sis, 40, 41; evolves Easter date formula there, 69

RESEARCH

Discovery of method of least squares, 16–21; discovery of heptadecagon, 28; Gaussian theory of cyclotomic or circle-division equations, 28–30; discovery of new proof of Lagrange's theorem, 37; publication of first major work, *Disquisitiones arithmeticae*, 37–45; Fundamental Theorem of Gauss, 39, 45, 46n; paper on "Descartes' Rule of Signs," 45; mastery of elliptic modular functions, 47; interpretation of mathematical theory of binary and ternary quadratic forms (in review of Seeber's *Works*), 48; two memoirs on theory of biquadratic residues, 48

Astronomical calculations to determine location of new planets, 52–54, 55, 72, 75, 76, 79; rediscovery of Ceres, 56–58; discovery of the Easter date formula, 69, 70 (see also *Braunschweigisches Magazin; Monatliche Correspondenz*); general research in correction of lunar tables, 70, 71, 72; Gauss' work in same, 71; orbital corrections and calculations of Pallas, Juno, and Vesta, 75, 76; publication of second major work, *Theoria motus*, 71, 90–93; impetus to astronomical research given by new observatory at Göttingen, 97, 98–112

Investigation of validity of an infinite series, 104; memoir on above presented to Royal Society of Sciences in Göttingen, 104; observations at the meridian circles, 110; collaborative observations for determination of differences in longitude, 111 (see also Nicolai; Soldner; Encke); explorations in theoretical astronomy, 111–112; enunciation of theorems on so-called arithmetico-geometric mean, 112; table of logarithms produced from earlier idea of Leonelli, 112

Application of Gauss' calculations to practical geodetic work, 114–117; *Supplementum theoriae combinationis observationum erroribus minimis obnoxiae*, published results of these calculations, 114; publication of *Methodus peculiaris elevationem poli determinandi*, a previous treatise, 116; invention of the heliotrope, 122–123; direction, by Gauss, of Hanoverian survey, 116–138 *passim* (see also "Gauss survey"); study of higher geodesy, 134; memoir on curved surfaces, 135; invention, by Gauss, of geodetic formulas used in calculating coordinates, 137; two memoirs on topics in higher geodesy, 138

Development by Gauss of principle of the needle telegraph, 150; invention (with Weber) of bifilar magnetometer, 158; co-editor (with Weber) of research periodical for Magnetic Association, 158–159 (see also *Resultate aus den Beobachtungen des magnetischen Vereins im Jahre . . .*); stimulated by others to undertake study of magnetism, 153, 154; completion of the theory of the intensity of terrestrial magnetism, 154; presentation of memoir on, to Royal Society of Göttingen, 154, 155; activity of research at the Göttingen center, 157, 158; completion of experimental data on Gauss' general theory of terrestrial magnetism, 158, 159; "Allgemeine Theorie des Erdmagnetismus," published data on same, 158, 159; collaborative publication of *Atlas*

Gauss, Carl Friedrich (*cont'd.*)
des Erdmagnetismus, 159 (*see also* Weber W. *and* Goldschmidt*)*; last memoir on work in magnetism, 160; law of induction formulated, 161; "Gauss' Fundamental Law," 162

Steps leading to the generalization of his theory of curved surfaces, 163, 164; published result, 164–166; effect of, upon subsequent development of theory, 164, 165; pioneer character of Gauss' work in surface theory, 166; Gauss interest in crystallography, 166, 167; work of devising system of crystallographic notation by Gauss and by Miller, 167, 168; devised formula for Gauss objective, 168; published article on achromatic objects, 168, 169; relative value of Gauss and Fraunhofer objectives, 169; Gauss' skepticism relative to the value of the dialytic telescope, 169, 170; use of the Gauss eye piece, 170; prize proposed by Gauss relative to method for photometry, 170; publication of *Dioptrische Untersuchungen,* 171, 172

Controversy over Gauss' work in non-Euclidean geometry, 174, 175; Gauss' early meditations on the idea, 175; interest at Göttingen in question of parallels, 176; continued efforts of Gauss and others to prove the eleventh axiom, 180; Schweikart's "theory of astral magnitudes" submitted to Gauss for an opinion, 180; interchange of ideas with Taurinus, 181, 182; research on "the metaphysics of space doctrine," 182; publication of notes on same, 183, 184, 186; interchange of ideas with the Bolyais, 183, 184; examination of Lobachevsky's works on the parallel theory, 185, 186, 187; Gauss first to free himself from Euclidean tradition, 188; independence of the respective researches, 188, 189

AND HIS FAMILY

Establishment of home, 57; marriage to Johanna Osthoff, 62–66; birth and christening of Joseph, 77; call to Göttingen and removal of family from Brunswick, 85; getting acquainted in new home, 85, 86; birth of daughter, Minna, 87; death of Gauss' father, 87; birth of second son, Louis, and early death of, 93; death of Johanna, 93; Gauss' grief, 93–95

Gauss' second marriage to Minna Waldeck, 100–103; births of sons Eugene and Wilhelm, and of daughter Therese recorded, 102; illness of Minna, 102, 103, 128, 131, 134, 144; her death, 103; Gauss settles paternal estate and assumes care of mother, 203; death of mother, 204; marriage of son Joseph, 205; Gauss' grandchildren in America, 205; administrator of estate of wife, Minna, 236

See also Gauss, Charles Frederick; Gauss, Dorothea (Benze); Gauss, Eugene; Gauss, Gebhard Dietrich; Gauss, Johanna; Gauss Johann Georg; Gauss, Carl (Joseph); Gauss, Louis; Gauss, Minna (Waldeck); Gauss, (Minna) Wilhelmine; Gauss, Robert; Gauss, Sophie; Gauss, Therese; Gauss, Wilhelm

LATER YEARS

Interests: special attention to topology, 221–223; continuation of astronomical observations, 223; calculations and notebooks, 227, 228; pleasure in music, 228, 229; director of the study and reorganization of the fund for professors' widows (Göttingen), 232–

INDEX 517

Gauss, Carl Friedrich (cont'd.) 234; attention to public finance, 234, 235; Gauss' financial status, 237; mastery of the Russian language, 238, 239; intensive reading, 240–243; interest in politics and reactions thereto, 243–246; interest in American progress, 287; prominence as a faculty member, 288; recreation, 318; interest in railroads, 319

Philosophy: Gauss' philosophy of life, 298; on doctrinal questions, 300; belief in immortality, 300, 301; constant effort by Gauss to harmonize his views on the human soul with the principles of mathematics, 310. *See also* Wagner, Rudolf

Last Days: illness, 321; writes will, 321; death, 322, 323; autopsy, 323, 324; burial, 324–326

Estate: inheritance of children, 356, 357; disposition of his library, 358; disposition of his medals, 359

FRIENDSHIPS

With Bartels, 14; with Bessel, 76–77, 226; with Bolyai, 26–30; with Eisenstein, 75, 274; with Eschenburg, 278; with Ewald, 145, 266, 364; with Gerling, 143, 166, 192; with Goldschmidt, 226; with Harding, 77, 78, 85; with Hassler, 286; with the Humboldts, Alexander and Wilhelm, 89, 192, 194; with Laplace, 87, 231; with Lindenau, 117, 226, 319; with Olbers, 54, 61, 62, 74–76 *passim*, 85, 87, 90, 92, 93, 111, 113, 115, 130, 132–134 *passim*, 147, 153, 205, 226; with Repsold, 89; with Seyffer, 26, 30, 176; with Schumacher, 89, 93, 116, 130, 163, 169, 208, 209, 215, 226, 238, 252, 367; with Rudolf Wagner, 301–310; with Weber, 141, 142, 199, 266; with von Zach, 75, 115; with Zimmermann, 16, 17; friends in Great Britain, 280; in America, 281–286

CORRESPONDENCE

With General Baeyer, 138; with Bessel, 106, 120, 160, 208, 212, 213, 214, 216, 217, 224, 413, 415, 417; with Bolyai, 30, 32–34, 35–37, 61, 62, 65, 77, 176, 177, 184, 215, 247, 255, 312, 413, 416; with von Buch, 293; with Drobisch, 314; with Encke, 169, 208, 210, 212, 245, 275; with Ewald, 206, 265; with daughter Minna Ewald, 201; with Fries, 314; with von Fuss, 58; with son Eugene Gauss in America, 369, 370; with son Joseph Gauss, 375; with wife Minna Gauss on Easter trip, 107, 108, 109; with Gerling, 144, 166, 180, 186, 214, 233, 234, 247, 287, 316; with Gould, 284; with Hassler, 286; with Hoppenstedt, 139, 140, 141; with Humboldt, 89, 195, 256; with Lobachevsky, 187; with nephew, 320; with Olbers, 54, 67, 68, 74, 75, 76, 147, 153, 176, 193, 199*n*, 205, 213, 214, 216, 221, 254, 256, 261, 262, 290, 295, 311, 413, 416; with Parrot, 289, 290; with Praël (Gauss' last note), 321; with Schumacher, 71, 88, 112, 116, 126, 129, 131, 148, 149, 168, 170, 175, 183, 186, 201, 208, 209–212, 215, 216, 218, 219, 221, 244, 255, 256, 257, 261, 293*n*, 296, 297, 312, 313, 314, 316, 348, 411, 412, 413, 414, 416, 417, 418; with Seyffer, 26, 30, 176; with Taurinus, 181; with Weber, 416; with von Zach, 73, 83, 114; with Zimmermann, 16, 17

Correspondence with British scientists, 280; with American friends, 282

PERSONAL GLIMPSES

Request of Gauss relative to his tombstone, 28; sorrow at death of

Gauss, Carl Friedrich (cont'd.)
Duke Ferdinand, 82, 83; letter of Johanna Gauss to Dorothea Köppe, 85, 86; letter of Johanna to her mother and to Dorothea, 87; noble conduct of Gauss under grief, 95; family life, 103; irritation of Gauss, 111; letter to Gerling concerning wife's illness, 144; fear of losing family maid, 145; condition of Gauss' mother, 199; health of household, 199, 202; trial to Gauss of being separated from daughter Minna, 200; Gauss treats self for deafness, 201, 202; Gauss' care of his mother, 203, 204; Gauss' worry over son in America, 204; grief over death of Olbers, 205; grief at death of daughter Minna, 206; Gauss in a social group of friends, 219; Gauss' pride in his invention of the heliotrope, 220; Gauss reading the newspapers, 225, 226; Gauss' enjoyment of whist and chess, 227; Gauss' physical appearance, 229, 260; letter from Gauss to Schumacher relative to insecurity, 235; Gauss' happiness at return of Ewald and Weber to Göttingen, 244; Gauss' views on "table rapping," 247–249; Cantor's reminiscences of Gauss, 257; Dedekind's reminiscences, 259–261; family scene at christening of Ewald's daughter, 265, 266; Gauss lighting his pipe, 276; letter of Therese to Eugene giving details of the jubilee celebration, 277; Gauss as a host, 285; professional bickering, 292; Gauss' emotional reactions during the visits of Rudolf Wagner, 303–310; Gauss' health, 318, 319, 321; letter to nephew on death of Gauss' brother, 320; Therese and family friends at Gauss' death, 323; family relationship revealed in letter of Therese to her brother Eugene on father's death, 326–330

ANECDOTES

Gauss' childhood skill with numbers, 12; his brother's reception of the duke's messenger, 15; conversation leading to discovery of Ceres, 58; Gauss' care of instruments, 110; Gauss with his grandson, 205; tale about "Hansi," 219; the Ribbentrop anecdotes, 219, 220; the Teipel incidents, 220; Eugene's version of his father's idea for invention of the heliotrope, 221; illustrating Gauss' power of concentration, 230; Gauss' amusement over the moon rise in a passage from Sir Walter Scott, 243; reminiscence concerning Gauss in the Revolution of 1848, 244; Gauss unknown to the English Admiralty, 250; concerning proofreading of *Theoria motus*, 252; concerning Fraunhofer's accident, 307; conversation between Gauss and student concerning Fries' book, 315; of little Minna, aged five, 363; concerning Ewald's absentmindedness, 364

MINOR INCIDENTS

Three little trips with friends of Gauss, 74, 75; Gauss' visit, after Johanna's death, with Olbers and Schumacher, 93; Gauss' Easter trip to Bavaria, 107; Bessel's displeasure with Gauss and Gauss' reply, 120; visit of Olbers, Schumacher, and Repsold with Gauss in Zeven, 130; competition with Legendre, 176; Gauss sends to Ewald souvenirs of the jubilee, 202; pony riding of grandson in America, 204; Therese visits her brother Joseph, 207; Encke accuses Chevalier d'Angos of decep-

Gauss, Carl Friedrich (cont'd.)
tion, 253, 254; episode with Schöpffer, 262-264; Dühring's attack, 273, 274; episode concerning publication of a memoir by Abel, 273; moon hoax in N. Y. *Sun*, 294, 295

PORTRAITS AND LIKENESSES

In pastel crayon, 75; in oil, 206, 207, 336, 337; portraits of grandchildren, 205; medallion of Gauss, 322; daguerreotypes, 324; busts, 324, 336, 337, 338; monuments, 335, 336, 337; picture on 1955 commemorative postage stamp, 340

PERSONAL CHARACTERISTICS

Sensitive, 174; noble in bearing, 192; thoroughly conservative, 199, 243; values not appraised from utilitarian viewpoint, 217; slow in passing judgment, 217; dislike of travel, 228; wise investor, 235; aristocratic, 245; thorough, 296; despised pretense, 298; religious in nature, 298; practical in concept of religion, 301; kind, but austere, 312; accepted misfortune, 313

PROFESSIONAL CHARACTERISTICS

Rapidity of calculations, 72, 76; noble attitude to popular sneers at scientific activity, 73, 74; method of approach to research, 153; new truth first concern, 174, 175; rigorous standards of perfecting research that hampered publication, 209-214; low opinion of majority of mathematicians, 215; compactness in presentation of his material, 216, 217; unwilling to turn over results to others without completion, 216, 217; style of writing, 217, 218; unique power in number calculating, 219; clearness in condensation, 227, 228, 296; affected by distractions, 230; entirely a mathematician, 231; deeply meditative, 231; conscious of high position, but modest, 252; abominated polemic natures, 253; satisfied only with whole proof, 254; isolated position, 225; methods of work, 295; purely mathematical thinking, 303

Gauss, (Carl) Joseph, son of C. F. and Johanna (Osthoff) Gauss, 86, 87, 108, 170n, 202, 266, 330; birth and christening of, 77; as assistant to his father on the Hanover survey, 126, 127, 135, 136, 137, 138; as administrator of grandmother's estate, 237; references to his American tour, 281, 282, 286, 287; reference to military career, 293; attends father's funeral, 325; his inheritance from father, 357; as executor of father's estate, 357-359; summary of his life, 361-363

Gauss, Caroline Magdalene Dorothee, niece of Gauss, 9

Gauss, Charles Frederick, son of Wilhelm Gauss, grandson of C. F. Gauss, birth of, 204; named for C. F. Gauss, 204; financial success of, 205; death of, 205

Gauss, Dorothea (Benze), second wife of Gebhard Dietrich, and mother of C. F. Gauss, 8-10, 30, 203, 204

Gauss, Dorothea Emerenzia (Warnecke), first wife of Gebhard Dietrich Gauss, 7, 8

Gauss [usually Gaus], Emil, 7

Gauss, Engel, 4

Gauss, Eugene [Peter Samuel Marius Eugenius], son of C. F. and Minna Gauss, birth of, 102; incident relative to his companionship with father, 221; his inheritance from mother's estate, 236; letters concerning his life in America, 311; his inheritance from father, 357;

Gauss, Eugene (*cont'd.*)
  summary of his life and experiences in America, 365-372
Gauss, Gebhard Dietrich, father of Gauss, 5, 7-10, 61, 88
Gauss, Georg Gebhard Albert, nephew of Gauss, 9, 203
Gauss, Hans, great-great-grandfather of Gauss, 4
Gauss, Heinrich Engel, 4, 5
Gauss, Johanna (Osthoff), first wife of C. F. Gauss, Gauss introduction to, 62; Gauss' description of, 62, 63; Gauss' letter of proposal to, 63, 64; Gauss' letter to Bolyai concerning acceptance by, 65, 66; marriage of, 66; birth of children to, 77, 87, 93; her life in Göttingen, 85-87; death of, 93; tribute from husband to, 93-95
Gauss, Johann Franz Heinrich, uncle of Gauss, 7
Gauss, Johann Georg Heinrich, half-brother of Gauss, 7-9, 320
Gauss, Ludwig (Louis), son of C. F. and Johanna Gauss, birth of, 93; burial place of, 326; sketch of, 365
Gauss, Minna (Waldeck), second wife of C. F. Gauss, family background of, 100, 102; marriage of, 102; children of, 102; ill health of, 102, 103, 128, 131, 134, 144; death of, 103; family relics of, 103; estate of, 236
Gauss, (Minna) Wilhelmine, second child of C. F. and Johanna Gauss, birth of, 87; named for Olbers, 88; personal appearance and characteristics of, 145; marriage of, to Heinrich Ewald, 145; ill health of, 146; her homesickness in Tübingen, 205, 206; death of, 206; biographical sketch of, 363-365
Gauss, Peter Heinrich, uncle of Gauss, 7
Gauss, Robert ("Colonel"), grandson of C. F. Gauss, 324

Gauss, Sophie (Erythropel), wife of Joseph Gauss, marriage into and congeniality with the Gauss family, 205, 362; birth of son to, 205
Gauss [usually Gaus], Theodore, 7
Gauss, Theodore, son of Eugene Gauss of St. Charles, Mo., 371
Gauss, Therese [Henriette Wilhelmine Caroline Therese], daughter of C. F. and Minna Gauss, 205, 207, 219, 240, 266, 277, 364; birth of, 102; devotion of, to family, 145; receives inheritance from mother, 236; godmother to Ewald's daughter, 265; her grief at death of father, 323; attends father's funeral, 324; her letter to brother Eugene relative to loss of father, 326-328; reference to, in Gauss' will, 356; her inheritance from father, 357; gift by, to University library, 358; her life after father's death, 374; summary, 373-375
Gauss, Wilhelm [Wilhelm August Carl Matthias], second son of C. F. and Minna Gauss, birth of, 102; as assistant to father, 156; his marriage to Bessel's niece and migration to America, 194; his father's worry over safe arrival, 204; birth of first son to, 205; receives inheritance from mother, 236; mentioned as farmer in Missouri, 356; his inheritance from father, 357; summary of life and experiences in America of, 272-273
"Gauss' Analogies," 90
Gauss Archive, 270, 337
Gaussberg, a volcanic mountain, 338
Gaussberg, knoll in Brunswick, 335
Gauss Centenary, 255
"Gauss curvature," 189
"Gauss' equations," 90*n*
Gauss family addresses, 6, 11, 35, 66, 86, 88, 119, 266, 374*n*
Gauss Fund, 338

INDEX 521

"Gauss Fundamental Law," 162
Gauss House, building at Technische Hochschule in Hanover, 338
Gaussiana, notes on conversations between Gauss and Schumacher, 89
Gaussian heliotrope, 123, 221, 286
Gaussian logarithms, so-called. *See* Leonelli, Z.
Gaussian school in astronomy, 267
Gauss jubilee, published accounts of, 275n; description of, 276; Jacobi's opinion on, 276, 277; Theresa Gauss' letter to brother Eugene concerning, 277; honors received on occasion of, 278
Gauss letter seal, 208
Gauss name, used in science and mathematics, 338–340
Gauss (schooner), 338
"Gauss stone," 338
"Gauss survey" ("Hanover survey"), origin of idea of, 116; Gauss' interest in, 116; reasons for Gauss' hesitancy in promoting project of, 117; Schumacher's initiative in, 118, 119; steps leading to Gauss' appointment by the King as director of, 118, 119, 120, 121, 135; beginnings of, 122; difficulties encountered in, 121, 125, 126, 133; accelerated by Gauss' invention of the heliotrope, 122, 123; actual triangulation period of, 124; principal points of, determined, 124; Gauss' assistants at, 118, 121, 124, 126, 129, 131, 135; collaboration of Encke and Schumacher with Gauss in the actual, 124, 134, 135; allowances paid to Gauss and assistants for, 135; amount of records pertaining to, 136; Gauss' contribution to science as result of, 117, 122, 123, 127, 138
Gauss Tower on Mount Hohenhagen, 337, 338
Gauss–Weber exhibit, 336

Gauss–Weber experiment with telegraph, description of device, 147; first public notice of, 147; first words sent by, 148; lines of, destroyed by lightning, 148, 149; inventions in equipment for, 149, 150; its success and disappointment, 150–152
Gauting, near Munich, 320
Geermans, Ilse, second wife of Hinrich Gooss [Gauss], 5
Geheimer Hofrat, title granted to Gauss by King Ernst August, 288
Geismarstrasse, Gauss address (at Volbaum's), 86
Geismar Tor, in Göttingen, 244
Gellert, C. F., 17
Gelpke, August Heinrich Christian, assistant principal of Martineum, 83
Genealogy of Gauss family, 376–390
Gentleman's Magazine, 242
Geodesy. *See* Gauss, Carl Friedrich, Research
Geometriae prima elementa, 182
Geometria magnitudinis, 222
Geometria situs, 221, 222
Géometrie de position, Carnot's, translation of, 221
Geometrische Untersuchungen zur Theorie der Parallellinien, memoir by Lobachevsky, 185; praised by Gauss, 186; by Wolfgang Bolyai, 187
Geometry. *See* Gauss, Carl Friedrich, Research
Geophysical Institute of Göttingen University, 123
George II, King of England, founder of Göttingen University, 23
George III, King of Great Britain, gift of, to Göttingen Observatory, 105
George IV, King of Great Britain and Hanover, 121, 135, 288
George V, King of Hanover, 322, 337
Georgplatz, Gauss survey of, 175
Gerhard, Paul, 308

Gerling, C. L., pupil of Gauss, 20, 91, 128, 134, 137, 140, 141, 166, 170, 171, 175, 180, 181, 194, 226, 247, 267, 272, 276, 280, 287, 316. *See also* Gauss, Carl Friedrich, Correspondence *and* Friendships
Germain, Sophie, anecdote concerning, 67, 68; pseudonym for, 68; academic data on, 68n, 93, 192
German literature, 239, 242, 243
German Oriental Society, 146
Germany, 279, 330
Gersdorf's *Repertorium der gesammten deutschen Literatur*, journal, 185
Gervinus, Georg Gottfried, signer of Göttingen protest to King over revocation of liberal constitution, 197; dismissed from university, 197, 198; to Heidelberg, 198
Geschichte der Philosophie, by J. F. Fries, 314
Gesner, Johann Matthias, founder of first philological seminar in Germany, 23, 24
Gibbon's *Decline and Fall of the Roman Empire*, 243
Gibraltar, magnetic station at, 157
Gildemeister, astronomer, 50
Gildersleeve, Basil L., 282
Gilliss, J. M., 283
Gnarrenburg, Gauss survey of, 130, 133
Goethe, Johann Wolfgang von, 3, 239, 242, 267, 302, 317, 416
Goldschmidt, C. W. B., 259, 284, 285, 288, 320; assistant to Gauss, 137; collaborator with Gauss, 159; winner of prize question, 165; his death, 226
Goldsmith's *Vicar of Wakefield*, 241
Goos *and* Gooss. *See also* Gauss
Goos, Jürgen, grandfather of Gauss, 4–7
Gooss, Engel, 6
Gooss, Hinrich, great-grandfather of Gauss, 4, 5, 6

Göschen, Johann Friedrich Ludwig, 193
Goslar, Germany, 34, 37
Gotha, Germany, 49, 60, 62, 75, 78, 90, 106, 108, 109, 115, 135
Gothmarstrasse 11, Gauss address, 86
Göttingen, *passim;* point of Gauss survey at, 122, 129, 135, 137, 138; revolution of January 18, 1831, in, 143; jubilee celebration at, 192, 207; as radical city, 244
Göttingen Observatory, rebuilt under Gauss, 95, 96; instruments in, 96, 97, 106–111 *passim;* description of building of, 97, 98; record of activity of, 105, 267
Göttingen, Royal Society of Sciences in. *See* Royal Society of Sciences in Göttingen
Göttingen, University of, *passim;* founding of, 23; characteristics of its greatness, 23, 139; noted men of its faculty, 23, 24; character of its student body, 23, 24; conditions of, under French occupation, 24, 95; improvement of Observatory of, 156; as center of magnetic research, 157; its centennial jubilee, 191–193, 256; resignation of seven professors of, 197–198; resignation effect upon, 198, 199; the seven recalled to, 244; its progressive attitude in granting degrees, 267, 268; honors bestowed on Gauss by, 336, 337; Gauss library acquired by, 358, 359. *See also* Gauss, Carl Friedrich, Education *and* Research
Göttingen magnetic observatory, Gauss' influence in obtaining, 156; building and equipment of, 156; as impetus to research, 157
Göttingen Seven, signers of protest at University (1837), 197, 235, 244, 279
"Göttingen theory of parallels," a treatise by Wolfgang Bolyai, 177
Göttingen widows' fund, 234

Göttingische gelehrte Anzeigen, Göttingen magnetic observatory described in, 156, 210n; mention in, of a Gauss discussion of topology, 222

Gould, Benjamin Apthorp, disciple of Gauss, 283–284, 285, 286

Graham's planet. *See* Metis

Grassmann, Hermann, 161, 167

Gremmel, Konrad, 5

Green, George, 153, 160

Gresy, Chevalier Cisa. *See* Jewish calendar

Grimm, Jakob Ludwig Carl, signer of Göttingen protest to King over revocation of liberal constitution, 197; dismissed from university, 197, 198; to Berlin, 198; suspicious of Gauss, 200

Grimm, Wilhelm Carl (brother of Jakob Grimm), signer of protest to King over revocation of liberal constitution, 197; dismissed from university, 197; to Berlin, 198; suspicious of Gauss, 200

Gross-Schwülper, Germany, 4

Grove, Anna, widow of Hinrich Wehrmann and first wife of Hinrich Gooss [Gauss], 5

Gruithuisen, Franz von Paula, 253

Grünen Jäger, Zum, a forest tavern, 279

Guelph Order, Knight's Cross of the, 252, 288

Guichard, Colonel K. G., 272

Haase, Carl, German translator of *Theoria motus*, 90n

Haassel, Gauss survey of, 127

Hagen, 20

Hague, The, magnetic station at, 157

Halifax, magnetic station at, 157

Halle, Germany, 82, 157; University of, 23, 140, 268

Haller, Albrecht von, 23, 302

Halstead, G. B., 251

Hamburg, Germany, 90, 107, 118, 119, 121, 127, 129, 221, 269, 293, 362

Hamburg Fire Insurance Company, 237

Hamburg optical works, 168

Hamburger, M., 69n

Hamburger Correspondent, 246

Hamilton, Sir William Rowan, 254

Hamlin, Germany, 93, 205, 244

Hamm, Westphalia, 267

Hankel, Hermann, 218

Hanover, Germany, 4, 90n, 115, 119, 121, 122, 124, 128, 136, 143, 194, 225, 236, 262, 271, 272, 277, 280, 320, 322, 324; Kingdom of, 107, 197, 198, 235, 338, 361

Hanoverian bonds, 237

Hanoverian constitution of 1833, granted, 143; revoked, 195; restored, 265

Hanoverian government, 85, 120, 142, 293, 357, 358

"Hanover Survey." *See* "Gauss survey"

Hänselmann, Ludwig, Brunswick librarian and archivist, 4

Hansen, Peter Andreas, 20, 132, 227, 276

"Hansi," name of Gauss' pet bird, 219

"Hans Pfaall," story by E. A. Poe, 295

Hansteen, Christoph, 157, 261, 262, 273

Hanstein, Germany, 81, 116

Harburg, Germany, 262

Harding, Carl Ludwig, 50, 56, 57, 292; as astronomer and tutor, 74; as discoverer of asteroid Juno, 74; visits Gauss, 78; Gauss' son named for 93; activities of, at Göttingen Observatory, 96, 105

Hardy clock, 111, 135

Harpke, Germany, 35

Harris, Snow, British physicist, 158

Harrisonburg, Va., 370

Härtel, printer, 252

Hartmann, F., Gauss' assistant, 121, 122, 125, 126, 129, 135, 136
Hartzer, F., sculptor, 337
Harvard University, 283, 284; catalogue of, 242
Harz, Germany, 137, 176
Hassler, F. R., friend of Gauss, 286
Hauck, G., 349
Hauerschildt, Miss, Gauss' domestic, 145
Haugwitz, C. A. H. K., 79
Hausblätter, edited by E. F. A. Hoefer, 243
Hauselberg, Gauss survey of, 126
Hausmann, J. F. L., 253, 263
Haverloh, Gauss survey of, 128
Heeren, Arnold, German historian, 24, 87, 280
Hegel, G. W. F., 313, 314
Heidelberg, University of, 198, 248, 257
Heinrich, Father Placidus, 109
Heliometer, Fraunhofer, 96, 106
Heliometer. *See* Gauss, Carl Friedrich Research
Heliotrope, Gaussian, 122, 221, 286
Hellwig, teacher at Carolineum, 25, 59, 60
Helmert, 20
Helmholtz, Hermann, 162, 273, 296, 311, 337, 347
"Helmklotz," derisive name, 273
Helmstedt, Germany, 33, 34, 35, 115, 278; University of, 35, 40, 41, 84, 198, 276
Henle, Jakob, 323
Heptadecagon, Gauss' discovery of, 28
Herbart, J. F., 253
Hermannstadt, Hungary, 26
Herodotus, 241
Herschel, Caroline, sister of William Herschel, 195, 280
Herschel, Sir John, 73, 254, 261, 280, 294, 295
Herschel, William, 280
Herschel's mirror telescope, 96
Hesemann, C. H., Hanoverian court sculptor, 322, 324

Hesiod, 241
Hesperides, 215
Hessel, J. F. C., 166
Hessen, survey of, 121, 128, 137
Heusinger, Conrad, 15
Heyne, Christian Gottlob, German classicist, 24
Hildebrand, L., student of Gauss, 258
Hildesheim, Germany, 128, 244, 320
Hillebrand, Karl, 347
Hils, Gauss survey of, 122, 124, 125
Himalayas, magnetic station in, 157
Hindenburg, C. F., 45
Hirst, Thomas Archer, 280
History of the Inductive Sciences, by William Whewell, 302
Hoefer, Edmund Franz Andreas, German writer and relative of Gauss, 236, 243
Hof, Germany, school in, 270
Hofrat, title granted to Gauss, 274, 288
Hohenhagen, mountain near Dransfeld, Germany, 122, 124, 125, 129, 136, 337
Hohenlohe-Ingelfingen, Prince of, 79, 80
Holberg, Ludvig, 241
Hollmann, S. C., influence of, at Göttingen, 24
Hollmannsegg, Count, 59
Hoppenstedt, Georg Ernst Friedrich, 139n, 197
Horace, 241
Horack, Johanna, second wife of C. W. Staufenau, 375
Hospitalstrasse 1, address of the "literary museum," Göttingen, 225
Howaldt, stonemason, 28
Hoyer, Peter, 6
Huber, Daniel, Basel mathematician, 19
Humboldt, Friedrich Heinrich Alexander von, influential and lifelong friend of Gauss, 89, 136, 145, 153, 157, 192, 194, 195, 200, 218, 226, 249, 250, 251, 256, 263, 266, 274,

INDEX 525

Humboldt, A. von (cont'd.) 275, 278, 281, 282, 283, 292, 293, 317, 318, 348
Humboldt, Wilhelm von, friend of Gauss, 89, 291
Hume, David, 313
Huth, J. S., 290
Huygens, Christian, 222
Hyperion, orbit of, 284

Icilius, E. W. G. von Quintus, student under Gauss, 271
Ide, Johann Joseph Anton, 18; biographical data on, 25; fellow student with Gauss at Göttingen, 25; in Moscow, 26; letter of, to Gauss, 31
Ilfeld, home of Count von Laffert, 199
Illiger, Johann Carl Wilhelm, 18; academic data on, 59, 60
Immortality, Gauss' belief in, 300, 301
"Imposture astronomique grossière du Chevalier d'Angos," article by Encke, 253
Induction, Law of. *See* Gauss, Carl Friedrich, Research
Infinite series. *See* Gauss, Carl Friedrich, Research
Inselsberg, Germany, 124, 125, 129, 338
Institut de France, Gauss honored with Lalande prize by, 93
Institute of Technology. *See* Collegium Carolinum
Intensitas vis magneticae terrestris ad mensuram absolutam revocata, memoir by Gauss, 154
Ionian Isles, magnetic station on, 157
Iris, planet, 223
Ivory, Sir James, 20

Jacobi, C. G. J., German-Jewish mathematician, 46, 47, 112, 210, 211, 245, 251n, 255, 273, 275, 276, 277
Jamaica, magnetic station on, 157
James, G. P. R., 241

Janensch, Gerhard, sculptor, 337
Jean Paul. *See* Richter, J. P. F.
Jefferson, Thomas, 286
Jena, Germany, 27
Jensen, Christian Albrecht, painter, 207
Jever, Gauss survey of, 133
Jewish calendar, Gaussian rule for, proved by Gresy, 69
Joachim, Joseph, 229
Johns Hopkins University, 282
Johnson, Dr. Samuel. *See* Dictionaries
Jordan, 20
Journal of Science, Edinburgh, 294
Julius Caesar, 241
Jung Deutschland movement, 225
Juno, planetoid, discovery of, 57, 74, 282
Jutland, Gauss survey of, 116
Juvenal, 241

Kalberlah, 4
Kampaner Tal, by Jean Paul, 307
Kant, Immanuel, 313, 314, 315, 316, 317
Kant's *Critique of Pure Reason*, 315
Karnlastrasse 4, Dresden address, 374n
Kasan, magnetic station at, 157
Kasan *Messenger*, 185
Kasan, University of, 185, 278
Kassel, Germany, 319
Kästner, Abraham Gotthelf, 17, 24, 30, 176, 296, 304
Kemény, Baron Simon, 27
"Kennst Du das Land," Mignon's song, 228
Kepler, Johann, 4, 55, 83, 91, 211, 303
Keyes, Lucretia (Fawcett), mother of Henrietta Fawcett, wife of Eugene Gauss, 370
Kharkov, Ukraine, 290
Kiel, Germany, 269
Kircher, printer, 37
Kirchoff, G., physicist, 161
Kirkland, J. T., 281
Klassiker der exakten Wissenschaften, Ostwald's, 154

Klausenburg, Transylvania, 27
Klein, Felix, 174, 254, 346
Klein-Paris (now Turmstrasse), street in Göttingen, 86
Kleist, Colonel von, 80
Klinger, von, university curator at Dorpat, 290n
Klinkerfues, Wilhelm, pupil of Gauss, 92, 223, 226, 289; prepared catalogue of Gauss library, 357
Klopstock, F. G., 83
Klügel, pupil of Kästner, research on theory of parallels by, 176
Klüver, Gauss' assistant, 128, 129
Knesebeck, von der, 143
Knorr, Ernst, 185
Köhne, Dorothee Elisabeth, 55n
Königsberg heliometer objective, 171
Konversationsblatt, journal, 247
Köppe, Dorothea (Müller), 85
Köppe, Karl, friend of Gauss, 93
Kramer, Dr., Berlin physician, 201
Krayenhoff, measurements of, 120, 134
Kritische Geschichte der allgemeinen Prinzipien der Mechanik, essay by Eugen Dühring, 273n
Kronecker, L., 45, 217
Krüksberg, Gauss survey of, 122
Kummer, E. E., 104, 348
Kupfer, Adolf Theodor, 261, 262
Kurzestrasse 15, Gauss address, 86

Laffert, Count Friedrich von, 199
Lagrange, J. L., 18, 27, 36, 39, 40, 42, 44, 45, 48, 57, 68n, 227, 256, 273, 293, 297, 347; Gauss' new proof of his theorem, 37
Lake Geneva, Switzerland, 374
Lalande, J. J. le Français de, 49, 50, 51
Lambert, J. H., 177, 293
Lamont, Johann von, 262
Landry, J. H., portraitist, 206
Landshut, Germany, 109
Lane, G. M., 282
Langwarden, Gauss survey of, 133

Laplace, Pierre Simon, 20, 44, 57, 70, 71, 72, 83, 87, 107, 114, 231, 251, 255, 256, 258, 282, 323
Lauenburg, Germany, 74, 116, 118, 119, 120
Least squares. See Method of least squares
Le Blanc. See Germain, Sophie
Lecoq, Colonel, 115
Legendre, Adrien-Marie, 19, 20, 39, 40, 42, 45, 48, 68n, 113, 165, 251, 347
Lehmann, J. W. H., 177
"Lehrsätze in Beziehung auf die im verkehrten Verhältnisse des Quadrats der Entfernungen wirkenden Anziehungs- und Abstossungskräfte," treatise by Gauss, as result of research in magnetism, 160, 161
Leibniz, Gottfried Wilhelm, 222, 232, 303, 310
Leipzig, Germany, 26, 42, 93, 140, 199, 285, 291; University of, 198, 290
Leonelli, Z., originator of table of logarithms, 112
Lesage, Alain René, 241
Leue, F. L. K., 287
Levern, Germany, 372
Leverrier, U. J. J., 285
Liagre, mathematician, 20
Library of C. F. Gauss, 241, 242, 285n, 358, 359
Lichtenberg, Georg Christoph, 26, 77, 304
Lichtenburg, Gauss survey of, 122, 125, 126, 127
Lie, Sophus, 166
Liebherr, mechanic, 108
Liebherr clock, 111
Liebig, Justus von, 253, 274, 311, 337
Liebner, Karl Theodor Albert, 191, 192
Life of Sir Humphry Davy, 309
Lilienthal, Germany, 50, 62, 107
Lilienthal telescope mirror: purchase of, 77; dispute over, 83, 84

Lind, Jenny, 229
Lindenau, Bernhard August von, 108, 115, 117, 134, 163, 226, 292, 319
Lion, J. C., student of Gauss, 258
Lippe-Detmold, Germany, 279
Listing, Johann Benedikt, student and assistant of Gauss, 200, 223, 261, 263, 323, 357
Liszt, Franz, 229
Litberg, Gauss survey of, 131
"Literary museum," an organization in Göttingen, 225, 259
Littrow, quotation of, on Olbers, 77
Livy, 241
Lloyd, Humphrey, 158, 262, 280
Lobachevsky, Nikolai Ivanovitch, 174, 278; contributions by, to research on the parallel axiom, 184, 185; works of, examined by Gauss, 186, 187; elected to Royal Society of Sciences in Göttingen as corresponding member, 187; possible influence of research by, upon Gauss, 188; contribution by, to liberation from Euclidean tradition, 189
Locke, John, 313
Locke, R. A., perpetrator of hoax, 294; published stories of, 295
Logarithms, so-called Gaussian. See Leonelli, Z.
Lohlein's *Klavierschule*, 229
London, England, 199, 262
London *Athenaeum*. See Gauss jubilee
Longfellow, H. W., 281
Loomis, Captain S., Eugene Gauss' superior officer in U. S. Army, 311, 367
Loomis, Elias, 282
Lotze, R. H., philosophy of, as quoted from his *Microcosmus*, 317
Lübsen, H. B., pupil of Gauss, 269
Luc, de. See De Luc
Lucchesini, Prussian diplomat, 79
Lüchow (Hanover), Germany, 270
Lucian, 241
Ludwig, Prince of Prussia, 80
Lueder, August Ferdinand, 18, 25

Lunar research, encouragement of, by Paris Academy, 70, 71. See also Gauss, Carl Friedrich, Research
Lüneburg, Germany, 110, 118, 125, 128, 137, 259, 270
Lürssen, Anna Adelheid, 55n
Luther's "Ein' feste Burg ist unser Gott," 325
Lütke, Katherine, third wife of Hinrich Gooss [Gauss], 5

Macaulay's *History of England*, 243
Mack, Austrian general, 79
Mack, Dr. Heinrich, Brunswick librarian, archivist and Gauss scholar, 16
Madras, magnetic station at, 157
Magendies, physician to Laplace, 323
Magnetic Association, 157, 262
Magnetic Congress, 262
Magnetic stations, sites of, 156, 157
Magnetism and electricity. See Gauss, Carl Friedrich, Research
Magnetometer, bifilar, 158
Malebranche, Nicolas, 313
Malmsten, Maria, wife of Ernst Schering, 271
Malus, 172
Mannheim, Germany, 134
Marburg, Germany, 134, 140, 144; University of, 180, 198, 277, 280
Margary's (?) *Anleitung zum Klavierspielen*, 229
Marie, M., 345
Maros-Vásárhely, Hungary, 177
Martial, 241
Marx, Karl Friedrich Heinrich, 201, 244
Maskelyne, Nevil, 58
Mason's Lunar Tables, 70, 72
Massenbach, Colonel von, 79
Mathematical research. See Gauss, Carl Friedrich, Research *and* Philosophy
Mathematische Abhandlungen, by F. G. M. Eisenstein, 275

Mathematische Abhandlungen, by J. W. H. Lehmann, 177
"Mathematische Brouillons," collection of notes by Gauss, 179
Mathematische Naturlehre, by J. F. Fries, 315
Maupertuis, P. L. M. de, 241
Mauritius, magnetic station on, 157
Maury, Commodore M. F., 283
Max Joseph, king of Bavaria, 307
Maxwell, J. Clerk, 162
Mayer, Robert, 273
Mayer, Tobias, astronomer, 51, 113, 139, 217, 267; Lunar Tables of, 70, 71; activity of, at Göttingen, 95–96
Mécanique céleste, Laplace publication, 44; translation of, 285
Mechain, astronomer, 58
Mecklenburg Credit Union bonds, 237
Medallion of Gauss, 332
Méhes, George, 27
Meine, Germany, 4
Meiningen, Germany, 109
Meissner, Germany, 138
Menges, Christian, student of Gauss, 259
Menschenschöpfung und Seelensubstanz, pamphlet by Rudolf Wagner, 309
Mental Physiology, by Carpenter, 230
Mercury, 223
Meridian circle, Repsold, 97, 107, 109, 110; Reichenbach, 97, 98, 110, 223, 227
Merry Mount, a history of environs of Boston, 285
Mertens, ship captain, 204
Merz, refractor, 97
Méthode des moindres quarrés. See Method of least squares
Method of least squares, definition of, 19; research of various scientists on, 19, 20; fully developed by Gauss, 20; importance of Gauss' contribution to, 20; its acceptance by other scientists and extension of literature on the subject, 20; Gauss' publication concerning, 21; adaption of, to Gauss' own research, 113, 114; Gauss' course in, described by Cantor, 258
Methodus peculiaris elevationem poli determinandi, 116
Metis (Graham's planet), 223
Metropulos, C. P., 287
Meyerhoff, Johann Heinrich, 15, 37–39
Meyerstein, Moritz, mechanic, 157, 272
Michelmann, servant to Gauss and Weber, 148
Michigan, University of, 288
Microcosmus, by R. H. Lotze, quotation from, 317
Mignon's song, 228
Miller, William Hallows, 167, 276
Milton, John, 241
Miquel, Johann, revolutionary student leader, 244
Mississippi River, 204, 294, 372
Missouri, 204, 326, 370, 372
Missouri River, 369
Mitchel, General O. MacK., 283
Mitchell, Maria, 283
Möbius, A. F., pupil of Gauss, 215, 223, 265
Moesta, Carlos Guillelmo, Chilean astronomer, 89n
Möllendorf, Field Marshal von, 80
Mollweide, Carl Brandon, 91n
Monatliche Correspondenz zur Beförderung der Erd- und Himmelskunde, magazine, purpose of, 49; discussion of important articles published therein, 49–54, 69, 71, 76, 306
Monge, Gaspard, 166
Montaigne, Michel de, 241
Montesquieu, Charles Louis, 241
Montreal, magnetic station at, 157
Montreux, Switzerland, 374
Monuments to Gauss, 335–337
Moore, Thomas, 241
Morgan, De. *See* De Morgan
Mortality tables of Brune, 234

INDEX 529

Motley, J. L., 281
Mozart-Dirichlet, 347
Müffling, Friedrich Ferdinand Carl, Baron von, 115, 116, 121, 292
Mühlhausen, Germany, 109
Müller, Captain G. W., Gauss' assistant, 121, 122, 126, 128, 129, 135, 136, 137, 138
Müller, Carl Otfried, 192, 225
Müller, Johann, 90
Münchhausen, Gerlach Adolph, first curator at Göttingen University, 23, 24
Münden, Germany, 116, 137
Munich, Germany, 96, 106, 107, 108, 109, 157, 169, 253, 255, 269, 270, 307, 338
Münster, Count, minister of Hanoverian affairs in London, 119, 121, 129, 143
Murawjeff, Count, 33
Murgtal, 134
Murhard, Friedrich Wilhelm August, biographical identification of, 33n; Gauss' postscript in letter to Bolyai concerning, 33, 34
Music. See Gauss, Carl Friedrich, Later Years

Nagy-Enyed, seat of Evangelical Reformed College, 26
Naples, magnetic station at, 157
Napoleon, 44, 78, 83, 251, 258
Napoleonic period, 231, 251, 296
Nathusius, editor of the *Volksblatt*, 229
Nautical Almanac Ephemerides, 70
Naval College, Royal, 280
Nehus, von, Gauss assistant, 135
Neptune, 223, 284, 285
Nertschinck, Russia, 157
Nether-Saxony, Gauss Scholarship Fund in, 338
Neubert, Dr. C., 309n
Neumann, Franz, 161, 162, 167
Newfoundland, magnetic station on, 157
New Orleans, La., 204, 372

Newton, Sir Isaac, 18, 42, 57, 58, 111, 162, 174, 211, 226, 232, 275, 299, 302, 303, 311
New York, 367, 368; City College of, 282
New York *Daily Advertiser*, 295
New York Review, 241
Nicolai, Bernhard, pupil of Gauss, 110, 272; as Mannheim collaborator with Gauss in lunar observations, 111, 134; follower of Gaussian School in astronomy, 267
Nicollet, Jean-Nicolas, 294
Niendorf, Gauss survey of, 128
Noah, of the Bible; 313
Non-Euclidean geometry, 174, 180, 184, 214, 253, 274, 316
Non-Euclidean trigonometry, 180
Nordheim, Germany, 34
Nordic and Nether-Saxon, racial type of Gauss, 229
Northampton, Mass., 219
North Pacific coast, 338
North Sea islands, 138
Norwegian bonds, 237
Nottingham, England, 153
Nouvelles méthodes pour la détermination des orbites, treatise on the method of least squares, by Adrien-Marie Legendre, 19
Nova theoria musicae, Euler's, 229
Numbers, theory of. See Gauss, Carl Friedrich, Research
Nürnberg, Germany, 109

Oak Grove Cemetery, St. Charles, Mo., 372
Ober Ohe, living quarters of Gauss, 126
Oersted, Hans Christian, 153, 157
Oertling, mechanic, 227
Of the Plurality of Worlds, by William Whewell, 303
Ohm, Georg Simon, 153, 161
Olbers, Doris, 55n
Olbers, Georg Heinrich, 55n
Olbers, Heinrich Wilhelm Matthias, physician and astronomer, 50-53,

Olbers, Heinrich (cont'd.)
287, 290, 292, 293, 304, 309, 311, 318, 319, 367, 373; biographical data on, 54–56n; as discoverer of Pallas, 57, 72, 73; portrait of, 75; importance of, as astronomer, 77; death of, 205, 226; honored, 281; his friendship and association with Gauss, 54, 61, 62, 74, 75, 76, 85, 87, 90, 92, 93, 111, 113, 115, 130, 132, 133, 134, 147, 153, 231, 236, 251, 256. See also Gauss, Carl Friedrich, Correspondence and Friendships

Oldenburg, Germany, 269
Olympian characterization, 231
Optics. See Gauss, Carl Friedrich, Research
Order of Henry the Lion, Commander's Cross of, bestowed on Gauss, 278
Oriani, Milan astronomer, 49, 50, 51, 52
Oslo, Norway, 262
Osnabrück, Gauss survey of, 137
Osterholz, Gauss survey of, 130, 133
Osterode, Germany, 145
Osthoff, Christian Ernst, father of Johanna Gauss, 62
Osthoff, Johanna Elizabeth Rosina. See Gauss, Johanna
Osthoff, Johanna Maria Christine, mother-in-law of Gauss, estate of, 236; Gauss vexed by, 237
Ostwald's *Klassiker der exakten Wissenschaften*, 154
Ottensen, site of death of Duke of Brunswick, Gauss' patron, 83
Ovid, 241
Oxford, Bodleian Library, 199

Paganini, Nicolo, 229
Palermo, Sicily, 49
Pallas, planetoid, 57, 72, 73, 282, 310
Papen, Captain August, 361
Parallels, Theory of. See Gauss, Carl Friedrich, Research
Paranatta, magnetic station at, 157

Paris, France, 68, 251
Paris Academy, encouragement of lunar research through prizes established by, 70, 71
Paris Revolution of 1830, 142
Parrot, G. F., 289
Parthenope, planet, 223
Pascal, Blaise, 311
"Pathfinder of the seas." See Maury, Commodore M. F.
Paul, Jean. See Richter, Jean Paul Friedrich
Peirce, Benjamin, 283, 285, 286
Pennsylvania, University of, 280
Pernety, French general, 67
Personalist and Emancipator, magazine, 273
Perthes, Friedrich Christoph, German publisher, 90, 252, 296
Pestalozzi, J. H., 242
Petri, Philipp, Göttingen photographer, 324
Petri, philologist, 259
Pfaff, J. F., 34, 35, 36, 45, 84, 104, 176, 255n, 289
Pfaff, J. W. A., brother of J. F. Pfaff, 289
Phaedrus, 241
Philadelphia, Penn., 367
Philosophy of Gauss. See Gauss, Carl Friedrich, Later Years
Photometry, prize for method for, 170
Piazzi, Joseph, Italian astronomer, 49–53, 306; godfather to Gauss' son Joseph, 77
Pindar, 241
Pisces, constellation, 73
Plana, astronomer, 71, 227
Planets. See Gauss, Carl Friedrich, Research
Plato, 241, 314
Plautus, 241
Pleissenburg tower, 290
Pliny, 241
Plössl, Simon, 169
Plurality of Worlds, Of the, by William Whewell, 303

INDEX   531

Poe, Edgar Allan, 295
Poggendorff's *Annalen*, 154
Poisson, Siméon Denis, 162
Pope, author of *Rape of the Lock*, 241
Poppe, a signature, 247
Portraits. See Gauss, Carl Friedrich, Portraits
Posen, Germany, 268
Posselt, Johannes Friedrich, pupil of Gauss, 267
Postage stamp, Gauss commemorative, 340
Potsdam, Germany, 372
Potsdam Bridge, Berlin, 337
Praël, Otto, 321
Prague, Czechoslovakia, 256
Prairie du Chien, Wis., 369
Prevorst, von, seeress, 307
Probability of error, mathematical. See Gauss, Carl Friedrich, Research
Prussia, 198
Prussian Academy of Sciences, 50, 151, 254, 275, 291, 296, 347, 348
Prussian bonds, 237
Puissant, 20
Pulkowa, Observatory of, 88*n*, 185
Pushkin, Alexander, 238
Pütter, Johann Stephan, 23

Quatre Bras, battle site, 82
Quedlinburg, Germany, 262
Querfeld, Gauss' assistant, 131
Quetelet, L. A. J., 20, 154
Quetelet's *Annuaire*, 234
Quincy, Josiah, 281

Raabe, Wilhelm, 243
Radowitz, Joseph Maria von, Prussian general and statesman, 302, 307
Ramsden, English instrument producer, 106, 119
Ramsden zenith sector, 111, 119, 121, 125
Recherches sur la coubure des surfaces, treatise by L. Euler, 165. See also Gauss, Carl Friedrich, Research
Redtenbacher, 337
Regensburg, Germany, 109
Rehberg, A. W., secretary of Hanover cabinet, 85
Reichenbach, Georg von, instrument maker, 106, 107, 108, 110
Reichenbach meridian circle, 97, 98, 110, 223, 227
Reichenbach passage instrument, 97, 109, 110
Reichenbach theodolite, 166
Reichenhall, Germany, 109
Reichspostmuseum, Berlin, Gauss portrait at, 337
Reidl, *Beiträge zur Theorie des Sehnenwinkels* (Wien, 1827), topology item, 222
Reiff, C. P. See Dictionaries
Reinhard, Rev. Dr. F. V., 291
Remer, teacher, 18
Repertorium der gesammten deutschen Literatur, 185
Repsold, Johann Georg, friend of Gauss, 89, 96, 97, 107, 109, 119, 123, 130, 168
Repsold meridian circle, 97, 107, 109, 110
Repsold theodolite, 119
Research of Gauss. See Gauss, Carl Friedrich, Research
Resultate aus den Beobachtungen des Magnetischen Vereins im Jahre . . . , research periodical of the Magnetic Association, 150, 157, 159, 160, 161
Rethen, Hanover, probable ancestral home of elder Hinrich Gooss [Gauss], 5, 6
Revolution of 1848, 244
Ribbentrop, Georg Julius, anecdotes concerning, 219
Richardson, author of *Clarissa Harlowe*, 241
Richter, Jean Paul Friedrich, favorite song of, 228, 229; as Gauss'

Richter, Jean Paul (cont'd.)
favorite German author, 239, 240; views of, on immortality, 307, 308
Riemann, Bernhard, pupil of Gauss, 104, 162, 174, 186, 189, 223, 266, 274, 311, 350
Ritter, August, student of Gauss, 258, 259, 260
Ritter, Georg Karl, godfather of C. F. Gauss, 61
Ritterbrunnen, Brunswick site of a Gauss home, 6
Robertson, author of *History of Charles V*, 241
Rockingham County, Virginia, 370
Röntgen, W. K., 337
Ross, Captain, 158
Rotenkirchen, castle of king of Hanover, 197
Rotenkreuz, Geheimrat Feronçe von, 16
Rothenburg, Germany, 128, 268; Gauss survey of, 130
Rousseau, J. J., 241
Royal Astronomical Society, 345
Royal Irish Academy, Dublin, 346
Royal Naval College, Greenwich, 280
Royal Prussian Academy of Sciences, 50, 151, 254, 275, 291, 296, 347, 348
Royal Society in Copenhagen, 127, 129, 164
Royal Society in London, 92, 157, 202
Royal Society of Sciences in Göttingen, 104, 114, 139, 149, 154, 155, 164, 165, 170, 187, 192, 154, 237, 275, 276, 278, 288, 337
Rüchel, General, 80
Rüdiger, C. F., 290
Rudloff, telescope maker, 78
Rule of signs. *See* Descartes
Rumpf, Philipp, mechanic, 111, 123, 125
Russian Academy of Sciences, 151, 289
Russian bonds, 237
Russian Empire, 261
Russian papers, 235

Russian topographical bureau, memoirs of, 238

Saalfeld, Germany, 80
Sabine, General Edward, 262, 280
Sabine, Mrs., wife of Edward, 262
Saccheri, Girolamo, research on theory of parallels by, 177
Sachsen-Gotha, Germany, 292
St. Albans Cemetery, Göttingen, burial place of Gauss, 326
St. Albans Church, Göttingen, 300
St. Charles, Mo., 204, 277, 369, 370, 371, 373
St. Helena, magnetic station on, 157
St. John's Church, Göttingen, 102, 191
St. Katharine's Church, Brunswick, 66, 77
St. Louis, Mo., 205, 285, 368, 373
St. Michael's Church, Lüneburg, 221
St. Paul's Church, Frankfort, 244
St. Petersburg, 58, 78, 238, 261, 306; Academy of Sciences in, 151, 289
St. Thomas School, 23
Sallust, 241
Sandbergen, Germany, 271
Sanskrit, 238
Sarnighausen, Rev. Dr., 326
Sartorius von Waltershausen, Baron Wolfgang, 3, 29, 92, 154, 207, 242, 254, 302, 320, 321, 322, 326; funeral sermon of, for Gauss, 334–335
Saturn, planet, 284, 299
Sauensieck, Germany, 131
Saussure, Swiss physicist, 241
Savart, 153
Schaper, Fritz, sculptor, 335
Scharnhorst, Gauss survey of, 127
Scharnhorst, Colonel von, 80
Schelling, F. W. J., 253, 313, 314
Schering, Ernst, 75, 255, 271
Scherk, Heinrich Ferdinand, pupil of Gauss, 268
Schiaparelli, Giovanni Virginio, 20

INDEX 533

Schiller, J. C. F. von, 317; Gauss' opinion on works by, 239, 242; weight of brain of, 324
Schinkel, K. F., 337
Schleiden, M. J., 315
Schleiermacher, Auguste. *See* Ewald, Auguste
Schleiermacher, Ludwig, uncle of Auguste Ewald, 265
"Schleswig-Holstein, meerumschlungen," a song, 229
Schliephacke, Heinrich, friend of Eugene Gauss in America, 367
Schlözer, A. L. von, 23
Schmidt, K. A., teacher at Carolineum, 17
Schmidt's Russian-German. *See* Dictionaries
Schneidewin, F. W., 278
Schneidewin, Max, 244, 245
Schnürlein, Ludwig Christoph, pupil of Gauss, 223, 269, 270
Schoenlein, Philipp, 288
Schoon, Christine Magdalene von, mother of H. C. Schumacher, 88n
Schöpffer, Carl, 263
Schott, 20
Schröder, landlord of Gauss, 35
Schröder, lens maker in Gotha, 78
Schrödinger, E., 171
Schröter, Johann Hieronymus, 50, 58, 62, 74, 107
Schröter telescope, 96
Schumacher, Andreas Anthon Friedrich, 88n
Schumacher, Christian Andreas, 88n
Schumacher, Heinrich Christian, student in astronomy under Gauss, 89, 116–134 *passim*, 163, 207, 238, 273, 278, 283, 284, 290, 293, 296, 297, 312, 313, 314, 316, 318, 348; biographical data on family of, 88n; correspondence between, and Gauss, 89, 226; Gauss visits, 93; interest of, in Gauss' work; 169, 170, 209, 210, 211; known as follower of Gaussian school in astronomy, 267; as editor, 125, 127, 164, 157; as translator, 221; death of, 226. *See also* Gauss, Carl Friedrich, Correspondence *and* Friendships; *Astronomische Nachrichten;* Carnot's *Géométrie de position*
Schumacher, Richard, 89n
Schur, Wilhelm, 96
Schütting inn, Lüneburg, Gauss address, 119
Schwarz, Christian August, portraits of Gauss and Olbers by, 75
Schwarz, Hermann Amandus, 266
Schweiger, Ludwig, appraiser of Gauss library, 357
Schweikart, F. C., on parallel theory, 180; on "theory of astral magnitudes," 181; role of, in liberation from Euclidean tradition, 188
Schwerd instrument, 173
Scientific research, Gauss'. *See* Gauss, Carl Friedrich, Research
Scientific Society of Brunswick, 336
Scott, General Winfield, 368
Scott, Sir Walter, Gauss' opinion on works of, 243
Seeber, L., 48, 140, 141; *Works of,* reviewed by Gauss, 48
Seeberg, Germany, 108, 117, 292
Seelenlehre, by Braubach, 310
Seneca, 241, 256
Serret, J. A., 48
Seven, The. *See* Göttingen Seven
Seyffer, Carl Felix, faculty friend of Gauss at Göttingen and later correspondent, 26, 30, 176
Shakespeare, 240
Shelton clock, 104, 110
Sheridan, author of *School for Scandal,* 241
Siemens, Wilhelm, 337
Silliman, Benjamin, editor and geologist of Yale, 283, 287
Silliman's Journal. See Silliman, Benjamin
Simon, Heinrich, 104, 114n
Skagen, Denmark, 116
Slonimsky, Ch. Z., 69n

Smith, H. J. S., appraisal of Gauss, by, 343
Smolensk, Russia, 98
Smollett, Tobias George, Gauss' opinion on works of, 240
Society. *See* Royal Society
Society, Scientific, of Brunswick, 336
"Socrates," name of parrot, 219
Soldner, Johann Georg von, Munich collaborator with Gauss on lunar observations, 111
Sollerich, or Sollicher. *See* Warnecke, Dorothea Emerenzia
Solling, Germany, 197
South America, 284
South Pole, 338
Spandau, Germany, 275
Spazier, Richard Otto, 240
Spehr, Friedrich Wilhelm, 137
Spinoza, Benedict, 244n
Spittler, L. T., 24
Springfield, Mass., 294
Stade, girlhood home of Sophie Gauss, 205, 207, 361, 362
Stamford, Major-General von, 33n
Stanley, A. D., of Yale University, 287
Staudt, Karl Georg Christian von, pupil of Gauss, 268, 270
Staufenau, Constantin Wilhelm, husband of Therese Gauss, 374, 375
Stein, Baron vom, 79
Steinburg, Gauss survey of, 130
Steiner, Jakob, 245
Steinheil, Carl August, 169, 262
Steinweg 22, Gauss address in Brunswick, 66, 82
Stern, Moritz A., pupil of Gauss, 222, 225, 259, 270, 275, 348
Stockholm, Sweden, 271
"Storm king." *See* Espy, James P.
Stowe, H. B., author of *Uncle Tom's Cabin*, 241
Stralenheim, Minister, von, 191, 192
Strassburg, 271
Streckverse, by Jean Paul, 240
Stromeyer, Friedrich, 87
Struve, Friedrich Georg Wilhelm, 185

Struve, Otto, grandson of Bartels, 187, 188
Stuttgart, Germany, 106, 134, 243
Suetonius, 241
Supplementum theoriae combinationis observationum erroribus minimis obnoxiae, treatise by C. F. Gauss, 114
Surface theory. *See* Gauss, Carl Friedrich, Research
Survey. *See* "Gauss survey"
Süssmilch's *Göttliche Ordnung gerettet*, 310
Swedish bonds and mining stock, 237
Swift's *Gulliver's Travels*, 241
Sylvester, J. J., 347
Szathmáry, Michael, theologian, 27

"Table rapping," 247
Tacitus, 241
Tait, P. G., 251, 254
Taurinus, F. A., research on theory of parallels by, 181; correspondence of, with Gauss on theory of parallels, 181, 182
Taylor, General Zachary, 368
Technical Institute of Brunswick, 336, 338
Technical Institute at Charlottenburg, 337
Technical Institute in Hanover, 259, 338
Teipel, J. H., anecdotes concerning, 220
Telegraph, needle, Gauss', 150
Telescope, dialytic, 169, 170; Herschel's mirror, 96; by Ahort, 75; Lilienthal's mirror, 77, 83, 84; Schröter, 96; factory in Wolfenbüttel, 78, 115
"Tell me the tales that to me were so dear, long long ago," a favorite song of Gauss, 228
Tempelhoff, General Georg Friedrich von, 36
Tentamen, by Johann Bolyai, 183
Terence, 241

Terrestrial magnetism. *See* Gauss, Carl Friedrich, Research
The Hague, magnetic station at, 157
"The Lay of the Lone Fishball," ballad, 282
Theodolite, Repsold, 119; Reichenbach, 166
Theophrastus, 241
Theoria motus, by C. F. Gauss, 71, 90–93, 252, 281, 296, 304, 345
Théorie des nombres, by A. M. Legendre, 251
Theorie der Parallellinien, 182
"Theory of astral magnitudes," 181
Theory of numbers. *See* Gauss, Carl Friedrich, Research
Thibaut, B. F., 266, 267
Thomson, James, 241, 299
Thucydides, 241
Thune, Copenhagen astronomer, 132
Thuringia, Germany, 121, 374
Tibullus, 241
Ticknor, George, 281
Timpenberg, Gauss survey of, 127, 128
Tittel, P., 70, 108
Todhunter, Isaac, 345
Topographical bureau, memoirs of Russian, 238
Topology. *See* Gauss, Carl Friedrich, Later Years
Transylvania, 177
Triangulation. *See* "Gauss survey"
Trigonometric survey of Brunswick, 137, 361
Troughton sextant, 116
Tübingen, University of, 134, 140, 198, 269, 374
Tucker, Robert, 39, 40
Turmstrasse. *See* Klein-Paris
Tychsen, Adelheid, 87
Tychsen, Cäcilie, 87
Tychsen, Thomas Christian, 87

Über die beste Art, Klavier zu spielen, by Bach, 229
Über die Hypothesen, welche der Geometrie zu Grunde liegen, 189
Über Wissen und Glauben, by Ruldolf Wagner, 309
"Union." *See* "Literary Museum"
Union College, 286
United States, bureau of weights and measures, 286; Coast and Geodetic Survey, 283, 286; government, 338
"Unser Leben gleicht der Reise eines Wanderers in der Nacht," song, 229
Uppsala, University of, 288; magnetic station at, 157
Uranus, planet, 51, 55, 56
Ursin, Georg Friedrich Krüger, assistant to Gauss in the Lüneburg survey, 118
Uslar, Gauss survey of, 137
Uslar, Lieutenant von, 258
Utzschneider, Joseph von, partner of Fraunhofer, 108
Uylenbrock, Dutch mathematician, 222

Valentia, Ireland, 283
Valett, A., student of Gauss, 258
Vandermonde, geometer, 221
Van Diemen's Land, magnetic station in, 157
Vassar College, 283
Vegesack, Gauss survey of, 130
Velpke, native village of Gauss' mother, 8, 13, 203
Veltheim, Count, benefactor to Ide, 25
Venus, planet, 223
Vesta, planetoid, 76, 224, 282, 419
Vevey, Switzerland, 374
Victoria, planet, 223
Vienna, 27, 157, 178
Vienna bank stock, 237
Vieweg, Friedrich, 37
Villers, de. *See* De Villers
Virgil, 241
Virginia, 370; University of, 282
Virgo, constellation, 72
Visselhövede, Gauss survey of, 129
Vogel, Göttingen barber, 260
Vogt, Karl, 302

Völkenrode, village home of branch of Gauss family, 4, 5
Völker, military physician, 81
Volksblatt. See Nathusius
Voltaire, 241
Voss, Aurel Edmund, 255

Wachter, Friedrich Ludwig, pupil of Gauss, 179, 267, 268
Wagener, G., student of Gauss, 258, 344
Wagner, Richard, 3
Wagner, Rudolf, 247, 301; essay by, on religious views of Gauss, 302; conversations of, with Gauss on philosophy, 302–310; report by, on Gauss autopsy, 323
Waisenhaus 271, Dresden address, 374n
Waldeck, Charlotte A. W. (Wyneken), Gauss' second mother-in-law, 237, 292
Waldeck, Johann Peter, father of Minna W. Gauss, 100, 236
Waldeck, Luise Christiane Sophie, sister of Minna W. Gauss, 236
Waldeck, Minna. See Gauss, Minna
Walker, Sears, 284
Waltershausen. See Sartorius
Wangeroog, Gauss survey of, 133
Warnecke, Dorothea Emerenzia (also known as Sollerich or Sollicher). See Gauss, Dorothea Emerenzia (Warnecke)
Warnstedt, Adolf von, 325
Washington, D. C., 286
Weavers' song copied by Gauss, 229
Weber, Carl Maria von, 229
Weber, Ernst Heinrich, brother of Wilhelm Weber, 142
Weber, Wilhelm, 263, 266, 271, 288, 302, 357; first meeting of, with Gauss, 136; recommendation of, by Gauss, 140, 141; called to Göttingen University, 141; friendship between Gauss and, 141, 142, 148; experiment of, with the telegraph, 147–151; as Gauss' associate in magnetic research, 153, 156, 157, 158, 194; one of the Göttingen Seven, 197; dismissed from university, 197; to Leipzig, 198; on tour of scientific study, 199; recalled to Göttingen, 244; last years of, 266; portrait of, 337; monument to, 337. See also Gauss, Carl Friedrich, Research *and* Friendships; Göttingen Seven; Göttingen, University of
Weende, Germany, 121
Wehrtmann, Hinrich. See Grove, Anna
Weierstrass, Karl W. T., 104
Weimar, Germany, 80, 306
Wellington, Duke of, 121
Wendeburg, home of Hans Gauss, 4
Wendengraben, Brunswick site of Gauss' birth house, 6
Wends' tower, a beer garden, 279
Weser River, Gauss survey of, 130 137
Western Reserve University, 282
Westphal, Justus Georg, pupil of Gauss, 223, 270
Westphalia, Germany, 115, 137
Westphalian government, 85, 95
West Point, N.Y., 286
Whewell, William, 167, 302, 304
Wildt, J., 30, 176
Wilhelm, Duke of Brunswick, honors Gauss, 278
Wilhelm, King of Hanover (William IV of England), 143; unveiling of monument to, 191; death of, and its effect upon Hanover, 195
Wilhelmsplatz, plaza in Göttingen, 191, 193
Wilhelmstrasse in Brunswick, street site of Gauss' birthplace, 6, 11
Wilhelmstrasse, street in Hanover, 266
Wilkes, Captain Charles, 158
Wilmerding, mayor of Brunswick, 6, 7

INDEX 537

Wilsede, Gauss survey of, 126–128 *passim*, 131
Wimmer, R., portraitist, 336
Winnecke, F. A. T., pupil of Gauss, 271
Wittstein, pupil of Gauss, 272
Wolfenbüttel, seat of telescope factory, 78, 115
Wolff, Christian, 313, 314
Wolffradt, Privy Councillor von, 83
Wollaston, W. H., 166
Wolzogen, Frau von, 418
Woodward, R. S., 154
Works of L. Seeber, reviewed by Gauss, 48
Wulfsode, 126, 127
Wurm, 134
Württemberg, Germany, 4, 134
Württemberg bonds, 237
Württemberg Credit Union, 236
Würzburg, Germany, 134
Wyneken, Charlotte Auguste Wilhelmine, mother of Minna W. Gauss, 100

Xenophon, 241

Yale University, 282, 283, 287

Zach, Franz Xavier G. von, 49–54 *passim*, 60, 70, 72, 73, 75, 115, 116, 163, 306
Zachariä, teacher, 17
"Zeitungstiger," Gauss nickname, 226n
Zenith sector, Ramsden, 111, 119, 121, 125
Zeven, Gauss survey of, 130, 133, 134
Zeyk, Baron von, 183
Zimmermann, E. A. W., mathematician and teacher, 14; as patron and friend of Gauss, 14–16; 28, 37, 39, 58, 281; as friend of Ide, 25; promotion of, to privy councillor, 58, 59
Zöpflein, professor, 248
Zugspitze, Bavarian mountain, Gauss compared to, 346
Zurich, Switzerland, 271